FUNCTIONAL IMAGING BY CONTROLLED NONLINEAR OPTICAL PHENOMENA

WILEY SERIES IN BIOMEDICAL ENGINEERING AND MULTI-DISCIPLINARY INTEGRATED SYSTEMS

KAI CHANG, SERIES EDITOR

Advances in Optical Imaging for Clinical Medicine
Nicusor Iftimia, William R. Brugge, and Daniel X. Hammer (Editors)

Antigen Retrieval Immunohistochemistry Based Research and Diagnostics
Shan-Rong Shi and Clive R. Taylor

Introduction to Nanomedicine and Nanobioengineering
Paras N. Prasad

Functional Imaging by Controlled Nonlinear Optical Phenomena
Keisuke Isobe, Wataru Watanabe, and Kazuyoshi Itoh

FUNCTIONAL IMAGING BY CONTROLLED NONLINEAR OPTICAL PHENOMENA

KEISUKE ISOBE
RIKEN Center for Advanced Photonics, 2-1 Hirosawa,
Wako, Saitama 351-0198, Japan

WATARU WATANABE
Department of Electrical & Electronic Engineering
College of Science and Engineering
Ritsumeikan University
1-1-1 Noji-Higashi, Kusatsu, Shiga 525-8577 Japan

KAZUYOSHI ITOH
Science Technology Entrepreneurship Laboratory (E-Square),
Osaka University, Yamada-oka 2-1, Suita, Osaka 565-0871, Japan

Published by John Wiley & Sons, Inc., Hoboken, New Jersey.
Published simultaneously in Canada.

Library of Congress Cataloging-in-Publication Data:

Isobe, Keisuke, author.
Functional imaging by controlled nonlinear optical phenomena / Keisuke Isobe, Wataru Watanabe, Kazuyoshi Itoh.
p. ; cm.
Includes bibliographical references and index.
ISBN 978-1-118-09144-9 (cloth)
I. Watanabe, Wataru, author. II. Itoh, Kazuyoshi, author. III. Title.
[DNLM: 1. Imaging, Three-Dimensional–methods. 2. Microscopy, Fluorescence, Multiphoton. 3. Lasers–diagnostic use. 4. Nonlinear Dynamics. 5. Optical Processes. WN 180]
RB43
616.07′58–dc23

2013017208

Printed in the United States of America

10 9 8 7 6 5 4 3 2 1

CONTENTS

LIST OF FIGURES

LIST OF TABLES

PREFACE

Since the advent of laser light, we have enjoyed its highly coherent properties due to the very narrow bandwidth hitherto impossible in the wide range of application fields, such as interferometry, spectroscopy, and holography. Recently, however, many researchers are examining pulsed light from lasers with very wide bandwidths. Ultrashort laser pulses can be generated from the wideband laser, if all the modes in the laser cavity are locked by some means. A typical commercially available mode-locked laser is a Ti:Sapphire or an Er-doped fiber laser. Subpicosecond laser pulses with more than 10 GW are readily available from these lasers. No less important, the spatial coherence inherent to these laser pulses allows us to readily focus them into an area smaller than 1 μm^2, where the filed strength can be extremely high. These laser pulses are called ultrashort or ultrafast laser pulses.

Natural light usually interacts with matter in a gentle manner. Its magnitude is attenuated, and/or its phase is simply delayed, depending on its wavelength and the materials it propagates. The wavelength-dependent attenuation of the magnitude gives light the color, and the tapered phase delay across the wavefront bends or disperses the light beam. Because of these interactions we can enjoy brilliant colors in the world. We can use glasses and camera lenses because of the refraction due to the phase shift in glass. If we use ultrafast laser pulses with a high peak power, the interaction of light with matter becomes much stronger and dependent on the field strength. Then the interaction regime gets into the nonlinear regime, and light that has different frequency components other than the incident one that is generated. The generated frequency components usually include those of the sum and difference of the input pulses. In extreme cases, the generated components span from terahertz to X-ray regions.

In the range of moderate peak power of ultrafast laser pulses, we can utilize the optical nonlinear effects for new modalities of imaging instruments, the optical nonlinear microscopy. The optical nonlinear effects include the two-photon absorption, high harmonics generation, and four-wave mixing. Nonlinear optical microscopy is a rapidly emerging research area with widespread clinical applications. Nonlinear optical effects allow both structural and functional imaging with cellular-level resolution. The introduction of endogenous or exogenous probes can selectively enhance contrast for molecular targets in a living cell, as well as supply functional information on processes. To obtain functional images, nonlinear optical processes can be controlled by photo probes and/or parameters of ultrafast laser pulses, such as time, space, polarization, and phase.

This book gives an overview of the nonlinear optical process caused by ultrafast laser pulses, and explains how the basics of nonlinear optical microscopy led to the most advanced techniques of photo-controlled nonlinear optical microscopy. If the incident pulses have higher peak power and total energy, much stronger interaction occurs. The ultrafast laser pulses can form localized plasmas and generate heat and shock waves in the sample and may cause ablation or explosion of a micro or nanometer scale. Such phenomena can be used for manipulation of biological samples. Thus, ultrafast laser pulses allow high-precision imaging and simultaneously the manipulation of the cells and intracellular substances for biological and medical applications. The latter issues of molecular control and laser surgery are dealt with in separate chapters.

The first author, Keisuke Isobe, who was one of my students, invented the concept of the stimulated parametric emission microscopy described in this book. The second author, Wataru Watanabe, also a student of mine, started the intracellular nanosurgery and single-organelle tracking in the early stages as an assistant professor of my laboratory. We have enjoyed frequent discussions in our laboratory on varieties of possible and impossible methods of nonlinear optical microscopy and manipulations of intracellular substances. Those days of stimulating discussions formed some of the best times of my life.

Kazuyoshi Itoh

On the 10th floor of GSE Common East,
Suita Campus, Osaka University

ACKNOWLEDGMENTS

Our work described in this book would not have been possible without collaboration and support from a large number of persons. We are pleased to take this opportunity to express our sincere gratitude. We would like to express our thanks to Professor Sachihiro Matsunaga (Tokyo University of Science), Assistant Professor Tsunehito Higashi (Hokkaido University), and Professor Kiichi Fukui (Osaka University) for their collaboration in the research in Chapters 4.2.2, 4.3.1, 5.1.2, and 6.1. We gratefully acknowledge Naomi Arakawa and Tomoko Shimada from Osaka University for their support of the work in Chapters 5.1.2 and 6.1. Next, we would like to thank Professor Nobuhiro Tsutsumi and Associate Professor Shin-ichi Arimura from the University of Tokyo for the preparation of Kaede in the experimental results of Chapter 5.1.2. K. Itoh and K. Isobe would like to express our appreciation to Shogo Kataoka (Osaka University), Takehito Kawasumi (Osaka University), Rena Murase (Osaka University) and Assistant Professor Takayuki Tamaki (Nara National College of Technology) for their important contributions to the study in Chapters 4.3.1 and 4.6.2. K. Itoh and K. Isobe would like to express our gratitude to Assistant Professor Yasuyuki Ozeki (Osaka University) and Associate Professor Shin'ichiro Kajiyama (Kinki University) for their significant contributions to the development of the research in Chapters 4.3.1, 4.4.2, and 4.6.2. K. Isobe gratefully acknowledges Professor Akira Suda (Tokyo University of Science), Masahiro Tanaka (Keio University), Hiroshi Hashimoto (Keio University), Professor Fumihiko Kannari (Keio University), Dr. Hiroyuki Kawano (RIKEN Brain Science Institute), Professor Hideaki Mizuno (RIKEN Brain Science Institute), Professor Atsushi Miyawaki (RIKEN Brain Science Institute), and Professor Katsumi Midorikawa (RIKEN Advanced Science Institute) for their collaboration and support in the research in Chapters 3.3, 4.2.3, 4.4.1, 5.1.2, and 5.2. K. Isobe would further like to acknowledge Dr. Takako Kogure, Dr. Tetsuya

Kitaguchi, and Yoshiko Wada from RIKEN Brain Science Institute for the preparation of the biological sample in the experimental results of Chapter 5.2.

We would moreover like to acknowledge the financial support of the work in Chapters 4.3.1 and 4.4.2 by SENTAN, JST (Japan Science and Technology Agency), and the work in Chapters 4.2.2 and 4.4.2 by the Northern Osaka (Saito) Biomedical Cluster of Science and Technology Forming Project. The work in Chapters 3.3, 4.4.1, 5.1.2, and 5.2 was supported by the Special Postdoctoral Researchers Program of RIKEN.

Finally, we would like to thank our families for their loving support and patience.

ACRONYMS

AC	Autocorrelation
ADC	Analog-to-Digital Converter
AOD	Acousto-optic Deflector
AOM	Acoustic Optical Modulator
BBO	Barium Borate
CALI	Chromophore-Assisted Laser Inactivation
CARS	Coherent Anti-Stokes Raman Scattering
CCD	Charged Coupled Device
CMOS	Complementary Metal Oxide Semiconductor
CSRS	Coherent Stokes Raman Scattering
CW	Continuous Wave
DF	Difference frequency
DFG	Difference-Frequency Generation
DOE	Diffractive Optical Element
EOM	Electro-optical Modulator
ESA	Excited-State Absorption
FCS	Fluorescence Correlation Spectroscopy
FCCS	Fluorescence Cross-correlation Spectroscopy
FL	Fluorescence
FLIM	Fluorescence Lifetime Imaging Microscopy
FP	Fluorescent protein
FT	Fourier Transform
FOD	Fourth-Order Dispersion
FPALM	Fluorescence Photoactivation Localization Microscopy
FRAP	Fluorescence Recovery after Photobleaching

FRET	Fluorescence Resonance Energy Transfer
FROG	Frequency-Resolved Optical Gating
FTL	Fourier Transform-Limited
FWHM	Full-Width at Half-Maximum
FWM	Four-Wave Mixing
GD	Group Delay
GDD	Group Delay Dispersion
GSD	Ground-State Depletion
IAC	Interferometric Autocorrelation
IC	Internal Conversion
IR	Infrared
ISC	Intersystem Crossing
LASIK	Laser-Assisted In situ Keratomileusis
LC	Liquid Crystal
MPEF	Multiphoton Excited Fluorescence
NA	Numerical Aperture
NFWM	Nonresonant Four-Wave Mixing
NIR	Near Infrared
NLOM	Nonlinear Optical Microscopy
NLOS	Nonlinear Optical Spectroscopy
NSOM	Near-Field Scanning Optical Microscopy
OL	Objective Lens
OPA	Optical Parametric Amplifier
OPE	One-Photon Excitation
OPEF	One-Photon Excited Fluorescence
OPO	Optical Parametric Oscillator
OR	Optical Rectification
OTF	Optical Transfer Function
PALM	Photoactivated Localization Microscopy
PCF	Photonic Crystal Fiber
PSF	Point Spread Function
PMT	Photomultiplier Tube
PPLN	Periodically Poled Lithium Niobate
QD	Quantum Dot
QPM	Quasi-phase Matching
RESOLFT	Reversible Saturable Optical Linear Fluorescence Transitions
SAX	Saturated Excitation
SC	Supercontinuum
SFG	Sum-Frequency Generation
SH	Second Harmonic
SHG	Second-Harmonic Generation
SIM	Structured Illumination Microscopy
SLM	Spatial Light Modulator
SSIM	Saturated Structured Illumination Microscopy
STED	Stimulated Emission Depletion

STORM	Stochastic Optical Reconstruction Microscopy
SPIDER	Spectral Phase Interferometry for Direct Electric-Field Reconstruction
SPE	Stimulated Parametric Emission
SPM	Self-phase Modulation
SRS	Stimulated Raman Scattering
SVEA	Slowly Varying Envelope Approximation
TAC	Time-to-Amplitude Converter
TCSPC	Time-Correlated Single-Photo Counting
TMP	Trans-membrane Potential
TOD	Third-Order Dispersion
THG	Third-Harmonic Generation
TPA	Two-Photon Absorption
TPE	Two-Photon Excitation
TPEF	Two-Photon Excited Fluorescence
UV	Ultraviolet
XPM	Cross-phase Modulation

CHAPTER 1

ULTRAFAST OPTICS FOR NONLINEAR OPTICAL MICROSCOPY

1.1 NONLINEAR OPTICAL PHENOMENA

1.1.1 Introduction to Nonlinear Optics

1.1.1.1 The Wave Equation For understanding the nonlinear optical phenomena, it is necessary to make use of the theory of electronic wave propagation in a nonlinear optical medium. Beginning with the Maxwell's equations, we can obtain the optical wave equations that govern the propagation of a light pulse through a nonlinear optical medium. In the International System of Units, the Maxwell equations take the form

$$\nabla \times \mathbf{E}(\mathbf{r}, t) = -\frac{\partial \mathbf{B}(\mathbf{r}, t)}{\partial t}, \tag{1.1.1}$$

$$\nabla \times \mathbf{H}(\mathbf{r}, t) = \frac{\partial \mathbf{D}(\mathbf{r}, t)}{\partial t} + \mathbf{J}(\mathbf{r}, t), \tag{1.1.2}$$

$$\nabla \cdot \mathbf{D}(\mathbf{r}, t) = \rho_e(\mathbf{r}, t), \tag{1.1.3}$$

$$\nabla \cdot \mathbf{B}(\mathbf{r}, t) = 0, \tag{1.1.4}$$

where $\mathbf{E}$ and $\mathbf{H}$ are electric and magnetic field vectors, respectively. $\mathbf{D}$ and $\mathbf{B}$ are corresponding electric and magnetic flux densities. The current density vector $\mathbf{J}$ and the charge density ρ_e represent the sources for the electromagnetic field. We are

Functional Imaging by Controlled Nonlinear Optical Phenomena, First Edition.
Keisuke Isobe, Wataru Watanabe and Kazuyoshi Itoh.

primarily interested in the solution of these equations in regions of space that contain no free charges, so that

$$\rho_e(\mathbf{r}, t) = \mathbf{0}, \tag{1.1.5}$$

and that contain no free currents, so that

$$\mathbf{J}(\mathbf{r}, t) = \mathbf{0}. \tag{1.1.6}$$

We assume that the material is nonmagnetic, so that

$$\mathbf{B}(\mathbf{r}, t) = \mu_0 \mathbf{H}(\mathbf{r}, t), \tag{1.1.7}$$

where μ_0 is the vacuum permeability. The electric flux density $\mathbf{D}$ arises in nonlinear response to the electric field $\mathbf{E}$ propagating inside the nonlinear medium and is related to $\mathbf{E}$ through the following relation:

$$\mathbf{D}(\mathbf{r}, t) = \varepsilon_0 \mathbf{E}(\mathbf{r}, t) + \mathbf{P}(\mathbf{r}, t), \tag{1.1.8}$$

where ε_0 is the vacuum permittivity and $\mathbf{P}$ is the induced electric polarization. To derive the optical wave equation from Maxwell's equations, we eliminate $\mathbf{B}$ and $\mathbf{D}$ in favor of $\mathbf{E}$ and $\mathbf{P}$ by taking the curl of Eq. (1.1.1) and using Eqs. (1.1.2), (1.1.6), (1.1.7), and (1.1.8). As a result, we obtain the wave equation

$$\nabla \times \nabla \times \mathbf{E}(\mathbf{r}, t) = -\frac{1}{c^2}\frac{\partial^2 \mathbf{E}(\mathbf{r}, t)}{\partial t^2} - \mu_0 \frac{\partial^2 \mathbf{P}(\mathbf{r}, t)}{\partial t^2}, \tag{1.1.9}$$

where c is the speed of light in vacuum and the relation $\mu_0 \varepsilon_0 = 1/c^2$ was used. By using an identity from vector calculus, we can rewrite the left-hand side of Eq. (1.1.9) as

$$\nabla \times \nabla \times \mathbf{E}(\mathbf{r}, t) = \nabla(\nabla \cdot \mathbf{E}(\mathbf{r}, t)) - \nabla^2 \mathbf{E}(\mathbf{r}, t) = -\nabla^2 \mathbf{E}(\mathbf{r}, t), \tag{1.1.10}$$

where we used the relation $\nabla \cdot \mathbf{E}(\mathbf{r}, t) = 0$ from $\nabla \cdot \mathbf{D}(\mathbf{r}, t) = 0$, which is obtained by inserting Eq. (1.1.5) into (1.1.3). From Eqs. (1.1.9) and (1.1.10), we obtain the complete description

$$\nabla^2 \mathbf{E}(\mathbf{r}, t) - \frac{1}{c^2}\frac{\partial^2 \mathbf{E}(\mathbf{r}, t)}{\partial t^2} = \mu_0 \frac{\partial^2 \mathbf{P}(\mathbf{r}, t)}{\partial t^2}. \tag{1.1.11}$$

The polarization plays a key role in the description of nonlinear optical phenomena because Eq. (1.1.11) indicates that a time-varying polarization can act as the source of new components of the electromagnetic field. In the nonlinear optics, the polarization

$\mathbf{P}$ induced by electric dipoles is not linear in the electric field $\mathbf{E}$ and satisfies the general relation

$$\mathbf{P}(\mathbf{r},t) = \mathbf{P}^{(1)}(\mathbf{r},t) + \mathbf{P}^{(2)}(\mathbf{r},t) + \mathbf{P}^{(3)}(\mathbf{r},t) + \cdots, \tag{1.1.12}$$

$$\mathbf{P}^{(1)}(\mathbf{r},t) = \varepsilon_0 \int_{-\infty}^{t} \chi^{(1)}(t-t_1)\cdot \mathbf{E}(\mathbf{r},t_1)dt_1, \tag{1.1.13}$$

$$\mathbf{P}^{(2)}(\mathbf{r},t) = \varepsilon_0 \int_{-\infty}^{t}\int_{-\infty}^{t} \chi^{(2)}(t-t_1, t_1-t_2) : \mathbf{E}(\mathbf{r},t_1)\mathbf{E}(\mathbf{r},t_2)dt_1dt_2, \tag{1.1.14}$$

$$\mathbf{P}^{(3)}(\mathbf{r},t) = \varepsilon_0 \int_{-\infty}^{t}\int_{-\infty}^{t}\int_{-\infty}^{t} \chi^{(3)}(t-t_1, t_1-t_2, t_2-t_3) \vdots \mathbf{E}(\mathbf{r},t_1)\mathbf{E}(\mathbf{r},t_2)\mathbf{E}(\mathbf{r},t_3)dt_1dt_2dt_3, \tag{1.1.15}$$

where $\chi^{(1)}$ and $\chi^{(j)}$ $(j = 2, 3, \ldots)$ are known as the linear susceptibility and the jth-order nonlinear optical susceptibilities, respectively [1–3]. In general, $\chi^{(j)}$ is a tensor of rank $j+1$ and depends on the history of the electric fields [1]. The first term in Eq. (1.1.12) represents the linear response and the effects of $\chi^{(1)}$ are included through the linear refractive index $\mathbf{n}^{(1)}$ (see Eq. (1.1.25)) and linear absorption coefficient $\alpha^{(1)}$ (see Eq. (1.1.26)). The remaining terms in Eq. (1.1.12) represent the nonlinear response and are known as the nonlinear polarization [2, 3]. The second-order susceptibility induces the physical process such as second-harmonic generation (SHG), sum-frequency generation (SFG), difference-frequency generation (DFG), and optical rectification (OR). However, the second-order susceptibility $\chi^{(2)}$ is nonzero only for media that lack an inversion symmetry at the molecular level. In addition, $\chi^{(2)}$ vanishes for amorphous materials and solutions with random structure such as silica glass and water even when the symmetry at the molecular level is broken. The third-order susceptibility is related to the physical process such as third-harmonic generation (THG), four-wave mixing (FWM), optical Kerr effect, Raman scattering, and two-photon absorption (TPA). The third-order susceptibility $\chi^{(3)}$ is nonzero for all materials.

We now consider the wave equation in the frequency domain. $\mathbf{E}(\mathbf{r},t)$, $\mathbf{P}^{(1)}(\mathbf{r},t)$, and $\mathbf{P}^{(j)}(\mathbf{r},t)$ can be written in the frequency domain as

$$\tilde{\mathbf{E}}(\mathbf{r},\omega) = FT_t\,[\mathbf{E}(\mathbf{r},t)] = \int_{-\infty}^{\infty} \mathbf{E}(\mathbf{r},t)e^{i\omega t}dt, \tag{1.1.16}$$

$$\tilde{\mathbf{P}}^{(1)}(\mathbf{r},\omega) = \int_{-\infty}^{\infty} \mathbf{P}^{(1)}(\mathbf{r},t)e^{i\omega t}dt = \varepsilon_0\tilde{\chi}^{(1)}(\omega)\cdot\tilde{\mathbf{E}}(\mathbf{r},\omega), \tag{1.1.17}$$

$$\tilde{\mathbf{P}}^{(j)}(\mathbf{r},\omega) = \int_{-\infty}^{\infty} \mathbf{P}^{(j)}(\mathbf{r},t)e^{i\omega t}dt, \tag{1.1.18}$$

where $FT_t\,[\mathbf{E}(\mathbf{r}, t)]$ indicates the Fourier transform of $\mathbf{E}(\mathbf{r}, t)$ along the time coordinate. By using $\tilde{\mathbf{E}}(\mathbf{r}, \omega)$, $\tilde{\mathbf{P}}^{(1)}(\mathbf{r}, \omega)$ and $\tilde{\mathbf{P}}^{(j)}(\mathbf{r}, \omega)$ in the frequency domain, $\mathbf{E}(\mathbf{r}, t)$, $\mathbf{P}^{(1)}(\mathbf{r}, t)$, and $\mathbf{P}^{(j)}(\mathbf{r}, t)$ are expressed as

$$\mathbf{E}(\mathbf{r}, t) = FT_{\omega}^{-1}\,[\mathbf{E}(\mathbf{r}, \omega)] = \frac{1}{2\pi}\int_{-\infty}^{\infty} \tilde{\mathbf{E}}(\mathbf{r}, \omega) e^{-i\omega t} d\omega, \tag{1.1.19}$$

$$\mathbf{P}^{(1)}(\mathbf{r}, t) = \frac{1}{2\pi}\int_{-\infty}^{\infty} \tilde{\mathbf{P}}^{(1)}(\mathbf{r}, \omega) e^{-i\omega t} d\omega, \tag{1.1.20}$$

$$\mathbf{P}^{(j)}(\mathbf{r}, t) = \frac{1}{2\pi}\int_{-\infty}^{\infty} \tilde{\mathbf{P}}^{(j)}(\mathbf{r}, \omega) e^{-i\omega t} d\omega, \tag{1.1.21}$$

where $FT_{\omega}^{-1}\left[\tilde{\mathbf{E}}(\mathbf{r}, \omega)\right]$ represents the inverse Fourier transform of $\tilde{\mathbf{E}}(\mathbf{r}, \omega)$ along the frequency coordinate. Substituting Eqs. (1.1.19), (1.1.20), and (1.1.21) into Eq. (1.1.11), we obtain

$$\nabla^2 \tilde{\mathbf{E}}(\mathbf{r}, \omega) + \frac{\omega^2}{c^2}\tilde{\mathbf{E}}(\mathbf{r}, \omega) = -\mu_0 \omega^2 \left\{ \tilde{\mathbf{P}}^{(1)}(\mathbf{r}, \omega) + \sum_{j=2}^{\infty} \tilde{\mathbf{P}}^{(j)}(\mathbf{r}, \omega) \right\}. \tag{1.1.22}$$

From Eqs. (1.1.17) and (1.1.22), the wave equation can be rewritten as

$$\nabla^2 \tilde{\mathbf{E}}(\mathbf{r}, \omega) + \frac{\omega^2}{c^2}\left\{\mathbf{1} + \tilde{\chi}^{(1)}(\omega)\right\} \cdot \tilde{\mathbf{E}}(\mathbf{r}, \omega) = -\mu_0 \omega^2 \sum_{j=2}^{\infty} \tilde{\mathbf{P}}^{(j)}(\mathbf{r}, \omega). \tag{1.1.23}$$

The frequency-dependent dielectric constant $\tilde{\varepsilon}^{(1)}(\omega)$ is defined as

$$\tilde{\varepsilon}^{(1)}(\omega) = \mathbf{1} + \tilde{\chi}^{(1)}(\omega) = \left\{ \tilde{\mathbf{n}}^{(1)}(\omega) + i\frac{c}{2\omega}\tilde{\alpha}^{(1)}(\omega) \right\}^2, \tag{1.1.24}$$

where $\tilde{\mathbf{n}}^{(1)}(\omega)$ and $\tilde{\alpha}^{(1)}(\omega)$ are the linear refractive index and the linear absorption coefficient, respectively. From Eq. (1.1.24), we see that $\tilde{\mathbf{n}}^{(1)}(\omega)$ and $\tilde{\alpha}^{(1)}(\omega)$ are related to $\tilde{\chi}^{(1)}(\omega)$ by the relations

$$\tilde{\mathbf{n}}^{(1)}(\omega) = \mathbf{1} + \frac{1}{2}\mathrm{Re}\left\{\tilde{\chi}^{(1)}(\omega)\right\}, \tag{1.1.25}$$

$$\tilde{\mathbf{n}}^{(1)}(\omega) \cdot \tilde{\alpha}^{(1)}(\omega) = \frac{\omega}{c}\mathrm{Im}\left\{\tilde{\chi}^{(1)}(\omega)\right\}, \tag{1.1.26}$$

where Re $\{\tilde{\chi}^{(1)}(\omega)\}$ and Im $\{\tilde{\chi}^{(1)}(\omega)\}$ denote the real and imaginary parts of the complex $\tilde{\chi}^{(1)}(\omega) = \text{Re}\{\tilde{\chi}^{(1)}(\omega)\} + i\text{Im}\{\tilde{\chi}^{(1)}(\omega)\}$, respectively. By replacing $\mathbf{1} + \tilde{\chi}^{(1)}(\omega)$ in Eq. (1.1.23) by $\{\tilde{\mathbf{n}}^{(1)}(\omega) + i\frac{c}{2\omega}\tilde{\alpha}^{(1)}(\omega)\}^2$, we obtain

$$\nabla^2\tilde{\mathbf{E}}(\mathbf{r},\omega) + \frac{\omega^2}{c^2}\left\{\tilde{\mathbf{n}}^{(1)}(\omega) + i\frac{c}{2\omega}\tilde{\alpha}^{(1)}(\omega)\right\}^2 \cdot \tilde{\mathbf{E}}(\mathbf{r},\omega) = -\mu_0\omega^2\sum_{j=2}^{\infty}\tilde{\mathbf{P}}^{(j)}(\mathbf{r},\omega). \tag{1.1.27}$$

Using the relation $\tilde{\mathbf{k}}(\omega) = \tilde{\mathbf{n}}^{(1)}(\omega)\omega/c$, Eq. (1.1.27) we can express by

$$\nabla^2\tilde{\mathbf{E}}(\mathbf{r},\omega) + \left\{\tilde{\mathbf{k}}(\omega) + i\frac{1}{2}\tilde{\alpha}^{(1)}(\omega)\right\}^2 \cdot \tilde{\mathbf{E}}(\mathbf{r},\omega) = -\mu_0\omega^2\sum_{j=2}^{\infty}\tilde{\mathbf{P}}^{(j)}(\mathbf{r},\omega), \tag{1.1.28}$$

where $\tilde{\mathbf{k}}(\omega)$ is the wave vector.

1.1.1.2 Light Pulses Propagating as Plane Waves To solve Eq. (1.1.28), we make several simplifying assumptions [4]. First, $\tilde{\mathbf{P}}^{(j)}(\mathbf{r},\omega)$ is treated as a small perturbation to $\tilde{\mathbf{P}}^{(1)}(\mathbf{r},\omega)$. Second, the electric field is assumed to be linearly polarized in x direction, to propagate in the $+z$ direction and to maintain its polarization along z axis so that a scalar approach is valid. Third, we also assume that the field is uniform in the transverse x, y direction, forming a plane wave. Finally, we adopt the slowly varying envelope approximation (SVEA), which is used for separating the rapidly varying part of the electric field (Fig. 1.1). Then, the solution for a plane wave at a carrier frequency of ω_c can be written as

$$E_x(\mathbf{r},t) = \frac{1}{2}\left\{A(z,t)e^{i(k_c z-\omega_c t)} + c.c.\right\}, \tag{1.1.29}$$

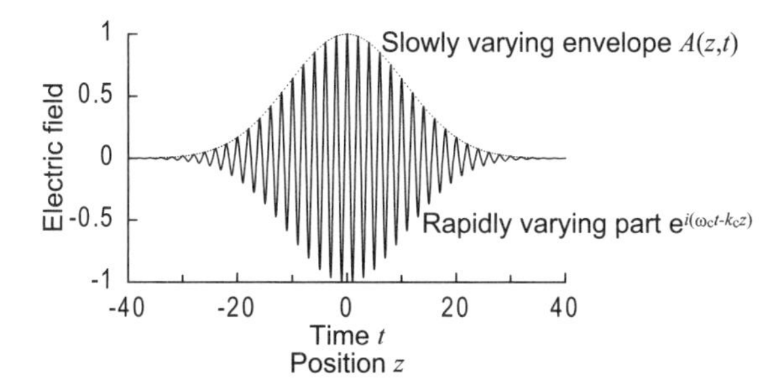

FIGURE 1.1 Slowly varying envelope approximation is used for separating the rapidly varying part (dotted line) of the electric field (solid line).

where $k_c = \tilde{k}(\omega_c) = \tilde{n}^{(1)}(\omega_c)\omega_c/c$ is the wave number at a carrier frequency of ω_c. $A(z,t)$ is the slowly varying function of z and t. The Fourier transform of the electric field of $E_x(\mathbf{r},t)$ is expressed by

$$\tilde{E}_x(\mathbf{r},\omega) = \frac{1}{2}\left\{\tilde{A}(z,\omega-\omega_c)e^{ik_c z} + c.c.\right\}. \tag{1.1.30}$$

The polarization component oscillating with the frequency ω_c can also be expressed in a similar way by writing

$$P_x^{(j)}(\mathbf{r},t) = \frac{\varepsilon_0\chi^{(j)}}{2}\left\{p^{(j)}(z,t)e^{i\left(k_p^{(j)}z-\omega_c t\right)} + c.c.\right\}, \tag{1.1.31}$$

$$\tilde{P}_x^{(j)}(\mathbf{r},\omega) = \frac{\varepsilon_0\chi^{(j)}}{2}\left\{\tilde{p}^{(j)}(z,\omega-\omega_c)e^{ik_p^{(j)}z} + c.c.\right\}, \tag{1.1.32}$$

where $k_p^{(j)}$ is the wave number of jth-order nonlinear polarization. By substituting Eqs. (1.1.30) and (1.1.32) into Eq. (1.1.28), we obtain the following equation for $A(z,\omega)$:

$$\begin{aligned}
&\frac{\partial^2\tilde{A}(z,\omega-\omega_c)}{\partial z^2} + i2k_c\frac{\partial\tilde{A}(z,\omega-\omega_c)}{\partial z} \\
&\quad + \left[\left\{\tilde{k}(\omega) + i\frac{1}{2}\tilde{\alpha}^{(1)}(\omega)\right\}^2 - k_c^2\right]\tilde{A}(z,\omega-\omega_c) \\
&\quad = -\frac{\omega^2}{c^2}\sum_{j=2}^{\infty}\chi^{(j)}\tilde{p}^{(j)}(z,\omega-\omega_c)e^{i\Delta k^{(j)}z},
\end{aligned} \tag{1.1.33}$$

where the relation $\mu_0\varepsilon_0 = 1/c^2$ was used and we have introduced the quantity $\Delta k^{(j)} = k_p^{(j)} - k_c$, which is called the wave vector mismatch. The second derivative $\partial^2 A/\partial z^2$ in the left-hand side of Eq. (1.1.33) can be neglected since $\tilde{A}(z,\omega)$ is assumed to be a slowly varying function of z. Thus, Eq. (1.1.33) is rewritten as

$$\begin{aligned}
&\frac{\partial\tilde{A}(z,\omega-\omega_c)}{\partial z} + \left[-i\left\{\tilde{k}(\omega) - k_c\right\} + \frac{1}{2}\tilde{\alpha}^{(1)}(\omega)\right]\tilde{A}(z,\omega-\omega_c) \\
&\quad = i\frac{\omega}{2\tilde{n}^{(1)}(\omega_c)c}\sum_{j=2}^{\infty}\chi^{(j)}\tilde{p}^{(j)}(z,\omega-\omega_c)e^{i\Delta k^{(j)}z},
\end{aligned} \tag{1.1.34}$$

where $\{\tilde{k}(\omega) + i\tilde{\alpha}^{(1)}(\omega)/2\}^2$, $\tilde{k}^2(\omega) - k_c^2$, $\tilde{n}(\omega)/\tilde{n}(\omega_c)$, and ω/ω_c were approximated by $\tilde{k}^2(\omega) + i\tilde{k}(\omega)\tilde{\alpha}^{(1)}(\omega)$, $2k_c\{\tilde{k}(\omega) - k_c\}$, 1 and 1, respectively. For further consideration we expand $\tilde{k}(\omega)$ and $\tilde{\alpha}^{(1)}(\omega)$ in a Taylor series around the carrier frequency ω_c as

$$\tilde{k}(\omega) = \sum_{m=0}^{\infty} \frac{1}{m!} (\omega - \omega_c)^m \left(\frac{d^m \tilde{k}}{d\omega^m} \right)_{\omega=\omega_c}, \tag{1.1.35}$$

$$\tilde{\alpha}^{(1)}(\omega) = \sum_{m=0}^{\infty} \frac{1}{m!} (\omega - \omega_c)^m \left(\frac{d^m \tilde{\alpha}^{(1)}}{d\omega^m} \right)_{\omega=\omega_c}. \tag{1.1.36}$$

By inserting Eqs. (1.1.35) and (1.1.36) into Eq. (1.1.34), we obtain

$$\begin{aligned} &\frac{\partial \tilde{A}(z, \omega - \omega_c)}{\partial z} - i \sum_{m=1}^{\infty} \frac{(\omega - \omega_c)^m}{m!} \left(\frac{d\tilde{k}}{d\omega} \right)_{\omega=\omega_c} \tilde{A}(z, \omega - \omega_c) \\ &+ \frac{1}{2} \sum_{m=0}^{\infty} \frac{(\omega - \omega_c)^m}{m!} \left(\frac{d\tilde{\alpha}^{(1)}}{d\omega} \right)_{\omega=\omega_c} \tilde{A}(z, \omega - \omega_c) \\ &= i \frac{\omega}{2\tilde{n}^{(1)}(\omega_c) c} \sum_{j=2}^{\infty} \chi^{(j)} \tilde{p}^{(j)}(z, \omega - \omega_c) e^{i \Delta k^{(j)} z}. \end{aligned} \tag{1.1.37}$$

Now we take the inverse Fourier transform of Eq. (1.1.37) using the following two equations:

$$A(z, t) = \frac{1}{2\pi} \int_{-\infty}^{\infty} \tilde{A}(z, \omega - \omega_c) e^{-i(\omega - \omega_c)t} d\omega, \tag{1.1.38}$$

$$p^{(j)}(z, t) = \frac{1}{2\pi} \int_{-\infty}^{\infty} \tilde{p}^{(j)}(z, \omega - \omega_c) e^{-i(\omega - \omega_c)t} d\omega. \tag{1.1.39}$$

During the Fourier-transform operation, $\omega - \omega_c$ is replaced by the differential operator $i\,(\partial/\partial t)$. The resulting equation for $A(z, t)$ becomes

$$\begin{aligned} &\frac{\partial A(z, t)}{\partial z} - i \sum_{m=1}^{\infty} i^m \frac{1}{m!} \left(\frac{d^m \tilde{k}}{d\omega^m} \right)_{\omega=\omega_c} \frac{\partial^m A(z, t)}{\partial t^m} \\ &+ \frac{1}{2} \sum_{m=0}^{\infty} i^m \frac{1}{m!} \left(\frac{d^m \tilde{\alpha}^{(1)}}{d\omega^m} \right)_{\omega=\omega_c} \frac{\partial^m A(z, t)}{\partial t^m} \\ &= i \frac{1}{2\tilde{n}^{(1)}(\omega_c) c} \left(\omega_c + i \frac{\partial}{\partial t} \right) \sum_{j=2}^{\infty} \chi^{(j)} p^{(j)}(z, t) e^{i \Delta k^{(j)} z}. \end{aligned} \tag{1.1.40}$$

The second and third terms in the left-hand side of Eq. (1.1.40) govern the effects of the chromatic dispersion and the linear absorption whereas the term in the right-hand side has the origin in nonlinear optical phenomena.

A monochromatic wave with a frequency of ω_c travels at the phase velocity $v_p(\omega_c) = c/\tilde{n}(\omega_c)$, while an ultrashort pulse with a center frequency of ω_c is the superposition of many such waves with different phase velocities that leads to a wave packet propagating with the group velocity $v_g(\omega_c) = 1/(d\tilde{k}/d\omega)_{\omega=\omega_c}$. In dealing with ultrashort pulses, the appropriate retarded frame of reference is moving at the group velocity rather than at the phase velocity. In dealing with ultrashort pulses, we transfer to a coordinate system (Z, T),

$$Z = z, \tag{1.1.41}$$

$$T = t - \frac{z}{v_g} = t - z\left(\frac{d\tilde{k}}{d\omega}\right)_{\omega=\omega_c}, \tag{1.1.42}$$

so that the retarded frame of reference is moving at the group velocity rather than at the phase velocity. Using the following relation obtains

$$\frac{\partial}{\partial z} = \frac{\partial Z}{\partial z}\frac{\partial}{\partial Z} + \frac{\partial T}{\partial z}\frac{\partial}{\partial T} = \frac{\partial}{\partial Z} - \left(\frac{d\tilde{k}}{d\omega}\right)_{\omega=\omega_c}\frac{\partial}{\partial T}, \tag{1.1.43}$$

$$\frac{\partial}{\partial t} = \frac{\partial Z}{\partial t}\frac{\partial}{\partial Z} + \frac{\partial T}{\partial t}\frac{\partial}{\partial T} = \frac{\partial}{\partial T}. \tag{1.1.44}$$

Eq. (1.1.40) is rewritten as

$$\begin{aligned}&\frac{\partial A(Z,T)}{\partial Z} - i\sum_{m=2}^{\infty} i^m \frac{1}{m!}\left(\frac{d^m\tilde{k}}{d\omega^m}\right)_{\omega=\omega_c}\frac{\partial^m A(Z,T)}{\partial T^m}\\ &+\frac{1}{2}\sum_{m=0}^{\infty} i^m \frac{1}{m!}\left(\frac{d^m\tilde{\alpha}^{(1)}}{d\omega^m}\right)_{\omega=\omega_c}\frac{\partial^m A(Z,T)}{\partial T^m}\\ &= i\frac{1}{2\tilde{n}^{(1)}(\omega_c)c}\left(\omega_c + i\frac{\partial}{\partial T}\right)\sum_{j=2}^{\infty}\chi^{(j)}p^{(j)}(Z,T)e^{i\Delta k^{(j)}Z}.\end{aligned} \tag{1.1.45}$$

If the pulse duration is much longer than an optical period, the first derivative $\partial p^{(j)}/\partial T$ in the right-hand side of Eq. (1.1.45) may be neglected. Then, Eq. (1.1.45) is rewritten as

$$\begin{aligned}&\frac{\partial A(Z,T)}{\partial Z} - i\sum_{m=2}^{\infty} i^m \frac{1}{m!}\left(\frac{d^m\tilde{k}}{d\omega^m}\right)_{\omega=\omega_c}\frac{\partial^m A(Z,T)}{\partial T^m}\\ &+\frac{1}{2}\sum_{m=0}^{\infty} i^m \frac{1}{m!}\left(\frac{d^m\tilde{\alpha}^{(1)}}{d\omega^m}\right)_{\omega=\omega_c}\frac{\partial^m A(Z,T)}{\partial T^m}\\ &= i\frac{\omega_c}{2\tilde{n}^{(1)}(\omega_c)c}\sum_{j=2}^{\infty}\chi^{(j)}p^{(j)}(Z,T)e^{i\Delta k^{(j)}Z}.\end{aligned} \tag{1.1.46}$$

If the linear optical loss and the chromatic dispersion can be neglected, we obtain

$$\frac{\partial A(Z,T)}{\partial Z} = i\frac{\omega_c}{2\tilde{n}^{(1)}(\omega_c)c}\sum_{j=2}^{\infty}\chi^{(j)}p^{(j)}(Z,T)e^{i\Delta k^{(j)}Z}. \tag{1.1.47}$$

We will encounter Eq. (1.1.47) again in Sections 1.1.2 and 1.1.3 when we discuss in detail second-order and third-order nonlinear optical phenomena. If the linear optical loss and the nonlinear optical phenomena can be neglected, we obtain

$$\frac{\partial A(Z,T)}{\partial Z} = \sum_{m=2}^{\infty} i^{m+1}\frac{1}{m!}\left(\frac{d^m\tilde{k}}{d\omega^m}\right)_{\omega=\omega_c}\frac{\partial^m A(Z,T)}{\partial T^m}, \tag{1.1.48}$$

By employing Eq. (1.1.48), the effect of dispersion is discussed in Section 1.3.2.

1.1.1.3 Light Pulse Propagating as Gaussian Beams

So far we have described the time-varying field with only one spatial coordinate by assuming light pulse propagating as plane waves. Here we will discuss solely the situation where the change of pulse characteristics can be separated from the spatial beam profile. Again, we assume that the electric field is linearly polarized in the x direction, to propagate in the $+z$ direction and to maintain its polarization along z-axis so that a scalar approach is valid. Then, the solution at a carrier frequency of ω_c can be written as

$$E_x(\mathbf{r},t) = \frac{1}{2}\left\{U(x,y,z)A(t)e^{i(k_c z-\omega_c t)} + c.c.\right\}, \tag{1.1.49}$$

where $U(x,y,z)$ is the complex amplitude function of spatial coordinates of x, y, and z and is the slowly varying with z. $A(t)$ is the slowly varying complex amplitude function of time. The Fourier transform of the electric field of $E_x(\mathbf{r},t)$ along the time t is expressed by

$$\tilde{E}_x(\mathbf{r},\omega) = \frac{1}{2}\left\{U(x,y,z)\tilde{A}(\omega-\omega_c)e^{ik_c z} + c.c.\right\}. \tag{1.1.50}$$

The polarization component oscillating with the frequency ω_c can also be expressed by

$$P_x^{(j)}(\mathbf{r},t) = \frac{\varepsilon_0\chi^{(j)}}{2}\left\{U_p^{(j)}(x,y,z)p^{(j)}(t)e^{i\left(k_p^{(j)}z-\omega_c t\right)} + c.c.\right\}, \tag{1.1.51}$$

$$\tilde{P}_x^{(j)}(\mathbf{r},\omega) = \frac{\varepsilon_0\chi^{(j)}}{2}\left\{U_p^{(j)}(x,y,z)\tilde{p}^{(j)}(\omega-\omega_c)e^{ik_p^{(j)}z} + c.c.\right\}, \tag{1.1.52}$$

where $k_p^{(j)}$ is the wave number of jth-order nonlinear polarization. Insertion of Eqs. (1.1.49) and (1.1.52) into Eq. (1.1.28) yields

$$\left[\frac{\partial^2}{\partial x^2}+\frac{\partial^2}{\partial y^2}+i2k_c\frac{\partial}{\partial z}+\left\{\tilde{k}(\omega)+i\frac{1}{2}\tilde{\alpha}^{(1)}(\omega)\right\}^2-k_c^2\right]U(x,y,z)\tilde{A}(\omega-\omega_c)$$
$$=-\frac{\omega^2}{c^2}\sum_{j=2}^{\infty}\chi^{(j)}U_p^{(j)}(x,y,z)\tilde{p}^{(j)}(\omega-\omega_c)e^{i\Delta k^{(j)}z}, \qquad (1.1.53)$$

where the second derivative $\partial^2 U/\partial z^2$ was neglected because $U(x, y, z)$ is the slowly varying with z. Here we also used the relation $\varepsilon_0\mu_0 = 1/c^2$ and have introduced the quantity $\Delta k^{(j)} = k_p^{(j)} - k_c$. If $\{\tilde{k}(\omega) + i\tilde{\alpha}^{(1)}(\omega)/2\}^2$, $\tilde{k}^2(\omega) - k_c^2$, $\tilde{n}(\omega)/\tilde{n}(\omega_c)$ and ω/ω_c can be approximated by $\tilde{k}^2(\omega) + i\tilde{k}(\omega)\tilde{\alpha}^{(1)}(\omega)$, $2k_c\{\tilde{k}(\omega) - k_c\}$, 1 and 1, respectively Eq. (1.1.53) can be written as

$$\left[-\frac{i}{2k_c}\left(\frac{\partial^2}{\partial x^2}+\frac{\partial^2}{\partial y^2}\right)+\frac{\partial}{\partial z}-i\left\{\tilde{k}(\omega)-k_c\right\}+\frac{1}{2}\tilde{\alpha}^{(1)}(\omega)\right]U(x,y,z)\tilde{A}(\omega-\omega_c)$$
$$=i\frac{\omega}{2\tilde{n}^{(1)}(\omega_c)c}\sum_{j=2}^{\infty}\chi^{(j)}U_p^{(j)}(x,y,z)\tilde{p}^{(j)}(\omega-\omega_c)e^{i\Delta k^{(j)}z}, \qquad (1.1.54)$$

When $\tilde{k}(\omega)$ and $\tilde{\alpha}^{(1)}(\omega)$ can be approximated by k_c and 0, respectively, Eq. (1.1.54) can be rewritten as

$$\left[-\frac{i}{2k_c}\left(\frac{\partial^2}{\partial x^2}+\frac{\partial^2}{\partial y^2}\right)+\frac{\partial}{\partial z}\right]U(x,y,z)\tilde{A}(\omega-\omega_c)$$
$$=i\frac{\omega}{2\tilde{n}^{(1)}(\omega_c)c}\sum_{j=2}^{\infty}\chi^{(j)}U_p^{(j)}(x,y,z)\tilde{p}^{(j)}(\omega-\omega_c)e^{i\Delta k^{(j)}z}. \qquad (1.1.55)$$

Now we take the inverse Fourier transform of Eq. (1.1.55). The resulting equation becomes

$$\left[-\frac{i}{2k_c}\left(\frac{\partial^2}{\partial x^2}+\frac{\partial^2}{\partial y^2}\right)+\frac{\partial}{\partial z}\right]U(x,y,z)A(t)$$
$$=i\frac{1}{2\tilde{n}^{(1)}(\omega_c)c}\left(\omega_c+i\frac{\partial}{\partial t}\right)\sum_{j=2}^{\infty}\chi^{(j)}U_p^{(j)}(x,y,z)p^{(j)}(t)e^{i\Delta k^{(j)}z}. \qquad (1.1.56)$$

In the case of the free propagation, as $\chi^{(j)} = 0$, Eq. (1.1.56) can be rewritten by

$$\left(\frac{\partial^2}{\partial x^2} + \frac{\partial^2}{\partial y^2} + 2ik_c\frac{\partial}{\partial z}\right) U(x, y, z) = 0. \tag{1.1.57}$$

This equation is known as the paraxial wave equation. By taking the Fourier transform of Eq. (1.1.57) along the space coordinates x and y, we obtain

$$\frac{\partial \tilde{U}(k_x, k_y, z)}{\partial z} = -\frac{i}{2k_c}\left(k_x^2 + k_y^2\right) \tilde{U}(k_x, k_y, z), \tag{1.1.58}$$

where

$$\tilde{U}(k_x, k_y, z) = \int_{-\infty}^{\infty}\int_{-\infty}^{\infty} U(x, y, z)e^{-i(k_x x + k_y y)} dxdy. \tag{1.1.59}$$

By integrating Eq. (1.1.58), we can obtain the integral form of Fresnel equation:

$$\tilde{U}(k_x, k_y, z) = \tilde{U}(k_x, k_y, 0) \exp\left\{-\frac{i}{2k_c}\left(k_x^2 + k_y^2\right) z\right\}. \tag{1.1.60}$$

The paraxial wave equation has many known solutions. An important particular solution of the wave equation within the paraxial approximation is the Gauusian beam [5], which can be written in the form

$$U(x, y, z) = \frac{U_c}{\sqrt{1 + z^2/\rho_c^2}} e^{-(x^2+y^2)/W_c^2(z)} e^{ik_c(x^2+y^2)/2R_c(z)} e^{-i\Phi_c(z)}, \tag{1.1.61}$$

where

$$W_c(z) = w_c\sqrt{\frac{1 + z^2}{\rho_c^2}} \tag{1.1.62}$$

$$R_c(z) = z\left[\frac{1 + \rho_c^2}{z^2}\right] \tag{1.1.63}$$

$$\Phi_c(z) = \arctan\left(\frac{z}{\rho_c}\right) \tag{1.1.64}$$

$$\rho_c = \frac{n\pi w_c^2}{\lambda_c} = \frac{1}{2}k_c w_c^2. \tag{1.1.65}$$

The origin of z-axis is chosen to be the position of the beam waist $w_c = W_c(z = 0)$. The radius curvature of planes of constant phase is $R_c(z)$. $\Phi_c(z)$ is the phase shift of π radians that any beams of light experiences in passing through its focus, and is called the Gouy phase shift [6]. The length ρ_c is called the Rayleigh range, and $2\rho_c$ is the confocal parameter. The angular divergence of the beam in the far field is given by

$$\theta_{\mathrm{ff}} = \frac{\lambda_c}{n\pi w_c}. \tag{1.1.66}$$

Sometimes it is convenient to represent the Gaussian beam in the more compact form:

$$U(x, y, z) = \frac{U_c}{1 + i\zeta_c} e^{-(x^2+y^2)/w_c^2(1+i\zeta_c)}, \tag{1.1.67}$$

where

$$\zeta_c = z \overline{\rho_c}. \tag{1.1.68}$$

1.1.2 Second-Order Nonlinear Optical Phenomena

Let us discuss the second-order nonlinear optical phenomena. The second order contribution to the nonlinear polarization is expressed by

$$\begin{aligned} P_i^{(2)}(\mathbf{r}, t) = \varepsilon_0 \int_{-\infty}^{t} \int_{-\infty}^{t} \sum_{j,k} \chi_{ijk}^{(2)}(t - t_1, t_1 - t_2) \\ \times E_j(\mathbf{r}, t_1) E_k(\mathbf{r}, t_2) dt_1 dt_2, \end{aligned} \tag{1.1.69}$$

where $\chi_{ijk}^{(2)}$ are the components of the second-order susceptibility tensor. Here the indexes ijk refer to the cartesian components (x, y, and z). Now we consider the second-order nonlinear optical phenomena in a centrosymmetric medium. If the sign of the applied electric field is changed, the sign of the induced nonlinear polarization must also changed:

$$\begin{aligned} -P_i^{(2)}(\mathbf{r}, t) = \varepsilon_0 \int_{-\infty}^{t} \int_{-\infty}^{t} \sum_{j,k} \chi_{ijk}^{(2)}(t - t_1, t_1 - t_2) \\ \times \left[-E_j(\mathbf{r}, t_1)\right] \left[-E_k(\mathbf{r}, t_2)\right] dt_1 dt_2. \end{aligned} \tag{1.1.70}$$

In order to satisfy Eqs. (1.1.69) and (1.1.70) simultaneously, $\chi_{ijk}^{(2)}$ must be zero. Thus, the second-order susceptibility $\chi_{ijk}^{(2)}$ vanishes for centrosymmetric media. The second-order susceptibility is nonzero only for media that lack an inversion symmetry at the molecular level. However, even when the symmetry at the molecular level is broken, the second-order susceptibility vanishes for amorphous materials and solutions with random structure such as silica glass and water.

By taking directly the Fourier transform of Eq. (1.1.69), we find that the second-order nonlinear polarization in the frequency domain is

$$
\begin{aligned}
\tilde{P}_i^{(2)}(\mathbf{r},\omega) &= \varepsilon_0 \int_{-\infty}^{t}\int_{-\infty}^{t}\sum_{j,k}\int_{-\infty}^{\infty} \chi_{ijk}^{(2)}(t-t_1,t_1-t_2)e^{i\omega t}dt \\
&\quad \times E_j(\mathbf{r},t_1)E_k(\mathbf{r},t_2)dt_1dt_2 \\
&= \frac{\varepsilon_0}{(2\pi)^2}\int_{-\infty}^{t}\int_{-\infty}^{t}\sum_{j,k}\int_{-\infty}^{\infty}\int_{-\infty}^{\infty}\int_{-\infty}^{\infty}\tilde{\chi}_{ijk}^{(2)}(\Omega_1,\Omega_2) \\
&\quad \times E_j(\mathbf{r},t_1)e^{i(\Omega_1-\Omega_2)t_1}E_k(\mathbf{r},t_2)e^{i\Omega_2 t_2}e^{i(\omega-\Omega_1)t}dt_1dt_2dtd\Omega_1d\Omega_2 \\
&= \frac{\varepsilon_0}{2\pi}\sum_{j,k}\int_{-\infty}^{\infty}\int_{-\infty}^{\infty}\tilde{\chi}_{ijk}^{(2)}(\Omega_1,\Omega_2) \\
&\quad \times \tilde{E}_j(\mathbf{r},\Omega_1-\Omega_2)\tilde{E}_k(\mathbf{r},\Omega_2)\delta(\omega-\Omega_1)d\Omega_1d\Omega_2 \\
&= \frac{\varepsilon_0}{2\pi}\sum_{j,k}\int_{-\infty}^{\infty}\tilde{\chi}_{ijk}^{(2)}(\omega,\Omega_2)\tilde{E}_j(\mathbf{r},\omega-\Omega_2)\tilde{E}_k(\mathbf{r},\Omega_2)d\Omega_2, \qquad (1.1.71)
\end{aligned}
$$

where

$$\tilde{\chi}_{ijk}^{(2)}(\Omega_1,\Omega_2) = \int_{-\infty}^{\infty}\int_{-\infty}^{\infty}\chi_{ijk}^{(2)}(t_1,t_2)e^{i\Omega_1 t_1}e^{i\Omega_2 t_2}dt_1dt_2, \qquad (1.1.72)$$

$$\chi_{ijk}^{(2)}(t_1,t_2) = \frac{1}{(2\pi)^2}\int_{-\infty}^{\infty}\int_{-\infty}^{\infty}\tilde{\chi}_{ijk}^{(2)}(\Omega_1,\Omega_2)e^{-i\Omega_1 t_1}e^{-i\Omega_2 t_2}d\Omega_1d\Omega_2. \qquad (1.1.73)$$

If the second-order susceptibility is not frequency dependent, we obtain

$$\tilde{P}_i^{(2)}(\mathbf{r},\omega) = \frac{\varepsilon_0}{2\pi}\sum_{j,k}\chi_{ijk}^{(2)}\int_{-\infty}^{\infty}\tilde{E}_j(\mathbf{r},\Omega)\tilde{E}_k(\mathbf{r},\omega-\Omega)d\Omega. \qquad (1.1.74)$$

If a light pulse propagates along z with only a field component along x, second-order nonlinear polarization is described by

$$P_x^{(2)}(\mathbf{r},t) = \varepsilon_0\int_{-\infty}^{t}\int_{-\infty}^{t}\chi_{xxx}^{(2)}(t-t_1,t_1-t_2)\,E_x(\mathbf{r},t_1)E_x(\mathbf{r},t_2)dt_1dt_2. \qquad (1.1.75)$$

This convolution takes a simple form in the case of an instantaneous nonlinearity:

$$\chi_{xxx}^{(2)}(t-t_1,t_1-t_2) = \chi_0^{(2)}\delta(t-t_1)\delta(t-t_2). \qquad (1.1.76)$$

Then, the second-order nonlinear polarization is written as

$$P_x^{(2)}(\mathbf{r}, t) = \varepsilon_0 \chi_0^{(2)} E_x^2(\mathbf{r}, t). \tag{1.1.77}$$

Fourier-transforming Eq. (1.1.77) yields

$$P_x^{(2)}(\mathbf{r}, \omega) = \frac{\varepsilon_0 \chi_0^{(2)}}{2\pi} \int_{-\infty}^{\infty} E_x(\mathbf{r}, \Omega) E_x(\mathbf{r}, \omega - \Omega) d\Omega. \tag{1.1.78}$$

The result of Eq. (1.1.78) corresponds to that of Eq. (1.1.74). This is a manifestation of the fact that an instantaneous response is characterized by nondispersive material properties.

Now we consider the circumstance where the incident optical field consists of two distinct frequency components ω_1 and ω_2 ($\omega_1 \geq \omega_2$), which we represent in the form

$$E_x(\mathbf{r}, t) = \sum_{m=1}^{2} E_{mx}(\mathbf{r}, t), \tag{1.1.79}$$

$$E_{mx}(\mathbf{r}, t) = \frac{1}{2} \left\{ A_m(z, t) e^{i(k_m z - \omega_m t)} + c.c. \right\}. \tag{1.1.80}$$

Here $A_m(z, t)$ is a slowly varying function of z. ω_m and k_m are the carrier frequency and the wave number, respectively. Substitution of Eq. (1.1.79) into Eq. (1.1.77) leads to the result

$$\begin{aligned} P_x^{(2)}(\mathbf{r}, t) = &\frac{1}{4} \varepsilon_0 \chi_0^{(2)} \\ &\times \left[A_1^2(z, t) e^{i(2k_1 z - 2\omega_1 t)} + A_2^2(z, t) e^{i(2k_2 z - 2\omega_2 t)} \right. \\ &+ 2A_1(z, t) A_2(z, t) e^{i\{(k_1 + k_2)z - (\omega_1 + \omega_2)t\}} \\ &+ 2A_1(z, t) A_2^*(z, t) e^{i\{(k_1 - k_2)z - (\omega_1 - \omega_2)t\}} \\ &\left. + 2\left|A_1(z, t)\right|^2 + 2\left|A_2(z, t)\right|^2 \right] \\ &+ c.c. \end{aligned} \tag{1.1.81}$$

We find that $P_x^{(2)}(\mathbf{r}, t)$ is composed of the terms oscillating with frequencies of $2\omega_1$, $2\omega_2$, $\omega_1 + \omega_2$, $\omega_1 - \omega_2$, and 0. The first two terms containing $e^{-i2\omega_1 t}$ and $e^{-i2\omega_2 t}$ in Eq. (1.1.81) result in second-harmonic generation (SHG). The third term including $e^{-i(\omega_1 + \omega_2)t}$ leads to sum-frequency generation (SFG). The fourth term involving $e^{-i(\omega_1 - \omega_2)t}$ contributes to difference-frequency generation (DFG). The remaining two terms at a frequency of 0 are responsible for optical rectification (OR). These physical processes are described in this section.

Next we consider the incident optical field consisting of two distinct frequency components ω_1 and ω_2 ($\omega_1 \geq \omega_2$), that are expressed in the Gaussian form

$$E_{mx}(\mathbf{r}, t) = \frac{1}{2}\left\{U_m(x, y, z)A_m(t)e^{i(k_m z - \omega_m t)} + c.c.\right\}, \tag{1.1.82}$$

where

$$U_m(x, y, z) = \frac{U_m}{1 + i\zeta_m} e^{-(x^2+y^2)/w_m^2(1+i\zeta_m)}. \tag{1.1.83}$$

Insertion of Eqs. (1.1.79) and (1.1.82) into Eq. (1.1.77) yields

$$\begin{aligned} P_x^{(2)}(\mathbf{r}, t) = \frac{1}{4}\varepsilon_0\chi_0^{(2)} & \\ \times \big[&U_1^2(x, y, z)A_1^2(t)e^{i(2k_1 z - 2\omega_1 t)} + U_2^2(x, y, z)A_2^2(t)e^{i(2k_2 z - 2\omega_2 t)} \\ &+ 2U_1(x, y, z)U_2(x, y, z)A_1(t)A_2(t)e^{i\{(k_1+k_2)z - (\omega_1+\omega_2)t\}} \\ &+ 2U_1(x, y, z)U_2^*(x, y, z)A_1(t)A_2^*(t)e^{i\{(k_1-k_2)z - (\omega_1-\omega_2)t\}} \\ &+ 2\,|U_1(x, y, z)A_1(t)|^2 + 2\,|U_2(x, y, z)A_2(t)|^2\big] \\ &+ c.c. \end{aligned} \tag{1.1.84}$$

$P_x^{(2)}(\mathbf{r}, t)$ for Gaussian beams also consists of the terms oscillating with frequencies of $2\omega_1$, $2\omega_2$, $\omega_1 + \omega_2$, $\omega_1 - \omega_2$, and 0.

1.1.2.1 Properties of Nonlinear Susceptibility Let us study some of the formal symmetry properties of the nonlinear susceptibility. Because j, k, 1, and 2 are dummy indexes, we could just as well have written the nonlinear susceptibility expressed by Eq. (1.1.71) with j interchanged with k and with 1 interchanged with 2, that is, as

$$\begin{aligned} \tilde{P}_i^{(2)}(\mathbf{r}, \omega) &= \varepsilon_0 \int_{-\infty}^{t}\int_{-\infty}^{t}\sum_{j,k}\int_{-\infty}^{\infty}\chi_{ikj}^{(2)}(t - t_2, t_2 - t_1)e^{i\omega t}dt \\ &\quad \times E_k(\mathbf{r}, t_2)E_j(\mathbf{r}, t_1)dt_1 dt_2 \\ &= \frac{\varepsilon_0}{2\pi}\sum_{j,k}\int_{-\infty}^{\infty}\tilde{\chi}_{ikj}^{(2)}(\omega, \omega - \Omega_1) \\ &\quad \times \tilde{E}_k(\mathbf{r}, \Omega_1)\tilde{E}_j(\mathbf{r}, \omega - \Omega_1)d\Omega_1. \end{aligned} \tag{1.1.85}$$

Since Eq. (1.1.71) is equal to Eq. (1.1.85), we see that the nonlinear susceptibility is unchanged by the simultaneous interchange of its last two frequency arguments and its last two cartesian indexes:

$$\tilde{\chi}_{ijk}^{(2)}(\omega, \Omega) = \tilde{\chi}_{ikj}^{(2)}(\omega, \omega - \Omega). \tag{1.1.86}$$

This property is known as "intrinsic permutation symmetry." The intrinsic permutation symmetry is written in an easy to understand form as

$$\tilde{\chi}_{ijk}^{(2)}(\omega, \Omega, \omega - \Omega) = \tilde{\chi}_{ikj}^{(2)}(\omega, \omega - \Omega, \Omega), \tag{1.1.87}$$

where the first frequency argument is always the sum of the latter two frequency arguments.

The additional symmetries of the nonlinear susceptibility are valid for the case of a lossless nonlinear medium. This condition is satisfied whenever all of optical frequencies are detuned from the resonance frequencies of the medium. Under this condition, all of the frequency arguments of the nonlinear susceptibility can be freely interchanged, as long as the corresponding cartesian indexes are interchanged simultaneously:

$$\tilde{\chi}_{ijk}^{(2)}(\omega, \Omega, \omega - \Omega) = \tilde{\chi}_{jki}^{(2)}(\Omega, \Omega - \omega, \omega) = \tilde{\chi}_{kij}^{(2)}(\omega - \Omega, \omega, -\Omega), \tag{1.1.88}$$

where the signs of the frequencies are adjusted to maintain the convention that the first argument is always the sum of the latter two when the first frequency is interchanged with either of the later two [3]. This property is known as "full permutation symmetry," which is satisfied for a lossless medium. Finally, we consider the condition that all of optical frequencies are much smaller than the lowest resonance frequency of the material system. Then, the nonlinear susceptibility is essentially independent of frequency. Under this condition, the medium is necessarily lossless. Thus, the full permutation symmetry is valid:

$$\begin{aligned}
\tilde{\chi}_{ijk}^{(2)}(\omega, \Omega, \omega - \Omega) &= \tilde{\chi}_{jki}^{(2)}(\Omega, \Omega - \omega, \omega) = \tilde{\chi}_{kij}^{(2)}(\omega - \Omega, \omega, -\Omega) \\
&= \tilde{\chi}_{ikj}^{(2)}(\omega, \omega - \Omega, \Omega) = \tilde{\chi}_{jik}^{(2)}(\Omega, \omega, \Omega - \omega) \\
&= \tilde{\chi}_{kji}^{(2)}(\omega - \Omega, -\Omega, \omega).
\end{aligned} \tag{1.1.89}$$

However, under the present condition, the nonlinear susceptibility is independent of frequency. Thus, we can freely permute the cartesian indexes without permuting the frequencies:

$$\begin{aligned}
\tilde{\chi}_{ijk}^{(2)}(\omega, \Omega, \omega - \Omega) &= \tilde{\chi}_{jki}^{(2)}(\omega, \Omega, \omega - \Omega) = \tilde{\chi}_{kij}^{(2)}(\omega, \Omega, \omega - \Omega) \\
&= \tilde{\chi}_{ikj}^{(2)}(\omega, \Omega, \omega - \Omega) = \tilde{\chi}_{jik}^{(2)}(\omega, \Omega, \omega - \Omega) \\
&= \tilde{\chi}_{kji}^{(2)}(\omega, \Omega, \omega - \Omega).
\end{aligned} \tag{1.1.90}$$

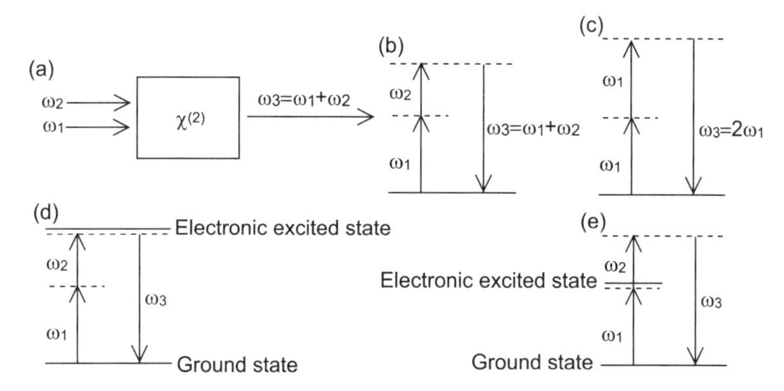

FIGURE 1.2 Sum-frequency generation (SFG) and second-harmonic generation (SHG). (a) Geometry of the interaction for SFG. (b, c) Energy diagrams describing the SFG process (b) and SHG process (c) under nonresonant condition. (d, e) Energy diagrams for the SFG processes under two-photon resonant (d) and one-photon resonant (e) conditions.

This result is known as the Kleinman symmetry condition, which is valid whenever dispersion of the susceptibility can be neglected.

1.1.2.2 Second Harmonic and Sum-Frequency Generation Let us now consider the processes of second-harmonic generation (SHG) and sum-frequency generation (SFG), which are illustrated in Fig. 1.2. In the SFG process, two photons of frequencies ω_1 and ω_2 are destroyed and one photon of frequency of $\omega_3 = \omega_1 + \omega_2$ is created as shown in Fig. 1.2(a) and (b). SHG is a special case of SFG where the two photons have the same frequency, $\omega_1 = \omega_2$ as shown in Fig. 1.2(c). SHG was first observed in 1961 by Franken et al. [7]. When the two-photon and/or one-photon transitions are nearly resonant as shown in Fig. 1.2(d) and (e), the efficiencies of SFG and SHG are enhanced. For the cases of the one-photon resonance, the incident fields experience linear absorption. However, for the two-photon resonance, the generated fields experience linear absorption. Thus, the incident and/or generated fields are rapidly attenuated as these fields propagate through the medium. Therefore, in the case of thick media, nonresonant SFG and SHG are typically preferred.

According to Eq. (1.1.81), the polarization representing the SFG process under nonresonant condition is given by the expression

$$P_{\rm SFG}^{(2)}(\mathbf{r}, t) = \frac{1}{2}\varepsilon_0 \chi_0^{(2)} A_1(z, t) A_2(z, t) e^{i\{(k_1+k_2)z-(\omega_1+\omega_2)t\}} + c.c. \quad (1.1.91)$$

From Eqs. (1.1.31) and (1.1.91), the complex amplitude $p_{\rm SFG}^{(2)}(z, t)$ and the wave number $k_{\rm SFG}^{(2)}$ of the polarization of SFG are written as

$$p_{\rm SFG}^{(2)}(z, t) = A_1(z, t) A_2(z, t) \quad (1.1.92)$$

$$k_{\rm SFG}^{(2)} = k_1 + k_2. \quad (1.1.93)$$

Let us assume low incident intensities and/or short length of the nonlinear medium. Under these circumstances, we may assume that the incident pulses don't suffer losses because of low conversion efficiencies. If we assume in addition that the linear absorption and the chromatic dispersion can be neglected, the distortionless propagation of the incident pulses is permitted. Then, the wave equation describing in Eq. (1.1.47) can be written as

$$\frac{\partial A_3(z,t)}{\partial z} = i\frac{\omega_3 \chi_0^{(2)}}{2\tilde{n}^{(1)}(\omega_3)c} p_{\mathrm{SFG}}^{(2)}(0,t) e^{i\Delta k_{\mathrm{SFG}}^{(2)} z}, \tag{1.1.94}$$

where $\Delta k_{\mathrm{SFG}}^{(2)} = k_{\mathrm{SFG}}^{(2)} - k_3 = \{\tilde{n}(\omega_1)\omega_1 + \tilde{n}(\omega_2)\omega_2 - \tilde{n}(\omega_3)\omega_3\}/c$ is called the wave vector mismatch. Here $A_3(z,t)$ and k_3 are the complex amplitude and the wave number of the SFG field, respectively. By integrating Eq. (1.1.94) from 0 to the integration length L, we obtain

$$\begin{aligned} A_3(L,t) &= i\frac{\omega_3 \chi_0^{(2)}}{2\tilde{n}^{(1)}(\omega_3)c} p_{\mathrm{SFG}}^{(2)}(0,t) \int_0^L e^{i\Delta k_{\mathrm{SFG}}^{(2)} z} dz \\ &= \frac{\omega_3 \chi_0^{(2)}}{2\tilde{n}^{(1)}(\omega_3)c} A_1(0,t) A_2(0,t) \left\{ \frac{e^{i\Delta k_{\mathrm{SFG}}^{(2)} L} - 1}{\Delta k_{\mathrm{SFG}}^{(2)}} \right\}, \end{aligned} \tag{1.1.95}$$

where we used Eq. (1.1.92). From Eq. (1.1.95), the SFG intensity $I_3(L,t)$ is obtained as

$$\begin{aligned} I_3(L,t) &= \frac{1}{2}\varepsilon_0 \tilde{n}^{(1)}(\omega_3)c\, |A_3(L,t)|^2 \\ &= \frac{\varepsilon_0 \left(\omega_3 \chi_0^{(2)} L\right)^2}{8\tilde{n}^{(1)}(\omega_3)c} |A_1(0,t)|^2 |A_2(0,t)|^2 \left\{ \frac{\sin\left(\Delta k_{\mathrm{SFG}}^{(2)} L/2\right)}{\Delta k_{\mathrm{SFG}}^{(2)} L/2} \right\}^2 \\ &\equiv \frac{\varepsilon_0 \left(\omega_3 \chi_0^{(2)} L\right)^2}{8\tilde{n}^{(1)}(\omega_3)c} |A_1(0,t)|^2 |A_2(0,t)|^2 \operatorname{sinc}^2\left(\frac{\Delta k_{\mathrm{SFG}}^{(2)} L}{2}\right). \end{aligned} \tag{1.1.96}$$

According to Eq. (1.1.96), we find that the SFG intensity is proportional to the product of the intensities of the two incident waves.

In the case of SHG at frequency $2\omega_1$, the wave equation describing in Eq. (1.1.47) can be written as

$$\frac{\partial A_4(z,t)}{\partial z} = i\frac{2\omega_1 \chi_0^{(2)}}{2\tilde{n}^{(1)}(2\omega_1)c} p_{\mathrm{SHG}}^{(2)}(0,t) e^{i\Delta k_{\mathrm{SHG}}^{(2)} z}, \tag{1.1.97}$$

where

$$p_{\mathrm{SHG}}^{(2)}(z,t) = \frac{1}{2}A_1^2(z,t), \tag{1.1.98}$$

$$\Delta k_{\mathrm{SHG}}^{(2)} = k_{\mathrm{SHG}}^{(2)} - k_4 = 2\omega_1 \frac{\tilde{n}(\omega_1) - \tilde{n}(2\omega_1)}{c}, \tag{1.1.99}$$

$$k_{\mathrm{SHG}}^{(2)} = 2k_1. \tag{1.1.100}$$

Here $p_{\mathrm{SHG}}^{(2)}(z,t)$ and $k_{\mathrm{SHG}}^{(2)}$ are the complex amplitude and the wave number of the polarization of SHG, and $A_4(z,t)$ and k_4 are the complex amplitude and the wave number of the SHG field, respectively. From Eqs. (1.1.97) and (1.1.98), the complex amplitude of SHG is obtained as

$$\begin{aligned} A_4(L,t) &= i\frac{2\omega_1\chi_0^{(2)}}{2\tilde{n}^{(1)}(2\omega_1)c}p_{\mathrm{SHG}}^{(2)}(0,t)\int_0^L e^{i\Delta k_{\mathrm{SHG}}^{(2)}z}dz \\ &= \frac{\omega_1\chi_0^{(2)}}{2\tilde{n}^{(1)}(2\omega_1)c}A_1^2(0,t)\left\{\frac{e^{i\Delta k_{\mathrm{SHG}}^{(2)}L}-1}{\Delta k_{\mathrm{SHG}}^{(2)}}\right\}. \end{aligned} \tag{1.1.101}$$

Therefore, the SHG intensity is given by

$$I_4(L,t) = \frac{\varepsilon_0\left(\omega_1\chi_0^{(2)}L\right)^2}{8\tilde{n}^{(1)}(2\omega_1)c}\left|A_1^2(0,t)\right|^2 \operatorname{sinc}^2\left(\frac{\Delta k_{\mathrm{SHG}}^{(2)}L}{2}\right). \tag{1.1.102}$$

According to Eq. (1.1.102), we see that the SHG intensity is proportional to the square of the incident intensity.

From Eq. (1.1.96), we see that the SFG intensity decreases as $|\Delta k_{\mathrm{SFG}}^{(2)}|L$ increases, with some oscillations occurring. We give an explanation of this behavior. When $\Delta k_{\mathrm{SFG}}^{(2)}$ is equal to zero, the relative phase between the SFG wave $E_{3x}(\mathbf{r},t)$ and its driving polarization $P_{\mathrm{SFG}}^{(2)}(\mathbf{r},t)$ is zero everywhere. Thus, because of constructive interference between the SFG wave and its driving polarization, the SFG field increases linearly with z. Therefore, the SFG intensity increases as the square of the propagation distance. This condition is known as the condition of perfect phase matching. In contrast, if $\Delta k_{\mathrm{SFG}}^{(2)}$ is not equal to zero, the relative phase difference between the SFG wave and its driving polarization increases as the propagation distance increases. Thus, due to destructive interference between the SF wave and its driving polarization, the SFG field decreases. Therefore, the SFG intensity decreases as $|\Delta k_{\mathrm{SFG}}^{(2)}|L$ increases. By satisfying the phase-matching condition $\Delta k_{\mathrm{SFG}}^{(2)} = 0$, the efficiency of SFG can be enhanced. However, the phase-matching condition $\Delta k_{\mathrm{SFG}}^{(2)} = 0$ is often difficult to achieve. This is because in the case of normally dispersive materials, the refractive index of the materials is an increasing function of the frequency. In order to achieve the phase-matching condition, birefringent crystals whose the refractive

TABLE 1.1 Phase-Matching Methods for Uniaxial Crystals

	Positive Uniaxial $\tilde{n}_e(\omega) > \tilde{n}_o(\omega)$	Negative Uniaxial $\tilde{n}_e(\omega) < \tilde{n}_o(\omega)$
Type I	$\tilde{n}_o(\omega_3)\omega_3 = \tilde{n}_e(\omega_1)\omega_1 + \tilde{n}_e(\omega_2)\omega_2$	$\tilde{n}_e(\omega_3)\omega_3 = \tilde{n}_o(\omega_1)\omega_1 + \tilde{n}_o(\omega_2)\omega_2$
Type II	$\tilde{n}_o(\omega_3)\omega_3 = \tilde{n}_e(\omega_1)\omega_1 + \tilde{n}_o(\omega_2)\omega_2$	$\tilde{n}_e(\omega_3)\omega_3 = \tilde{n}_o(\omega_1)\omega_1 + \tilde{n}_e(\omega_2)\omega_2$

index depends on the direction of polarization of the optical radiation are often used. The SFG wave is polarized in the direction that gives it the lower of the two possible refractive indexes. There are two choices for the polarizations of two incident waves [8]. One is type-I phase matching where the two incident waves have the same polarization. The other is type-II phase matching where the two incident waves have orthogonal polarizations. The phase-matching methods using birefringence are summarized in Table 1.1. Careful control of the refractive indexes at each of the three frequencies is required in order to establish the phase-matching condition, Typically phase matching is achieved by one of two methods: angle tuning and temperature tuning [3].

Now we describe the example of type-I SHG phase matching in a positive uniaxial crystal ($\tilde{n}_e(\theta, \omega) < \tilde{n}_o(\omega)$). Here $\tilde{n}_e(\theta, \omega)$ is the refractive index for the extraordinary (e) wave, which is polarized parallel to the plane containing the optic axis of the crystal and the propagation direction, and $\tilde{n}_o(\omega)$ is the refractive index for the ordinary (o) wave, which is polarized perpendicular to the extraordinary wave. θ is the angle between the direction of the wave and the optic axis of the crystal. The refractive indexes for the ordinary and extraordinary waves are related to the following equation:

$$\frac{1}{\tilde{n}^2(\theta, \omega)} = \frac{\cos^2\theta}{\tilde{n}_o^2(\omega)} + \frac{\sin^2\theta}{\tilde{n}_e^2(\omega)}, \tag{1.1.103}$$

which is represented graphically by an ellipse. Type-I SHG phase matching can be achieved by using a combination of two extraordinary fundamental waves and an ordinary second-harmonic (SH) wave. Then, the angle θ is selected according to the following equation:

$$\tilde{n}(\theta, \omega_1) = \tilde{n}_o(2\omega_1). \tag{1.1.104}$$

This is illustrated graphically in Fig. 1.3, which displays the ordinary and extraordinary refractive indexes (a circle and an ellipse) at ω_1 (solid curves) and at $2\omega_1$ (dashed curves). The angle at which phase matching is satisfied is that at which the circle at $2\omega_1$ intersects the ellipse at ω_1.

There is a technique known as quasi-phase matching (QPM) to enhance the frequency conversion efficiency when normal phase matching cannot be implemented [9]. The idea of QPM proposed by Armstrong et al. in 1962 [10]. As shown in Fig. 1.4(b), in QPM, periodic reversal of the sign of the nonlinear susceptibility of the material occurs in the period of Λ ($= 2\pi/\Delta k = 2d$). In each region with a width of d,

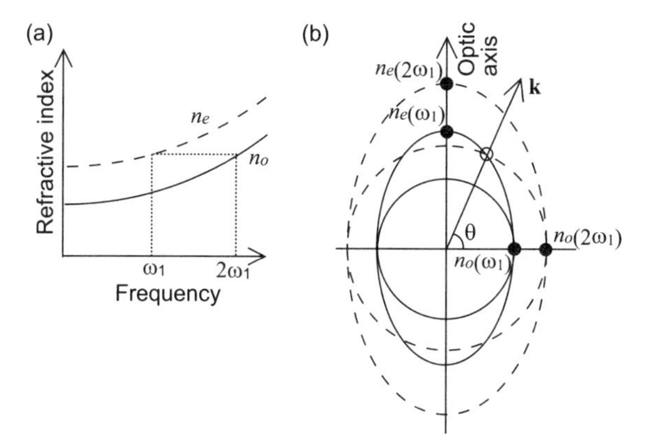

FIGURE 1.3 Type-I SHG phase-matching angle in a positive uniaxial crystal. (a) Matching the index of the extraordinary wave (dashed curves) at ω_1 with that of the ordinary wave (solid curves) at $2\omega_1$. (b) Index surfaces at ω_1 (solid curves) and $2\omega_1$ (dashed curves) for an uniaxial crystal. The wave is chosen to travel at an angle θ with respect to the crystal optic axis, such that the extraordinary refractive index $\tilde{n}_e(\theta, \omega_1)$ of the ω_1 wave equals the ordinary refractive index $\tilde{n}_o(2\omega 1)$ of the $2\omega_1$ wave.

the relative phase difference between the generated wave and its driving polarization is smaller than π. Thus, the frequency conversion efficiency is enhanced because of constructive interference. The most challenging aspect of QPM is the fabrication of the periodic nonlinear structure. Yamada et al. developed the fabrication technique where the nonlinear crystal is lithographically exposed to a periodic electric field that reverses the direction of the crystal's permanent electric polarization [11]. This approach, which is called poling, has been applied to ferroelectric crystals such as $LiTaO_3$, KTP, and $LiNbO_3$; the latter has spawned a technology known as periodically poled lithium niobate (PPLN).

So far we have considered SHG excited by plane waves. Let us now treat SHG using a focused Gaussian beam. From Eqs. (1.1.52), and (1.1.84), the complex amplitudes of the polarization of SHG $U^{(2)}_{\rm SHG}(x, y, z)$ is written as

$$U^{(2)}_{\rm SHG}(x, y, z) = \frac{1}{2} U_1^2(x, y, z). \tag{1.1.105}$$

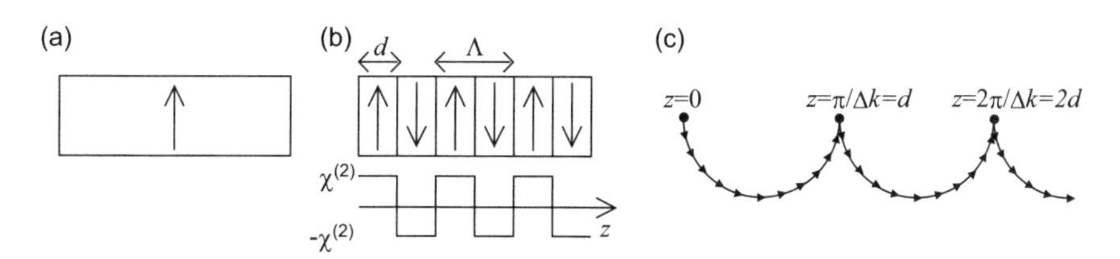

FIGURE 1.4 (a) Homogeneous single crystal. (b) Periodically poled material in which the positive c-axis alternates in orientation with period Λ $(= 2d)$. Arrows in (b) indicate the direction of the c-axis. (c) Phasors of the waves radiated by incremental elements at different positions z in the QPM medium.

By assuming distortionless propagation of the incident pulses, we can rewrite the wave equation described in Eq. (1.1.56) as

$$\left[-\frac{i}{2k_4}\left(\frac{\partial^2}{\partial x^2}+\frac{\partial^2}{\partial y^2}\right)+\frac{\partial}{\partial z}\right]U_4(x,y,z)$$
$$=i\frac{2\omega_1}{2\tilde{n}^{(1)}(2\omega_1)c}\chi_0^{(2)}U_{\mathrm{SHG}}^{(2)}(x,y,z)e^{i\Delta k_{\mathrm{SHG}}^{(2)}z}. \quad (1.1.106)$$

Here $U_4(x, y, z)$ is the complex amplitude of the SHG field. We solve Eq. (1.1.106) by adopting the trial solution

$$U_4(x,y,z)=\frac{U_4(z)}{1+i\zeta_1}e^{-2(x^2+y^2)/w_1^2(1+i\zeta_1)}, \quad (1.1.107)$$

where $U_4(z)$ is a function of z. Because the second-harmonic (SH) wave is generated coherently over a region whose axial extent is equal to the fundamental wave, we have assumed that the trial solution corresponds to a beam with the same confocal parameter as the fundamental wave. Substitution of Eqs. (1.1.107), (1.1.105), and (1.1.83) into Eq. (1.1.106) leads to the result

$$\frac{\partial U_4(z)}{\partial z}=i\frac{\omega_1}{2\tilde{n}^{(1)}(2\omega_1)c}\chi_0^{(2)}U_1^2\frac{e^{i\Delta k_{\mathrm{SHG}}^{(2)}z}}{1+i\zeta_1}. \quad (1.1.108)$$

By integrating this equation, we obtain

$$U_4(z)=i\frac{\omega_1}{2\tilde{n}^{(1)}(2\omega_1)c}\chi_0^{(2)}U_1^2\int_{z_0}^{z}\frac{e^{i\Delta k_{\mathrm{SHG}}^{(2)}z'}}{1+iz'/\rho_1}dz', \quad (1.1.109)$$

where z_0 indicates the value of z at the entrance to the nonlinear medium. If ρ_1 is much larger than $|z_0|$ and $|z|$, the integral in Eq. (1.1.109) is expressed by

$$J_{\mathrm{SHG}}^{(2)}(\Delta k_{\mathrm{SHG}}^{(2)},z_0,z)\equiv\int_{z_0}^{z}\frac{e^{i\Delta k_{\mathrm{SHG}}^{(2)}z'}}{1+iz'/\rho_1}dz'\approx\frac{e^{i\Delta k_{\mathrm{SHG}}^{(2)}z}-e^{i\Delta k_{\mathrm{SHG}}^{(2)}z_0}}{i\Delta k_{\mathrm{SHG}}^{(2)}}. \quad (1.1.110)$$

From Eq. (1.1.110), we obtain

$$\left|J_{\mathrm{SHG}}^{(2)}(\Delta k_{\mathrm{SHG}}^{(2)},z_0,z)\right|^2=(z-z_0)^2\,\mathrm{sinc}^2\left\{\frac{\Delta k_{\mathrm{SHG}}^{(2)}(z-z_0)}{2}\right\}, \quad (1.1.111)$$

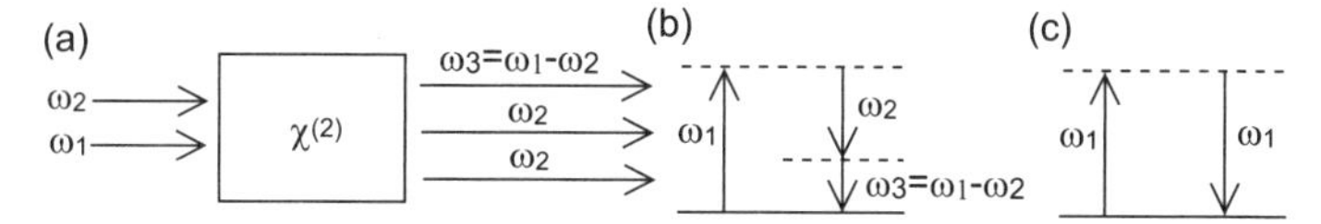

FIGURE 1.5 Difference-frequency generation (DFG) and optical rectification (OR). (a) Geometry of the interaction for DFG. (b, c) Energy diagrams describing the DFG process (b) and OR process.

which corresponds to Eq. (1.1.102). This equation indicates a propagation of plane waves. Next, we consider the situation where ρ_1 is much smaller than $|z_0|$ and $|z|$. The wave propagation occurs under a tight-focusing condition. Then, the integral in Eq. (1.1.109) can be expressed by

$$J_{\mathrm{SHG}}^{(2)}(\Delta k_{\mathrm{SHG}}^{(2)}, z_0, z) \approx \int_{-\infty}^{\infty} \frac{e^{i\Delta k_{\mathrm{SHG}}^{(2)} z'}}{1 + iz'/\rho_1} dz' = \begin{cases} 0, & \Delta k_{\mathrm{SHG}}^{(2)} \leq 0, \\ 2\pi\rho_1 e^{-\rho_1 \Delta k_{\mathrm{SHG}}^{(2)}}, & \Delta k_{\mathrm{SHG}}^{(2)} > 0. \end{cases} \quad (1.1.112)$$

According to Eq. (1.1.112), we find that with perfect phase matching, $\Delta k_{\mathrm{SHG}}^{(2)} = 0$, no efficiency of SHG under the tight-focusing condition is achieved. On the contrary, efficiency is maximized by the use of a positive wave vector mismatch. This behavior can be understood from the Gouy phase shift [3]. The SH wave experiences phase shift of π radians in passing through its focus, while its driving polarization experiences a phase shift of 2π radians. This phase shift is twice as large as the phase shift of the fundamental wave as is evident from Eq. (1.1.105). Consequently, even though the perfect phase-matching condition is satisfied, the relative phase shift between the SH wave and its driving polarization is induced by the Gouy phase shift. That is to say perfect constructive interference between the SH wave and its driving polarization can be achieved not by satisfying the perfect phase-matching condition but by compensating for this relative phase shift through the use of a positive wave vector mismatch. Therefore, the SHG efficiency is maximized in the tight-focusing limit for the case of positive wave vector mismatch.

1.1.2.3 Deference-Frequency Generation Next we consider the processes of difference-frequency generation (DFG) and optical rectification (OR), which are shown in Fig. 1.5. The DFG process is induced by two photons of frequencies ω_1 and ω_2 and then generation of one photon at a new frequency of $\omega_3 = \omega_1 - \omega_2$ is accompanied by the amplification of the lower-frequency photon of ω_2 as shown in Fig. 1.5(c). This amplification, which is known as optical parametric amplification, can be used to construct devises known as an optical parametric oscillator (OPO) and an optical parametric amplifier (OPA). DFG was first observed by Smith and Braslau in 1962 [12]. Optical parametric amplification was first observed in KH2PO4 by Wang and Racette [13], and the first tunable infrared optical source based on optical parametric oscillation in LiNbO3 was demonstrated by Giordmaine and Miller [14] in 1965. Since then, OPOs and OPAs have become widely used as sources of tunable radiation. Although the efficiencies of DFG and OR are enhanced by the two-photon resonance and/or one-photon resonance, the incident and/or generated fields experience linear absorption. Thus, in general, DFG and OR are not employed under resonant condition.

According to Eq. (1.1.81), the polarization representing the DFG process under nonresonant condition is described by the expression

$$P_{\mathrm{DFG}}^{(2)}(\mathbf{r}, t) = \frac{1}{2}\varepsilon_0 \chi_0^{(2)} A_1(z, t) A_2^*(z, t) e^{i\{(k_1 - k_2)z - (\omega_1 - \omega_2)t\}} + c.c. \tag{1.1.113}$$

From Eqs. (1.1.31), and (1.1.113), the complex amplitude $p_{\mathrm{DFG}}^{(2)}(z, t)$ and the wave number $k_{\mathrm{DFG}}^{(2)}$ of the polarization of DFG are expressed by

$$p_{\mathrm{DFG}}^{(2)}(z, t) = A_1(z, t) A_2^*(z, t), \tag{1.1.114}$$

$$k_{\mathrm{DFG}}^{(2)} = k_1 - k_2. \tag{1.1.115}$$

We will assume distortionless propagation of the incident pulses, so that the wave equation describing in Eq. (1.1.47) can be written as

$$\frac{\partial A_3(z, t)}{\partial z} = i \frac{\omega_3 \chi_0^{(2)}}{2\tilde{n}^{(1)}(\omega_3) c} p_{\mathrm{DFG}}^{(2)}(0, t) e^{i \Delta k_{\mathrm{DFG}}^{(2)} z}, \tag{1.1.116}$$

where $\Delta k_{\mathrm{DFG}}^{(2)} = k_{\mathrm{DFG}}^{(2)} - k_3 = \{\tilde{n}(\omega_1)\omega_1 - \tilde{n}(\omega_2)\omega_2 - \tilde{n}(\omega_3)\omega_3\}/c$ is the wave vector mismatch. Here $A_3(z, t)$ and k_3 are the complex amplitude and the wave number of the DFG field, respectively. By integrating Eq. (1.1.116) from 0 to the integration length L, we obtain

$$A_3(L, t) = \frac{\omega_3 \chi_0^{(2)}}{2\tilde{n}^{(1)}(\omega_3) c} A_1(0, t) A_2^*(0, t) \left\{ \frac{e^{i \Delta k_{\mathrm{DFG}}^{(2)} L} - 1}{\Delta k_{\mathrm{DFG}}^{(2)}} \right\}. \tag{1.1.117}$$

From Eq. (1.1.117), the DFG intensity $I_3(L, t)$ is obtained by

$$I_3(L, t) = \frac{\varepsilon_0 \left(\omega_3 \chi_0^{(2)} L\right)^2}{8\tilde{n}^{(1)}(\omega_3) c} |A_1(0, t)|^2 |A_2(0, t)|^2 \operatorname{sinc}^2 \left(\frac{\Delta k_{\mathrm{DFG}}^{(2)} L}{2} \right). \tag{1.1.118}$$

This equation indicates that the DFG intensity is also proportional to the product of the intensities of the two incident waves. This result is similar to Eq. (1.1.118). However, the phase-matching conditions $\Delta k^{(2)}_{\mathrm{SFG}} = 0$ and $\Delta k^{(2)}_{\mathrm{DFG}} = 0$ are different. If one phase-matching condition is satisfied, typically the other phase-matching condition cannot be satisfied. Thus, we can selectively enhance one of the second-order nonlinear optical phenomena by satisfying a specific phase-matching condition.

1.1.3 Third-Order Nonlinear Optical Phenomena

The third-order contribution to the nonlinear polarization is

$$P_i^{(3)}(\mathbf{r}, t) = \varepsilon_0 \int_{-\infty}^{t} \int_{-\infty}^{t} \int_{-\infty}^{t} \sum_{j,k,l} \chi^{(3)}_{ijkl}(t - t_1, t_1 - t_2, t_2 - t_3)$$
$$\times E_j(\mathbf{r}, t_1) E_k(\mathbf{r}, t_2) E_l(\mathbf{r}, t_3) dt_1 dt_2 dt_3, \quad (1.1.119)$$

where $\chi^{(3)}_{ijkl}(t - t_1, t_1 - t_2, t_2 - t_3)$ is the components of the third-order susceptibility tensor. Here the indexes i, j, k, l refer to the cartesian components. The symmetric properties of the fourth-rank tensors $\chi^{(3)}_{ijkl}$ for all 32 crystallographic point groups have been known since the early days of nonlinear optics. In 1965, Midwinter and Warner obtained the compact matrices of for uniaxial and isotropic crystals by simply comparing $\chi^{(3)}_{in}$ terms with the table of elasto-optic tensor elements [15]. Butcher has also shown the independent elements of $\chi^{(3)}_{ijkl}$ for all 32 crystallographic point groups [16]. In 1987, Shang and Hsu tested Butcher's results and made some modifications using a computer program [17]. The revised table of the independent elements of $\chi^{(3)}_{ijkl}$ is published by Butcher [18]. In 1995, Yang and Xie contracted the independent elements of $\chi^{(3)}_{ijkl}$ to the compact matrices form $\chi^{(3)}_{in}$ for all 32 crystallographic point groups [19]. We will consider the third-order nonlinear polarization in an isotropic medium such as a glass, a liquid, or a vapor. Since each of the coordinate axes must be equivalent in an isotropic medium, it is clear that susceptibility possesses the following symmetry properties:

$$\chi^{(3)}_{xxxx} = \chi^{(3)}_{yyyy} = \chi^{(3)}_{zzzz} = \chi^{(3)}_{xxyy} + \chi^{(3)}_{xyxy} + \chi^{(3)}_{xyyx},$$
$$\chi^{(3)}_{xxyy} = \chi^{(3)}_{xxzz} = \chi^{(3)}_{yyxx} = \chi^{(3)}_{yyzz} = \chi^{(3)}_{zzxx} = \chi^{(3)}_{zzyy},$$
$$\chi^{(3)}_{xyxy} = \chi^{(3)}_{xzxz} = \chi^{(3)}_{yxyx} = \chi^{(3)}_{yzyz} = \chi^{(3)}_{zxzx} = \chi^{(3)}_{zyzy},$$
$$\chi^{(3)}_{xyyx} = \chi^{(3)}_{xzzx} = \chi^{(3)}_{yxxy} = \chi^{(3)}_{yzzy} = \chi^{(3)}_{zxxz} = \chi^{(3)}_{zyyz}. \quad (1.1.120)$$

There are 21 nonzero elements, of which only 3 are independent. If a light pulse propagates along $+z$ with only a field component along x, the third-order nonlinear polarization is given by

$$P_x^{(3)}(\mathbf{r},t) = \varepsilon_0 \int_{-\infty}^{t}\int_{-\infty}^{t}\int_{-\infty}^{t} \chi_{xxxx}^{(3)}(t-t_1, t_1-t_2, t_2-t_3) \times E_x(\mathbf{r},t_1)E_x(\mathbf{r},t_2)E_x(\mathbf{r},t_3)dt_1dt_2dt_3. \quad (1.1.121)$$

Now we consider the circumstance in which the incident optical field consists of three distinct frequency components ω_1, ω_2, and ω_3 ($\omega_1 \geq \omega_2 \geq \omega_3$), which we represent in the form

$$E_x(\mathbf{r},t) = \sum_{m=1}^{3} E_{mx}(\mathbf{r},t), \quad (1.1.122)$$

$$E_{mx}(\mathbf{r},t) = \frac{1}{2}\left\{A_m(z,t)e^{i(k_m z-\omega_m t)} + c.c.\right\}. \quad (1.1.123)$$

Here $A_m(z,t)$ is a slowly varying function of z. ω_m and k_m are the carrier frequency and the wave number, respectively. When the two-photon transition is nearly resonant, the time-delayed transition involving a real level can be separated from the instantaneous transition between virtual levels, and $\chi_{xxxx}^{(3)}(t-t_1, t_1-t_2, t_2-t_3)$ is described as

$$\chi_{xxxx}^{(3)}(t-t_1, t_1-t_2, t_2-t_3) = \chi^{(3)}\delta(t-t_1)R(t-t_2)\delta(t_2-t_3). \quad (1.1.124)$$

The contributions of the nonresonance, two-photon electronic resonance and two-photon Raman resonance can be included by assuming the following form for the time-delayed response:

$$R(t) = (1-f_{\mathrm{E}}-f_{\mathrm{R}})\delta(t) + f_{\mathrm{E}}h_{\mathrm{ER}}(t) + f_{\mathrm{R}}h_{\mathrm{RR}}(t). \quad (1.1.125)$$

$R(t)$ is normalized such that $\int_{-\infty}^{\infty} R(t)dt = 1$. Here, $h_{\mathrm{ER}}(t)$ and $h_{\mathrm{RR}}(t)$ are the electronic and Raman responses under the resonant conditions, respectively. The first term in Eq. (1.1.125) represents the nonresonant contribution. If we assume that $\tilde{h}_{\mathrm{ER}}(\omega) = FT_t[h_{\mathrm{ER}}(t)]$ and $\tilde{h}_{\mathrm{RR}}(\omega) = FT_t[h_{\mathrm{RR}}(t)]$ are described by Lorentzian line shapes with bandwidths (full widths at half maximum) of $2\Gamma_{\mathrm{ER}}$ and $2\Gamma_{\mathrm{RR}}$ at the resonant frequencies of ω_{ER} and ω_{RR} such that

$$\tilde{h}_{\mathrm{ER}}(\omega) = \frac{1}{2}\frac{\omega_{\mathrm{ER}}^2+\Gamma_{\mathrm{ER}}^2}{\omega_{\mathrm{ER}}} \times \left[\frac{1}{\omega_{\mathrm{ER}}-\omega-i\Gamma_{\mathrm{ER}}} + \frac{1}{\omega_{\mathrm{ER}}+\omega+i\Gamma_{\mathrm{ER}}}\right], \quad (1.1.126)$$

$$\tilde{h}_{RR}(\omega) = \frac{1}{2}\frac{\omega_{RR}^2 + \Gamma_{RR}^2}{\omega_{RR}} \times \left[\frac{1}{\omega_{RR} - \omega - i\Gamma_{RR}} + \frac{1}{\omega_{RR} + \omega + i\Gamma_{RR}}\right], \quad (1.1.127)$$

$h_{ER}(t)$ and $h_{RR}(t)$ are given by

$$h_{ER}(t) = \frac{\omega_{ER}^2 + \Gamma_{ER}^2}{2i\omega_{ER}} u(t) e^{-\Gamma_{ER} t}\left(e^{i\omega_{ER} t} - e^{-i\omega_{ER} t}\right), \quad (1.1.128)$$

$$h_{RR}(t) = \frac{\omega_{RR}^2 + \Gamma_{RR}^2}{2i\omega_{RR}} u(t) e^{-\Gamma_{RR} t}\left(e^{i\omega_{RR} t} - e^{-i\omega_{RR} t}\right). \quad (1.1.129)$$

Here $u(t)$ is the unit step function. By substituting Eq. (1.1.124) into Eq. (1.1.121), the third-order polarization can be written as

$$\begin{aligned} P_x^{(3)}(\mathbf{r}, t) &= \varepsilon_0 \chi^{(3)} E_x(\mathbf{r}, t) \int_{-\infty}^{\infty} R(t - t_2) E_x^2(\mathbf{r}, t_2) dt_2 \\ &= \varepsilon_0 (1 - f_E - f_R) \chi^{(3)} E_x^3(\mathbf{r}, t) \\ &\quad + \varepsilon_0 f_E \chi^{(3)} E_x(\mathbf{r}, t) \int_{-\infty}^{t} h_{ER}(t - t_2) E_x^2(\mathbf{r}, t_2) dt_2 \\ &\quad + \varepsilon_0 f_R \chi^{(3)} E_x(\mathbf{r}, t) \int_{-\infty}^{t} h_{RR}(t - t_2) E_x^2(\mathbf{r}, t_2) dt_2 \\ &= P_{NR}^{(3)}(\mathbf{r}, t) + P_{ER}^{(3)}(\mathbf{r}, t) + P_{RR}^{(3)}(\mathbf{r}, t). \end{aligned} \quad (1.1.130)$$

$P_{NR}^{(3)}(\mathbf{r}, t)$, $P_{ER}^{(3)}(\mathbf{r}, t)$, and $P_{RR}^{(3)}(\mathbf{r}, t)$ are the contributions of the nonresonance, the electronic resonance and Raman resonance, respectively and are described by

$$P_{NR}^{(3)}(\mathbf{r}, t) = \varepsilon_0 (1 - f_E - f_R) \chi^{(3)} E_x^3(\mathbf{r}, t), \quad (1.1.131)$$

$$P_{ER}^{(3)}(\mathbf{r}, t) = \varepsilon_0 f_E \chi^{(3)} E_x(\mathbf{r}, t) \int_{-\infty}^{t} h_{ER}(t - t_2) E_x^2(\mathbf{r}, t_2) dt_2, \quad (1.1.132)$$

$$P_{RR}^{(3)}(\mathbf{r}, t) = \varepsilon_0 f_R \chi^{(3)} E_x(\mathbf{r}, t) \int_{-\infty}^{t} h_{RR}(t - t_2) E_x^2(\mathbf{r}, t_2) dt_2. \quad (1.1.133)$$

After inserting Eqs. (1.1.122) and (1.1.123) into Eq. (1.1.131), we obtain

$$
\begin{aligned}
P_{\mathrm{NR}}^{(3)}(\mathbf{r}, t) = & \frac{1}{8}\varepsilon_0(1 - f_{\mathrm{E}} - f_{\mathrm{R}})\chi^{(3)} \\
& \times \left[A_1^3(z,t)e^{i\{3k_1z-3\omega_1t\}}\right. \\
& + A_2^3(z,t)e^{i\{3k_2z-3\omega_2t\}} \\
& + A_3^3(z,t)e^{i\{3k_3z-3\omega_3t\}} \\
& + 3A_1^2(z,t)A_2(z,t)e^{i\{(2k_1+k_2)z-(2\omega_1+\omega_2)t\}} \\
& + 3A_1^2(z,t)A_3(z,t)e^{i\{(2k_1+k_3)z-(2\omega_1+\omega_3)t\}} \\
& + 3A_1(z,t)A_2^2(z,t)e^{i\{(k_1+2k_2)z-(\omega_1+2\omega_2)t\}} \\
& + 3A_1(z,t)A_3^2(z,t)e^{i\{(k_1+2k_3)z-(\omega_1+2\omega_3)t\}} \\
& + 3A_2^2(z,t)A_3(z,t)e^{i\{(2k_2+k_3)z-(2\omega_2+\omega_3)t\}} \\
& + 3A_2(z,t)A_3^2(z,t)e^{i\{(k_2+2k_3)z-(\omega_2+2\omega_3)t\}} \\
& + 3A_1^2(z,t)A_2^*(z,t)e^{i\{(2k_1-k_2)z-(2\omega_1-\omega_2)t\}} \\
& + 3A_1^2(z,t)A_3^*(z,t)e^{i\{(2k_1-k_3)z-(2\omega_1-\omega_3)t\}} \\
& + 3A_1^*(z,t)A_2^2(z,t)e^{i\{(2k_2-k_1)z-(2\omega_2-\omega_1)t\}} \\
& + 3A_2^2(z,t)A_3^*(z,t)e^{i\{(2k_2-k_3)z-(2\omega_2-\omega_3)t\}} \\
& + 3A_1^*(z,t)A_3^2(z,t)e^{i\{(2k_3-k_1)z-(2\omega_3-\omega_1)t\}} \\
& + 3A_2^*(z,t)A_3^2(z,t)e^{i\{(2k_3-k_2)z-(2\omega_3-\omega_2)t\}} \\
& + 6A_1(z,t)A_2(z,t)A_3(z,t)e^{i\{(k_1+k_2+k_3)z-(\omega_1+\omega_2+\omega_3)t\}} \\
& + 6A_1(z,t)A_2(z,t)A_3^*(z,t)e^{i\{(k_1+k_2-k_3)z-(\omega_1+\omega_2-\omega_3)t\}} \\
& + 6A_1(z,t)A_2^*(z,t)A_3(z,t)e^{i\{(k_1+k_3-k_2)z-(\omega_1+\omega_3-\omega_2)t\}} \\
& + 6A_1^*(z,t)A_2(z,t)A_3(z,t)e^{i\{(k_2+k_3-k_1)z-(\omega_2+\omega_3-\omega_1)t\}} \\
& + 3\left|A_1(z,t)\right|^2 A_1(z,t)e^{i(k_1z-\omega_1t)} \\
& + 6\left\{\left|A_2(z,t)\right|^2 + \left|A_3(z,t)\right|^2\right\} A_1(z,t)e^{i(k_1z-\omega_1t)} \\
& + 3\left|A_2(z,t)\right|^2 A_2(z,t)e^{i(k_2z-\omega_2t)} \\
& + 6\left\{\left|A_1(z,t)\right|^2 + \left|A_3(z,t)\right|^2\right\} A_2(z,t)e^{i(k_2z-\omega_2t)} \\
& + 3\left|A_3(z,t)\right|^2 A_3(z,t)e^{i(k_3z-\omega_3t)} \\
& \left. + 6\left\{\left|A_1(z,t)\right|^2 + \left|A_2(z,t)\right|^2\right\} A_3(z,t)e^{i(k_3z-\omega_3t)}\right] \\
& + c.c.
\end{aligned}
\tag{1.1.134}
$$

Note that the third-order polarization $P_{NR}^{(3)}(\mathbf{r}, t)$ consists of terms oscillating with frequencies of $3\omega_1$, $3\omega_2$, $3\omega_3$, $2\omega_1 + \omega_2$, $2\omega_1 + \omega_3$, $\omega_1 + 2\omega_2$, $\omega_1 + 2\omega_3$, $2\omega_2 + \omega_3$, $\omega_2 + 2\omega_3$, $2\omega_1 - \omega_2$, $2\omega_1 - \omega_3$, $2\omega_2 - \omega_1$, $2\omega_2 - \omega_3$, $2\omega_3 - \omega_1$, $2\omega_3 - \omega_2$, $\omega_1 + \omega_2 + \omega_3$, $\omega_1 + \omega_2 - \omega_3$, $\omega_1 + \omega_3 - \omega_2$, $\omega_2 + \omega_3 - \omega_1$, ω_1, ω_2, and ω_3. The contributions at the frequencies that are different from the incident field frequencies ω_1, ω_2, and ω_3 can lead to the generation of radiation at new frequencies. The first three terms called containing $e^{-i3\omega_1 t}$, $e^{-i3\omega_2 t}$, and $e^{-i3\omega_3 t}$ in Eq. (1.1.134) are responsible for the physical process called third-harmonic generation (THG). The next 16 terms result in the physical process called four-wave mixing (FWM). The remaining six terms at the frequencies ω_1, ω_2, and ω_3 lead to change in the refractive index experienced by light pulses with carrier frequencies ω_1, ω_2, and ω_3. The change in the refractive index is called the optical Kerr effect, and it contributes to the physical process called self-phase modulation (SPM) and cross-phase modulation (XPM). These physical processes are discussed in detail later in this section.

We next consider the contribution of electronic resonance $P_{\text{ER}}^{(3)}(\mathbf{r}, t)$ under the condition where the electronic resonant frequency ω_{ER} is nearly equal to $\omega_1 + \omega_2$, and is far from $2\omega_1$, $2\omega_2$, $2\omega_3$, $\omega_1 + \omega_3$, $\omega_2 + \omega_3$, $\omega_1 - \omega_2$, $\omega_1 - \omega_3$, $\omega_2 - \omega_3$, and 0. From Eqs. (1.1.122) and (1.1.123), $E_x^2(\mathbf{r}, t)$ is written by

$$
\begin{aligned}
E_x^2(\mathbf{r}, t) = \frac{1}{4} \big[& A_1^2(z, t) e^{i(2k_1 z - 2\omega_1 t)} \\
& + A_2^2(z, t) e^{i(2k_2 z - 2\omega_2 t)} \\
& + A_3^2(z, t) e^{i(2k_3 z - 2\omega_3 t)} \\
& + 2A_1(z, t) A_2(z, t) e^{i\{(k_1 + k_2) z - (\omega_1 + \omega_2) t\}} \\
& + 2A_1(z, t) A_3(z, t) e^{i\{(k_1 + k_3) z - (\omega_1 + \omega_3) t\}} \\
& + 2A_2(z, t) A_3(z, t) e^{i\{(k_2 + k_3) z - (\omega_2 + \omega_3) t\}} \\
& + 2A_1(z, t) A_2^*(z, t) e^{i\{(k_1 - k_2) z - (\omega_1 - \omega_2) t\}} \\
& + 2A_1(z, t) A_3^*(z, t) e^{i\{(k_1 - k_3) z - (\omega_1 - \omega_3) t\}} \\
& + 2A_2(z, t) A_3^*(z, t) e^{i\{(k_2 - k_3) z - (\omega_2 - \omega_3) t\}} \\
& + 2 |A_1(z, t)|^2 + 2 |A_2(z, t)|^2 + 2 |A_3(z, t)|^2 \big] \\
& + c.c. \qquad (1.1.135)
\end{aligned}
$$

Because only the fourth term in Eq. (1.1.135) is found to satisfy the electronic resonant condition, $P_{\text{ER}}^{(3)}(\mathbf{r}, t)$ becomes

$$
\begin{aligned}
P_{\text{ER}}^{(3)}(\mathbf{r}, t) = & \frac{1}{2} \varepsilon_0 f_{\text{E}} \chi^{(3)} E_x(\mathbf{r}, t) \int_{-\infty}^{t} h_{\text{ER}}(t - t_2) \\
& \times \left\{ A_1(z, t_2) A_2(z, t_2) e^{i\{(k_1 + k_2) z - (\omega_1 + \omega_2) t_2\}} \right. \\
& \left. + A_1^*(z, t_2) A_2^*(z, t_2) e^{-i\{(k_1 + k_2) z - (\omega_1 + \omega_2) t_2\}} \right\} dt_2. \qquad (1.1.136)
\end{aligned}
$$

Substituting Eqs. (1.1.122), (1.1.123) and (1.1.128) into Eq. (1.1.136), we obtain

$$
\begin{aligned}
P_{\mathrm{ER}}^{(3)}(\mathbf{r}, t) &\cong i\frac{\omega_{\mathrm{ER}}^2 + \Gamma_{\mathrm{ER}}^2}{4\omega_{\mathrm{ER}}}\varepsilon_0 f_{\mathrm{E}}\chi^{(3)}E_x(\mathbf{r}, t) \\
&\times \left[e^{i\{(2k_1+k_2)z-(2\omega_1+\omega_2)t\}} A_{12(\mathrm{ER})}^{(2)}(z, t) \right. \\
&\left. - e^{-i\{(2k_1+k_2)z-(2\omega_1+\omega_2)t\}} A_{12(\mathrm{ER})}^{(2)*}(z, t) \right] \\
&= i\frac{\omega_{\mathrm{ER}}^2 + \Gamma_{\mathrm{ER}}^2}{8\omega_{\mathrm{ER}}}\varepsilon_0 f_{\mathrm{E}}\chi^{(3)} \\
&\times \left[A_1(z, t)A_{12(\mathrm{ER})}^{(2)}(z, t)e^{i\{(2k_1+k_2)z-(2\omega_1+\omega_2)t\}} \right. \\
&+ A_2(z, t)A_{12(\mathrm{ER})}^{(2)}(z, t)e^{i\{(k_1+2k_2)z-(\omega_1+2\omega_2)t\}} \\
&+ A_3(z, t)A_{12(\mathrm{ER})}^{(2)}(z, t)e^{i\{(k_1+k_2+k_3)z-(\omega_1+\omega_2+\omega_3)t\}} \\
&+ A_3^*(z, t)A_{12(\mathrm{ER})}^{(2)}(z, t)e^{i\{(k_1+k_2-k_3)z-(\omega_1+\omega_2-\omega_3)t\}} \\
&+ A_2^*(z, t)A_{12(\mathrm{ER})}^{(2)}(z, t)e^{i(k_1z-\omega_1t)} \\
&\left. + A_1^*(z, t)A_{12(\mathrm{ER})}^{(2)}(z, t)e^{i(k_2z-\omega_2t)} - c.c. \right]. \qquad (1.1.137)
\end{aligned}
$$

Here we introduce $A_{12(\mathrm{ER})}^{(2)}$ using

$$
A_{13(\mathrm{ER})}^{(2)}(z, t) = \int_{-\infty}^{t} u\,(t - t_2)\, e^{[-i\{\omega_{\mathrm{ER}}-(\omega_1+\omega_2)\}-\Gamma_{\mathrm{ER}}](t-t_2)} A_1(z, t_2)A_2(z, t_2)dt_2. \qquad (1.1.138)
$$

We find that $P_{\mathrm{ER}}^{(3)}(\mathbf{r}, t)$ is composed of the terms oscillating with frequencies of $2\omega_1 + \omega_2$, $\omega_1 + 2\omega_2$, $\omega_1 + \omega_2 + \omega_3$, $\omega_1 + \omega_2 - \omega_3$, ω_1, and ω_2. The first four terms containing $e^{-i(2\omega_1+\omega_2)t}$, $e^{-i(\omega_1+2\omega_2)t}$, $e^{-i(\omega_1+\omega_2+\omega_3)t}$, and $e^{-i(\omega_1+\omega_2-\omega_3)t}$ in Eq. (1.1.137) are responsible for the electronic resonant FWM, and are enhanced by the two-photon electronic resonance. Especially, the contribution at frequency $\omega_1 + \omega_2 - \omega_3$ is called a simulated parametric emission (SPE). The remaining two terms, including $e^{-i\omega_1 t}$ and $e^{-i\omega_2 t}$ lead to change the refractive index and the absorption coefficient experienced by light pulses with carrier frequencies ω_1 and ω_2. The change in the absorption coefficient contributes to the physical process called two-photon absorption (TPA).

If the electronic resonant frequency ω_{ER} is nearly equal to $2\omega_1$, and is far from $2\omega_2, 2\omega_3, \omega_1+\omega_2, \omega_1+\omega_3, \omega_2+\omega_3, \omega_1-\omega_2, \omega_1-\omega_3, \omega_2-\omega_3$, and 0, we can set $P_{ER}^{(3)}(\mathbf{r}, t)$ to

$$
\begin{aligned}
P_{ER}^{(3)}(\mathbf{r}, t) &\cong i\frac{\omega_{ER}^2+\Gamma_{ER}^2}{8\omega_{ER}}\varepsilon_0 f_E \chi^{(3)} E_x(\mathbf{r}, t) \\
&\quad \times \left[e^{i(2k_1 z-2\omega_1 t)} A_{11(ER)}^{(2)}(z, t) - e^{-i(2k_1 z-2\omega_1 t)} A_{11(ER)}^{(2)*}(z, t)\right] \\
&= i\frac{\omega_{ER}^2+\Gamma_{ER}^2}{16\omega_{ER}}\varepsilon_0 f_E \chi^{(3)} \\
&\quad \times \left[A_1(z, t) A_{11(ER)}^{(2)}(z, t) e^{i(3k_1 z-3\omega_1 t)}\right. \\
&\quad + A_2(z, t) A_{11(ER)}^{(2)}(z, t) e^{i\{(2k_1+k_2)z-(2\omega_1+\omega_2)t\}} \\
&\quad + A_2^*(z, t) A_{11(ER)}^{(2)}(z, t) e^{i\{(2k_1-k_2)z-(2\omega_1-\omega_2)t\}} \\
&\quad \left. + A_1^*(z, t) A_{11(ER)}^{(2)}(z, t) e^{i(k_1 z-\omega_1 t)} - c.c.\right].
\end{aligned}
\tag{1.1.139}
$$

where we also assumed that ω_2 is equal to ω_3. We see that $P_{ER}^{(3)}(\mathbf{r}, t)$ is composed of the terms oscillating with frequencies of $3\omega_1$, $2\omega_1+\omega_2$, $2\omega_1-\omega_2$, and ω_1. The first term containing $e^{-i3\omega_1 t}$ in Eq. (1.1.139) results in electronic resonant THG, which are enhanced by the two-photon electronic resonance. The next two terms involving $e^{-i(2\omega_1+\omega_2)t}$ and $e^{-i(2\omega_1-\omega_2)t}$ lead to electronic resonant FWM. The final term including $e^{-i\omega_1 t}$ contributes to TPA. These physical processes are also discussed in detail later in this section.

Under the condition that the Raman resonant frequency ω_{RR} is nearly equal to $\omega_1-\omega_3$, and is far from $2\omega_1$, $2\omega_2$, $2\omega_3$, $\omega_1+\omega_2$, $\omega_1+\omega_3$, $\omega_2+\omega_3$, $\omega_1-\omega_2$, $\omega_2-\omega_3$, and 0, the contribution of the Raman resonance $P_{RR}^{(3)}(\mathbf{r}, t)$ can be written as

$$
\begin{aligned}
P_{RR}^{(3)}(\mathbf{r}, t) &= \frac{1}{2}\varepsilon_0 f_R \chi^{(3)} E_x(\mathbf{r}, t) \\
&\quad \times \int_{-\infty}^{t} h_{RR}(t-t_2)\left[A_1(z, t_2) A_3^*(z, t_2) e^{i\{(k_1-k_3)z-(\omega_1-\omega_3)t_2\}}\right. \\
&\quad \left. + A_1^*(z, t_2) A_3(z, t_2) e^{-i\{(k_1-k_3)z-(\omega_1-\omega_3)t_2\}}\right] dt_2.
\end{aligned}
\tag{1.1.140}
$$

Here we used Eqs. (1.1.133) and (1.1.135). Inserting Eqs. (1.1.122), (1.1.123) and (1.1.129) into Eq. (1.1.140), we can express $P_{\mathrm{RR}}^{(3)}(\mathbf{r}, t)$ by

$$\begin{aligned} P_{\mathrm{RR}}^{(3)}(\mathbf{r}, t) &\cong i \frac{\omega_{\mathrm{RR}}^2 + \Gamma_{\mathrm{RR}}^2}{4\omega_{\mathrm{RR}}} \varepsilon_0 f_{\mathrm{R}} \chi^{(3)} E_x(\mathbf{r}, t) \\ &\quad \times \left[e^{i\{(k_1 - k_3)z - (\omega_1 - \omega_3)t\}} A_{13(\mathrm{RR})}^{(2\#)} - e^{-i\{(k_1 - k_3)z - (\omega_1 - \omega_3)t\}} A_{13(\mathrm{RR})}^{(2\#)*} \right] \\ &= i \frac{\omega_{\mathrm{RR}}^2 + \Gamma_{\mathrm{RR}}^2}{8\omega_{\mathrm{RR}}} \varepsilon_0 f_{\mathrm{R}} \chi^{(3)} \\ &\quad \times \left[A_1(z, t) A_{13(\mathrm{RR})}^{(2\#)}(z, t) e^{i\{(2k_1 - k_3)z - (2\omega_1 - \omega_3)t\}} \right. \\ &\quad + A_2(z, t) A_{13(\mathrm{RR})}^{(2\#)}(z, t) e^{i\{(k_1 + k_2 - k_3)z - (\omega_1 + \omega_2 - \omega_3)t\}} \\ &\quad - A_3(z, t) A_{13(\mathrm{RR})}^{(2\#)*}(z, t) e^{i\{(2k_3 - k_1)z - (2\omega_3 - \omega_1)t\}} \\ &\quad - A_2(z, t) A_{13(\mathrm{RR})}^{(2\#)*}(z, t) e^{i\{(k_2 + k_3 - k_1)z - (\omega_2 + \omega_3 - \omega_1)t\}} \\ &\quad + A_3(z, t) A_{13(\mathrm{RR})}^{(2\#)}(z, t) e^{i(k_1 z - \omega_1 t)} \\ &\quad \left. - A_1(z, t) A_{13(\mathrm{RR})}^{(2\#)*}(z, t) e^{i(k_3 z - \omega_3 t)} - c.c. \right]. \end{aligned} \tag{1.1.141}$$

Here $A_{13(\mathrm{RR})}^{(2\#)}$ is introduced by using

$$A_{13(\mathrm{RR})}^{(2\#)}(z, t) = \int_{-\infty}^{t} u\,(t - t_2)\, e^{[-i\{\omega_{\mathrm{RR}} - (\omega_1 - \omega_3)\} - \Gamma_{\mathrm{RR}}](t - t_2)} A_1(z, t_2) A_3^*(z, t_2) dt_2. \tag{1.1.142}$$

The first four terms containing $e^{-i(2\omega_1 - \omega_3)t}$, $e^{-i(\omega_1 + \omega_2 - \omega_3)t}$, $e^{-i(2\omega_3 - \omega_1)t}$, and $e^{-i(\omega_2 + \omega_3 - \omega_1)t}$ in Eq. (1.1.141) result in Raman resonant FWM, which are enhanced by the two-photon Raman resonance. The first two terms and next two terms are called coherent anti-Stokes Raman scattering (CARS) and coherent Stokes Raman scattering (CSRS), respectively. The remaining two terms involving $e^{-i\omega_1 t}$ and $e^{-i\omega_3 t}$ are responsible for the physical process called stimulated Raman scattering (SRS). These physical processes are described in detail later in this section.

Next we consider the incident optical field consists of three distinct frequency components ω_1, ω_2, and ω_3 ($\omega_1 \geq \omega_2 \geq \omega_3$), which are expressed in the Gaussian form

$$E_{mx}(\mathbf{r}, t) = \frac{1}{2} \left\{ U_m(x, y, z) A_m(t) e^{i(k_m z - \omega_m t)} + c.c. \right\}, \tag{1.1.143}$$

where

$$U_m(x, y, z) = \frac{U_m}{1 + i\zeta_m} e^{-(x^2 + y^2)/w_m^2(1 + i\zeta_m)}. \tag{1.1.144}$$

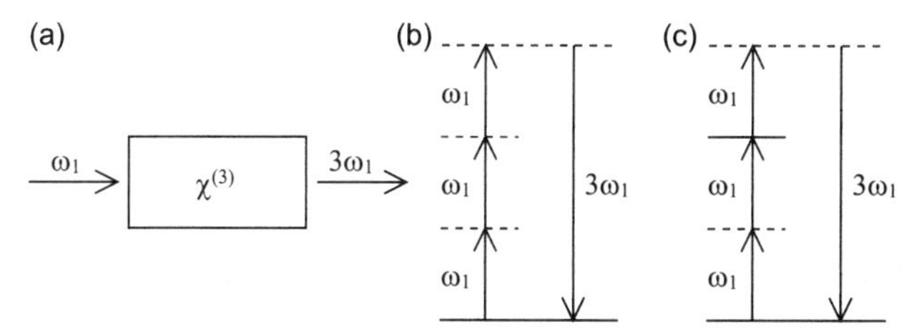

FIGURE 1.6 Third-harmonic generation. (a) Geometry of the interaction. (b, c) Energy diagrams describing the third-harmonic generation processes under nonresonant (b) and two-photon resonant (c) conditions.

The third-order nonlinear polarizations $P^{(3)}_{\mathrm{NR}}(\mathbf{r}, t)$, $P^{(3)}_{\mathrm{ER}}(\mathbf{r}, t)$, and $P^{(3)}_{\mathrm{RR}}(\mathbf{r}, t)$ for Gaussian beams can be obtained by replacing $A_m(z, t)$ in Eqs. (1.1.134), (1.1.137), (1.1.139), and (1.1.141) with $U_m(x, y, z)A_m(t)$, and these polarizations are also composed of the terms oscillating with frequencies of $3\omega_1$, $3\omega_2$, $3\omega_3$, $2\omega_1 + \omega_2$, $2\omega_1 + \omega_3$, $\omega_1 + 2\omega_2$, $\omega_1 + 2\omega_3$, $2\omega_2 + \omega_3$, $\omega_2 + 2\omega_3$, $2\omega_1 - \omega_2$, $2\omega_1 - \omega_3$, $2\omega_2 - \omega_1$, $2\omega_2 - \omega_3$, $2\omega_3 - \omega_1$, $2\omega_3 - \omega_2$, $\omega_1 + \omega_2 + \omega_3$, $\omega_1 + \omega_2 - \omega_3$, $\omega_1 + \omega_3 - \omega_2$, $\omega_2 + \omega_3 - \omega_1$, ω_1, ω_2, and ω_3.

1.1.3.1 Third-Harmonic Generation Let us now consider the process of third-harmonic generation (THG), which is illustrated in Fig. 1.6. In the THG process, three photons of frequency ω_1 are destroyed and one photon of frequency of $3\omega_1$ is created as shown in Fig. 1.6(b). THG was first observed in gasses by New et al. in 1967 [20]. When the one-photon, two-photon, and/or three-photon transitions are nearly resonant, the efficiency of THG is enhanced. However, for the cases of one-photon resonance and three-photon resonance, the incident and/or generated fields experience linear absorption and are rapidly attenuated as these fields propagate through the medium. In contrast, for the case of two-photon resonance, there is no linear absorption to limit the efficiency of THG. Thus, the enhancement through the two-photon resonance shown in Fig. 1.6(c) is usually the preferred way.

According to Eqs. (1.1.134) and (1.1.139), the polarizations describing the THG process under nonresonant and two-photon resonant conditions are given by the expression

$$P^{(3)}_{\mathrm{THG(NR)}}(\mathbf{r}, t) = \frac{1}{8}\varepsilon_0(1 - f_{\mathrm{E}})\chi^{(3)} \times A_1^3(z, t)e^{i(3k_1z - 3\omega_1 t)} + c.c., \tag{1.1.145}$$

$$P^{(3)}_{\mathrm{THG(ER)}}(\mathbf{r}, t) = \frac{\omega_{\mathrm{ER}}^2 + \Gamma_{\mathrm{ER}}^2}{16\omega_{\mathrm{ER}}}\varepsilon_0 f_{\mathrm{E}}\chi^{(3)}A_1(z, t) \times A^{(2)}_{11(ER)}(z, t)e^{i(3k_1z - 3\omega_1 t + (\pi/2))} + c.c. \tag{1.1.146}$$

From Eqs. (1.1.31), (1.1.145), and (1.1.146), the slowly varying amplitude $p_{\mathrm{THG}}^{(3)}(z,t)$ and the wave number $k_{\mathrm{THG}}^{(3)}$ of the polarization of THG are expressed by

$$p_{\mathrm{THG}}^{(3)}(z,t) = \frac{1}{4}(1-f_{\mathrm{E}})A_1^3(z,t) + i\frac{\omega_{\mathrm{ER}}^2+\Gamma_{\mathrm{ER}}^2}{8\omega_{\mathrm{ER}}} f_{\mathrm{E}} A_1(z,t) A_{11(\mathrm{ER})}^{(2)}(z,t), \tag{1.1.147}$$

$$k_{\mathrm{THG}}^{(3)} = 3k_1. \tag{1.1.148}$$

If the lifetime of the electronic excited state, which is expressed by $1/\Gamma_{\mathrm{ER}}$, is much shorter than the pulse duration, $A_{11(\mathrm{ER})}^{(2)}(z,t)$ is approximated by

$$\begin{aligned} A_{11(\mathrm{ER})}^{(2)}(z,t) &\approx A_1^2(z,t)\int_{-\infty}^{t} u\,(t-t_2)\,e^{\{-i(\omega_{\mathrm{ER}}-2\omega_1)-\Gamma_{\mathrm{ER}}\}(t-t_2)}dt_2 \\ &= -\frac{i}{\omega_{\mathrm{ER}}-2\omega_1-i\Gamma_{\mathrm{ER}}}A_1^2(z,t). \end{aligned} \tag{1.1.149}$$

Substituting Eq. (1.1.149) into Eq. (1.1.147), we obtain

$$\begin{aligned} p_{\mathrm{THG}}^{(3)}(z,t) &= \frac{1}{4}\left\{(1-f_{\mathrm{E}}) + \frac{\omega_{\mathrm{ER}}^2+\Gamma_{\mathrm{ER}}^2}{2\omega_{\mathrm{ER}}\,(\omega_{\mathrm{ER}}-2\omega_1-i\Gamma_{\mathrm{ER}})} f_{\mathrm{E}}\right\} A_1^3(z,t) \\ &= \frac{1}{4}\left\{(1-f_{\mathrm{E}}) + f_{\mathrm{E}}\tilde{h}_{\mathrm{ER}}\,(2\omega_1)\right\} A_1^3(z,t). \end{aligned} \tag{1.1.150}$$

Let us assume low incident intensities and/or a short length of the nonlinear medium. Under these circumstances, we may assume that the incident pulses don't suffer losses because of low conversion efficiencies. If we assume in addition that the linear absorption and the chromatic dispersion can be neglected, the distortionless propagation of the incident pulses is permitted. Then, the wave equation describing in Eq. (1.1.47) can be written as

$$\frac{\partial A_4(z,t)}{\partial z} = i\frac{3\omega_1\chi^{(3)}}{2\tilde{n}^{(1)}(3\omega_1)c} p_{\mathrm{THG}}^{(3)}(0,t) e^{i\Delta k_{\mathrm{THG}}^{(3)} z}, \tag{1.1.151}$$

where $\Delta k_{\mathrm{THG}}^{(3)} = k_{\mathrm{THG}}^{(3)} - k_4 = 3\omega_1\{\tilde{n}(\omega_1)-\tilde{n}(3\omega_1)\}/c$ is the wave vector mismatch. Here $A_4(z,t)$ and k_4 are the complex amplitude and the wave number of the third-harmonic field, respectively. By integrating Eq. (1.1.151) from 0 to the integration length L, we obtain

$$\begin{aligned} A_4(L,t) &= i\frac{3\omega_1\chi^{(3)}}{2\tilde{n}^{(1)}(3\omega_1)c} p_{\mathrm{THG}}^{(3)}(0,t)\int_0^L e^{i\Delta k_{\mathrm{THG}}^{(3)} z}dz \\ &= \frac{3\omega_1\chi^{(3)}}{2\tilde{n}^{(1)}(3\omega_1)c} p_{\mathrm{THG}}^{(3)}(0,t)\left\{\frac{e^{i\Delta k_{\mathrm{THG}}^{(3)} L}-1}{\Delta k_{\mathrm{THG}}^{(3)}}\right\}, \end{aligned} \tag{1.1.152}$$

From Eq. (1.1.152), we obtain the THG intensity $I_4(L,t)$

$$I_4(L,t) = \frac{9\varepsilon_0\left(\omega_1\chi^{(3)}L\right)^2}{8\tilde{n}^{(1)}(3\omega_1)c}\left|p_{\mathrm{THG}}^{(3)}(0,t)\right|^2 \operatorname{sinc}^2\left(\Delta k_{\mathrm{THG}}^{(3)}L/2\right). \quad (1.1.153)$$

According to Eq. (1.1.150), we find that $|p_{\mathrm{THG}}^{(3)}(0,t)|^2$ is related to the incident intensity by the following form:

$$\left|p_{\mathrm{THG}}^{(3)}(0,t)\right|^2 \propto \left\{|A_1(0,t)|^2\right\}^3 \propto I_1^3(0,t). \quad (1.1.154)$$

From Eqs. (1.1.153) and (1.1.154), we see that the THG intensity is proportional to the third power of the incident intensity, and the THG efficiency is maximized under the perfect phase-matching condition.

Next we treat nonresonant THG using a focused Gaussian beam. From Eqs. (1.1.52), (1.1.143), and (1.1.134), the complex amplitudes of the polarization of THG $U_{\mathrm{THG}}^{(3)}(x,y,z)$ is expressed by

$$U_{\mathrm{THG}}^{(3)}(x,y,z) = \frac{1}{4}U_1^3(x,y,z). \quad (1.1.155)$$

From Eqs. (1.1.155) and (1.1.143), we see that the polarization of THG experiences a phase shift of 3π radians, which is 3 times larger than the phase shift of the fundamental wave in passing through its focus due to the Gouy phase shift. If we assume a distortionless propagation of the incident pulses, the wave equation describing in Eq. (1.1.56) can be written as

$$\left[-\frac{i}{2k_4}\left(\frac{\partial^2}{\partial x^2}+\frac{\partial^2}{\partial y^2}\right)+\frac{\partial}{\partial z}\right]U_4(x,y,z)$$
$$= i\frac{3\omega_1}{2\tilde{n}^{(1)}(3\omega_1)c}\chi^{(3)}U_{\mathrm{THG}}^{(3)}(x,y,z)e^{i\Delta k_{\mathrm{THG}}^{(3)}z}. \quad (1.1.156)$$

Here $U_4(x,y,z)$ is the complex amplitude of the THG field. As shown in Section 1.12, we solve Eq. (1.1.156) by adopting the trial solution

$$U_4(x,y,z) = \frac{U_4(z)}{1+i\zeta_1}e^{-3(x^2+y^2)/w_1^2(1+i\zeta_1)}, \quad (1.1.157)$$

where $U_4(z)$ is a function of z. Insertion of Eqs. (1.1.157), (1.1.155), and (1.1.144) into Eq. (1.1.156) leads to the result

$$\frac{\partial U_4(z)}{\partial z} = i\frac{3\omega_1}{8\tilde{n}^{(1)}(3\omega_1)c}\chi^{(3)}U_1^3\frac{e^{i\Delta k_{\mathrm{THG}}^{(3)}z}}{(1+i\zeta_1)^2}. \quad (1.1.158)$$

By integrating this equation, we obtain

$$U_4(z) = i\frac{3\omega_1}{8\tilde{n}^{(1)}(3\omega_1)c}\chi^{(3)}U_1^3\int_{z_0}^{z}\frac{e^{i\Delta k_{\mathrm{THG}}^{(3)}z'}}{(1+iz'/\rho_1)^2}dz', \tag{1.1.159}$$

where z_0 indicates the value of z at the entrance to the nonlinear medium. If ρ_1 is much larger than $|z_0|$ and $|z|$, the integral in Eq. (1.1.159) is expressed by

$$J_{\mathrm{THG}}^{(3)}(\Delta k_{\mathrm{THG}}^{(3)}, z_0, z) \equiv \int_{z_0}^{z}\frac{e^{i\Delta k_{\mathrm{THG}}^{(3)}z'}}{(1+iz'/\rho_1)^2}dz' \approx \frac{e^{i\Delta k_{\mathrm{THG}}^{(3)}z} - e^{i\Delta k_{\mathrm{THG}}^{(3)}z_0}}{i\Delta k_{\mathrm{THG}}^{(3)}}. \tag{1.1.160}$$

From Eq. (1.1.160), we obtain

$$\left|J_{\mathrm{THG}}^{(3)}(\Delta k_{\mathrm{THG}}^{(3)}, z_0, z)\right|^2 = (z-z_0)^2\,\mathrm{sinc}^2\left\{\frac{\Delta k_{\mathrm{THG}}^{(3)}(z-z_0)}{2}\right\}, \tag{1.1.161}$$

which corresponds to Eq. (1.1.153). This condition indicates the propagation of plane waves. Next we consider the situation where ρ_1 is much smaller than $|z_0|$ and $|z|$. This situation represents the propagation under tight-focusing condition. The integral in Eq. (1.1.159) can now be expressed by

$$\begin{aligned} J_{\mathrm{THG}}^{(3)}(\Delta k_{\mathrm{THG}}^{(3)}, z_0, z) &\approx \int_{-\infty}^{\infty}\frac{e^{i\Delta k_{\mathrm{THG}}^{(3)}z'}}{(1+iz'/\rho_1)^2}dz' \\ &= \begin{cases} 0, & \Delta k_{\mathrm{THG}}^{(3)} \le 0, \\ 2\pi\rho_1^2\Delta k_{\mathrm{THG}}^{(3)}e^{-\rho_1\Delta k_{\mathrm{THG}}^{(3)}}, & \Delta k_{\mathrm{THG}}^{(3)} > 0. \end{cases} \end{aligned} \tag{1.1.162}$$

From Eq. (1.1.162), we see that similarly to SHG in tight-focusing limit, the THG efficiency is also maximized in the tight-focusing limit for the case of positive wave vector mismatch. This is because a positive wave vector mismatch can be used to compensate for the relative phase shift between the THG wave and the its driving polarization, which originate from the Gouy phase shift.

1.1.3.2 Four-Wave Mixing We discuss the process of four-wave mixing (FWM), which is shown in Fig. 1.7. In the FWM processes, a new photon with a frequency of ω_4 is generated by the interaction among three incident photons with frequencies of ω_1, ω_2, and ω_3. Although ω_4 is one of frequencies $2\omega_1+\omega_2$, $2\omega_1+\omega_3$, $\omega_1+2\omega_2$, $\omega_1+2\omega_3$, $2\omega_2+\omega_3$, $\omega_2+2\omega_3$, $2\omega_1-\omega_2$, $2\omega_1-\omega_3$, $2\omega_2-\omega_1$, $2\omega_2-\omega_3$, $2\omega_3-\omega_1$, $2\omega_3-\omega_2$, $\omega_1+\omega_2+\omega_3$, $\omega_1+\omega_2-\omega_3$, $\omega_1+\omega_3-\omega_2$, and $\omega_2+\omega_3-\omega_1$, an explanation of the generation of frequency components at $\omega_1+\omega_2+\omega_3$ and $\omega_1+\omega_2-\omega_3$ can easily extend to learning the FWM processes for other frequency components. Since the FWM process at a frequency of $\omega_1+\omega_2+\omega_3$ in a special case of $\omega_1=\omega_2=\omega_3$ corresponds to the THG process, we consider the FWM process at a frequency of $\omega_1+\omega_2-\omega_3$. The efficiency of FWM is also enhanced by the one-photon, two-photon, and/or three-photon resonances. To avoid the linear absorption of the

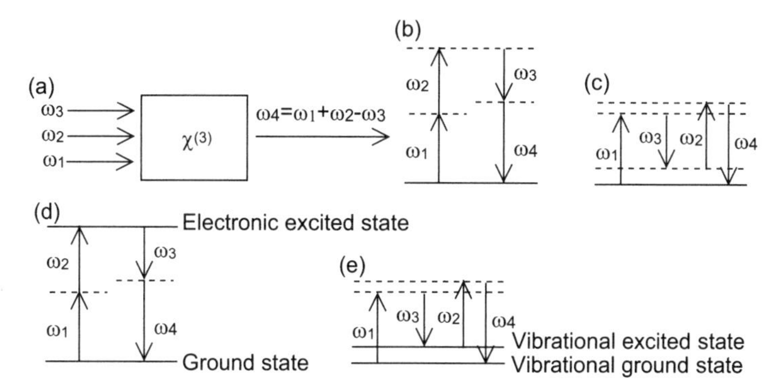

FIGURE 1.7 Four-wave mixing. (a) Geometry of the interaction. (b–e) Energy diagrams describing the four-wave mixing processes under nonresonant (b, c) and two-photon electronic resonant (d), two-photon vibrational resonant (e) conditions.

incident and/or generated fields, the enhancement technique by the two-photon resonance shown in Fig. 1.7(d) and (e) is commonly used. The FWM process in which the frequency difference between ω_1 and ω_2 coincides with the Raman resonant frequency is called coherent anti-Stokes Raman (CARS) as shown in Fig. 1.7(d). The FWM process using the enhancement technique by tuning sum frequency of ω_1 and ω_2 to the electronic resonance frequency is called stimulated parametric emission (SPE) as shown in Fig. 1.7(e). FWM process was theoretically studied by Armstrong et al. in 1962 [10]. The first comprehensive experiments of the FWM process was carried out by Maker and Terhune in 1965 [21]. The Raman resonance including the nonresonant electronic contribution to the FWM process was observed by Levenson and Bloembergen [22]. The FWM signal enhanced by the two-photon electronic resonance was also observed at extremely low temperatures by Kramer and Bloembergen [23].

According to Eqs. (1.1.134), (1.1.137), and (1.1.141), the polarizations representing the FWM process at a frequency of $\omega_1 + \omega_2 - \omega_3$ are given by the expression

$$\begin{aligned}
P^{(3)}_{\mathrm{FWM}}(\mathbf{r}, t) = {} & \frac{1}{4}\varepsilon_0(3 - f_{\mathrm{E}} - f_{\mathrm{R}})\chi^{(3)} \\
& \times A_1(z,t)A_2(z,t)A_3^*(z,t)e^{i\{(k_1+k_2-k_3)z-(\omega_1+\omega_2-\omega_3)t\}} \\
& + \frac{\omega_{\mathrm{ER}}^2 + \Gamma_{\mathrm{ER}}^2}{8\omega_{\mathrm{ER}}}\varepsilon_0 f_{\mathrm{E}}\chi^{(3)} \\
& \times A_3^*(z,t)A^{(2)}_{12(\mathrm{ER})}(z,t)e^{i\frac{\pi}{2}}e^{i\{(k_1+k_2-k_3)z-(\omega_1+\omega_2-\omega_3)t\}} \\
& + \frac{\omega_{\mathrm{RR}}^2 + \Gamma_{\mathrm{RR}}^2}{8\omega_{\mathrm{RR}}}\varepsilon_0 f_{\mathrm{R}}\chi^{(3)} \\
& \times A_2(z,t)A^{(2\#)}_{13(\mathrm{RR})}(z,t)e^{i\frac{\pi}{2}}e^{i\{(k_1+k_2-k_3)z-(\omega_1+\omega_2-\omega_3)t\}} \\
& + c.c. \qquad (1.1.163)
\end{aligned}$$

From Eqs. (1.1.31), and (1.1.163), the complex amplitude $p^{(3)}_{\mathrm{FWM}}(z,t)$ and the wave number $k^{(3)}_{\mathrm{FWM}}$ of the polarization of FWM are described by

$$\begin{aligned} p^{(3)}_{\mathrm{FWM}}(z,t) &= \frac{1}{2}(3 - f_{\mathrm{E}} - f_{\mathrm{R}})A_1(z,t)A_2(z,t)A_3^*(z,t) \\ &\quad + i\frac{\omega_{\mathrm{ER}}^2 + \Gamma_{\mathrm{ER}}^2}{4\omega_{\mathrm{ER}}} f_{\mathrm{E}} A_3^*(z,t)A^{(2)}_{12(\mathrm{ER})}(z,t) \\ &\quad + i\frac{\omega_{\mathrm{RR}}^2 + \Gamma_{\mathrm{RR}}^2}{4\omega_{\mathrm{RR}}} f_{\mathrm{R}} A_2(z,t)A^{(2\#)}_{13(\mathrm{RR})}(z,t), \end{aligned} \tag{1.1.164}$$

$$k^{(3)}_{\mathrm{FWM}} = k_1 + k_2 - k_3. \tag{1.1.165}$$

If we assume the distortionless propagation of the incident pulses, the wave equation described in Eq. (1.1.47) can be written as

$$\frac{\partial A_4(z,t)}{\partial z} = i\frac{\omega_4\chi^{(3)}}{2\tilde{n}^{(1)}(\omega_4)c} p^{(3)}_{\mathrm{FWM}}(0,t)e^{i\Delta k^{(3)}_{\mathrm{FWM}} z}, \tag{1.1.166}$$

where $\Delta k^{(3)}_{\mathrm{FWM}}$ is the wave vector mismatch. Here $A_4(z,t)$ and k_4 are the complex amplitude and the wave number of the FWM field, respectively. By integrating Eq. (1.1.166) from 0 to the integration length L, we obtain

$$A_4(L,t) = \frac{\omega_4\chi^{(3)}}{2\tilde{n}^{(1)}(\omega_4)c} p^{(3)}_{\mathrm{FWM}}(0,t)\left\{\frac{e^{i\Delta k^{(3)}_{\mathrm{FWM}} L} - 1}{\Delta k^{(3)}_{\mathrm{FWM}}}\right\}. \tag{1.1.167}$$

From Eq. (1.1.167), the FWM intensity $I_4(L,t)$ is obtained by

$$I_4(L,t) = \frac{\varepsilon_0\left(\omega_4\chi^{(3)}L\right)^2}{8\tilde{n}^{(1)}(\omega_4)c}\left|p^{(3)}_{\mathrm{FWM}}(0,t)\right|^2 \mathrm{sinc}^2\left(\Delta k^{(3)}_{\mathrm{FWM}} L/2\right). \tag{1.1.168}$$

In the frequency domain, Eq. (1.1.167) is written as

$$\tilde{A}_4(L,\omega-\omega_4) = \frac{\omega_4\chi^{(3)}}{2\tilde{n}^{(1)}(\omega_4)c}\left\{\frac{e^{i\Delta k^{(3)}_{\mathrm{FWM}} L} - 1}{\Delta k^{(3)}_{\mathrm{FWM}}}\right\}\tilde{p}^{(3)}_{\mathrm{FWM}}(0,\omega-\omega_4). \tag{1.1.169}$$

Thus, the spectral intensity of FWM is proportional to

$$\left|\tilde{A}_4(L,\omega-\omega_4)\right|^2 = \left\{\frac{\omega_4\chi^{(3)}L}{2\tilde{n}^{(1)}(\omega_4)c}\mathrm{sinc}\left(\frac{\Delta k^{(3)}_{\mathrm{FWM}} L}{2}\right)\right\}^2\left|\tilde{p}^{(3)}_{\mathrm{FWM}}(0,\omega-\omega_4)\right|^2. \tag{1.1.170}$$

If the frequency difference $\omega_1 - \omega_3$ is nearly equal to the Raman resonance frequency and the frequency sum $\omega_1 + \omega_2$ is far from the electronic resonance frequency,

$p_{\mathrm{FWM}}^{(3)}(z,t)$ is expressed by

$$p_{\mathrm{FWM}}^{(3)}(z,t) = \frac{1}{4}A_2(z,t)\int_{-\infty}^{\infty}\left[(6-2f_{\mathrm{R}})\delta(t-t_2) + if_{\mathrm{R}}\frac{\omega_{\mathrm{RR}}^2+\Gamma_{\mathrm{RR}}^2}{\omega_{\mathrm{RR}}}u(t-t_2)e^{[-i\{\omega_{\mathrm{RR}}-(\omega_1-\omega_3)\}-\Gamma_{\mathrm{RR}}](t-t_2)}\right] \times A_1(z,t_2)A_3^*(z,t_2)dt_2. \tag{1.1.171}$$

In the frequency domain, $\tilde{p}_{\mathrm{FWM}}^{(3)}(z,\omega)$ is given by

$$\tilde{p}_{\mathrm{FWM}}^{(3)}(z,\omega) = \frac{1}{4}\int_{-\infty}^{\infty}\tilde{A}_2(z,\omega-\Omega)\left[(6-2f_{\mathrm{R}}) + f_{\mathrm{R}}\frac{\omega_{\mathrm{RR}}^2+\Gamma_{\mathrm{RR}}^2}{\omega_{\mathrm{RR}}}\frac{1}{\{\omega_{\mathrm{RR}}-(\omega_1-\omega_3)\}-\Omega-i\Gamma_{\mathrm{RR}}}\right] \times \tilde{A}_{13}^{(2\#)}(z,\Omega)d\Omega, \tag{1.1.172}$$

where $\tilde{A}_{13}^{(2\#)}(z,\omega)$ is introduced using

$$\tilde{A}_{13}^{(2\#)}(z,\omega) = \int_{-\infty}^{\infty}\tilde{A}_1(z,\Omega)\tilde{A}_3^*(z,\Omega-\omega)d\Omega. \tag{1.1.173}$$

If we assume that the incident field with a carrier frequency of ω_2 is described as a narrowband pulse $\tilde{A}_2(z,\omega) = A_2(z)\delta(\omega)$, $\tilde{p}_{\mathrm{FWM}}^{(3)}(z,\omega)$ is written as

$$\begin{aligned}\tilde{p}_{\mathrm{FWM}}^{(3)}(z,\omega) &= \frac{1}{4}A_2(z)\tilde{A}_{13}^{(2\#)}(z,\omega)\left[(6-2f_{\mathrm{R}}) + f_{\mathrm{R}}\frac{\omega_{\mathrm{RR}}^2+\Gamma_{\mathrm{RR}}^2}{\omega_{\mathrm{RR}}}\frac{1}{\{\omega_{\mathrm{RR}}-(\omega_1-\omega_3)\}-\omega-i\Gamma_{\mathrm{RR}}}\right]\\ &= \frac{1}{4}A_2(z)\tilde{A}_{13}^{(2\#)}(z,\omega)\left[[(6-2f_{\mathrm{R}}) + 2f_{\mathrm{R}}\tilde{h}_{\mathrm{RR}}(\omega_1-\omega_3+\omega)\right].\end{aligned} \tag{1.1.174}$$

From Eqs. (1.1.174) and (1.1.170), we see that the spectral information about the Raman response can be acquired from the FWM spectrum, which is measured by employing the narrowband pulse at ω_2:

$$\left|\tilde{A}_4(L,\omega)\right|^2 = \frac{1}{16}\left\{\frac{\omega_4 L}{2\tilde{n}^{(1)}(\omega_4)c}\operatorname{sinc}\left(\frac{\Delta k_{\mathrm{FWM}}^{(3)}L}{2}\right)\right\}^2\left|A_2(0)\tilde{A}_{13}^{(2\#)}(0,\omega)\right|^2 \times\left|(6-2f_{\mathrm{R}})\chi^{(3)} + 2f_{\mathrm{R}}\tilde{\chi}_{\mathrm{RR}}^{(3)}(\omega_1-\omega_3+\omega)\right|^2, \tag{1.1.175}$$

where

$$\tilde{\chi}_{\mathrm{RR}}^{(3)}(\omega_1 - \omega_3 + \omega) = \chi^{(3)}\tilde{h}_{\mathrm{RR}}(\omega_1 - \omega_3 + \omega). \tag{1.1.176}$$

When $\tilde{A}_{13}^{(2\#)}(z, \omega)$ can be approximated by $A_1(z)A_3^*(z)\delta(\omega)$, we obtain

$$|A_4(L,t)|^2 = \frac{1}{16}\left\{\frac{\omega_4 L}{2\tilde{n}^{(1)}(\omega_4)c}\operatorname{sinc}\left(\frac{\Delta k_{\mathrm{FWM}}^{(3)}L}{2}\right)\right\}^2 |A_1(0)|^2 |A_2(0)|^2$$
$$\times |A_3(0)|^2 \left|\chi_{\mathrm{R}}^{(\mathrm{NR})} + \tilde{\chi}^{(\mathrm{RR})}(\omega_1 - \omega_3)\right|^2, \tag{1.1.177}$$

where

$$\chi_{\mathrm{R}}^{(\mathrm{NR})} = (6 - 2f_{\mathrm{R}})\chi^{(3)}, \tag{1.1.178}$$

$$\chi^{(\mathrm{RR})}(\omega_1 - \omega_3) = 2f_{\mathrm{R}}\tilde{\chi}_{\mathrm{RR}}^{(3)}(\omega_1 - \omega_3). \tag{1.1.179}$$

In contrast, when the frequency difference $\omega_1 - \omega_3$ is far from the Raman resonance frequency and the frequency sum $\omega_1 + \omega_2$ coincides with the electronic resonance frequency, $p_{\mathrm{FWM}}^{(3)}(z, t)$, which is given by

$$p_{\mathrm{FWM}}^{(3)}(z,t) = \frac{1}{4}A_3(z,t)\int_{-\infty}^{\infty}\left[(6 - 2f_{\mathrm{E}})\delta(t - t_2)\right.$$
$$\left. + if_{\mathrm{E}}\frac{\omega_{\mathrm{ER}}^2 + \Gamma_{\mathrm{ER}}^2}{\omega_{\mathrm{ER}}}u(t - t_2)e^{[-i\{\omega_{\mathrm{ER}} - (\omega_1 + \omega_2)\} - \Gamma_{\mathrm{ER}}](t - t_2)}\right]$$
$$\times A_1(z, t_2)A_2(z, t_2)dt_2. \tag{1.1.180}$$

In the frequency domain, $\tilde{p}_{\mathrm{FWM}}^{(3)}(z, \omega)$ is expressed as

$$\tilde{p}_{\mathrm{FWM}}^{(3)}(z,\omega) = \frac{1}{4}\int_{-\infty}^{\infty}\tilde{A}_3(z, \omega - \Omega)\left[(6 - 2f_{\mathrm{E}})\right.$$
$$\left. + f_{\mathrm{E}}\frac{\omega_{\mathrm{ER}}^2 + \Gamma_{\mathrm{ER}}^2}{\omega_{\mathrm{ER}}}\frac{1}{\{\omega_{\mathrm{ER}} - (\omega_1 + \omega_2)\} - \Omega - i\Gamma_{\mathrm{ER}}}\right]\tilde{A}_{12}^{(2)}(z, \Omega)d\Omega, \tag{1.1.181}$$

where $\tilde{A}_{12}^{(2)}(z, \omega)$ is introduced by using

$$\tilde{A}_{12}^{(2)}(z,\omega) = \int_{-\infty}^{\infty}\tilde{A}_1(z, \Omega)\tilde{A}_2(z, \omega - \Omega)d\Omega. \tag{1.1.182}$$

We can assume that the incident field with a carrier frequency of ω_3 is described as a narrowband pulse $\tilde{A}_3(z,\omega) = A_3(z)\delta(\omega)$, so we can write $\tilde{p}^{(3)}_{\mathrm{FWM}}(z,\omega)$ as

$$\begin{aligned}\tilde{p}^{(3)}_{\mathrm{FWM}}(z,\omega) &= \frac{1}{4}A_3(z)\tilde{A}^{(2)}_{12}(z,\omega)\Bigg[(6-2f_{\mathrm{E}}) \\ &\quad + f_{\mathrm{E}}\frac{\omega_{\mathrm{ER}}^2+\Gamma_{\mathrm{ER}}^2}{\omega_{\mathrm{ER}}}\frac{1}{\{\omega_{\mathrm{ER}}-(\omega_1+\omega_2)\}-\omega-i\Gamma_{\mathrm{ER}}}\Bigg] \\ &= \frac{1}{4}A_3(z)\tilde{A}^{(2)}_{12}(z,\omega)\left[(6-2f_{\mathrm{E}})+2f_{\mathrm{E}}\tilde{h}_{\mathrm{ER}}(\omega_1+\omega_2+\omega)\right].\end{aligned} \tag{1.1.183}$$

From Eq. (1.1.183), we see that the spectral information about the electronic response can be obtained from the FWM spectrum, which is recorded by using the narrowband pulse at ω_3:

$$\begin{aligned}\left|\tilde{A}_4(L,\omega)\right|^2 &= \frac{1}{16}\left\{\frac{\omega_4 L}{2\tilde{n}^{(1)}(\omega_4)c}\operatorname{sinc}\left(\frac{\Delta k^{(3)}_{\mathrm{FWM}}L}{2}\right)\right\}^2\left|A_3(0)\tilde{A}^{(2)}_{12}(0,\omega)\right|^2 \\ &\quad\times\left|(6-2f_{\mathrm{E}})\,\chi^{(3)}+2f_{\mathrm{E}}\tilde{\chi}^{(\mathrm{ER})}(\omega_1+\omega_2+\omega)\right|^2,\end{aligned} \tag{1.1.184}$$

where

$$\tilde{\chi}^{(3)}_{\mathrm{ER}}(\omega_1+\omega_2+\omega) = \chi^{(3)}\tilde{h}_{\mathrm{ER}}(\omega_1+\omega_2+\omega). \tag{1.1.185}$$

If we assume that $\tilde{A}^{(2)}_{12}(z,\omega)$ can be described as $A_1(z)A_2(z)\delta(\omega)$, we obtain

$$\begin{aligned}|A_4(L,t)|^2 &= \frac{1}{16}\left\{\frac{\omega_4 L}{2\tilde{n}^{(1)}(\omega_4)c}\operatorname{sinc}\left(\frac{\Delta k^{(3)}_{\mathrm{FWM}}L}{2}\right)\right\}^2|A_1(0)|^2\,|A_2(0)|^2 \\ &\quad\times|A_3(0)|^2\left|\tilde{\chi}^{(\mathrm{NR})}_{\mathrm{E}}+\tilde{\chi}^{(\mathrm{ER})}(\omega_1+\omega_2)\right|^2,\end{aligned} \tag{1.1.186}$$

where

$$\tilde{\chi}^{(\mathrm{ER})}(\omega_1+\omega_2+\omega) = 2f_{\mathrm{E}}\tilde{\chi}^{(3)}_{\mathrm{ER}}(\omega_1+\omega_2+\omega) \tag{1.1.187}$$

and

$$\tilde{\chi}^{(\mathrm{NR})}_{\mathrm{E}} = (6-2f_{\mathrm{E}})\chi^{(3)}. \tag{1.1.188}$$

From Eqs. (1.1.177) and (1.1.186), we known that the FWM intensity is proportional to the product of the incident intensities I_1, I_2 and I_3.

Next we discuss nonresonant FWM using a focused Gaussian beam. Equations (1.1.52), (1.1.143), and (1.1.134) show that the complex amplitudes of the polarization of FWM $U_{\mathrm{FWM}}^{(3)}(x, y, z)$ can be expressed by

$$U_{\mathrm{FWM}}^{(3)}(x, y, z) = \frac{1}{2} U_1(x, y, z) U_2(x, y, z) U_3^*(x, y, z). \tag{1.1.189}$$

According to Eqs. (1.1.189) and (1.1.144), because of the Gouy phase shift, the polarization of FWM experiences a phase shift of $\pi(\approx 2\pi - \pi)$ radians, which is similar to the phase shift of the fundamental wave in passing through its focus. Of course, the FWM wave experiences the phase shift of π radians. Thus, the relative phase between the FWM wave and its driving polarization, which originates from the Gouy phase shift, can be neglected. Therefore, the FWM intensity is also maximized under the perfect phase-matching condition in the tight-focusing limit.

1.1.3.3 Self-Phase Modulation Let us next consider the process of self-phase modulation (SPM). Thus is when the phenomenon when a phase modulation occurs due to the intensity-dependent refractive index induced by the intensity of the pulse itself. SPM was first observed as a modulated spectrum extending both above and below the laser frequency after self-focusing had occurred in a liquid-filled cell, and this was explained by Shimizu in 1967 as phase modulation due to the intensity-dependent refractive index [24]. According to Eq. (1.1.134), the polarization representing the SPM process at a frequency of ω_1 is given by the expression

$$\begin{aligned} P_{\mathrm{SPM}}^{(3)}(\mathbf{r}, t) = & \frac{3}{8} \varepsilon_0 \chi^{(3)} \\ & \times |A_1(z, t)|^2 A_1(z, t) e^{i(k_1 z - \omega_1 t)} + c.c. \end{aligned} \tag{1.1.190}$$

From Eqs. (1.1.31) and (1.1.190), the slowly varying amplitude $p_{\mathrm{SPM}}^{(3)}(z, t)$ and the wave number $k_{\mathrm{SPM}}^{(3)}$ of the polarization of SPM are expressed by

$$p_{\mathrm{SPM}}^{(3)}(z, t) = \frac{3}{4} |A_1(z, t)|^2 A_1(z, t) \tag{1.1.191}$$

$$k_{\mathrm{SPM}}^{(3)} = k_1. \tag{1.1.192}$$

If we assume that the linear absorption and the chromatic dispersion can be neglected, the wave equation describing in Eq. (1.1.47) can be written as

$$\frac{\partial A_1(z, t)}{\partial z} = i \frac{\omega_1 \chi^{(3)}}{2\tilde{n}^{(1)}(\omega_1) c} p_{\mathrm{SPM}}^{(3)}(z, t), \tag{1.1.193}$$

Substituting Eq. (1.1.191) into Eq. (1.1.193), we obtain

$$\begin{aligned}\frac{\partial A_1(z,t)}{\partial z} &= i\frac{3\omega_1\chi^{(3)}}{8\tilde{n}^{(1)}(\omega_1)c}\left|A_1(z,t)\right|^2 A_1(z,t)\\ &= i\frac{\omega_1\tilde{n}^{(2)}(\omega_1)}{c}\left|A_1(z,t)\right|^2 A_1(z,t).\end{aligned} \tag{1.1.194}$$

Here we have introduced the quantity

$$\tilde{n}^{(2)}(\omega_1) = \frac{3\chi^{(3)}}{8\tilde{n}^{(1)}(\omega_1)}, \tag{1.1.195}$$

which is called nonlinear index coefficient. Inserting $A_1(z,t) = |A_1(z,t)|e^{-i\varphi_1(z,t)}$ into Eq. (1.1.194) and separating the real and imaginary parts obtains the equations for the envelope $|A_1(z,t)|$ and the phase $\varphi_1(z,t)$:

$$\frac{\partial\left|A_1(z,t)\right|}{\partial z} = 0, \tag{1.1.196}$$

$$\frac{\partial\varphi_1(z,t)}{\partial z} = -\frac{\omega_1\tilde{n}^{(2)}(\omega_1)}{c}\left|A_1(z,t)\right|^2. \tag{1.1.197}$$

From Eq. (1.1.196), we know that the envelope is constant in the coordinate system traveling with the group velocity. Thus, the envelope remains unchanged, $|A_1(z,t)| = |A_1(0,t)|$. Taking this into consideration and integrating Eq. (1.1.197), we obtain the phase as

$$\varphi_1(z,t) = \varphi_1(0,t) - \frac{\omega_1}{c}\tilde{n}_{11}^{(2)}(\omega_1)I_1(0,t)z. \tag{1.1.198}$$

Here the quantity $\tilde{n}_{11}^{(2)}(\omega_1)$ is introduced by using

$$\tilde{n}_{11}^{(2)}(\omega_1) = \frac{2}{\varepsilon_0\tilde{n}^{(1)}(\omega_1)c}\tilde{n}^{(2)}(\omega_1). \tag{1.1.199}$$

This result implies that the refractive index increases with the increasing incident intensity. This refractive index change is called the optical Kerr effect. According to Eq. (1.1.198), the incident intensity varying in time results in the phase modulation given by

$$\frac{\partial\varphi_1(z,t)}{\partial t} = \frac{d\varphi_1(0,t)}{dt} - \frac{\omega_1}{c}\tilde{n}_{11}^{(2)}(\omega_1)\frac{dI_1(0,t)}{dt}z. \tag{1.1.200}$$

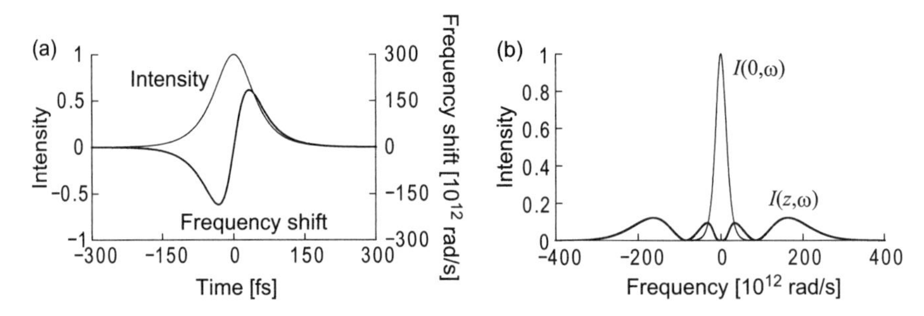

FIGURE 1.8 Self-phase modulation. (a) Instantaneous frequency shift. (b) Spectral broadening.

This phase modulation can be interpreted as an instantaneous frequency shift, which is given by

$$\delta\omega(t) = -\frac{\omega_1}{c}\tilde{n}_{11}^{(2)}(\omega_1)\frac{dI_1(0,t)}{dt}z. \tag{1.1.201}$$

From Eq. (1.1.201), we see that the instantaneous frequency shift is proportional to the first derivation of instantaneous intensity of the ultrashort pulse. The leading edge of the pulse has negative slope, the peak has a zero slope, and the tailing edge of the pulse has positive slope as shown in Fig. 1.8(a). Thus, the frequency shift at the peak of the pulse is zero, and the leading edge is red-shifted while the tailing edge is blue-shifted as shown in Fig. 1.8(a). As a result, SPM induces a spectral broadening as shown in Fig. 1.8(b).

1.1.3.4 Cross-Phase Modulation The process of cross-phase modulation (XPM), is similar to the SPM process. In the SPM process, the refractive index experienced by a light pulse with a carrier frequency of ω_1 is changed by the intensity of the pulse itself. In the XPM process, the intensity of the copropagating pulse with a carrier frequency of $\omega_2(\neq \omega_1)$ increases the refractive index at a frequency of ω_1. According to the 6th term to the last in Eq. (1.1.134), the XPM process at a frequency of ω_1 is given by the expression

$$P_{\mathrm{XPM}}^{(3)}(\mathbf{r},t) = \frac{3}{4}\varepsilon_0\chi^{(3)}\left|A_2(z,t)\right|^2 A_1(z,t)e^{i(k_1z-\omega_1t)} + c.c. \tag{1.1.202}$$

From Eqs. (1.1.31) and (1.1.202), the slowly varying amplitude $p_{\mathrm{XPM}}^{(3)}(z,t)$ and the wave number $k_{\mathrm{XPM}}^{(3)}$ of the polarization of XPM are expressed by

$$p_{\mathrm{XPM}}^{(3)}(z,t) = \frac{3}{2}\left|A_2(z,t)\right|^2 A_1(z,t) \tag{1.1.203}$$

$$k_{\mathrm{XPM}}^{(3)} = k_1. \tag{1.1.204}$$

We will assume that the linear absorption and the chromatic dispersion can be neglected, so that the wave equation describing in Eq. (1.1.47) can be written as

$$\frac{\partial A_1(z,t)}{\partial z} = i\frac{\omega_1 \chi^{(3)}}{2\tilde{n}^{(1)}(\omega_1)c} p_{\mathrm{XPM}}^{(3)}(z,t), \tag{1.1.205}$$

Inserting Eq. (1.1.203) into Eq. (1.1.205), we obtain

$$\begin{aligned}\frac{\partial A_1(z,t)}{\partial z} &= i\frac{3\omega_1 \chi^{(3)}}{4\tilde{n}^{(1)}(\omega_1)c} |A_2(z,t)|^2 A_1(z,t) \\ &= i\frac{2\omega_1 \tilde{n}^{(2)}(\omega_1)}{c} |A_2(z,t)|^2 A_1(z,t)\end{aligned} \tag{1.1.206}$$

Here we used Eq. (1.1.195). Inserting $A_1(z,t) = |A_1(z,t)|e^{-i\varphi_1(z,t)}$ into Eq. (1.1.206) and separating the real and imaginary parts, we obtain the equations for the envelope $|A_1(z,t)|$ and the phase $\varphi_1(z,t)$ as

$$\frac{\partial |A_1(z,t)|}{\partial z} = 0, \tag{1.1.207}$$

$$\frac{\partial \varphi_1(z,t)}{\partial z} = -\frac{2\omega_1 \tilde{n}^{(2)}(\omega_1)}{c} |A_2(z,t)|^2 . \tag{1.1.208}$$

From Eq. (1.1.207), we see that the envelope remains unchanged, $|A_1(z,t)| = |A_1(0,t)|$. Thus, we can obtain the phase from Eq. (1.1.208) in the following form:

$$\varphi_1(z,t) = \varphi_1(0,t) - \frac{2\omega_1}{c}\tilde{n}_{12}^{(2)}(\omega_1) I_2(0,t) z. \tag{1.1.209}$$

Here the quantity $\tilde{n}_{12}^{(2)}(\omega_1)$ is introduced by using

$$\tilde{n}_{12}^{(2)}(\omega_1) = \frac{2}{\varepsilon_0 \tilde{n}^{(1)}(\omega_2)c}\tilde{n}^{(2)}(\omega_1). \tag{1.1.210}$$

From Eq. (1.1.209), we see that the refractive index experienced by the pulse with a carrier frequency of ω_1 depends on the intensity of the copropagating pulse with a carrier frequency of ω_2. According to Eq. (1.1.209), the XPM process induces the phase modulation

$$\frac{\partial \varphi_1(z,t)}{\partial t} = \frac{d\varphi_1(0,t)}{dt} - \frac{2\omega_1}{c}\tilde{n}_{12}^{(2)}(\omega_1)\frac{dI_2(0,t)}{dt} z, \tag{1.1.211}$$

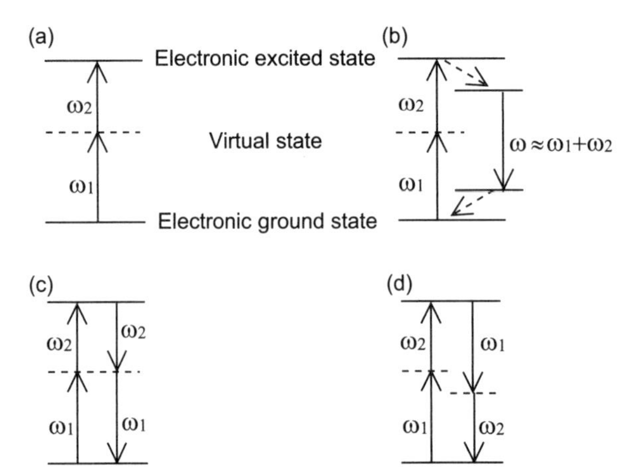

FIGURE 1.9 Energy diagrams describing the processes of two-photon absorption (TPA) (a) and two-photon excited fluorescence (b). (c, d) TPA can be interpreted as a special case of two-photon electronic resonant FWM where the frequency of the generated photon is equal to that of the incident photon at ω_1 (c) and ω_2 (d).

which results in the instantaneous frequency shift given by

$$\delta\omega(t) = -\frac{2\omega_1}{c}\tilde{n}_{12}^{(2)}(\omega_1)\frac{dI_2(0,t)}{dt}z. \tag{1.1.212}$$

From Eq. (1.1.212), we find that the spectrum of the pulse with ω_1 is broadened through XPM induced by the copropagating pulse with ω_2. The spectral broadening by XPM is similar to that obtained by SPM.

1.1.3.5 Two-Photon Absorption The process of two-photon absorption (TPA) is illustrated in Fig. 1.9(a). In the TPA process, an atom or a molecule makes a transition from its ground state to an excited state by the simultaneous absorption of two incident photons at frequencies of ω_1 and ω_2. Figure 1.9(b) shows the energy diagram describing the process of two-photon excited fluorescence (TPEF) in which fluorescence is emitted after the TPA process. TPA was predicted by Göppert-Meyer in 1931 [25] and TPEF was first observed in a europium-doped crystal by Kaiser et al. in 1961 [26].

To simplify the consideration of the TPA process, we first assume that ω_2 is equal to ω_2. According to Eq. (1.1.139), the nonlinear polarization including the TPA process at ω_1 is given by the expression

$$P_{\mathrm{TPA}}^{(3)}(\mathbf{r},t) = i\frac{\omega_{\mathrm{ER}}^2+\Gamma_{\mathrm{ER}}^2}{16\omega_{\mathrm{ER}}}\varepsilon_0\chi^{(3)} \times \left\{A_1^*(z,t)A_{11(\mathrm{ER})}^{(2)}(z,t)e^{i(k_1z-\omega_1t)} - c.c.\right\}. \tag{1.1.213}$$

From Eqs. (1.1.31) and (1.1.213), the slowly varying amplitude $p_{\mathrm{TPA}}^{(3)}(z,t)$ and the wave number $k_{\mathrm{TPA}}^{(3)}$ of the polarization of TPA are expressed by

$$p_{\mathrm{TPA}}^{(3)}(z,t) = i\frac{\omega_{\mathrm{ER}}^2 + \Gamma_{\mathrm{ER}}^2}{8\omega_{\mathrm{ER}}} A_1^*(z,t) A_{11(\mathrm{ER})}^{(2)}(z,t), \tag{1.1.214}$$

$$k_{\mathrm{TPA}}^{(3)} = k_1. \tag{1.1.215}$$

If we assume that the linear absorption and the chromatic dispersion can be neglected, the wave equation describing in Eq. (1.1.47) can be written as

$$\frac{\partial A_1(z,t)}{\partial z} = i\frac{\omega_1 \chi^{(3)}}{2\tilde{n}^{(1)}(\omega_1)c} p_{\mathrm{TPA}}^{(3)}(z,t). \tag{1.1.216}$$

Substituting Eq. (1.1.214) into Eq. (1.1.216), we obtain

$$\frac{\partial A_1(z,t)}{\partial z} = -\frac{\omega_1 \chi^{(3)}}{16\tilde{n}^{(1)}(\omega_1)c} \frac{\omega_{\mathrm{ER}}^2 + \Gamma_{\mathrm{ER}}^2}{\omega_{\mathrm{ER}}} A_1^*(z,t) A_{11(\mathrm{ER})}^{(2)}(z,t). \tag{1.1.217}$$

By assuming that the lifetime of the electronic excited state is much shorter than the pulse duration, $A_{11(\mathrm{ER})}^{(2)}(z,t)$ is approximated by Eq. (1.1.149). Inserting Eq. (1.1.149) into Eq. (1.1.217), we obtain

$$\begin{aligned}\frac{\partial A_1(z,t)}{\partial z} &= i\frac{\omega_1 \chi^{(3)}}{16\tilde{n}^{(1)}(\omega_1)c} \frac{\omega_{\mathrm{ER}}^2 + \Gamma_{\mathrm{ER}}^2}{\omega_{\mathrm{ER}}(\omega_{\mathrm{ER}} - 2\omega_1 - i\Gamma_{\mathrm{ER}})} |A_1(z,t)|^2 A_1(z,t) \\ &= \left\{ i\frac{\omega_1 \tilde{n}_{\mathrm{ER},1}^{(2)}(2\omega_1)}{c} - \frac{1}{8}\tilde{\alpha}_1^{(2)}(2\omega_1) \right\} |A_1(z,t)|^2 A_1(z,t).\end{aligned} \tag{1.1.218}$$

Here we have introduced the two quantities

$$\begin{aligned}\tilde{\alpha}_i^{(2)}(\omega_1+\omega_2) &= \frac{\omega_i \chi^{(3)}}{\tilde{n}^{(1)}(\omega_i)c} \frac{\omega_{\mathrm{ER}}^2 + \Gamma_{\mathrm{ER}}^2}{2\omega_{\mathrm{ER}}} \frac{\Gamma_{\mathrm{ER}}}{\{\omega_{\mathrm{ER}} - (\omega_1+\omega_2)\}^2 + \Gamma_{\mathrm{ER}}^2} \\ &= \frac{\omega_i}{\tilde{n}^{(1)}(\omega_i)c} \chi^{(3)} \mathrm{Im}\left\{\tilde{h}_{\mathrm{ER}}(\omega_1+\omega_2)\right\} \\ &= \frac{\omega_i}{\tilde{n}^{(1)}(\omega_i)c} \mathrm{Im}\left\{\tilde{\chi}_{\mathrm{ER}}^{(3)}(\omega_1+\omega_2)\right\},\end{aligned} \tag{1.1.219}$$

$$\begin{aligned}\tilde{n}_{\mathrm{ER},i}^{(2)}(\omega_1+\omega_2) &= \frac{\chi^{(3)}}{8\tilde{n}^{(1)}(\omega_i)} \frac{\omega_{\mathrm{ER}}^2 + \Gamma_{\mathrm{ER}}^2}{2\omega_{\mathrm{ER}}} \frac{\omega_{\mathrm{ER}} - (\omega_1+\omega_2)}{\{\omega_{\mathrm{ER}} - (\omega_1+\omega_2)\}^2 + \Gamma_{\mathrm{ER}}^2} \\ &= \frac{\chi^{(3)}}{8\tilde{n}^{(1)}(\omega_i)} \mathrm{Re}\left\{\tilde{h}_{\mathrm{ER}}(\omega_1+\omega_2)\right\} \\ &= \frac{1}{8\tilde{n}^{(1)}(\omega_i)} \mathrm{Re}\left\{\tilde{\chi}_{\mathrm{ER}}^{(3)}(\omega_1+\omega_2)\right\},\end{aligned} \tag{1.1.220}$$

where $i = 1, 2$. Here $\mathrm{Re}\{h\}$ and $\mathrm{Im}\{h\}$ denote the real and imaginary parts of the complex $h = \mathrm{Re}\{h\} + i\mathrm{Im}\{h\}$, respectively. The second term in Eq. (1.1.218) represents the SPM process enhanced by two-photon electronic resonance. Now we consider only the TPA process:

$$\frac{\partial A_1(z,t)}{\partial z} = -\frac{1}{8}\tilde{\alpha}_1^{(2)}(2\omega_1)\left|A_1(z,t)\right|^2 A_1(z,t). \tag{1.1.221}$$

From Eq. (1.1.221), the equation for the intensity $I_1(z, t)$ can be acquired

$$\begin{aligned}\frac{\partial I_1(z,t)}{\partial z} &= \frac{1}{2}\varepsilon_0\tilde{n}^{(1)}(\omega_1)c\frac{\partial\left|A_1(z,t)\right|^2}{\partial z} \\ &= \frac{1}{2}\varepsilon_0\tilde{n}^{(1)}(\omega_1)c\left\{\frac{\partial A_1(z,t)}{\partial z}A_1^*(z,t) + A_1(z,t)\frac{\partial A_1^*(z,t)}{\partial z}\right\} \\ &= -\tilde{\alpha}_{11}^{(2)}(2\omega_1)I_1^2(z,t),\end{aligned} \tag{1.1.222}$$

Here the quantity $\tilde{\alpha}_{ij}^{(2)}(\omega_1 + \omega_2)$ is called a two-photon absorption coefficient, and it is introduced by using

$$\tilde{\alpha}_{ij}^{(2)}(\omega_1+\omega_2) = \frac{\tilde{\alpha}_i^{(2)}(\omega_1+\omega_2)}{2\varepsilon_0\tilde{n}^{(1)}(\omega_j)c} = \frac{\omega_i\mathrm{Im}\left\{\tilde{\chi}_{\mathrm{ER}}^{(3)}(\omega_1+\omega_2)\right\}}{2\varepsilon_0\tilde{n}^{(1)}(\omega_i)\tilde{n}^{(1)}(\omega_j)c^2}, \tag{1.1.223}$$

where $i, j = 1, 2$. From Eq. (1.1.223), we find that the absorption loss by TPA is proportional to the square of the incident intensity. Using the relation

$$\frac{\partial}{\partial z}\frac{1}{I_1(z,t)} = -\frac{1}{I_1^2(z,t)}\frac{\partial I_1(z,t)}{\partial z}, \tag{1.1.224}$$

we obtain the solution of Eq. (1.1.223) in the form

$$I_1(z,t) = \frac{I_1(0,t)}{1+\tilde{\alpha}_{11}^{(2)}(2\omega_1)I_1(0,t)z}. \tag{1.1.225}$$

In the case of the weak absorption, it reduces to

$$I_1(z,t) = I_1(0,t)\left\{1 - \alpha_{11}^{(2)}(2\omega_1)I_1(0,t)z\right\}. \tag{1.1.226}$$

Next, we consider the TPA process under the condition that ω_2 is not equal to ω_1. According to Eq. (1.1.137), the polarization describing the TPA process with an electronic resonant frequency of $\omega_1 + \omega_2$,

$$P_{\mathrm{ER}}^{(3)}(\mathbf{r}, t) = i\frac{\omega_{\mathrm{ER}}^2 + \Gamma_{\mathrm{ER}}^2}{8\omega_{\mathrm{ER}}}\varepsilon_0\chi^{(3)} \times \Big[A_2^*(z, t)A_{12(\mathrm{ER})}^{(2)}(z, t)e^{i(k_1 z - \omega_1 t)} + A_1^*(z, t)A_{12(\mathrm{ER})}^{(2)}(z, t)e^{i(k_2 z - \omega_2 t)} - c.c.\Big]. \tag{1.1.227}$$

From Eqs. (1.1.47), (1.1.31) and (1.1.227), the equations for the slowly varying amplitudes at ω_1 and ω_2 are given by

$$\frac{\partial A_1(z, t)}{\partial z} = -\frac{\omega_1\chi^{(3)}}{8\tilde{n}^{(1)}(\omega_1)c}\frac{\omega_{\mathrm{ER}}^2 + \Gamma_{\mathrm{ER}}^2}{\omega_{\mathrm{ER}}}A_2^*(z, t)A_{12(\mathrm{ER})}^{(2)}(z, t), \tag{1.1.228}$$

$$\frac{\partial A_2(z, t)}{\partial z} = -\frac{\omega_2\chi^{(3)}}{8\tilde{n}^{(1)}(\omega_2)c}\frac{\omega_{\mathrm{ER}}^2 + \Gamma_{\mathrm{ER}}^2}{\omega_{\mathrm{ER}}}A_1^*(z, t)A_{12(\mathrm{ER})}^{(2)}(z, t). \tag{1.1.229}$$

If the lifetime of the electronic excited state is much shorter than the pulse duration, $A_{12(\mathrm{ER})}^{(2)}(z, t)$ is approximated by

$$A_{12(\mathrm{ER})}^{(2)}(z, t) \approx A_1(z, t)A_2(z, t)\int_{-\infty}^{t} u\,(t - t_2)\, e^{[-i\{\omega_{\mathrm{ER}} - (\omega_1 + \omega_2)\} - \Gamma_{\mathrm{ER}}](t - t_2)}dt_2 = -\frac{i}{\omega_{\mathrm{ER}} - (\omega_1 + \omega_2) - i\Gamma_{\mathrm{ER}}}A_1(z, t)A_2(z, t). \tag{1.1.230}$$

Using Eqs. (1.1.230) and (1.1.219), we can rewrite Eqs. (1.1.228) and (1.1.229) as

$$\frac{\partial A_1(z, t)}{\partial z} = -\frac{1}{4}\tilde{\alpha}_1^{(2)}(\omega_1 + \omega_2)\,|A_2(z, t)|^2\, A_1(z, t), \tag{1.1.231}$$

$$\frac{\partial A_2(z, t)}{\partial z} = -\frac{1}{4}\tilde{\alpha}_2^{(2)}(\omega_1 + \omega_2)\,|A_1(z, t)|^2\, A_2(z, t), \tag{1.1.232}$$

where we have neglect the contribution of the XPM process enhanced by two-photon electronic resonance. The right-hand sides of these equations indicate the FWM polarizations, which can act as the source of new photons whose phases are shifted by π relative to the incident photons. Thus, the incident photons vanish due to destructive interference between the incident photons and the generated photons. Therefore, we can interpret TPA as a special case of two-photon electronic resonant FWM where the frequency of the generated photon is equal to that of the incident photon as shown in Fig. 1.9(c) and (d). From Eqs. (1.1.231) and (1.1.232), the coupled equations for the intensities $I_1(z,t)$ and $I_2(z,t)$ can be obtained

$$\frac{\partial I_1(z,t)}{\partial z} = -2\tilde{\alpha}_{12}^{(2)}(\omega_1+\omega_2)I_1(z,t)I_2(z,t), \tag{1.1.233}$$

$$\frac{\partial I_2(z,t)}{\partial z} = -2\tilde{\alpha}_{21}^{(2)}(\omega_1+\omega_2)I_1(z,t)I_2(z,t), \tag{1.1.234}$$

From Eqs. (1.1.233) and (1.1.234), we see that

$$\frac{1}{\tilde{\alpha}_{12}^{(2)}(\omega_1+\omega_2)}\frac{\partial I_1(z,t)}{\partial z} = \frac{1}{\tilde{\alpha}_{21}^{(2)}(\omega_1+\omega_2)}\frac{\partial I_2(z,t)}{\partial z}. \tag{1.1.235}$$

The following relation is obtained from Eq. (1.1.235):

$$I_2(z,t) = \frac{\tilde{\alpha}_{21}^{(2)}(\omega_1+\omega_2)}{\tilde{\alpha}_{12}^{(2)}(\omega_1+\omega_2)}\{I_1(z,t)-I_1(0,t)\}+I_2(0,t). \tag{1.1.236}$$

Inserting Eq. (1.1.236) into (1.1.233), we have

$$\frac{\partial I_1(z,t)}{\partial z} = -2\tilde{\alpha}_{21}^{(2)}(\omega_1+\omega_2)I_1^2(z,t)+2K_{\mathrm{E}}(t)I_1(z,t), \tag{1.1.237}$$

where

$$K_{\mathrm{E}}(t) = \tilde{\alpha}_{21}^{(2)}(\omega_1+\omega_2)I_1(0,t)-\tilde{\alpha}_{12}^{(2)}(\omega_1+\omega_2)I_2(0,t). \tag{1.1.238}$$

Here we assume $I_1(0,t) > I_2(0,t)$. Employing Eq. (1.1.224), we can rewrite Eq. (1.1.237) as

$$\frac{\partial}{\partial z}\frac{1}{I_1(z,t)} = 2\tilde{\alpha}_{21}^{(2)}(\omega_1+\omega_2)-2K_{\mathrm{E}}(t)\frac{1}{I_1(z,t)}. \tag{1.1.239}$$

The solution of Eq. (1.1.239) takes the form

$$I_1(z,t) = \frac{I_1(0,t)K_{\mathrm{E}}(t)}{\tilde{\alpha}_{21}^{(2)}(\omega_1+\omega_2)I_1(0,t)-\tilde{\alpha}_{12}^{(2)}(\omega_1+\omega_2)I_2(0,t)\exp\{-2K_{\mathrm{E}}(t)z\}}. \tag{1.1.240}$$

Substituting Eq. (1.1.240) into Eq. (1.1.236), we obtain

$$I_2(z,t) = \frac{I_2(0,t)K_{\mathrm{E}}(t)\exp\{-2K_{\mathrm{E}}(t)z\}}{\tilde{\alpha}_{21}^{(2)}(\omega_1+\omega_2)I_1(0,t)-\tilde{\alpha}_{12}^{(2)}(\omega_1+\omega_2)I_2(0,t)\exp\{-2K_{\mathrm{E}}(t)z\}}. \tag{1.1.241}$$

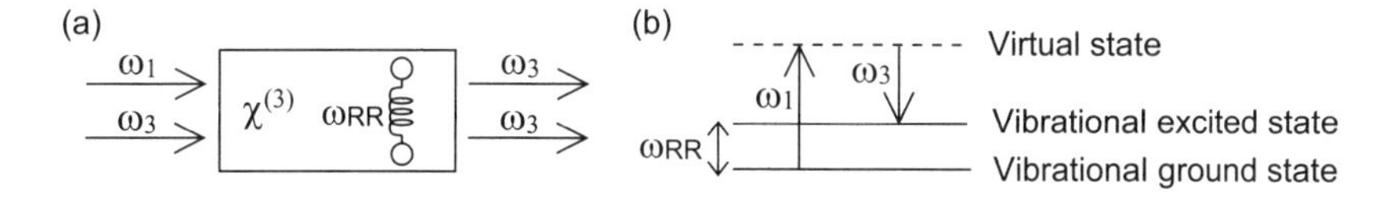

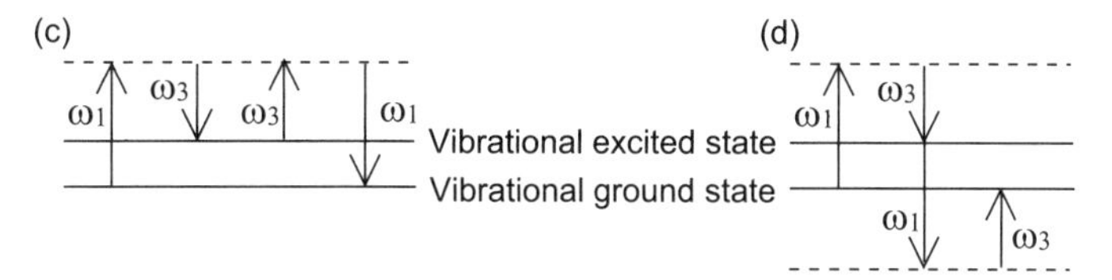

FIGURE 1.10 Stimulated Raman scattering. (a) Geometry of the interaction. (b) Energy diagram describing the stimulated Raman scattering process. (c, d) SRS can be interpreted as a special case of two-photon vibrational resonant FWM where the frequency of the generated photon is equal to that of the incident photon at ω_1 (c) and ω_3 (d).

1.1.3.6 Stimulated Raman Scattering We next consider the process of stimulated Raman scattering (SRS), which is illustrated in Fig. 1.10. In the SRS process, the molecule makes a transition from a vibrational ground state to a vibrational excited state by the two-photon transition in which a pump photon with a frequency of ω_1 induces an intermediate transition to a virtual state and then a lower energy Stokes photon with a frequency of ω_3 causes the stimulated emission of the Stokes photon. SRS was first observed in 1962 [27].

According to Eq. (1.1.141), the polarization representing the SRS process with a Raman resonant frequency of $\omega_{\text{RR}} = \omega_1 - \omega_3$ is given by the expression

$$P_{\text{SRS}}^{(3)}(\mathbf{r}, t) = i\frac{\omega_{\text{RR}}^2 + \Gamma_{\text{RR}}^2}{8\omega_{\text{RR}}}\varepsilon_0\chi^{(3)} \times \Big[A_3(z,t)A_{13(\text{RR})}^{(2\#)}(z,t)e^{i(k_1 z-\omega_1 t)} - A_1(z,t)A_{13(\text{RR})}^{(2\#)*}(z,t)e^{i(k_3 z-\omega_3 t)} - c.c.\Big]. \quad (1.1.242)$$

From Eqs. (1.1.31) and (1.1.242), the slowly varying amplitudes $p_{\text{SRS1}}^{(3)}(z,t)$ and $p_{\text{SRS3}}^{(3)}(z,t)$, and the wave numbers $k_{\text{SRS1}}^{(3)}$ and $k_{\text{SRS3}}^{(3)}$ of the polarization of SRS at ω_1 and ω_3 are expressed by

$$p_{\text{SRS1}}^{(3)}(z,t) = i\frac{\omega_{\text{RR}}^2 + \Gamma_{\text{RR}}^2}{4\omega_{\text{RR}}}A_3(z,t)A_{13(\text{RR})}^{(2\#)}, \quad (1.1.243)$$

$$p_{\text{SRS3}}^{(3)}(z,t) = -i\frac{\omega_{\text{RR}}^2 + \Gamma_{\text{RR}}^2}{4\omega_{\text{RR}}}A_1(z,t)A_{13(\text{RR})}^{(2\#)*}(z,t), \quad (1.1.244)$$

$$k_{\text{SRS1}}^{(3)} = k_1, \quad (1.1.245)$$

$$k_{\text{SRS3}}^{(3)} = k_3. \quad (1.1.246)$$

We can assume that the linear absorption and the chromatic dispersion can be neglected, so that the wave equation describing in Eq. (1.1.47) can be written as

$$\frac{\partial A_1(z,t)}{\partial z} = i\frac{\omega_1 \chi^{(3)}}{2\tilde{n}^{(1)}(\omega_1)c} p^{(3)}_{\mathrm{SRS1}}(z,t), \tag{1.1.247}$$

$$\frac{\partial A_3(z,t)}{\partial z} = i\frac{\omega_3 \chi^{(3)}}{2\tilde{n}^{(1)}(\omega_3)c} p^{(3)}_{\mathrm{SRS3}}(z,t). \tag{1.1.248}$$

Inserting Eqs. (1.1.243) and (1.1.244) into Eqs. (1.1.247) and (1.1.248), respectively, we obtain

$$\frac{\partial A_1(z,t)}{\partial z} = -\frac{\omega_1 \chi^{(3)}}{8\tilde{n}^{(1)}(\omega_1)c}\frac{\omega_{\mathrm{RR}}^2 + \Gamma_{\mathrm{RR}}^2}{\omega_{\mathrm{RR}}} A_3(z,t) A^{(2\#)}_{13(\mathrm{RR})}, \tag{1.1.249}$$

$$\frac{\partial A_3(z,t)}{\partial z} = \frac{\omega_3 \chi^{(3)}}{8\tilde{n}^{(1)}(\omega_3)c}\frac{\omega_{\mathrm{RR}}^2 + \Gamma_{\mathrm{RR}}^2}{\omega_{\mathrm{RR}}} A_1(z,t) A^{(2\#)*}_{13(\mathrm{RR})}. \tag{1.1.250}$$

If we assume that the lifetime of the vibrational excited state is much shorter than the pulse duration, $A^{(2\#)}_{13(\mathrm{RR})}$ is approximated by

$$\begin{aligned} A^{(2\#)}_{13(\mathrm{RR})}(z,t) &\approx A_1(z,t)A_3^*(z,t)\int_{-\infty}^{t} u\,(t-t_2)\, e^{[-i\{\omega_{\mathrm{RR}}-(\omega_1-\omega_3)\}-\Gamma_{\mathrm{RR}}](t-t_2)}dt_2 \\ &= -\frac{i}{\omega_{\mathrm{RR}}-(\omega_1-\omega_3)-i\Gamma_{\mathrm{RR}}} A_1(z,t)A_3^*(z,t). \end{aligned} \tag{1.1.251}$$

Substituting Eq. (1.1.251) into Eqs. (1.1.249) and (1.1.250), we obtain

$$\frac{\partial A_1(z,t)}{\partial z} = -\frac{1}{4}\tilde{g}_{\mathrm{R}1}(\omega_1-\omega_3)\,|A_3(z,t)|^2\,A_1(z,t), \tag{1.1.252}$$

$$\frac{\partial A_3(z,t)}{\partial z} = \frac{1}{4}\tilde{g}_{\mathrm{R}2}(\omega_1-\omega_3)\,|A_1(z,t)|^2\,A_3(z,t). \tag{1.1.253}$$

Here we have neglected the contribution of the XPM process enhanced by two-photon Raman resonance and have introduced the quantity

$$\begin{aligned} \tilde{g}_{\mathrm{R}i}(\omega_1-\omega_3) &= \frac{\omega_i \chi^{(3)}}{2\tilde{n}^{(1)}(\omega_i)c}\frac{\omega_{\mathrm{RR}}^2+\Gamma_{\mathrm{RR}}^2}{\omega_{\mathrm{RR}}}\frac{\Gamma_{\mathrm{RR}}}{\{\omega_{\mathrm{RR}}-(\omega_1-\omega_3)\}^2+\Gamma_{\mathrm{RR}}^2} \\ &= \frac{\omega_i}{\tilde{n}^{(1)}(\omega_i)c}\chi^{(3)}\mathrm{Im}\left\{\tilde{h}_{\mathrm{RR}}(\omega_1-\omega_3)\right\} \\ &= \frac{\omega_i}{\tilde{n}^{(1)}(\omega_i)c}\mathrm{Im}\left\{\tilde{\chi}^{(3)}_{\mathrm{RR}}(\omega_1-\omega_3)\right\}, \end{aligned} \tag{1.1.254}$$

which is called Raman gain coefficient. Here $\mathrm{Im}\{h\}$ denotes the imaginary part of the complex $h = \mathrm{Re}\{h\} + i\mathrm{Im}\{h\}$. The right-hand sides of Eqs. (1.1.252) and (1.1.253) indicate the FWM polarizations, which can act as the source of new photon. The phase of the generated photon in Eq. (1.1.252) is shifted by π relative to the incident photon, while that by Eq. (1.1.253) is not shifted relative to the incident photon. Thus, the incident photon in Eq. (1.1.252) vanishes due to destructive interference between the incident photons and the generated photons, whereas the incident photon in Eq. (1.1.253) is amplified due to constructive interference. Therefore, we can interpret SRS as a special case of two-photon vibrational resonant FWM where the frequency of the generated photon is equal to that of the incident photon as shown in Fig. 1.10(c) and (d). From Eqs. (1.1.252) and (1.1.253), we have the coupled equations for the intensities $I_1(z, t)$ and $I_3(z, t)$:

$$\frac{\partial I_1(z,t)}{\partial z} = -\tilde{g}_{13}^{(\mathrm{R})}(\omega_1 - \omega_3) I_1(z,t) I_3(z,t), \tag{1.1.255}$$

$$\frac{\partial I_3(z,t)}{\partial z} = \tilde{g}_{31}^{(\mathrm{R})}(\omega_1 - \omega_3) I_1(z,t) I_3(z,t), \tag{1.1.256}$$

where

$$\tilde{g}_{ij}^{(\mathrm{R})}(\omega_1 - \omega_3) = \frac{\tilde{g}_{\mathrm{R}i}(\omega_1 - \omega_3)}{\varepsilon_0 \tilde{n}^{(1)}(\omega_j) c} = \frac{\omega_i \mathrm{Im}\left\{\tilde{\chi}_{\mathrm{RR}}^{(3)}(\omega_1 - \omega_3)\right\}}{\varepsilon_0 \tilde{n}^{(1)}(\omega_i) \tilde{n}^{(1)}(\omega_j) c^2}. \tag{1.1.257}$$

From Eqs. (1.1.255) and (1.1.256), we see that

$$\frac{1}{\tilde{g}_{13}^{(\mathrm{R})}(\omega_1 - \omega_3)} \frac{\partial I_1(z,t)}{\partial z} = -\frac{1}{\tilde{g}_{31}^{(\mathrm{R})}(\omega_1 - \omega_3)} \frac{\partial I_3(z,t)}{\partial z}. \tag{1.1.258}$$

The following relation is obtained from Eq. (1.1.258):

$$I_3(z,t) = -\frac{\tilde{g}_{31}^{(\mathrm{R})}(\omega_1 - \omega_3)}{\tilde{g}_{13}^{(\mathrm{R})}(\omega_1 - \omega_3)} \{I_1(z,t) - I_1(0,t)\} + I_3(0,t). \tag{1.1.259}$$

Inserting Eq. (1.1.259) into (1.1.255), we have

$$\frac{\partial I_1(z,t)}{\partial z} = \tilde{g}_{31}^{(\mathrm{R})}(\omega_1 - \omega_3) I_1^3(z,t) - K_{\mathrm{R}}(t) I_1(z,t), \tag{1.1.260}$$

where

$$K_{\mathrm{R}}(t) = \tilde{g}_{31}^{(\mathrm{R})}(\omega_1 - \omega_3) I_1(0,t) + \tilde{g}_{13}^{(\mathrm{R})}(\omega_1 - \omega_3) I_3(0,t). \tag{1.1.261}$$

Employing Eq. (1.1.224), we can rewrite Eq. (1.1.260) as

$$\frac{\partial}{\partial z} \frac{1}{I_1(z,t)} = -\tilde{g}_{31}^{(\mathrm{R})}(\omega_1 - \omega_3) + K_{\mathrm{R}}(t) \frac{1}{I_1(z,t)}. \tag{1.1.262}$$

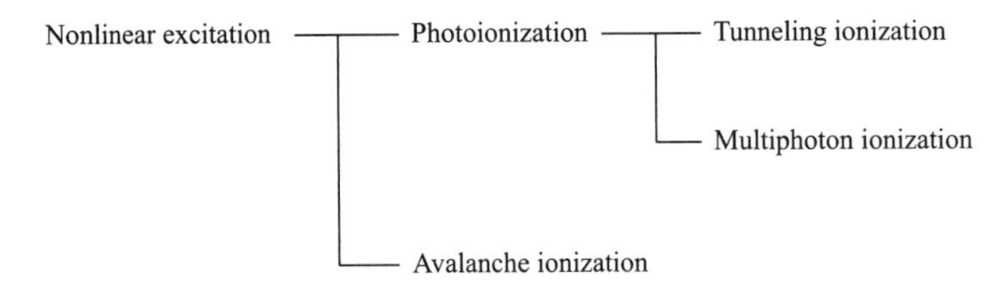

FIGURE 1.11 Type of nonlinear excitation.

The solution of Eq. (1.1.262) takes the form

$$I_1(z,t) = \frac{I_1(0,t)K_R(t)}{\tilde{g}_{31}^{(R)}(\omega_1-\omega_3)I_1(0,t)+\tilde{g}_{13}^{(R)}(\omega_1-\omega_3)I_3(0,t)\exp\{K_R(t)z\}}. \tag{1.1.263}$$

Substituting Eq. (1.1.263) into Eq. (1.1.259), we obtain

$$I_3(z,t) = \frac{I_3(0,t)K_R(t)}{\tilde{g}_{31}^{(R)}(\omega_1-\omega_3)I_1(0,t)\exp\{-K_R(t)z\}+\tilde{g}_{13}^{(R)}(\omega_1-\omega_3)I_3(0,t)}. \tag{1.1.264}$$

1.2 NONLINEAR IONIZATION

1.2.1 Nonlinear Optical Ionization

When intense ultrafast laser pulses are focused, the intensity in the focal volume can become high enough to cause nonlinear absorption, through photoionization and avalanche ionization, resulting in optical breakdown and the formation of a high-density plasma. There are two classes of nonlinear excitation mechanisms; photoionization and avalanche ionization as described in Fig. 1.11. In photoionization, electrons are directly excited from the valence to the conduction band by the laser field. Depending on the laser frequency and intensity, there are two different regimes of photoionization, the multiphoton ionization regime and the tunneling ionization regime. Keldysh showed that both multiphoton and tunneling ionization could be described within the same framework [28].

1.2.1.1 Tunneling Ionization For strong laser fields and low laser frequency, photoionization is mainly a tunneling process. In the tunneling ionization, the electric field of the laser suppresses the Coulomb well that binds a valence electron to its parent atom. When the electric field is very strong, the Coulomb well can be suppressed enough that the bound electron tunnels through the short barrier and becomes free, as shown schematically in the left panel of Fig. 1.12.

1.2.1.2 Multiphoton Ionization For higher laser frequencies photoionization is usually described in terms of the simultaneous absorption of several photons by

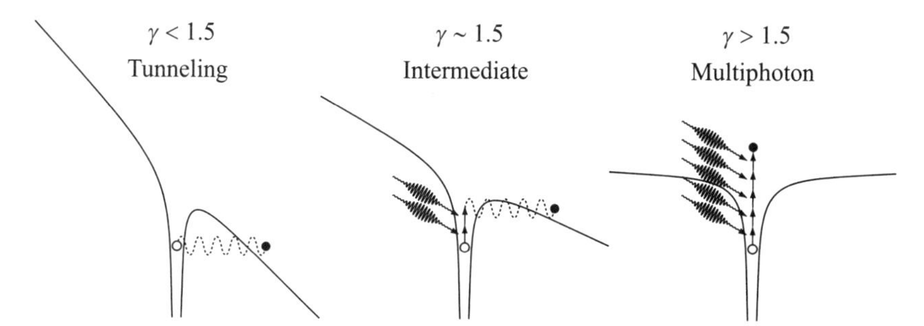

FIGURE 1.12 Schematic diagram of photoionization for different values of the Keldysh parameter [29].

an electron, as shown in the right panel of Fig. 1.12 [30, 31].This process is called multiphoton ionization. The number of photons necessary for the exciatationof the electron is determined by the energy difference between the electronic binding energy and the energy of the single photon. In order to be excited from the valence to the conduction band by multiphoton absorption, the electron must absorb enough photons so that total energy of photons absorbed is greater than the bandgap of the material. Because the energy of a single photon is smaller than the bandgap, the simultaneous absorption of several photons is required to ionize an electron. For example, water has a bandgap of 6.5 eV; therefore, five photons at a wavelength of 800 nm (1.5 eV) are necessary to ionize an electron.

The transition point between multiphoton ionization and tunneling ionization was defined by Keldysh [28]. The adiabatic parameter, also known as the Keldysh parameter γ is defined as

$$\gamma = \frac{\omega}{e}\sqrt{\frac{mcn\varepsilon_0 E_g}{I}}, \tag{1.2.1}$$

where ω is the laser frequency, I is the laser intensity at the focus, m and e are the reduced mass and charge of the electron, c is the velocity of light, n is refractive index of the material, E_g is the bandgap of the material, and ε_0 is the permittivity of free space. When the Keldysh parameter is smaller than approximately 1.5, photoionization is a tunneling process as shown in the left panel of Fig. 1.12. When the Keldysh parameter is larger than approximately 1.5, photoionization is a multiphoton process as shown in the right panel of Fig. 1.12. In the intermediate regime, the photoionization is a mixture between tunneling and multiphoton ionization as depicted in the middle panel of Fig. 1.12.

1.2.2 Avalanche Ionization

The avalanche ionization process is illustrated in Fig. 1.13 [30]. In order to initiate the avalanche process, there have to be some electrons already promoted to the bottom

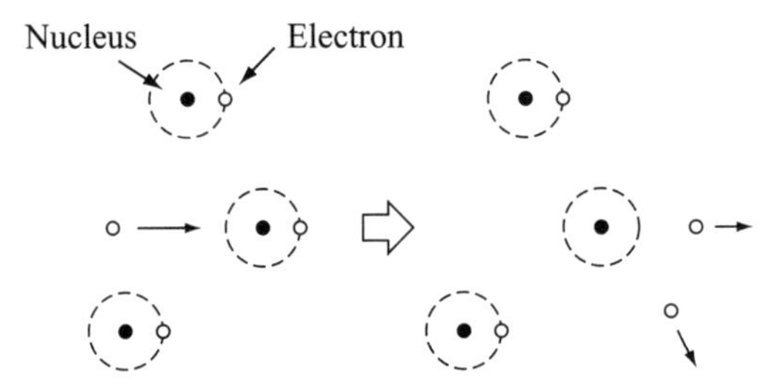

FIGURE 1.13 Schematic of electron avalanche by collisional impact ionization. Secondary free electrons are generated during collisions with electrons whose kinetic energy is grater the bound electrons binding energy.

of the conduction band. When ultrashort laser pulses are focused in a transparent material, the bound valence electrons have an ionization potential or bandgap greater than the laser photon energy. The bound electrons do not absorb the laser light at low intensities. However, when extremely high intensity of a laser pulse is applied, electrons in the conduction band are created through the photoionization as shown in Fig 1.12.

Once an electron is promoted to the bottom of the conduction band, it serves as a seed for avalanche ionization. The electron linearly absorbs photons, and once its energy has reached twice that of the bandgap, it can collisionally ionize with an electron in the valence band resulting in two electrons at the bottom of the conduction band. For electrons excited in the conduction band the scattering time is on the order of 1 fs, making this is a fast and efficient process [33]. Through this process, it is possible to create a free electron plasma, within the duration of the laser pulse.

1.2.3 Photodisruption/Optical Breakdown

1.2.3.1 Optical Breakdown When ultrashort laser pulses are tightly focused, the intensity in the focal volume can become high enough to cause nonlinear absorption, through the multiphoton, tunnel, and avalanche ionization process, as described in Fig. 1.13, resulting in optical breakdown and the formation of a high-density plasma. This high-density plasma leads to material removal. Through avalanche ionization the free electron density increases exponentially until a plasma is formed. The electron density grows until it approaches the critical density, where the plasma frequency matches that of the incident laser radiation. The plasma is strongly absorbing through free-carrier absorption, increasing the kinetic energy of the electrons. For femtosecond laser, the energy transfer from the pulse to the electrons is completed before any heating of the ions can start. Ultrashort laser pulses can produce localized energy absorption because the pulse width is shorter than the time required for heat to diffuse out of the focal volume, and material removal occurs only in and around the focal volume.

Multiphoton excitation is the lead mechanism of generating these electrons when using femtosecond laser pulses. Femtosecond laser optical breakdown is deterministic. In contrast, when longer pulses, such as nanosecond laser pulses are used, the laser intensity might not be high enough to generate seed electrons, and these electrons have to come from impurities and defects. Therefore, the threshold for femtosecond laser breakdown is deterministic, and this makes femtosecond pulses more favorable over longer laser pulses.

1.2.3.2 Bubble Formation and Shockwave Generation in Water Water is a good candidate for biomedical research because most biological cells and tissues are comprised of 80% water. Through nonlinear absorption and plasma generation, it is possible to vaporize water. Optical breakdown in water cause a pressure wave and the formation of a cavitation bubble. The size of the region affected by this mechanical disruption increases with the pulse energy and can exceed many times the focal volume [34]. The threshold for cavitation bubble is intensity dependent and the size of the bubble scales down with shorter pulses. For example, a 10-mJ nanosecond pulse produces a bubble with a half a millimeter radius [35]. For shorted pulse durations, the size of the bubble is much smaller. A 1-μJ, 100-fs pulse produces cavitation bubble of 11 μm [36]. Figure 1.14 shows a time-resolved image of the bubble and the pressure wave 35 ns after optical excitation of 100-fs laser pulses.

1.2.3.3 Laser Surgery If pulse intensities become large enough, high concentrations of free elections can be generated in the focal volume by nonlinear absorption and avalanche ionization. The optical breakdown generates a rapidly expanding high-density plasma that leads to either modification or complete ablation of tissue material in the focal region [37]. It has been shown that for pulses with durations on the order

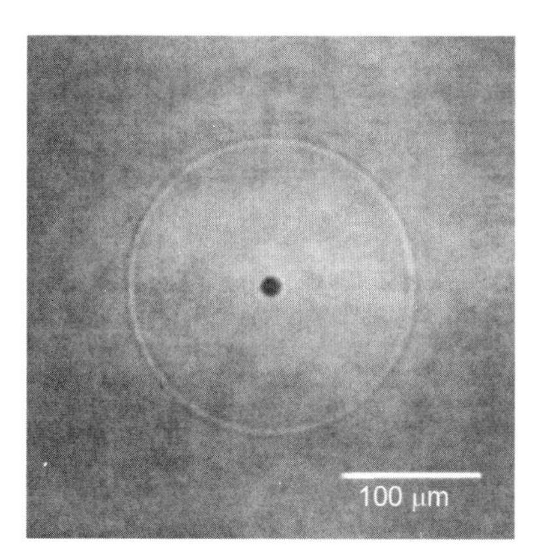

FIGURE 1.14 Time-resolved image for dynamics of laser-induced breakdown in water. Small vapor bubbles and expanding pressure waves photographed 35 ns after optical excitation by 100-fs, 800-nm laser pulses of 1-μJ. Reprinted with permission from [36]. Copyright 1997 Optical Society of America.

of 100 fs, only a few nanojoules of energy are required for tissue dissection. Since femtosecond pulses are one to two orders of magnitude shorter than thermal diffusion time, [38] heat transport is minimized and modification or ablation can remain highly localized, leaving the surrounding tissue intact. Ultrafast lasers have thus emerged as an important tool for not just high-resolution imaging but for precision surgical procedures as well.

1.3 LIGHT SOURCE

1.3.1 Ultrashort Laser Pulse

The practical applications of nonlinear opticl microscopy (NLOM) have been the result of dramatic improvements in the quality of laser sources [39–46]. Picosecond and femtosecond laser pulses have typically been employed as excitation light sources for NLOM because of the relatively low cross sections of the nonlinear optical processes that are used. In the case of SHG microscopy as first reported by Hellwarth et al. [47], a picosecond laser source was employed. The longer pulse duration reduced the signal generation efficiency [41]. The original TPEF work by Denk et al. was performed using a colliding pulse mode-locked dye laser operating at a pulse duration of 100 fs [48]. Although the shorter pulse duration increased the signal generation efficiency, the dyes limited the use of this technique due to several concerns such as toxicity and rapid aging. The capability of femtosecond lasers changed abruptly with the advent of the Kerr-lens mode locking technique, as reported by Spence et al. in 1991 [49]. A Ti:sapphire crystal, which is a mechanically durable laser medium with excellent thermal conduction properties and has no toxicity problems, was chosen as a medium for Kerr-lens mode-locked lasers. This work sparked a revolution in solid-state femtosecond oscillators. Due to their broad tuning range (700–1100 nm) and short pulse duration (100 fs or less), Ti:sapphire lasers have typically been used for the excitation light sources in NLOM applications. This breakthrough has lead to the demonstration of 5-fs pulses from a Ti:sapphire laser, which are the shortest pulses generated from a laser oscillator [50, 51]. The 5-fs pulses have ultrabroadband spectra over 400 nm.

For NLOM applications employing an excitation wavelength longer than 1100 nm, Cr:forsterite lasers [52–55] and optical parametric oscillators pumped by a Ti:sapphire oscillator [56] are available. The wavelength tuning ranges of a Cr:forsterite laser [53] and an optical parametric oscillator [56] are between 1130 to 1367 nm and 1000 to 2750 nm, respectively. Zhang et al. [54] and Chudoba et al. [55] have demonstrated the generation of sub-20-fs pulses using all-solid-state Kerr-lens mode-locked Cr:forsterite lasers.

Fiber lasers have attracted attention as practical alternatives to solid-state lasers, offering compact size, great stability, and convenience for alignment. Rare-earth doped fibers are attractive as laser gain media in mode-locked lasers because they possess the high single-pass gain combined with broad gain bandwidths and excellent beam quality. Passively mode-locked fiber lasers based on erbium-doped fibers has initially been developed as sources for 1.55 μm region [57, 58]. Recently, mode-locked fiber lasers based on ytterbium-doped fibers have attracted much attention, they have broad gain bandwidth [59], which enable ultrashort pulse generation. Zhou et al. generated 28 fs pulses from a mode-locked ytterbium fiber oscillator whose spectrum ranges from 980 nm to 1120 nm [60].

Several applications of the supercontinuum (SC) in the visible and infrared have been demonstrated, such as CARS microscopy and spectroscopy [62, 63] and TPEF microscopy [64, 65]. The SC generation is a complex nonlinear phenomenon that is characterized by the dramatic spectral broadening of intense light pulses passing through a nonlinear material [66, 67]. The spectral broadening takes place through the interaction of SPM, dispersion, self-steepening, SRS, and FWM. For the application of SC to NLOM, a high-power SC in the near infrared region operating at high repetition rate is preferred. Until recently, SC generation required a regeneratively amplified Ti:sapphire laser operating at a repetition rate of less than a few hundred kilohertz [66]. SC generation has been associated with the recent introduction of photonic crystal fibers (PCFs) [68]. By using PCFs (several milimeters to several tens of centimeters), the generation of broadband SC spectra that span more than an octave has been demonstrated with low-peak-power pulses from Ti:sapphire oscillators operating at a repetition rate of approximately 100 MHz.

It is important to acknowledge the relationship between spectral width and pulse duration when considering the generation of ultrashort pulses. The electric field $E(t)$ and the Fourier transforms $E(\omega)$ of $E(t)$ are described by

$$\tilde{E}(\omega) = \int_{-\infty}^{\infty} E(t)\exp(i\omega t)dt, \tag{1.3.1}$$

$$E(t) = \frac{1}{2\pi}\int_{-\infty}^{\infty} \tilde{E}(\omega)\exp(-i\omega t)d\omega. \tag{1.3.2}$$

The spectral profile $I(\omega)$ and the intensity profile $I(t)$ are expressed by

$$\tilde{I}(\omega) = \frac{1}{4\pi}\varepsilon_0\tilde{n}^{(1)}(\omega)c\,|E(\omega)|^2, \tag{1.3.3}$$

$$I(t) = \frac{1}{2}\varepsilon_0\tilde{n}^{(1)}(\omega_0)c\,|E(t)|^2. \tag{1.3.4}$$

The commonly used definition of pulse duration is based on the full-width at half-maximum (FWHM) principle of the optical power against time because it is easy to measure experimentally. The pulse duration and the spectral bandwidth can then be related by the following inequality:

$$\frac{\Delta t\Delta\omega}{2\pi} \geq K, \tag{1.3.5}$$

where $\Delta\omega$ is the spectral width that is FWHM of $I(\omega)$ and Δt is the pulse duration that is FWHM of $I(t)$. The value of K depends on the actual pulse shape. Some

TABLE 1.2 Examples of Standard Pulse Profiles

Shape	Temporal Profile	Pulse Dulation	Spectral Profile	Spectral Bandwidth	K
Gauss	$\exp\{-2(t/t_g)^2\}$	$1.177t_g$	$\exp\{-(\omega t_g)^2/2\}$	$2.335/t_g$	0.441
Sech	$\text{sech}^2(t/t_s)$	$1.763t_s$	$\text{sech}^2(\pi\omega t_s/2)$	$1.122/t_s$	0.315
Lorentz	$\{1+(t/t_l)^2\}^{-2}$	$1.287t_l$	$\exp\{-2\lvert\omega\rvert t_l)^2\}$	$0.693/t_l$	0.142
Square	$1, \lvert t/t_r\rvert \leq 1/2$ 0, elsewhere	t_r	$\text{sinc}^2(\omega t_r)$	$2.78/t_r$	0.443

examples are shown in Table 1.2 The most commonly cited are the Gaussian, for which the temporal dependence of the intensity is

$$I(t) = I_0 \exp\left\{-2\left(\frac{t}{t_g}\right)^2\right\} \tag{1.3.6}$$

and the secant hyperbolic,

$$I(t) = I_0 \text{sech}^2\left(\frac{t}{t_s}\right). \tag{1.3.7}$$

The parameters $t_g = \Delta t/\sqrt{2\ln 2}$ and $t_s = \Delta t/1.76$ are generally more convenient to use in theoretical calculations involving pulses with these assumed shapes than the FWHM of the intensity, Δt. The equality in Eq. (1.3.5) holds for pulses without frequency modulation (unchirped), which are called Fourier transform limited (FTL) or bandwidth limited. Such a pulse exhibits the shortest possible duration at a given spectral width and pulse shape. In order to generate an ultrashort pulse with a specific duration (Δt), a broad spectral bandwidth ($\Delta\omega$) is required because the minimum spectral width is inversely proportional to the pulse duration. For example, the spectral width of a 100-fs pulse at a wavelength of 1 μm is 10 nm. If the pulse width decreases to 10 fs, then the minimum spectral linewidth extends to 100 nm, which is as much as 10% of the wavelength itself. When we use ultrashort pulses, due to the broadband spectrum, we need to pay attention to the dispersion of the media or optical devices, because of the dependence of the refractive index n on the optical angular frequency ω. The effect of the dispersion is considered in Section 1.3.2.

1.3.2 Dispersion Management and Pulse Shaping

1.3.2.1 Pulse Distortions Induced by Linear Optical Elements Pulse distortions are induced by linear optical elements including lenses, mirrors, prisms, and gratings, and are found in nearly all optical setups. A linear optical element of this type can be characterized by a complex optical transfer function

$$\tilde{H}(\omega) = R(\omega)\exp\{i\phi(\omega)\}. \tag{1.3.8}$$

Here $R(\omega)$ is the amplitude response and $\phi(\omega)$ is the phase response. The field at the output from the linear optical element $\tilde{E}_{out}(\omega)$ is related to the incident field $\tilde{E}_{in}(\omega)$:

$$\tilde{E}_{\text{out}}(\omega) = \tilde{E}_{\text{in}}(\omega)\tilde{H}(\omega) = \tilde{E}_{\text{in}}(\omega)R(\omega)\exp\{i\phi(\omega)\}. \tag{1.3.9}$$

As can be seen from Eq. (1.3.9), the influence of $R(\omega)$ is that of a frequency filter. The phase factor $\phi(\omega)$ can be interpreted as the phase delay that a spectral component at a frequency of ω experiences. In order to understand how the phase response affects the light pulse, we assume that $R(\omega)$ does not change over the pulse spectrum whereas $\phi(\omega)$ does. Thus, the output field is expressed by

$$E_{\text{out}}(t) = \frac{R}{2\pi}\int_{-\infty}^{\infty} \tilde{E}_{\text{in}}(\omega)\exp\{i\phi(\omega)\}\exp(-i\omega t)d\omega. \tag{1.3.10}$$

To get insight as to how the phase response affects the light pulse, we will express the pulse spectral phase $\phi(\omega)$ in terms of a Taylor series as in Eq. (1.3.11):

$$\begin{aligned}\phi(\omega) = \phi(\omega_0) &+ \left(\frac{d\phi}{d\omega}\right)_{\omega=\omega_0}(\omega-\omega_0) + \frac{1}{2!}\left(\frac{d^2\phi}{d\omega^2}\right)_{\omega=\omega_0}(\omega-\omega_0)^2 \\ &+ \frac{1}{3!}\left(\frac{d^3\phi}{d\omega^3}\right)_{\omega=\omega_0}(\omega-\omega_0)^3 + \frac{1}{4!}\left(\frac{d^4\phi}{d\omega^4}\right)_{\omega=\omega_0}(\omega-\omega_0)^4 + \cdots,\end{aligned} \tag{1.3.11}$$

where ω_0 denotes the center angular frequency of the pulse. By substituting Eq. (1.3.11) into Eq. (1.3.10), we obtain

$$\begin{aligned}E_{\text{out}}(t) = \frac{R}{2\pi}\int_{-\infty}^{\infty} &\tilde{E}_{\text{in}}(\omega)\exp\left\{i\sum_{m=2}^{\infty}\frac{1}{m!}\left(\frac{d^m\phi}{d\omega^m}\right)_{\omega=\omega_0}(\omega-\omega_0)^m\right\} \\ &\times e^{i\phi(\omega_0)}e^{-i\omega_0 t}\exp\left[-i(\omega-\omega_0)\left\{t-\left(\frac{d\phi}{d\omega}\right)_{\omega=\omega_0}\right\}\right]d\omega.\end{aligned} \tag{1.3.12}$$

From Eq. (1.3.12), we can easily interpret the effects of the various expansion coefficients. The term $e^{i\phi(\omega_0)}$ is a constant phase shift having no effect on the pulse envelope. The first-order expansion term $(d\phi/d\omega)_{\omega=\omega_0}$ leads solely to a shift of the pulse on the time axis t; obviously the pulse will keep its position on a time scale $T = t - (d\phi/d\omega)_{\omega=\omega_0}$. The term $(d\phi/d\omega)_{\omega=\omega_0}$ determines the group delay (GD) of the wave packet. The higher order ($m \geq 2$) expansion terms produce a nonlinear behavior of the spectral phase, which changes the pulse envelope and chirp. The second-order expansion term $\left(d^2\phi/d\omega^2\right)_{\omega=\omega_0}$ describes the frequency dependence of the GD and is called group delay dispersion (GDD). The third-order expansion term $\left(d^3\phi/d\omega^3\right)_{\omega=\omega_0}$ and the fourth-order expansion term $\left(d^4\phi/d\omega^4\right)_{\omega=\omega_0}$ are called third order dispersion (TOD) and fourth order dispersion (FOD), respectively. While the GDD increases the pulse duration as shown in Fig. 1.15(a), the existence of the TOD

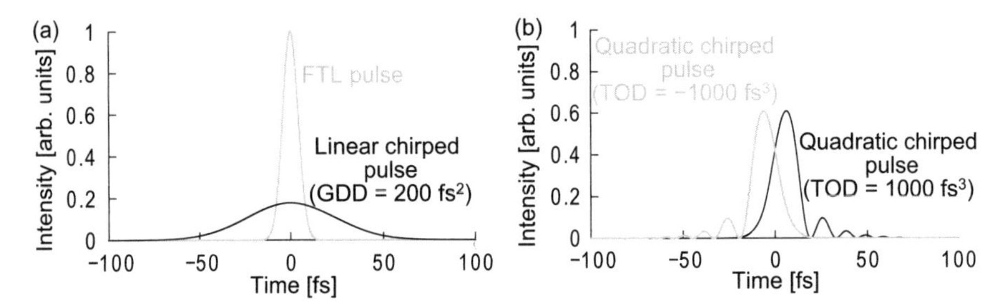

FIGURE 1.15 (a) Intensity profiles for the FTL Gaussian pulse with a pulse duration of 10 fs at a central wavelength of 800 nm (gray line) and the linearly chirped Gaussian pulse with a GDD of 200 fs^2 (black line). (b) Intensity profiles for the quadratic chirped Gaussian pulse with TODs of 1000 fs^3 (black line) and 1000 fs^3 (gray line).

distorts the pulse shape asymmetrically. The asymmetry of the pulse can be reversed by flipping the sign of the TOD as shown in Fig. 1.15(b). The FOD once again, relates to a pulse broadening.

For a quantitative picture of the influence that GDD has on the pulse duration, we analyze the effect of a quadratic spectral phase modulation

$$\phi(\omega) = \frac{1}{2}\left(\frac{d^2\phi}{d\omega^2}\right)_{\omega=\omega_0}(\omega - \omega_0)^2 \tag{1.3.13}$$

on an unchirped Gaussian pulse whose electric field is expressed by

$$\tilde{E}(\omega) = E_0 \exp\left[-\left\{\frac{t_g(\omega - \omega_0)}{2}\right\}^2\right]. \tag{1.3.14}$$

By inserting Eqs. (1.3.13) and (1.3.14) into Eq. (1.3.10), we obtain the electric field of the modulated pulse:

$$E(t) = E_0 \frac{1}{\sqrt{\pi\left[t_g^4 + 4\left\{\left(d^2\phi/d\omega^2\right)_{\omega=\omega_0}\right\}^2\right]}} \exp\left[-i\psi(t)\right]$$
$$\times \exp\left[-\left[\frac{t}{t_g\sqrt{1+\left\{2\left(d^2\phi/d\omega^2\right)_{\omega=\omega_0}/t_g^2\right\}^2}}\right]^2\right], \tag{1.3.15}$$

where

$$\psi(t) = \omega_0 t + \frac{2\left(d^2\phi/d\omega^2\right)_{\omega=\omega_0}}{t_g^2 + 4\left\{\left(d^2\phi/d\omega^2\right)_{\omega=\omega_0}\right\}^2} t^2 - \frac{1}{2}\arctan\left\{2\left(d^2\phi/d\omega^2\right)_{\omega=\omega_0}/t_g^2\right\}^2. \tag{1.3.16}$$

According to Eq. (1.3.15), we can express the pulse duration by

$$\Delta t_{\mathrm{GDD}} = \Delta t_{\mathrm{FL}} \sqrt{\left[1 + \left\{\frac{4\ln 2\left(d^2\phi/d\omega^2\right)_{\omega=\omega_0}}{\Delta t_{\mathrm{FL}}^2}\right\}^2\right]}, \tag{1.3.17}$$

where $\Delta t_{\mathrm{FL}} = t_g\sqrt{2\ln 2}$ is pulse duration assumed to be FTL. As can be seen from Eq. (1.3.17), a nonzero value for GDD indicates that the pulse duration is broadened from its transform limit. The time dependence of $\psi(t)$ implies that the instantaneous frequency differs across the pulse from the central frequency ω_0. The instantaneous frequency is given by

$$\omega(t) = \frac{d\psi(t)}{dt} = \omega_0 + \frac{4\left(d^2\phi/d\omega^2\right)_{\omega=\omega_0}}{t_g^2 + 4\left\{\left(d^2\phi/d\omega^2\right)_{\omega=\omega_0}\right\}^2} t. \tag{1.3.18}$$

This equation shows the frequency changes linearly across the pulse, namely that the GDD imposes linear frequency chirp on the pulse. The frequency chirp depends on the sign of GDD. In the positive GDD, the frequency chirp $\omega(t) - \omega_0$ is negative at the leading edge ($t < 0$) and increases linearly across the pulse. The opposite occurs in the negative GDD. The reason for the pulse broadening induced by GDD is that different frequency components of a pulse travel at slightly different speeds along the optical element. More specially, red components travel faster than blue components in the positive GDD, while the opposite occurs in the negative GDD. Any time delay in the arrival of different spectral components leads to pulse broadening.

1.3.2.2 Material Dispersion Effect Given the known functional dependence of the refractive index with respect to wavelength, $n(\lambda)$, it is usually convenient to express the different terms in expansion of Eq. (1.3.11) in terms of derivative of the refractive index with respect to the wavelength:

$$\phi(\omega) = \frac{L}{c} n(\lambda)\omega, \tag{1.3.19}$$

$$\left(\frac{d\phi}{d\omega}\right)_{\omega=\omega_0} = \frac{L}{c}\left[n(\lambda_0) - \lambda_0\left(\frac{dn}{d\lambda}\right)_{\lambda=\lambda_0}\right], \tag{1.3.20}$$

TABLE 1.3 Sellmeier Coefficients of Glass Materials

Material	B1 (μm^2)	B2 (μm^2)	B3 (μm^2)	C1 (μm^{-2})	C2 (μm^{-2})	C3 (μm^{-2})
Fused silica	0.6961663	0.4079426	0.8974794	0.004679148	0.013512063	97.9340025
BK7	1.03961212	0.231792344	1.01046945	0.006000699	0.020017914	103.560653
SF6	1.72448482	0.390104889	1.04572858	0.013487195	0.05693181	1.18557185
SF10	1.61625977	0.259229334	1.07762317	0.012753456	0.058198395	116.60768
SF11	1.73848403	0.311168974	1.17490871	0.01360686	0.061596046	121.922711
SFL6	1.78922056	0.328427448	2.01639441	0.013516354	0.06227296	168.014713
LAKN22	1.83021453	0.29156359	1.28544024	0.009048233	0.033075669	89.3675501
BAFN1	1.59034337	0.138464579	1.21988043	0.009327343	0.042749865	119.251777
LAKL21	0.996356844	0.651392837	1.22432622	0.014482159	0.001548264	89.9818604

$$\left(\frac{d^2\phi}{d\omega^2}\right)_{\omega=\omega_0} = \frac{L}{c}\left(\frac{\lambda_0}{2\pi c}\right)\left[\lambda_0^2\left(\frac{d^2n}{d\lambda^2}\right)_{\lambda=\lambda_0}\right], \tag{1.3.21}$$

$$\left(\frac{d^3\phi}{d\omega^3}\right)_{\omega=\omega_0} = -\frac{L}{c}\left(\frac{\lambda_0}{2\pi c}\right)^2\left[3\lambda_0^2\left(\frac{d^2n}{d\lambda^2}\right)_{\lambda=\lambda_0} + \lambda_0^3\left(\frac{d^3n}{d\lambda^3}\right)_{\lambda=\lambda_0}\right], \tag{1.3.22}$$

$$\left(\frac{d^4\phi}{d\omega^4}\right)_{\omega=\omega_0} = \frac{L}{c}\left(\frac{\lambda_0}{2\pi c}\right)^3\left[12\lambda_0^2\left(\frac{d^2n}{d\lambda^2}\right)_{\lambda=\lambda_0}, + 8\lambda_0^3\left(\frac{d^3n}{d\lambda^3}\right)_{\lambda=\lambda_0} + \lambda_0^4\left(\frac{d^4n}{d\lambda^4}\right)_{\lambda=\lambda_0}\right], \tag{1.3.23}$$

where L is the path length through the dispersive material and c is the speed of light. The degree to which each of mth ($m \geq 2$) order terms affects the pulse duration and shape is dependent on the pulse bandwidth and the length of the optical path over which the pulse propagates through the dispersive material. For the pulse durations and material path lengths considered here, it is important to know the amount and the sign of the dispersion through the fourth order for each element in the optical system.

Often a transparent material will be characterized by a fit of the refractive index n as a function of wavelength. A common form is the Sellmeier equation:

$$n^2(\lambda) = 1 + \sum_{j=1}^{3}\frac{B_j\lambda^2}{\lambda^2 - C_j}. \tag{1.3.24}$$

Some examples are shown in Table 1.3.

The dispersion of typical objectives has been measured by several groups [69,70]. From these measurements broadening of femtosecond pulses can be estimated. For example, for a 100-fs pulse at 800 nm, the measured GDD ranges from 580 to 1200 fs^2 for standard objectives. The net dispersion of the microscope, including beam expanders, filters, and dichroic mirrors, can then be on the order of 4500 fs^2.

This amount of dispersion broadens a 100-fs pulse to 270 fs, significantly impacting the efficiency of any nonlinear optical processes. The dispersion produced by the microscope can be precompensated through a variety of means.

1.3.2.3 Dispersion Compensation Next, we introduce the methods for dispersion compensation. In the near-infrared (NIR), the GDD, TOD, and FOD are all positive in the most common glasses. Thus, to compensate for a net positive dispersion, a dispersive system that is capable of producing negative GDD, TOD, and FOD must be used. A grating pair [71, 72] or a prism pair [73–76] arranged to provide negative GDD are the most common methods of dispersion precompensation. The negative GDD arises from the angular dispersion produced by these devices. The grating pairs have the advantage of high dispersion, making the system very compact, whereas the prism pairs have the advantage of high efficiency. A second advantage of the prism pairs over the grating pairs is the compensation for TOD and FOD, which is negative for the prism pairs and positive for the grating pairs. Thus, the higher order dispersions of a prism system is opposite to those of the glass material that is being compensated, whereas the grating's TOD and FOD adds to the TOD and FOD of the glass.

In 1994, Szipöecs et al. came up with a new idea of dispersion compensation, so-called chirped mirrors [77]. Similarly to other dielectric mirrors, chirped mirrors are composed of alternating high and low index quarter-wavelength thick layers, resulting in strong Bragg reflection. In chirped mirrors, the Bragg wavelength is chirped so that different wavelengths penetrate at different depths into the mirror upon reflection, giving rise to a wavelength-dependent GD.

Ultimately, it is the residual higher order dispersion of the compressor that limits the pulse duration. Careful attention to the higher order terms is obviously necessary to ensure optimum compensation. While the advantage of the prism pairs over the grating pairs may compensate for the higher order dispersion, care must also be taken with regard to the type of glass used in the prism pair.

Although highly dispersive glasses such as SF10 are often used in a compact prism pair, the negative TOD of such prism pairs is much larger than the positive TOD of glass materials used in lenses. Thus, such highly dispersive prism pairs result in high residual TOD that can limit pulse compression [70]. Fused silica is better matched with respect to balancing the higher order dispersion than highly dispersive prism pairs. Thus, fused silica is a better glass choice when either the shorter pulse duration or broader wavelength tenability is desired. Recently, a pulse shaper with a spatial light modulator (SLM) has been used to compensate for the residual higher order dispersion. Pulses as short as 5 fs have been generated at the focus of high-numerical aperture objective lenses by compensating for the GDD and the higher order dispersion through the use of a prism pair, and a pulse shaper with a liquid-crystal SLM [78]. Here we describe dispersion compensation with the prism pair, the grating pair, and the pulse shaper.

Prism Pair In dispersion compensation by using angular dispersion through a prism, a grating, or a grism, the first element creates an angular dispersed beam. The

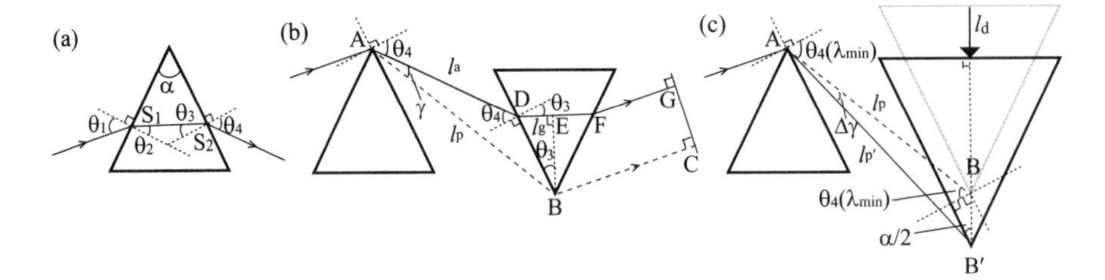

FIGURE 1.16 (a) The incidence and refraction in a prism. The successive angles of incidence-refraction are θ_1 and θ_2 at point S_1, and θ_3 and θ_4 at point S_2. (b) A typical prism pair arrangement. $l_p = \overline{\mathrm{AB}}$, $l_a = \overline{\mathrm{AD}}$, $l_g = \overline{\mathrm{DE}}$, $\gamma = \mathrm{BAD}$. (c) Prism insertion. $l_d = \overline{\mathrm{BB}}$, $l'_p = \overline{\mathrm{AB'}}$, $\Delta\gamma = \mathrm{BAB'}$.

second element will be associated to the first one, such that the angular dispersion introduced by the first element is compensated, and all frequency components of the beam are parallel again, as shown in Fig. 1.16(b). In order to remove the spatial chirp, the prism, grating, or grism pair systems must be double-passed. In two popular arrangements, the beam is sent through two elements, and retro-reflected by a plane mirror or sent directly through a sequence of four elements. To compensate for the angular dispersion in the prism pair system, the two prisms are put in opposition, in such a way that any face of one prism corresponds to a parallel face of the other prism (Fig. 1.16) [73–76]. In the first prism, the successive angles of incidence-refraction are θ_1 and θ_2 at point S_1, and θ_3 and θ_4 at point S_2. When the two prisms are identical, with equal apex angle α, and with pair of faces oriented parallel, in the second prism, the successive angles of incidence-refraction are θ_4 and θ_3 at point D, and θ_2 and θ_1 at point F. According to geometrical theorems and Snell's law, the relations between the successive angles are described by

$$\theta_2 + \theta_3 = \alpha, \tag{1.3.25}$$

$$\sin\theta_1 = n \sin\theta_2, \tag{1.3.26}$$

$$\sin\theta_4 = n \sin\theta_3. \tag{1.3.27}$$

The optical path ADEFG at a frequency ω (wavelength λ) and the optical path ABC at the maximum frequency $\omega_{\max}$ (the minimum wavelength $\lambda_{\min}$) are represented by the solid line and the dotted line in Fig. 1.16(b), respectively. We consider the optical path difference between ADEFG and ABC. Because a wavefront BE corresponds to a wavefront CG, the section of path indicated by rays EFG and BC makes no contribution to dispersion. Thus, we just have to calculate the optical path length of the ray ADE. The distance between D and E, l_g is expressed by

$$l_g = \overline{\mathrm{BD}} \sin\theta_3. \tag{1.3.28}$$

Using the notation in Fig. 1.16(b), from the law of sines in a triangle ADB,

$$\frac{l_p}{\sin(\pi/2+\theta_4)} = \frac{\overline{\mathrm{BD}}}{\sin\gamma} = \frac{l_a}{\sin(\pi/2-\theta_4-\gamma)}, \tag{1.3.29}$$

we obtain

$$\overline{\mathrm{BD}} = \frac{\sin\gamma}{\cos\theta_4} l_p \tag{1.3.30}$$

$$l_a = \frac{\cos(\theta_4+\gamma)}{\cos\theta_4} l_p, \tag{1.3.31}$$

where

$$\gamma(\lambda) = \theta_4(\lambda_{\min}) - \theta_4(\lambda). \tag{1.3.32}$$

Given Eq. (1.3.30), Eq. (1.3.28) can be rewritten as

$$l_g = \frac{\sin\gamma\,\sin\theta_3}{\cos\theta_4} l_p. \tag{1.3.33}$$

From Eqs. (1.3.31) and (1.3.33), we can describe the optical path length of the ray ADE, $P(\lambda)$ contributing the dispersion, by

$$P(\lambda) = l_a + n(\lambda) l_g = l_p \cos\gamma(\lambda), \tag{1.3.34}$$

where Eq. (1.3.27) is used. Thus, the phase $\phi(\omega)$ that passes through the prism pair can be expressed by

$$\phi(\omega) = \frac{\omega P(\lambda)}{c}. \tag{1.3.35}$$

Next we must tend to the finite beam size. The effect in the case of the finite beam size is equivalent to that in the case of prism insertion. When the second prism is inserted as shown in Fig. 1.16(c), l_p and γ are replaced by l'_p and $\gamma + \Delta\gamma'$, respectively:

$$P(\lambda) = l_{p'} \cos\{\gamma(\lambda) + \Delta\gamma\}. \tag{1.3.36}$$

Using the notation in Fig. 1.16(c), from the law of cosines in a triangle ABB$'$,

$$l_{p'}^2 = l_p^2 + l_d^2 - 2 l_p l_d \cos\left\{\theta_4(\lambda_{\min}) + \left(\frac{\pi}{2} - \frac{\alpha}{2}\right)\right\}, \tag{1.3.37}$$

we obtain

$$l_{p'} = \sqrt{l_p^2 + l_d^2 + 2 l_p l_d \sin\left\{\theta_4(\lambda_{\min}) - \frac{\alpha}{2}\right\}}. \tag{1.3.38}$$

From the law of sines in a triangle ABB′,

$$\frac{l_d}{\sin(\Delta\gamma)} = \frac{l_{p'}}{\sin\{\theta_4(\lambda_{\min}) + (\pi/2 - \alpha/2)\}}, \tag{1.3.39}$$

$\Delta\gamma$ can be written as

$$\Delta\gamma = \arcsin\left[\frac{l_d}{l_{p'}}\cos\left\{\theta_4(\lambda_{\min}) - \frac{\alpha}{2}\right\}\right]. \tag{1.3.40}$$

According to Eqs. (1.3.35) and (1.3.36), we can calculate the GD, GDD, TOD, and FOD of the prism pair using the following equations:

$$\frac{d\phi}{d\omega} = \frac{1}{c}\left(P - \lambda\frac{dP}{d\lambda}\right) \tag{1.3.41}$$

$$\frac{d^2\phi}{d\omega^2} = \frac{\lambda^3}{2\pi c^2}\frac{d^2P}{d\lambda^2}, \tag{1.3.42}$$

$$\frac{d^3\phi}{d\omega^3} = -\frac{\lambda^4}{4\pi^2c^3}\left(3\frac{d^2P}{d\lambda^2} + \lambda\frac{d^3P}{d\lambda^3}\right), \tag{1.3.43}$$

$$\frac{d^4\phi}{d\omega^4} = \frac{\lambda^5}{8\pi^3c^4}\left(12\frac{d^2P}{d\lambda^2} + 8\lambda\frac{d^3P}{d\lambda^3} + \lambda^2\frac{d^4P}{d\lambda^4}\right), \tag{1.3.44}$$

where

$$\frac{dP}{d\lambda} = -l_{p'}\sin\{\gamma + \Delta\gamma\}\frac{d\gamma}{d\lambda}, \tag{1.3.45}$$

$$\frac{d^2P}{d\lambda^2} = -l_{p'}\left[\cos\{\gamma + \Delta\gamma\}\left(\frac{d\gamma}{d\lambda}\right)^2 + \sin\{\gamma + \Delta\gamma\}\frac{d^2\gamma}{d\lambda^2}\right], \tag{1.3.46}$$

$$\frac{d^3P}{d\lambda^3} = l_{p'}\left[\sin\{\gamma + \Delta\gamma\}\left\{\left(\frac{d\gamma}{d\lambda}\right)^3 - \frac{d^3\gamma}{d\lambda^3}\right\} - 3\cos\{\gamma + \Delta\gamma\}\frac{d\gamma}{d\lambda}\frac{d^2\gamma}{d\lambda^2}\right], \tag{1.3.47}$$

$$\frac{d^4P}{d\lambda^4} = l_{p'}\left[\sin\{\gamma + \Delta\gamma\}\left\{6\left(\frac{d\gamma}{d\lambda}\right)^2\frac{d^2\gamma}{d\lambda^2} - \frac{d^4\gamma}{d\lambda^4}\right\}\right.$$
$$\left. + \cos\{\gamma + \Delta\gamma\}\left\{\left(\frac{d\gamma}{d\lambda}\right)^4 - 3\left(\frac{d^2\gamma}{d\lambda^2}\right)^2 - 4\frac{d\gamma}{d\lambda}\frac{d^3\gamma}{d\lambda^3}\right\}\right]. \tag{1.3.48}$$

Using Eq. (1.3.32), we obtain

$$\frac{d\gamma}{d\lambda} = -\frac{d\theta_4}{d\lambda} = -\frac{d\theta_4}{dn}\frac{dn}{d\lambda}, \tag{1.3.49}$$

$$\frac{d^2\gamma}{d\lambda^2} = -\frac{d^2\theta_4}{dn^2}\left(\frac{dn}{d\lambda}\right)^2 - \frac{d\theta_4}{dn}\frac{d^2n}{d\lambda^2}, \tag{1.3.50}$$

$$\frac{d^3\gamma}{d\lambda^3} = -\frac{d^3\theta_4}{dn^3}\left(\frac{dn}{d\lambda}\right)^3 - 3\frac{d^2\theta_4}{dn^2}\frac{dn}{d\lambda}\frac{d^2n}{d\lambda^2} - \frac{d\theta_4}{dn}\frac{d^3n}{d\lambda^3}, \tag{1.3.51}$$

$$\begin{aligned}\frac{d^4\gamma}{d\lambda^4} &= -\frac{d^4\theta_4}{dn^4}\left(\frac{dn}{d\lambda}\right)^4 - 6\frac{d^3\theta_4}{dn^3}\left(\frac{dn}{d\lambda}\right)^2\frac{d^2n}{d\lambda^2} \\ &\quad -\frac{d^2\theta_4}{dn^2}\left\{3\left(\frac{d^2n}{d\lambda^2}\right)^2 + 4\frac{dn}{d\lambda}\frac{d^3n}{d\lambda^3}\right\} - \frac{d\theta_4}{dn}\frac{d^4n}{d\lambda^4}.\end{aligned} \tag{1.3.52}$$

From Eqs. (1.3.27), (1.3.25), and (1.3.26), we obtain the following relationships:

$$\frac{d\theta_4}{dn} = \frac{\sin\theta_3 + n\cos\theta_3(d\theta_3/dn)}{\cos\theta_4} \tag{1.3.53}$$

$$\frac{d\theta_3}{dn} = -\frac{d\theta_2}{dn}, \tag{1.3.54}$$

$$\frac{d\theta_2}{dn} = -\frac{\tan\theta_2}{n}. \tag{1.3.55}$$

Given Eqs. (1.3.54), (1.3.55), and (1.3.25), Eq. (1.3.53) can be re-written as

$$\frac{d\theta_4}{dn} = \frac{\sin\alpha}{\cos\theta_2\cos\theta_4}. \tag{1.3.56}$$

Using Eq. (1.3.56), we obtain

$$\frac{d^2\theta_4}{dn^2} = \frac{d\theta_4}{dn}\left(-\frac{\tan^2\theta_2}{n} + \tan\theta_4\frac{d\theta_4}{dn}\right) \tag{1.3.57}$$

$$\begin{aligned}\frac{d^3\theta_4}{dn^3} &= \frac{d^2\theta_4}{dn^2}\left(-\frac{\tan^2\theta_2}{n} + 2\tan\theta_4\frac{d\theta_4}{dn}\right) \\ &\quad + \frac{d\theta_4}{dn}\left\{\frac{\tan^2\theta_2}{n^2}\left(1 + \frac{2}{\cos^2\theta_2}\right) + \frac{1}{\cos^2\theta_4}\left(\frac{d\theta_4}{dn}\right)^2\right\}\end{aligned} \tag{1.3.58}$$

$$\frac{d^4\theta_4}{dn^4} = \frac{d^3\theta_4}{dn^3}\left(-\frac{\tan^2\theta_2}{n} + 2\tan\theta_4\frac{d\theta_4}{dn}\right)$$
$$+\frac{d^2\theta_4}{dn^2}\left\{\frac{\tan^2\theta_2}{n^2}\left(1+\frac{4}{\cos^2\theta_2}\right)+\frac{5}{\cos^2\theta_4}\left(\frac{d\theta_4}{dn}\right)^2+2\tan\theta_4\frac{d^2\theta_4}{dn^2}\right\}$$
$$+\frac{d\theta_4}{dn}\left\{-\frac{2\tan^2\theta_2}{n^3\cos^2\theta_2}\left(1+\frac{2-2\sin\theta_2}{\cos^2\theta_2}\right)\right.$$
$$\left.-\frac{2\tan^2\theta_2}{n^3}\left(1+\frac{2}{\cos^2\theta_2}\right)-\frac{2}{\cos^3\theta_4}\left(\frac{d\theta_4}{dn}\right)^3\right\}. \tag{1.3.59}$$

By inserting Eqs. (1.3.46)–(1.3.52) and (1.3.56)–(1.3.59) into Eqs. (1.3.42)–(1.3.44), we can calculate the practical GDD, TOD, and FOD introduced by the prism pair systems. Here, the value of the GDD, TOD, and FOD doubles because prism pair systems must be double passed to remove spatial chirp.

Grating Pair Gratings can produce larger angular dispersion than prisms. In analogy with prisms, the simplest practical device consists of two identical gratings arranged as in Fig. 1.17 [71, 72]. The dispersion introduced by a pair of parallel gratings can be determined by tracing the optical path length ABC between A and an output wavefront CC′ which is frequency dependent. The optical path length ABC is expressed by

$$\overline{\text{ABC}} = \overline{\text{AB}} + \overline{\text{AB}}\ \cos(\theta_i - \theta_d) = \frac{L_g}{\cos\theta_d}\{1+\cos(\theta_i-\theta_d)\}, \tag{1.3.60}$$

where θ_i is the angle of incidence, θ_d is the diffraction angle at a frequency ω (wavelength λ) and L_g is the normal separation between the first grating and the second grating. Here we consider the first-order diffraction. Grating diffraction may be characterized in terms of phase matching by a 2π-phase jump at each ruling in first-order diffraction. Thus, the adding phase shift equals -2π times the

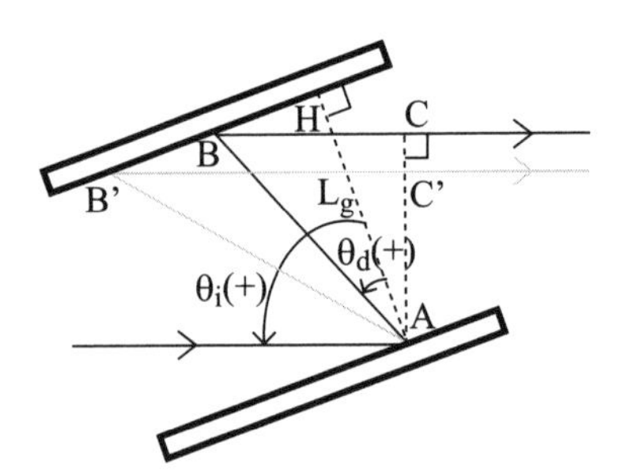

FIGURE 1.17 A typical grating pair arrangement. θ_i and θ_d are incident and diffraction angles at the first grating, respectively, $L_g = \overline{\text{AH}}$.

number of rulings between point B and the foot of the perpendicular from A to the second grating. The phase $\phi(\omega)$ after passing through the prism pair can be expressed by

$$\phi(\omega) = \frac{\omega}{c}\left(\overline{\mathrm{ABC}}\right) - 2\pi\frac{\overline{\mathrm{BH}}}{d} = \frac{\omega}{c}P(\lambda), \tag{1.3.61}$$

where

$$P(\lambda) = L_g\left[\frac{1+\cos(\theta_i - \theta_d)}{\cos\theta_d} - \frac{\lambda}{d}\tan\theta_d\right]. \tag{1.3.62}$$

Here d is the grating constant. If we obtain the derivative of Eq. (1.3.62), we can calculate the GDD, TOD, and FOD by using Eqs. (1.3.42) to (1.3.44). The angle of incidence and the diffraction angle are related through the grating equation

$$d\,(\sin\theta_i + \sin\theta_d) = \lambda. \tag{1.3.63}$$

From Eq. (1.3.63), we have

$$\frac{d\theta_d}{d\lambda} = \frac{1}{d\cos\theta_d}. \tag{1.3.64}$$

Using Eqs. (1.3.62) and (1.3.64), we obtain the derivative of Eq. (1.3.62):

$$\frac{dP}{d\lambda} = -\frac{L_g\tan\theta_d}{d}, \tag{1.3.65}$$

$$\frac{d^2P}{d\lambda^2} = -\frac{L_g}{d^2\cos^3\theta_d}, \tag{1.3.66}$$

$$\frac{d^3P}{d\lambda^3} = \frac{d^2P}{d\lambda^2}\frac{3\sin\theta_d}{d\cos^2\theta_d}, \tag{1.3.67}$$

$$\frac{d^4P}{d\lambda^4} = \frac{d^3P}{d\lambda^3}\frac{1+5\tan^2\theta_d}{d\sin\theta_d}. \tag{1.3.68}$$

By substituting Eqs. (1.3.66)–(1.3.68) into Eqs. (1.3.42)–(1.3.44), we calculate the GDD, TOD, and FOD introduced by a pair of gratings. We then double-pass a grating pair system, so that the value of the GDD, TOD, and FOD doubles.

Pulse Shaper It is difficult to compensate for the higher order dispersion by prism pairs, grating pairs, and grism pairs when either the shorter pulse duration or broader wavelength tenability is desired. A technique best suite for compensating for the higher order dispersion is the pulse-shaping technique using adaptive optics, by which we can manipulate the pulse spectrum in amplitude and phase [79–83]. The pulse-shaping technique was originally introduced for ps light pulses [84, 85]

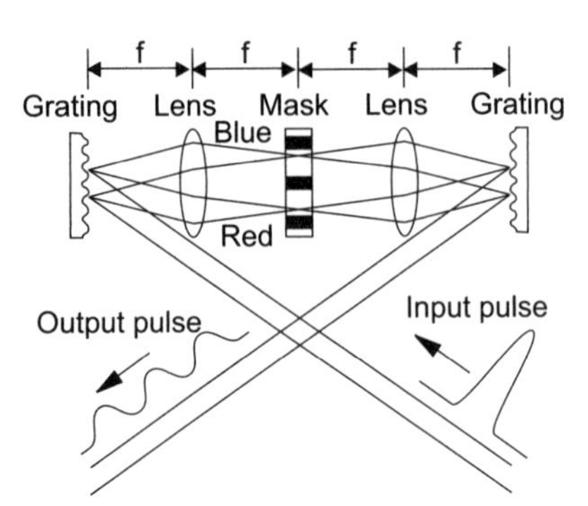

FIGURE 1.18 Typical grating-pair-formed pulse shaper.

and later improved for fs pulses [86]. A typical grating-pair-formed pulse shaper is illustrated in Fig. 1.18. The pulse to be shaped is spectrally dispersed using a first grating or a first prism. Each spectral component is focused at the position of the mask that spectrally filters the pulse. We assume that the frequency varies linearly in the focal plane of the lens. The complex transmission across the mask causes a transfer function of Eq. (1.3.8). The spectral components are recollimated into a beam by a second grating or a second prism. These masks can be adaptively controlled by liquid crystal (LC) based spatial light modulators (SLMs), acousto-optic based SLMs or deformable mirrors. Thus, by combining the pulse-shaping techniques with diagnostic methods for characterization of femtosecond pulses as described in Section 1.3.3, we can easily compensate not only for the GDD but also the higher order dispersions. In addition, the pulse shaping techniques can be used to control nonlinear optical phenomena as shown in Section 5.2.

1.3.3 Pulse Characterization in Nonlinear Optical Microscope

Pulse characterization and compression are of critical importance in ultrashort pulse science and technology. Especially in nonlinear optical microscopy, we need to characterize the pulse at the focal point in the microscope because the signal intensity decreases as the pulse duration is broadened by the dispersion produced by the microscope. The femtosecond time scale is beyond the reach of standard electronic detectors. Diagnostic methods for characterization of arbitrary femtosecond pulses have to be designed to freeze and time resolve events as short as a few optical cycles. Most diagnostic techniques are based on nonlinear optical phenomena, which may have an influence on the amplitude and phase of the pulse to be measured. The intensity autocorrelation (AC) technique is the most common method used to estimate the pulse duration of ultrashort pulses [87,88]. The AC width is related to the pulse duration when a particular pulse shape is assumed. However, the intensity AC measurement cannot give the phase of the pulse. The interferometric autocorrelation (IAC) technique has also been widely utilized for pulse characterization [89,90]. It was introduced by Jean-Claude Diels in 1983 [89]. Naganuma et al. have investigated the pulse reconstruction method for uniquely characterizing the pulse amplitude and phase, which is based on the IAC measurement [91]. The intensity AC, the second-harmonic (SH) AC, and the fundamental power spectrum can be used for unique pulse reconstruction. Frequency-resolved optical gating (FROG), which is a spectrally resolved autocorrelation, is also one of the most commonly used pulse characterization methods [92]. The spectrally resolved AC measurement gives a two-dimensional data set in the time–frequency domain, which is usually sufficient for unique reconstruction of the amplitude and phase of the ultrashort pulse via iterative phase retrieval algorithms. Spectral phase interferometry for direct electric-field reconstruction (SPIDER) is an interferometric pulse characterization technique in the frequency domain based on spectral shearing interferometry [93]. In SPIDER, the spectral phase of the ultrashort pulse can be extracted from the interferogram without iterative phase retrieval algorithms. Multiphoton intrapulse interference phase scan (MIIPS) is a method to characterize the pulse by manipulating the ultrashort pulse [94, 95]. In this section, we will describe IAC and MIIPS techniques, which have been typically used to confirm dispersion compensation at the focal point in the microscope.

1.3.3.1 Interferometric Autocorrelation A method that combines quantities related to the autocorrelation and spectrum in a single data trace is the IAC, often called phase-sensitive autocorrelation and the fringe-resolved autocorrelation (FRAC). The setup for the IAC measurement contains a Michelson interferometer with a variable delay line as shown in Fig. 1.19. A test pulse is split into two pulses by a first-beam splitter (BS1) and then recombined by a second-beam splitter (BS2). Next, the two collinear pulses from the Michelson interferometer are focused into a nonlinear medium. The resultant nonlinear signal is detected by a detector after passing through a short-pass filter and/or a band-pass filter, which are used to eliminate the fundamental. The detector only has to measure an average power. The delay time between the two pulses can be mechanically adjusted via the variable delay line. When the nonlinear signal intensity is proportional to the square of the excitation intensity, we can obtain the second-order IAC signal. The third-order IAC signal can be measured by using the nonlinear signal whose intensity is proportional to the cube of the excitation intensity. The SHG crystal such as a β-barium borate (BBO) crystal

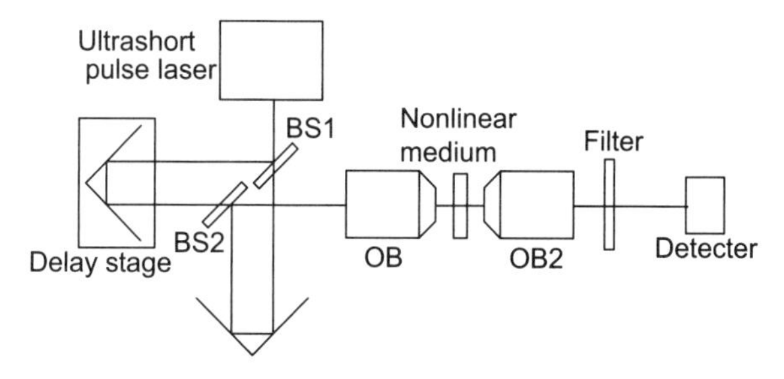

FIGURE 1.19 Setup for the IAC measurement.

has been widely used to measure the second-order IAC signal. For a simple implementation in the second-order IAC measurement, two-photon-induced free-carrier current in semiconductors such as GaAsP or GaP photodiodes has been employed [96]. We need to take care in using very short pulse durations (< 20 fs). The beam splitter may have a limited bandwidth, and its substrate and the other optical elements may introduce chromatic dispersion. Since the group velocity mismatch can restrict the phase-matching bandwidth, very thin nonlinear crystals with a thickness less than 10 μm have to be used.

The second-order IAC signal is proportional to the function

$$S^{(2)}(\tau) = \int_{-\infty}^{\infty} \left|\{E(t-\tau) + E(t)\}^2\right|^2 dt. \tag{1.3.69}$$

If the electric field is described by

$$E(t) = A(t)\exp(-i\omega_0 t), \tag{1.3.70}$$

where ω_0 is the central frequency of the electric field, the second-order IAC signal is written by

$$S^{(2)}(\tau) = \int_{-\infty}^{\infty} \left[F_0(\tau) + F_1(\tau)e^{-i\omega_0\tau} + F_1^*(\tau)e^{i\omega_0\tau}\right. \tag{1.3.71}$$

$$\left. + F_2(\tau)e^{-i2\omega_0\tau} + F_2^*(\tau)e^{i2\omega_0\tau}\right] dt. \tag{1.3.72}$$

Here $F_0(\tau)$, $F_1(\tau)$, and $F_2(\tau)$ are expressed by

$$F_0(\tau) = \int_{-\infty}^{\infty} \left[|A(t)|^2 + |A(t-\tau)|^2 + 4\,|A(t)|^2\,|A(t-\tau)|^2\right] dt, \tag{1.3.73}$$

$$F_1(\tau) = 2\int_{-\infty}^{\infty} \left[A^*(t)A^*(t-\tau)\left\{A^2(t) + A^2(t-\tau)\right\}\right.$$

$$\left. + A(t)A(t-\tau)\left\{A^{*2}(t) + A^{*2}(t-\tau)\right\}\right] dt, \tag{1.3.74}$$

$$F_2(\tau) = \int_{-\infty}^{\infty} A^2(t)A^2(t-\tau)dt. \tag{1.3.75}$$

According to Eq. (1.3.72), the second-order IAC signal is composed of fringe components at frequencies of around 0, ω_0, and $2\omega_0$. The peak $S^{(2)}(0)$ to background $S^{(2)}(\infty)$ ratio for the second IAC signal is 8 to 1. The terms $F_0(\tau)$, $F_1(\tau)$, and $F_2(\tau)$ can be extracted from a measurement by taking Fourier transform of the data, identifying the cluster of data near the three characteristic frequencies, and recovering by successive inverse Fourier transforms.

The dc term $F_0(\tau)$ indicates the sum of background term and background-free intensity AC, and is generally referred to as an intensity autocorrelation with the

TABLE 1.4 Examples of Deconvolution Factors for Standard Pulse Profiles

Shape	Intensity Profile $I(t)$	Pulse Dulation Δt	Deconvolution Factor $K_d = \Delta\tau/\Delta t$	Time − bandwidth Product K
Gauss	$\exp\{-2(t/t_g)^2\}$	$1.177t_g$	1.414	0.441
Sech	$\mathrm{sech}^2(t/t_s)$	$1.763t_s$	1.543	0.315
Lorentz	$\{1+(t/t_l)^2\}^{-1}$	$2t_l$	2.000	0.221
Square	$1, \lvert t/t_r\rvert \le 1/2$ 0, elsewhere	t_r	1.000	0.443

background. $F_0(\tau)$ has a peak $F_0(0)$ to background $F_0(\infty)$ of 3 to 1. The AC width of $F_0(\tau)$ depends on the temporal shape of the pulse intensity. If we can assume the pulse shape, the AC width $\Delta\tau$ is related to the pulse duration Δt (intensity width)

$$\frac{\Delta\tau}{\Delta t} = K_d, \tag{1.3.76}$$

where K_d is the deconvolution factor. Some examples are shown in Table 1.4. The deconvolution factor for a Gaussian pulse is 1.543, and that for the secant hyperbolic pulse is 1.414.

The second-harmonic (SH) term $F_2(\tau)$ shows an AC of the SH fields, which provides the SH power spectrum. In the absence of phase modulation—namely for FTL pulses—$F_2(\tau)$ is identical to the intensity AC. This property has been exploited to determine whether pulse phase is modulated [91]. The IAC contains phase information. The shape and phase sensitivity of the IAC can be also extracted to quantitatively measure a linear chirp.

Figure 1.20 shows the intensity profile, the IAC, and the extracted intensity AC with background for an FTL Gaussian pulse with a pulse duration of 10 fs at a central wavelength of 800 nm and a linearly chirped Gaussian pulse with a GDD of 100 fs^2. As shown in Fig. 1.20(c) pertaining to the FTL pulse, the lower and upper envelopes of the interference pattern split evenly from the background level. In the case of the chirped pulse, the width of interference pattern in Fig. 1.20(d) is much narrower than the width of the intensity AC in Fig. 1.20(f). The wings of the IAC are identical to those of the intensity AC. The level at which the interference pattern starts relative to the peak (2/8 in the case of 1.20(d)) indicates the magnitude of the chirp. The magnitude of the chirp can be estimated by comparison of upper and lower envelopes for the measured IAC and those for analytically calculated IACs of various chirped pulses.

1.3.3.2 Multiphoton Intrapulse Interference Phase Scan It has long been recognized that nonlinear optical processes are sensitive to the spectral phase. In the MIIPS method, the characterization of ultrashort pulses is achieved by scanning the spectral phase, and by detecting the spectrum of the nonlinear optical process. Figure 1.21 shows a typical setup for the MIIPS measurement. Ultrashort pulses pass

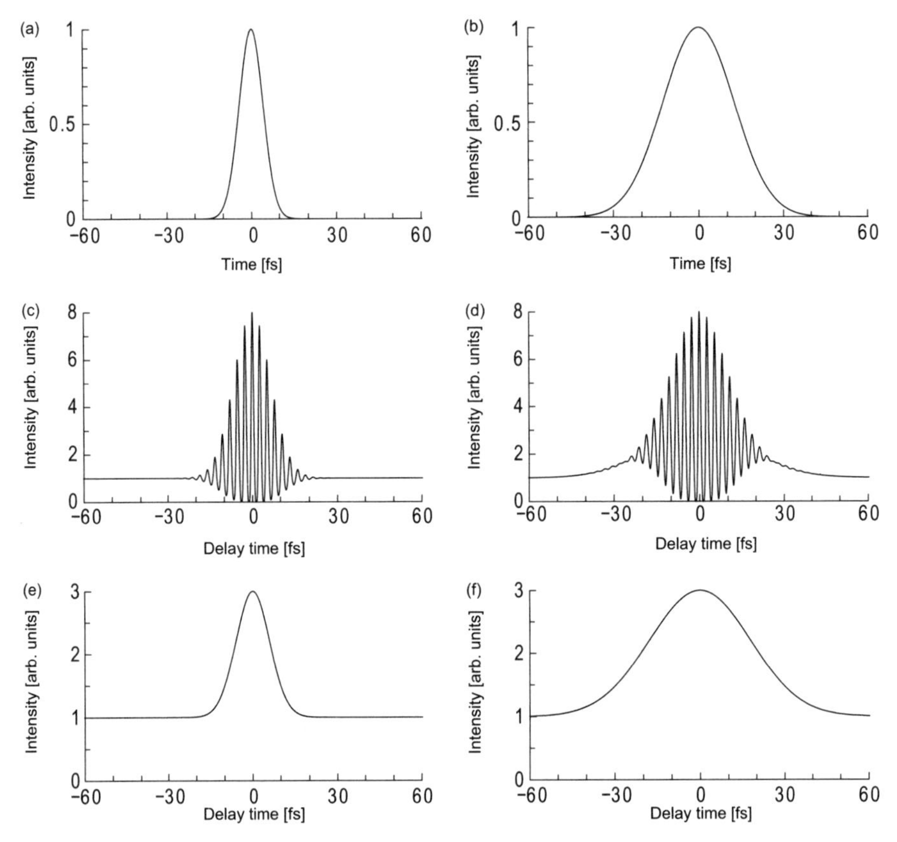

FIGURE 1.20 (a, b) The intensity profiles of the FTL Gaussian pulse with a pulse duration of 10 fs at a central wavelength of 800 nm (a) and the linearly chirped Gaussian pulse with a GDD of 100 fs^2 (b). (c, d) The calculated IAC signals for the FTL pulse (c) and the chirped pulse (d). (e, f) The extracted intensity AC with background for the FTL pulse (e) and the chirped pulse (f).

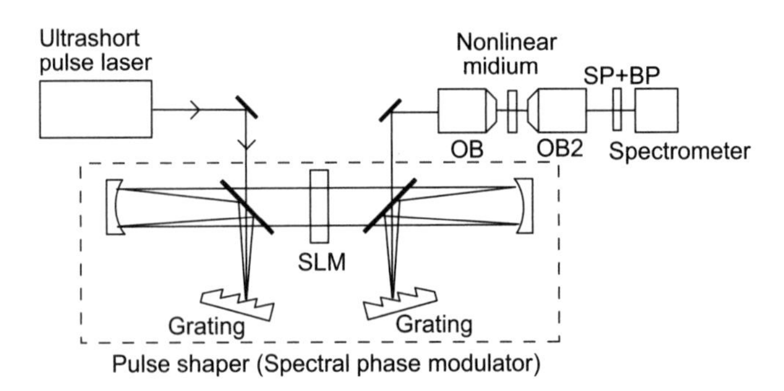

FIGURE 1.21 Setup for the MIIPS measurement.

through a pulse shaper with a liquid crystal based spatial light modulator (SLM), which is used to scan the spectral phase of the pulses, and then they are focused into a nonlinear medium. The resultant nonlinear signal is detected by a spectrometer after passing through a short-pass (SP) filter and/or a band-pass (BP) filter, which are used to eliminate the fundamental. Because the pulse shaper can be also used for dispersion precompensation, the MIIPS method provides not only pulse characteristics but also dispersion compensation. The second-harmonic (SH) spectra at each phase scan can be written as

$$\left|\tilde{E}^{(2)}(2\omega)\right|^2 = \left|\int \left|\tilde{E}(\omega+\Omega)\right| \left|\tilde{E}(\omega-\Omega)\right| \right. \\ \left. \times \exp\left[i\left\{\phi(\omega+\Omega)+\phi(\omega-\Omega)\right\}\right] d\Omega\right|^2, \qquad (1.3.77)$$

where $|E(\omega)|$ and $\phi(\omega)$ are the spectral amplitude and the spectral phase of the excitation pulse, respectively. Here the spectral phase is given by the sum of the unknown phase $\varphi(\omega)$ and a well-known reference function, $\psi(\omega)$, which is given by the phase scan in MIIPS. According to Eq. (1.3.77), the SHG intensity at a frequency of 2ω reaches maximum when $\phi(\omega+\Omega)+\phi(\omega-\Omega)$ is zero. This condition is satisfied when $\varphi(\omega) = -\psi(\omega)$. By expressing the pulse spectral phase $\phi(\omega)$ in terms of a Taylor series, we obtain

$$\phi(\omega+\Omega)+\phi(\omega-\Omega) = \sum_{m=0}^{\infty} \frac{2}{(2m!)} \frac{d^{2m}\phi}{d\omega^{2m}} \Omega^{2m} \\ = \sum_{m=0}^{\infty} \frac{2}{(2m!)} \left(\frac{d^{2m}\varphi}{d\omega^{2m}} + \frac{d^{2m}\psi}{d\omega^{2m}}\right) \Omega^{2m}. \qquad (1.3.78)$$

From Eq. (1.3.78), we see that the SHG intensity at a frequency of 2ω, $|E^{(2)}(2\omega)|^2$ is not affected by the odd terms $d^{2m+1}\phi/d\omega^{2m+1}$, and it is sensitive to the second derivative of the spectral phase $d^2\phi/d\omega^2$. By searching $d^2\psi/d\omega^2$ which produces maximum SHG signals, the unknown $d^2\varphi/d\omega^2$ can be obtained by $-d^2\psi/d\omega^2$. As the reference function,

$$\psi(\omega) = a\sin(b\omega - \delta) \qquad (1.3.79)$$

has been used [94]. Scanning the parameter δ and collecting the SHG spectrum for each value generates the two-dimensional MIIPS trace from which we can find the condition $\delta_{\max}(\omega)$ when the maximum SHG signal is obtained. From the second derivative of the reference function where the maximum signal is observed, we obtain

$$\frac{d^2\varphi(\omega)}{d\omega^2} = -\frac{d^2\psi(\omega)}{d\omega^2} = ab^2 \sin\left\{b\omega - \delta_{\max}(\omega)\right\}. \qquad (1.3.80)$$

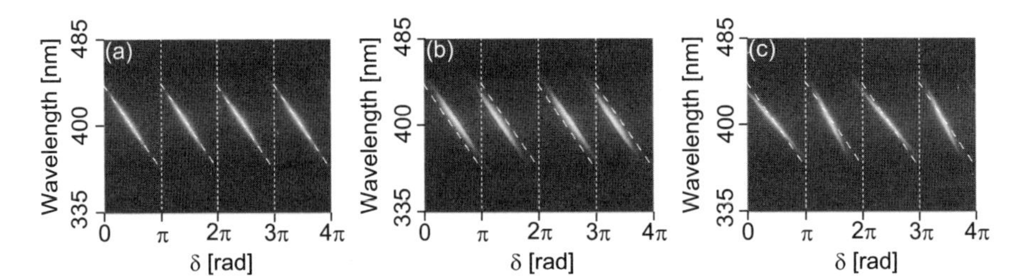

FIGURE 1.22 Calculated MIIPS traces for the FTL Gaussian pulse with a pulse duration of 10 fs at a central wavelength of 800 nm (a), the linearly chirped Gaussian pulse with a GDD of 200 fs^2 (b), and the quadratic chirped Gaussian pulse with a TOD of 1000 fs^3 (c). The dashed lines indicate $\delta_{\max}(\omega) = b\omega + N\pi$ for the FTL pulse.

The unknown spectral phase $\varphi(\omega)$ is directly obtained by double integration of $d^2\varphi/d\omega^2$ over frequency ω. Then, the unknown constants of integration are set to zero.

Once $\varphi(\omega)$ is determined, a compensation phase function equal to $-\varphi(\omega)$ is applied to obtain FTL pulses. For FTL pulses, the second derivative of the unknown phase $d^2\varphi/d\omega^2$ should be zero. Thus, from Eq. (1.3.80), we see that $\delta_{\max}(\omega)$ is given by $\delta_{\max}(\omega) = b\omega + N\pi$, where N is an integer. This fact indicates that two-dimensional MIIPS trace in the δ, ω plane for FTL pulses are characterized by parallel lines separated by π. If the only phase distortion is linear chirp, the parallel lines $\delta_{\max}(\omega)$ are no longer spaced by π. Cubic phase modulation (quadratic chirp) induces a change in the slope of $\delta_{\max}(\omega)$. Therefore, we can quickly and intuitively determine the spectral phase of ultrashort pulses. In addition, by applying an iterative procedure of measurement and compensation, the absolute accuracy of MIIPS can improve.

Figure 1.22 shows the calculated MIIPS traces for the FTL Gaussian pulse with a pulse duration of 10 fs at a central wavelength of 800 nm, the linearly chirped Gaussian pulse with a GDD of 200 fs^2 and the quadratic chirped Gaussian pulse with a TOD of 1000 fs^3. From these MIIPS traces, we see that the FTL pulse gives parallel lines $\delta_{\max}(\omega) = b\omega + N\pi$, the linearly chirped pulse produces unequally spaced parallel lines of $\delta_{\max}(\omega)$, and the quadratic chirped pulse causes a change in the slope of $\delta_{\max}(\omega)$.

1.3.4 Wavefront Compensation

Aberrations in a microscope are caused by optical elements quality and alignment, by the refractive index mismatch from the objective, and by the inhomogeneous structure of biological samples [97, 98]. These aberrations lead to a spreading of the focusing spot inside the sample, reducing the spatial resolution. Reduction of signal levels deteriorates the contrast of the image [99]. Nonlinear optical microscopy allows larger penetration depths compared with confocal microscopy (see Chapter 4) [100–102]; however, aberration effects and scattering limit.

Let us consider the image quality when focusing a light by an objective lens. In the absence of aberration, the focal spot is diffraction limited. When the light is focused through a planar interface between two media with different refractive indexes, refraction at the interface distorts the wavefronts. This effect occurs when the light is focused using an immersion objective, through a cover glass, or into a specimen-mounting medium [103, 104]. The refractive index mismatch causes focal shift and spherical aberration that are proportional to the focusing depth. These aberrations can be represented by a series of rotationally symmetric Zernike modes. Aberrations include spherical aberration (e.g., in the case of a sample whose index of refraction is different from that of the immersion medium), astigmatism, and coma (e.g., if the surface of the sample is curved or tilted). In some objective lenses, the spherical aberration induced by the refractive index mismatch of the cover glass can be removed with adjustable cover glass correction, to enable the use of different cover glass thicknesses [105]. A more significant problem is presented by the complex aberrations introduced by the optically inhomogeneous structure of biological samples. Spatial variations of the refractive indexes can lead to aberrations when focused through thick specimens. Aberration introduced by inhomogeneous samples distorts the focal spot. The detrimental effects of aberrations on image quality include reduced resolution and a decrease in image brightness and/or contrast.

Adaptive optics is used to restore a diffraction-limited focus within an inhomogeneous sample that introduces wavefront aberrations. A wavefront deformation of the excitation beam is introduced before the focusing objective compensates for aberrations created within the sample. This permits improving the quality of images obtained deep inside biological tissue in terms of resolution and signal level. Adaptive optical devices have been are used in adaptive microscopes such as deformable mirrors and liquid crystal spatial light modulators. Wavefront compensation can be categorized into (sensor based) direct sensing and (sensor less) indirect sensing. In direct sensing, the correction can be performed by measuring the aberrations through a sensing device such as a wavefront sensor [106, 107]. Several types of wavefront sensing have been developed for adaptive optics. The most common devices used are the Shack–Hartman wavefront sensor and the interferometric sensor. In indirect sensing, wavefront measurement is conducted with a sensorless adaptive optics system. In sensor less schemes, iterative algorithms are generally used to control an adaptive optical devices [108, 109].

In confocal microscopy, both the excitation and collected beams need to be compensated to improve the intensity of the reconstructed images [110]. In nonlinear microscopy, aberrated wavefronts can be corrected in the excitation beam [111]. Aberration correction can be used in nonlinear microscopy of a sensorless method relying on indirect measurement of the aberrations. This compensates the unknown existing optical aberrations without having to measure them. Wilson presented an analysis of the information contained in the image as a function of the applied aberration to determine the aberration initially introduced by the sample. Olivier et al. showed that correction could improve the signal by up to a factor of 2.7 in biological samples in third harmonic generation (THG) microscopy (Fig. 1.23) [112]. In widefield microscopy, only the correction of the collected signal is important [113, 114].

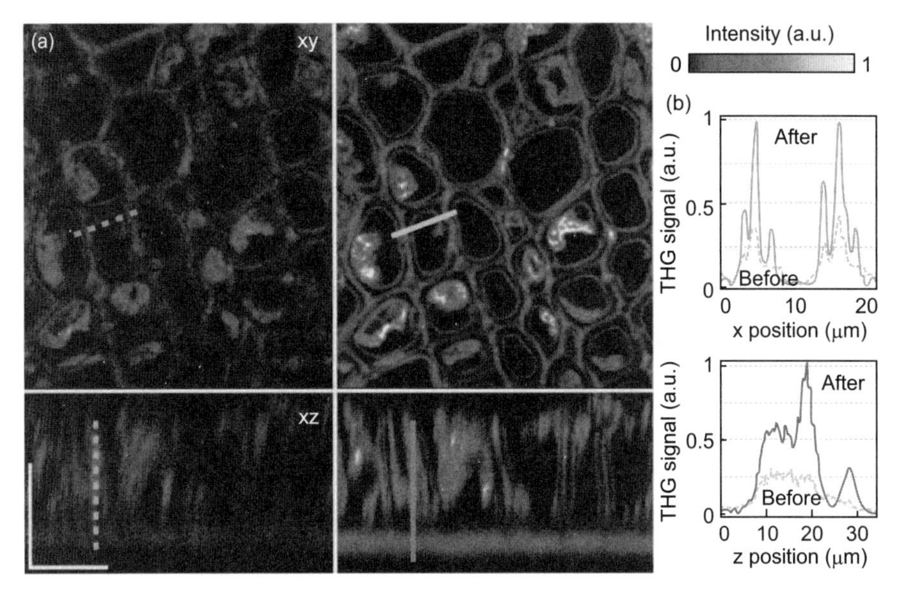

FIGURE 1.23 (a) THG images of an elderberry stem slice (left) before and (right) after aberration correction on the middle plane of the z stack. Scale bars, 20 μm. (b) Profiles along the lines in (a). Reprinted with permission from [112]. Copyright 2009 Optical Society of America. (*See insert for color representation of the figure.*)

Adaptive optics can reduce the time required for image acquisition by increasing signal generation and collection efficiency. Wavefront compensation with adaptive optics is a powerful method used to restore spatial resolution and signal levels within an inhomogeneous sample that introduces wavefront aberration in thick media.

REFERENCES

1. J.-C. Diels and W. Rudolph, *Ultrashort Laser Pulse Phenomena*, 2nd ed., Academic Press, 2006.
2. Y. R. Shen, *The Principle of Nonlinear Optics*, Wiley, NY, 1984.
3. R. Boyd, *Nonlinear Optics*, 2nd ed., Academic Press, San Diego, 2003.
4. G. Agrawal, *Nonlinear Fiber Optics*, 4th ed., Academic Press, San Diego, 2007.
5. A. Siegman, *Lasers*, University Science Books, Mill Valley, 1986.
6. L. G. Gouy, "Sur une propriete nouvelle des ondes lumineuses," *Acad. Sci. Paris*, **110**, 1251–1253 (1890).
7. P. A. Franken, A. E. Hill, C. W. Peters, and G. Weinreich, "Generation of optical harmonics," *Phys. Rev. Lett.*, **7**, 118–119 (1961).
8. J. E. Midwinter and J. Warner, "The effects of phase matching method and of uniaxial crystal symmetry on the polar distribution of second-order non-linear optical polarization," *Br. J. Appl. Phys.*, **16**, 1135–1142 (1965).
9. R. L. Byer, "Quasi-phasematched nonlinear interactions and devices," *J. Nonlinear Opt. Phys. Mater.*, **6**, 549–592 (1997).

10. J. A. Armstrong, N. Bloembergen, J. Ducuing, and P. S. Pershan, "Interactions between light waves in a nonlinear dielectric," *Phys. Rev.*, **127**, 1918–1939 (1962).
11. M. Yamada, N. Nada, M. Saitoh, and K. Watanabe, "First-order quasi-phase matched $LiNbO_3$ waveguide periodically poled by applying an external field for efficient blue second-harmonic generation," *Appl. Phys. Lett.*, **62**, 435–436 (1993).
12. A. W. Smith and N. Braslau, "Optical mixing of coherent and incoherent light," *IBM J. Res. Dev.*, **6**, 361–362 (1962).
13. C. C. Wang and G. W. Racette, "Measuement of parametric gain accompanying optical diffrence frequency generation," *Appl. Phys. Lett.*, **6**, 169–171 (1965).
14. J. A. Giordmaine and R. C. Miller, "Tunable coherent parametric oscillation in $LiNbO_3$ at optical frequencies,"*Phys. Rev. Lett.*, **14**, 973–976 (1965).
15. J. E. Midwinter and J. Warner, "The effects of phase matching method and of crystal symmetry on the polar dependence of third-order nonlinear optical polarization," *Br. J. Appl. Phys.*, **16**, 1665–1674 (1965).
16. P. N. Butcher, "Nonlinear optical phenomena," *Bulletin 200* (Engineer Experiment Station, Ohio State University, Columbus, 1965).
17. C. C. Shang and H. Hsu, "The spatial symmetric forms of the third-order nonlinear susceptibility," *IEEE J. Quantum Electron.*, **23**, 177–179 (1987).
18. P. N. Butcher and D. Cotter, *The Elements of Nonlinear Optics*, Cambridge U. Press, Cambridge, 1990.
19. X.-L. Yang and S.-W. Xie, "Expression of third-order effective nonlinear susceptibility for third-harmonic generation in crystals," *Appl. Opt.*, **34**, 6130–6135 (1995).
20. G. H. C. New and J. F. Ward, "Optical third-harmonic generation in gases," *Phys. Rev. Lett.*, **19**, 556–559 (1967).
21. P. D. Maker and R. W. Terhune, "Study of optical effects due to an induced polarization third order in the electric field strength," *Phys. Rev.*, **137**, A801–A819 (1965).
22. S. D. Kramer and N. Bloembergen, "Third-order nonlinear optical spectroscopy inCuCl," *Phys. Rev. B*, **14**, 4654–4669 (1976).
23. M. D. Levenson and N. Bloembergen, "Dispersion of the nonlinear optical susceptibility tensor in centrosymmetric media," *Phys. Rev. B*, **10**, 4447–4464 (1974).
24. F. Shimizu, "Frequency broadening in liquids by a short light pulse," *Phys. Rev. Lett.*, **19**, 1097–1100 (1967).
25. M. Goeppert-Mayer, "Über Elementarakte mit zwei Quantensprüngen," *Ann. Phys.*, **9**, 273–295 (1931).
26. W. Kaiser and C. G. B. Garrett, "Two-photon excitation in $CaF_2{:}Eu^{2+}$," *Phys. Rev. Lett.*, **7**, 229–232 (1961).
27. E. J. Woodbury and W. K. Ng, "Ruby laser operation in the Near IR*," *Proc. Inst. Radio Eng.*, **50**, 2367 (1962).
28. L. V. Keldysh, "Ionization in the field of a strong electromagnetic wave," *Sov. Phys. JETP*, **20**, 1307–1314 (1965) [L.V. Keldysh, J. Exptl. Theoret. Phys. (U. S. S. R) **47**, 1945–1957 (1964)].
29. C. B. Schaffer, A. Brodeur, and E. Mazur, "Laser-induced breakdown and damage in bulk transparent materials induced by tightly focused femtosecond laser pulses," *Meas. Sci. Technol.*, **12**, 1784–1794 (2001).
30. X. Liu, D. Du, and G. Mourou, "Laser ablation and micromachining with ultrashort laser pulses," *IEEE J. Quantum Electron.*, **33**, 1706–1716 (1997).

31. M. Lenzner, J. Krüger, S. Sartania, Z. Cheng, Ch. Spielmann, G. Mourou, W. Kautek, and F. Krausz, "Femtosecond optical breakdown in dielectrics," *Phys. Rev. Lett.*, **80**, 4076–4079 (1998).
32. M. Li, S. Menon, J. P. Nibarger, and G. N. Gibson, "Ultrafast electron dynamics in femtosecond optical breakdown of dielectrics," *Phys. Rev. Lett.*, **82**, 2394–2397 (1999).
33. N. Bloembergen, "Laser-induced electric breakdown in solids," *IEEE J. Quantum Electron.*, **QE10**, 375–386 (1974).
34. A. Vogel, J. Noack, G. Huttman, and G. Paltauf, "Mechanisms of femtosecond laser nanosurgery of cells and tissues," *Appl. Phys. B*, **81**, 1015–1047 (2005).
35. J. G. Fujimoto, W. Z. Lin, E. P. Ippen, C. A. Puliafito, and R. F. Steinert, "Time-resolved studies of nd-yag laser-induced breakdown - plasma formation, acoustic-wave generation, and cavitation," *Invest. Ophthalmol. Vis. Sci.*, **26**, 1771–1777 (1985).
36. E. N. Glezer, C. B. Schaffer, N. Nishimura, and E. Mazur, "Minimally disruptive laser-induced breakdown in water," *Opt. Lett.*, **22**, 1817–1819 (1997).
37. A. Vogel and V. Venugopalan, "Mechanisms of pulsed laser ablation of biological tissues,"*Chem. Rev.*, **103**, 577–644 (2003).
38. V. Kohli, A. Y. Elezzabi, and J. P. Acker, "Cell nanosurgery using ultrashort (femtosecond) laser pulses: applications to membrane surgery and cell isolation," *Laser Surg. Med.*, **37**, 227–230 (2005).
39. K. König, "Multiphoton microscopy in life sciences," *J. Microsc.*, **200**, 83–104 (2000).
40. W. R. Zipfel, R. M. Williams, and W. W. Webb, "Nonlinear magic: multiphoton microscopy in the biosciences," *Nat. Biotechnol.*, **21**, 1369–1377 (2003).
41. J. Squier and M. Müller, "High resolution nonlinear microscopy: A review of sources and methods for achieving optimal imaging," *Rev. Sci. Inst.*, **72**, 2855–2867 (2001).
42. P. J. Campagnola and L. M. Loew, "Second-harmonic imaging microscopy for visualizing biomolecular arrays in cells, tissues and organisms," *Nat. Biotechnol.*, **21**, 1356–1360 (2003).
43. J.-X. Cheng and X. S. Xie, "Coherent anti-Stokes Raman scattering microscopy: instrumentation, theory, and applications," *J. Phys. Chem. B*, **108**, 827–840 (2004).
44. C. L. Evans and X. S. Xie, "Coherent anti-Stokes Raman scattering microscopy: chemical imaging for biology and medicine," *Annu. Rev. Anal. Chem.*, **1**, 883–909 (2008).
45. F. Helmchen and W. Denk, "Deep tissue two-photon microscopy," Nat. Methods **2**, 932–940 (2005).
46. P. Theer and W. Denk, "On the fundamental imaging-depth limit in two-photon microscopy," *J. Opt. Soc. Am. A.*, **23**, 3139–3149 (2006).
47. R. Hellwarth and P. Christensen, "Nonlinear optical microscopic examination of structure in polycrystalline ZnSe," *Opt. Commun.*, **12**, 318–322 (1974).
48. W. Denk, J. H. Strickler, and W. W. Webb, "Two-photon laser scanning fluorescence microscopy," *Science*, **248**, 73–76 (1990).
49. D. E. Spence, P. N. Kean, and W. Sibbett, "60-fsec pulse generation from a self-modelocked Ti:sapphire laser," *Opt. Lett.*, **16**, 42–44 (1991).
50. U. Morgner, F. X. Krtner, S. H. Cho, Y. Chen, H. A. Haus, J. G. Fujimoto, E. P. Ippen, V. Scheuer, G. Angelow, and T. Tschudi, "Sub-two-cycle pulses from a Kerr-lens mode-locked Ti:sapphire laser," *Opt. Lett.*, **24**, 411–413 (1999).

51. D. H. Sutter, G. Steinmeyer, L. Gallmann, N. Matuschek, F. Morier-Genoud, U. Keller, V. Scheuer, G. Angelow, and T. Tschudi, "Semiconductor saturable-absorber mirror assisted Kerr-lens mode-locked Ti:sapphire laser producing pulses in the two-cycle regime," *Opt. Lett.*, **24**, 631–633 (1999).
52. A. Seas, V. Petricevic, and R. R. Alfano, "Generation of sub-100-fs pulses from a cw mode-locked chromium-doped forsterite laser," *Opt. Lett.*, **17**, 937–939 (1992).
53. J. M. Evans, V. Petricevic, A. B. Bykov, A. Delgado, and R. R. Alfano, "Direct diode-pumped continuous-wave near-infrared tunable laser operation of Cr^{4+}:forsterite and Cr^{4+}:Ca_2GeO_4," *Opt. Lett.*, **22**, 1171–1173 (1997).
54. Z. Zhang, K. Torizuka, T. Itatani, K. Kobayashi, T. Sugaya and T. Nakagawa, T. Sugaya, and T. Nakagawa, "Broadband semiconductor saturable-absorber mirror for a self-starting mode-locked Cr:forsterite laser," *Opt. Lett.*, **23**, 1465–1467 (1998).
55. C. Chudoba, J. G. Fujimoto, E. P. Ippen, H. A. Haus, U. Morgner, F. X. Kärtner, V. Scheuer, G. Angelow, and T. Tschudi, "All-solid-state Cr:forsterite laser generating 14-fs pulses at 1.3μ m," *Opt. Lett.*, **26**, 292–294 (2001).
56. W. S. Pelouch, P. E. Powers, and C. L. Tang, "Ti:sapphire-pumped, high-repetition-rate femtosecond optical parametric oscillator," *Opt. Lett.*, **17**, 1070–1072 (1992).
57. I. N. Duling III, "Subpicosecond all-fiber erbium laser," *Electron. Lett.*, **27**, 544–545 (1991).
58. K. Tamura, E. P. Ippen, H. A. Haus, and L. E. Nelson, "77-fs pulse generation from a stretched-pulse mode-locked all-fiber ring laser," *Opt. Lett.*, **18**, 1080–1082 (1993).
59. H. M. Pask, R. J. Carman, D. C. Hanna, A. C. Tropper, C. J. Mackechnie, P. R. Barber, and J. M. Dawes, "Ytterbium-doped silica fiber lasers: versatile sources for the 1-1.2 μm region," *IEEE J. Sel. Top. Quantum Electron.*, **1**, 2–13 (1995).
60. X. Zhou, D. Yoshitomi, Y. Kobayashi, and K. Torizuka, "Generation of 28-fs pulses from a mode-locked ytterbium fiber oscillator," *Opt. Express*, **16**, 7055–7059 (2008).
61. N. Kuse, Y. Nomura, A. Ozawa, M. Kuwata-Gonokami, S. Watanabe, and Y. Kobayashi, "Self-compensation of third-order dispersion for ultrashort pulse generation demonstrated in an Yb fiber oscillator," *Opt. Lett.*, **35**, 3868–3870 (2010).
62. H. N. Paulsen, K. M. Hilligse, J. Thgersen, S. R. Keiding, and J. J. Larsen, "Coherent anti-Stokes Raman scattering microscopy with a photonic crystal fiber based light source," *Opt. Lett.*, **28**, 1123–1125 (2003).
63. H. Kano and H. Hamaguchi, "Characterization of a supercontinuum generated from a photonic crystal fiber and its application to coherent Raman spectroscopy," *Opt. Lett.*, **28**, 2360–2362 (2003).
64. G. McConnell and E. Riis, "Two-photon laser scanning fluorescence microscopy using photonic crystal fiber," *J. Bio. Opt.*, **9**, 922–927 (2004).
65. K. Isobe, W. Watanabe, S. Matsunaga, T. Higashi, K. Fukui, and K. Itoh, "Multispectral two-photon excited fluorescence microscopy using supercontinuum light source," *Jpn. J. Appl. Phys.*, **44**, L167–L169 (2005).
66. R. R. Alfano ed., *The Suprecontinuum Light Source*, Springer-Verlag, New York, 1989.
67. G. P. Agrawal, *Nonlinear Fiber Optics*, 4th ed., Academic Press, San Diego, 2007.
68. J. K. Ranka, R. S. Windeler, and A. J. Stentz, "Visible continuum generation in airsilica microstructure optical fibers with anomalous dispersion at 800 nm," *Opt. Lett.*, **25**, 25–27 (2000).

69. J. B. Guild, C. Xu, and Watt W. Webb, "Measurement of group delay dispersion of high numerical aperture objective lenses using two-photon excited fluorescence," *Appl. Opt.*, **36**, 397–401 (1997).
70. M. Müller, J. Squier, R. Wolleschensky, U. Simons, and G. J. Brankenhoff, "Dispersion pre-compensation of 15 femtosecond optical pulses for high numerical aperture objectives," *J. Micrisc.*, **191**, 141–150 (1998).
71. M. B. Treacy, "Compression of picosecond light pulses," *Phys. Lett.*, **28A**, 34–35 (1968).
72. M. B. Treacy, "Optical pulse compression with difraction gratings," *IEEE J. Quantum Electron.*, **QE-5**, 454–458 (1969).
73. O. E. Martinez, J. P. Gordon, and R. L. Fork, "Negative group-velocity dispersion using refraction," *J. Opt. Soc. Am. A*, **1**, 1003–1006 (1984).
74. R. L. Fork, O. E. Martinez, and J. P. Gordon, "Negative dispersion using pairs of prisms," *Opt. Lett.*, **9**, 150–152 (1984).
75. Z. Zhang, and T. Yagi, "Observation of group delay dispersion as a function of the pulse width in a mode locked Ti:sapphire laser," *Appl. Phys. Lett.*, **63**, 2993–2995 (1993).
76. K. Osvay, P. Dombi, A. P. Kovács, and Z. Bor, "Fine tuning of the higher-order dispersion of a prismatic pulse compressor," *Appl. Phys. B*, **75**, 649–654 (2002).
77. R. Szipöcs, K. Ferencz, C. Spielmann, and F. Krausz, "Chirped multilayer coatings for broadband dispersion control in femtosecond lasers," *Opt. Lett.*, **19**, 201–203 (1994).
78. K. Isobe, A. Suda, M. Tanaka, F. Kannari, H. Kawano, H. Mizuno, A. Miyawaki, and K. Midorikawa, "Fourier transform spectroscopy combined with 5-fs broadband pulse for multispectral nonlinear microscopy," *Phys. Rev. A.*, **77**, 063832 (2008).
79. A. M. Weiner, "Femtosecond pulse shaping using spatial light modulators," *Rev. Sci. Instrum.*, **71**, 1929–1960 (2000).
80. A. Präkelt, M. Wollenhaupt, A. Assion, C. Horn, C.-S. Tudoran, M. Winter, and T. Baumert, "Compact, robust, and flexible setup for femtosecond pulse shaping," *Rev. Sci. Instrum.*, **74**, 4950–4953 (2003).
81. A. M. Weiner and J. P. Heritage, "Picosecond and femtosecond Fourier pulse shape synthesis," *Rev. Phys. Appl.*, **22**, 1619–1628 (1987).
82. M. M. Wefers and K. A. Nelson, "Analysis of programmable ultrashort waveform generation using liquid-crystal spatial light modulators," *J. Opt. Soc. Am. B*, **12**, 1343–1362 (1995).
83. H. Wang, Z. Zheng, D. E. Leaird, A. M. Weiner, T. A. Dorschner, J. J. Jijol, L. J. Friedman, H. Q. Nguyen, and L. A. Palmaccio, "20-fs pulse shaping with a 512-element phase-only liquid crystal modulator," *IEEE J. Select. Top. Quantum. Electron.*, **7**, 718–727 (2001).
84. J. Desbois, F. Gires, and P. Tournois, "A new approach to picosecond laser pulse analysis, shaping and coding," *IEEE J. Quantum Electron.*, **9**, 213–218 (1973).
85. J. Agostinelli, G. Harvey, T. Stone, and C. Gabel, "Optical pulse shaping with a grating pair," *Appl. Opt.*, **18**, 2500–2504 (1979).
86. A. M. Weiner, J. P. Heritage, and E. M. Kirschner, "High-resolution femtosecond pulse shaping," *J. Opt. Soc. Am. B*, **5**, 1563–1572 (1988).
87. J. A. Armstrong, "Measurement of picosecond laser pulse widths," *Appl. Phys. Lett.*, **10**, 16–18 (1967).
88. K. L. Sala, G. A. Kenney-Wallace, and G. E. Hall, "CW autocorrelation measurements of picosecond laser pulses," *IEEE J. Quantum Electron.*, **16**, 990–996 (1980).

89. J.-C. M. Diels, J. J. Fontaine, and F. Simoni, "Phase sensitive measurements of femtosecond laser pulses from a ring cavity," *Proceedings of the International Conference on Lasers*, STS Press: McLean, VA. 348–355 (1983).

90. J.-C. M. Diels, J. J. Fontaine, I. C. McMichael, and F. Simoni, "Control and measurement of ultrashort pulse shapes (in amplitute and phase) with femtosecond accuracy," *Appl. Opt.*, **24**, 1270–1282 (1985).

91. K. Naganuma, K. Mogi, and H. Yamada, "General-method for ultrashort light-pulse chirp measurement," *IEEE J. Quantum Electron.*, **25**, 1225–1233 (1989).

92. R. Trebino, and D. J. Kane, "Using phase retrieval to measure the intensity and phase of ultrashort pulses—frequency-resolved optical gating," *J. Opt. Soc. Am. A*, **10**, 1101–1111 (1993).

93. C. Iaconis, and I. A. Walmsley, "Spectral phase interferometry for direct electric-field reconstruction of ultrashort optical pulses," *Opt. Lett.*, **23**, 792–794 (1998).

94. V. V. Lozovoy, I. Pastirk, and M. Dantus, "Multiphoton intrapulse interference 4: characterization and compensation of the spectral phase of ultrashort laser pulses," *Opt. Lett.*, **29**, 775–777 (2004).

95. V. V. Lozovoy, Bingwei X. Y. Coello, and M. Dantus, "Direct measurement of spectral phase for ultrashort laser pulses," *Opt. Express*, **16**, 592–597 (2008).

96. J. K. Ranka, A. L. Gaeta, A. Baltuska, M. S. Pshenichnikov, and D. A. Wiersma, "Autocorrelation measurement of 6 fs pulses based on the two-photon-induced photocurrent in a GaAsP photodiode," *Opt. Lett.,* **22**, 1344–1366 (1997).

97. M. A. A. Neil, R. Juskaitis, M. J. Booth, T. Wilson, T. Tanaka, and S. Kawata, "Adaptive aberration correction in a two-photon microscope," *J. Microsc.*, **200**, 105–108 (2000).

98. D. Debarre, E. Botcherby, T. Watanabe, S. Srinivas, M. Booth, and T. Wilson, "Image-based adaptive optics for two-photon microscopy," *Opt. Lett.*, **34**, 2495–2497 (2009).

99. M. J. Booth and T. Wilson, "Refractive-index-mismatch induced aberrations in single-photon and two-photon microscopy and the use of aberration correction," *J. Biomed. Opt.*, **6**, 266–272 (2001).

100. M. J. Booth, "Adaptive optics in microscopy," *Philos. Transact. A Math. Phys. Eng. Sci.*, **365**, 2829–2843 (2007).

101. M. J. Booth, D. Debarre, and A. Jesacher, "Adaptive optics for biomedical microscopy," *Opt. Photonics News*, **23**, 22–29 (2012).

102. F. Helmchen and W. Denk, "Deep tissue two-photon microscopy," *Nat. Methods*, **2**, 932–940 (2005).

103. C. J. R. Sheppard and C. J. Cogswell, "Effects of aberrating layers and tube length on confocal imaging properties," *Optik*, **87**, 34–38 (1991).

104. C. J. R. Sheppard and M. Gu, "Aberration compensation in confocal microscopy," *Appl. Opt.*, **30**, 3563–3568 (1991).

105. M. Schwertner, M. J. Booth, and T. Wilson, "Simple optimization procedure for objective lens correction collar setting," *J. Microsc.*, **217**, 184–187 (2005).

106. J. W. Cha, J. Ballesta, and P. T. C. So, "Shack-Hartmann wavefront-sensor-based adaptive optics system for multiphoton microscopy," *J. Biomed. Opt.*, **15**, 046022 (2010).

107. M. Rueckel, J. A. Mack-Bucher, and W. Denk, "Adaptive wavefront correction in two-photon microscopy using coherence-gated wavefront sensing," *Proc. Natl. Acad. Sci. U. S. A.*, **103**, 17137–17142 (2006).

108. D. Debarre, M. J. Booth, and T. Wilson, "Image based adaptive optics through optimization of low spatial frequencies," *Opt. Express*, **15**, 8176–8190 (2007).
109. M. J. Booth, "Wave front sensor-less adaptive optics: a model-based approach using sphere packings," *Opt. Express*, **14**, 1339–1352 (2006).
110. M. J. Booth, M. Schwertner, and T. Wilson, "Specimen-induced aberrations and adaptive optics for microscopy," *Proc. SPIE*, **5894**, 26–34 (2005).
111. J. M. Bueno, E. J. Gualda, and P. Artal, "Adaptive optics multiphoton microscopy to study ex vivo ocular tissues," *J. Biomed. Opt.*, **15**, 066004 (2010).
112. N. Olivier, D. Debarre, and E. Beaurepaire, "Dynamic aberration correction for multiharmonic microscopy," *Opt. Lett.*, **34**, 3145–3147 (2009).
113. Z. Kam, P. Kner, D. Agard, and J. W. Sedat, "Modelling the application of adaptive optics to wide-field microscope live imaging," *J. Microsc.*, **226**, 33–42 (2007).
114. D. Debarre, E. J. Botcherby, M. J. Booth, and T. Wilson, "Adaptive optics for structured illumination microscopy," *Opt. Express*, **16**, 9290–9305 (2008).

CHAPTER 2

BASIC MICROSCOPIC TECHNIQUE

2.1 BASIC ARCHITECTURE OF A LASER SCANNING MICROSCOPE

In this section, we describe basic architecture of a laser scanning microscope (LSM). Figure 2.1(a) shows a schematic of a typical confocal laser scanning microscope (CLSM). CLSM has been summarized in a comprehensive book [1]. In CLSM, a specimen is generally illuminated with a point light source produced by focusing laser beams with a high numerical-aperture (NA) objective lens (OB). Scanning mirrors are used to change positions of the focal spot on the specimen. The emitted light or the scattered light from this point in the specimen is collected by the objective lens and its path retraced back through the scanning mirrors to a dichroic mirror (DM) or a beam splitter, either of which is used to separate the signal light from the excitation light. Then, the scanning mirrors are used for descanning the signal light. The signal light is imaged through a pinhole (PH) onto a detector and is detected with the detector. An image is obtained either by scanning the laser focal point by point across the specimen or by scanning the specimen across the fixed laser spot. Because the signal is detected through the pinhole, only light originating from the focal plane can reach the detector, while light coming from above or below the focal plane is not in focus in the pinhole plane as shown in Fig. 2.1(b) and does not contribute to the image. Thus, CLSM provides an inherent optical sectioning capability [1].

In the case of multicolor fluorescence the imaging of specimens is labeled with multiple fluorophores, as illustrated in Fig. 2.1(c); multiple lasers and detectors are typically employed because the use of multiple fluorophores in a single specimen

Functional Imaging by Controlled Nonlinear Optical Phenomena, First Edition.
Keisuke Isobe, Wataru Watanabe and Kazuyoshi Itoh.

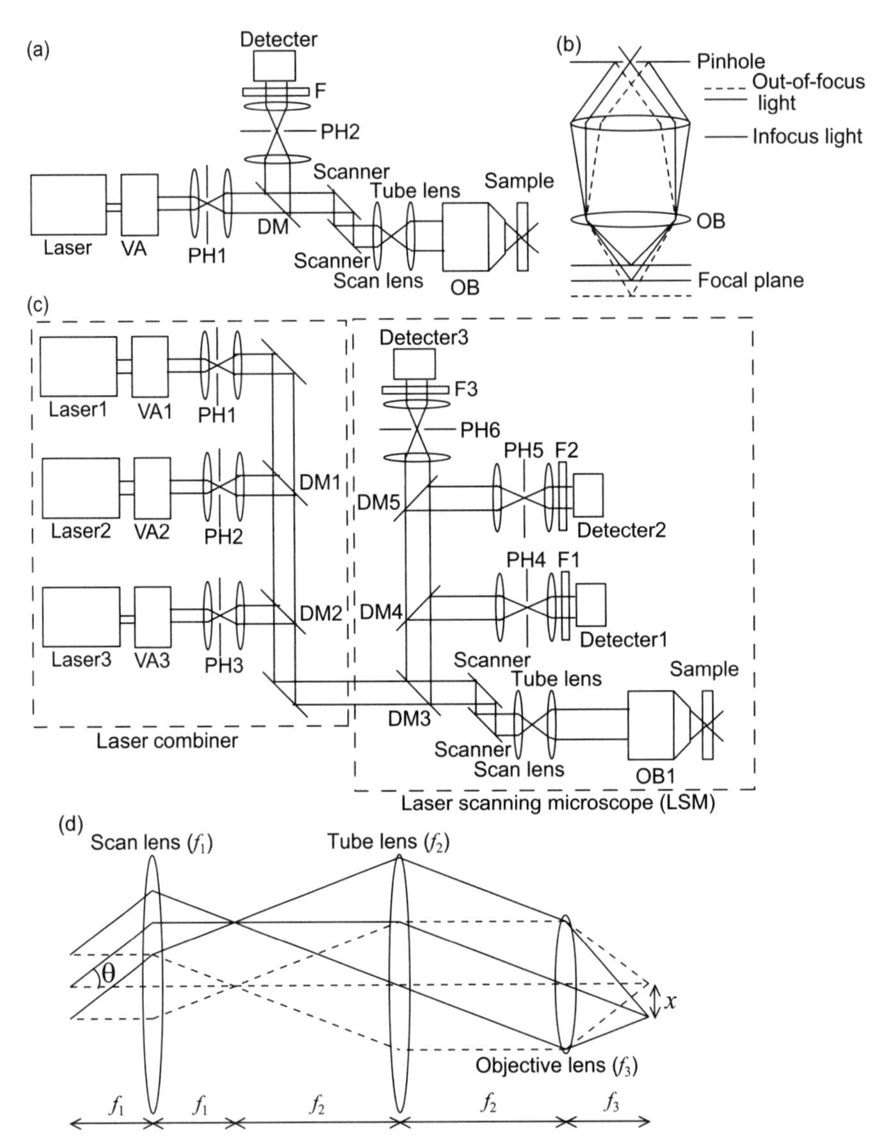

FIGURE 2.1 (a) Schematic of a typical confocal laser scanning microscope: VA, variable attenuator; PH, pinhole; DM, dichroic mirror; OB, objective lens; F, filter. (b) Schematic diagram of the confocal effect. (c) Schematic of the typical confocal laser scanning microscope for multicolor fluorescence imaging. (d) Beam scanning system used to translate an angular deflection θ of the excitation beam into a lateral translation of x of the focal spot in the sample.

requires the independent detection of signals from each fluorophore. Multiple laser beams are combined by a series of DMs and are directed into a CLSM. The signal light is divided into multiple wavelength bands with a series of DMs. The each signal light is imaged through a PH onto a detector and is detected with the detector. In the case where dispersive optical elements, such as a prism or a grating for dividing into multiple bands, instead of DMs, is used, the signal light is directed through a PH into the prism or grating.

2.1.1 Scanning Methods

The simplest way to achieve CLSM imaging is to scan the sample across the fixed laser spot. Because all the lenses work on an axis, lenses can easily be diffraction limited for the single on-axis focus. Piezoelectric-driven stages, which can be controlled within nanometer precision, can be used to scan the sample in the lateral directions. The stage movement is precise; however, it is slow. Thus, instead of sample scanning in the lateral directions, most CLSMs use a beam scanning method. As shown in Fig. 2.1(d), in the beam scanning method, the angular deflection of the excitation beam is translated to a lateral change in position of the focal spot on the sample by pivoting collimated beams at the back focal plane of the objective lens. The beam scanning is generally implemented using mirrors mounted on galvanometer motors. The galvanometer mirror system can control the scan angle with exquisite accuracy and has the additional advantage that the scanning angle is insensitive to the wavelength. For two-dimensional (2D) scanning in the lateral direction, the reflection at two orthogonally pivoting mirrors is used, and their angles are controlled with high precision by galvanometer motors. The spatial resolution in CLSM is basically proportional to the focal spot size. Therefore, the scanning methods attempt to create the smallest possible (diffraction-limited) illumination spot on the sample. Then, the diameter of a laser beam is expanded to fill the aperture at the back focal plane of the objective lens by using a combination of a scan lens and a tube lens. In the axial direction, the objective lens or the sample is typically scanned with motorized scanning stages.

In order to increase the beam scanning speed, resonant galvanometers that vibrate at a fixed frequency are often used. The mirror angle varies in a sinusoidal manner at a frequency determined by the mechanical resonance of the device, which is usually 4 kHz and above. By using acousto-optical deflectors (AODs), a laser beam can be scanned at up to 100 kHz, compared to about 4 to 16 kHz for resonant galvanometers. However, the AODs cannot be simply replaced with the galvanometer scanners. Because the deflection of the AOD is produced by diffraction, the scan angle depends on the wavelength. Thus, the signal light with different wavelengths from the excitation beam will be deflected by a different amount on its way back through the AOD and will fail to pass through the pinhole. Although this dispersion effect can be compensated for by a chromatic correction system, such systems are rarely employed because many additional expensive and lossy elements need of be used [1].

The spinning-disk CLSM enables us to increase the imaging speed not by increasing the scanning speed of a single beam but by increasing the number of focal

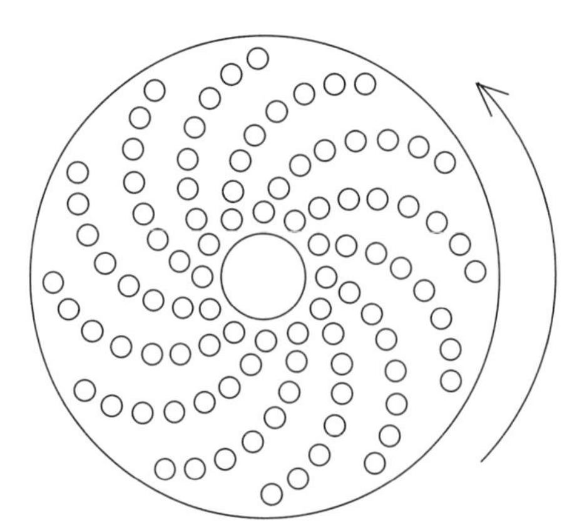

FIGURE 2.2 Nipkow disk.

spots. In the spinning-disk CLSM, rotation of a disk with a series of small holes arranged in an Archimedean spiral of constant pitch produces a raster scanning pattern (Fig. 2.2), thereby simultaneously illuminating many discrete points. The disk is called a Nipkow disk. It was developed by a young student in Berlin, Paul Nipkow, in 1884 as a means of dissecting an image for electromechanical transmission to a receiver. Egger and Petráň first used a Nipkow disc to implement a spinning-disk confocal microscope, in which illumination and detection were performed in tandem through pinholes on opposite sides of the disk [2]. Xiao et al. have designed a real-time spinning-disk confocal microscope using a Nipkow disk with the same pinholes for illumination and detection [3]. Spinning-disk confocal units have been vastly improved by adding microlenses to the pinholes [4,5].

2.1.2 Signal Detection

In CLSM, in order to pass the signal lights through the confocal pinholes in front of the detectors, the signal lights retrace its path back through the scanning mirrors to a dichromatic mirror (DM). DMs can be selected such that the fluorescence emitted light from the sample is divided into each expected signal wavelength band and is preferentially directed into the detectors corresponding to each signal wavelength band. Filters such as band-pass filters, short-pass filters, and long-pass filters are also typically in front of the detectors to further isolate the desired signal. In the single-point scanning method, the signals are detected by point detectors such as photomultiplier tubes (PMTs) and avalanche photodiodes (APDs). PMTs generally have higher internal gains but lower quantum efficiencies than APDs. The active areas of APDs are typically very small. APDs measuring 100 μm in diameter act as both detector and pinhole. Photon counting techniques can be used with PMTs and APDs. For photon counting, the detector circuit threshold is set to give a large response to any excitation by a photon, but to ignore the smaller pulses from internal random events. The photon counting technique effectively reduces dark noise, but requires substantially slower response than direct detection. In the spinning-disk method, imaging detectors such as charged coupled devices (CCDs) and complementary metal oxide semiconductors (CMOSs) are employed. CCDs have higher image uniformities but slower acquisition times than CMOSs.

2.2 FLUORESCENCE TECHNIQUE

2.2.1 Various Fluorescent Molecules and Multi-Color Fluorescence Imaging

Fluorescence microscopy is a versatile technique for visualizing biological phenomena such as the expression of genes, ionic concentration changes, and membrane potential changes. Fluorophores currently used in fluorescence microscopy are divided into small organic dyes, fluorescent proteins, semiconductor nanocrystals ("quantum dots"), and lanthanide chelates [6,7]. Multi-color fluorescence imaging using multiple fluorophores has become a popular way to discriminate among multiple proteins, organelles, or functions in a single cell or animal. Multiple fluorophores with different emission spectra are used to spectrally distinguish multiple biomolecules that overlap spatially. In this section, we discuss several types of fluorophores used in multi-color imaging.

2.2.1.1 Organic Dye Originally, fluorescence microscopy was used to visualize a specific protein or antigen in cells or tissue sections by binding a specific antibody chemically conjugated with an organic dye, as has been summarized in excellent review articles [1,8]. This technique is referred to immuno-fluorescence staining. Small organic dyes for immuno-fluorescence staining include fluorescein, rhodamine, borate-dipyrromethene (BODIPY) complexes [9,10], cyanine (Cy) dyes [11], and Alexa Fluor dyes [12]. However, immuno-fluorescence staining is performed in fixed and permeabilized cells and tissue sections. BODIPY, Cy, and Alexa Fluor dyes have been developed to complement the traditional fluorescein and rhodamine. The Cy and Alexa Fluor dyes have very high water solublility and excellent spectral properties (molar excitation coefficient, fluorescence quantum yield, and photostability). Later organic dyes could directly recognize organelles, nucleic acids, and certain important ions in living cells. Some examples of covalent labeling organic dyes and coupled targeting organic dyes are listed in Tables 2.1 and 2.2, respectively.

DAPI and Hoechst dyes preferentially bind to A-T clusters in DNA, and binding of these dyes to DNA produces an 20-fold fluorescence enhancement. Propidium iodide (PI) and SYTO series bind to DNA and RNA. Reduced MitoTracker® dyes are concentrated inside mitochondria by their negative mitochondrial membrane potential and do not emit fluoresce until they enter mitochondria in live cells, where they are oxidized to the corresponding fluorescent mitochondrion-selective probe and then sequestered in the mitochondria.

There are also selective environmental probes that produce a change of fluorescence intensity or wavelength in response to the interaction of the probe with its

TABLE 2.1 Spectroscopic Properties of Covalent Labeling Organic Dye

Organic Dye	Ex [nm]	Em [nm]	ϵ	QY	References[a]
Fluorescein	494	518	77	0.93	MP
Tetramethyl rhodamine	555	580	84		MP
Rhodamine 6G	525	555	108	0.95	MP
Texas Red	595	615	84		MP
BODIPY FL	505	513	80	0.94	MP
BODIPY R6G	528	550	70		MP
BODIPY TMR	542	574	60		MP
BODIPY TR	589	617	61		MP
BODIPY 630/650	625	640	101		MP
Cy2	489	506	150	>0.12	GE
Cy3	550	570	150	>0.15	GE
Cy3B	558	572	130	0.67	GE
Cy3.5	581	594	150	>0.15	GE
Cy5	649	670	250	>0.28	GE
Cy5.5	675	694	250	>0.28	GE
Cy7	743	767	250	0.28	GE
Alexa Fluor 350	346	442	19		MP
Alexa Fluor 405	401	421	34		MP
Alexa Fluor 430	433	541	16		MP
Alexa Fluor 488	496	519	71	0.92	MP
Alexa Fluor 532	532	553	81	0.61	MP
Alexa Fluor 546	556	573	104	0.79	MP
Alexa Fluor 555	555	565	150	0.1	MP
Alexa Fluor 568	578	603	91	0.69	MP
Alexa Fluor 594	590	617	73	0.66	MP
Alexa Fluor 610	612	628	138		MP
Alexa Fluor 633	632	647	239		MP
Alexa Fluor 647	650	665	239	0.33	MP
Alexa Fluor 660	663	690	132	0.37	MP
Alexa Fluor 680	679	702	184	0.36	MP
Alexa Fluor 700	702	723	192	0.25	MP
Alexa Fluor 750	749	775	240	0.12	MP
Alexa Fluor 790	784	814	270		MP

Note: Ex and Em denote the absorption and emission peak wavelengths, respectively. ϵ is the molar extinction coefficient; multiply listed values by 1000 to convert to units of $[M^{-1}cm^{-1}]$. QY represents Quantum yield.

[a] MP, Molecular Probes; GE, GE Healthcare Lifesciences.

target. The calcium ion (Ca^{2+}) is an important intracellular messenger ion responsible for the activation and deactivation of numerous biological events in cells. The binding of Ca^{2+} to the 1,2-bis(o-aminophenoxy)ethane-*N*,*N*,*N*′,*N*′-tetraacetic acid (BAPTA) targeting group such as the Fura series reconfigures the nitrogen lone pair of electrons, resulting in spectral shifts and increased fluorescence quantum yields. By measuring the spectral and/or intensity changes, the changes in concentration

TABLE 2.2 Spectroscopic Properties of Coupled Targeting Organic Dye

Organic Dye	Ex [nm]	Em [nm]	ϵ	Coupled Target	References[a]
DAPI	358	461	21	DNA	MP
Hoechst 33342	350	461	45	DNA	MP
Propidium iodide(PI)	535	617	5.4	DNA/RNA	MP
SYTO® 16	488	518	42	DNA/RNA	MP
MitoTracker® Green	490	516	119	Mitochondria	MP
MitoTracker® Orange	551	576	102	Mitochondria	MP
MitoTracker® Red	578	599	116	Mitochondria	MP
Rhodamine 123	507	529	101	Mitochondria	MP
LysoTracker® BlueDND − 22	373	422	9.6	Lysosome	MP
LysoTracker® GrrenDND − 26	504	511	54	Lysosome	MP
LysoTracker® RedDND − 99	577	590	48	Lysosome	MP
LysoSensor® GreenDND − 189	443	505	16	Lysosome	MP
Indo-1 (low Ca^{2+})	346	475	33	Calcium	MP
Indo-1 (high Ca^{2+})	330	400	33		
Fura-2 (low Ca^{2+})	363	512	28	Calcium	MP
Fura-2 (high Ca^{2+})	335	505	34		
Fluo-3 (low Ca^{2+})	503	-	92	Calcium	MP
Fluo-3 (high Ca^{2+})	505	526	102		
Fluo-4 (low Ca^{2+})	491	-	82	Calcium	MP
Fluo-4 (high Ca^{2+})	494	516	88		
BCECF (pH 5)	482	520	35	pH	MP
BCECF (pH 9)	503	528	90		
SNARF-1 (pH 6)	548	587	27	pH	MP
SNARF-1 (pH 10)	576	635	48		
SNARF-4F (pH 5)	552	589	27	pH	MP
SNARF-4F (pH 9)	581	652	48		
$DiBAC_4(3)$	493	516	140	Membrane potential	MP
FIERhR	490	520,573	70	cAMP	[13]

Note: Ex and Em denote the absorption and emission peak wavelengths, respectively. ϵ is the molar extinction coefficient; multiply listed values by 1000 to convert to units of $[M^{-1}cm^{-1}]$.
[a] MP, Molecular Probes.

of Ca^{2+} can be monitored. In the case of Fura-2, the Ca^{2+} concentration is determined from the fluorescence intensity ratio at an emission wavelength of 510 nm obtained at two excitation wavelengths of 340 nm and 380 nm. The modified fluorescein BCECF is the most widely used fluorescent indicator for intracellular pH. Intracellular pH measurements with BCECF are made by determining the ratio of fluorescence intensity at an emission wavelength of 535 nm when BCECF is excited at 490 nm versus the fluorescence intensity when excited at its isosbestic point of 440 nm. $DiBAC_4(3)$ is the slow-response potential-sensitive probe. When the cells are depolarized, more DiBAC enters the cells, and the increased concentration of DiBAC binding to intracellular proteins or membrane lipids causes an increase in fluorescence signal. When the cells are hyperpolarized, DiBAC exits the cells and the decreased concentration of DiBAC binding to proteins or membrane lipids results in a decrease of fluorescence signal. Protein-based fluorescent sensors have been developed [13,14]. Adams et al. have reported a fluorescence indicator, FIERhR, of a cyclic adenosine 3,5-monophosphate (cAMP), which is the important intracellular messenger [13]. Richieri et al. have developed a fluorescent indicator, ADIFAB, of fatty acids, which are of considerable importance in nutrition, membrane structure, protein modification, modulation of cell signaling, and eicosanoid formation [14].

2.2.1.2 Fluorescent Proteins Fluorescent proteins (FPs) from jellyfish *Aequorea victoria* and Anthozoa corals have revolutionized noninvasive imaging in living cells and organisms of reporter gene expression, protein trafficking, and many dynamic biochemical signals because they provide genetic encoding of strong fluorescence of a wide range of colors [15–19]. Since the wild type green fluorescent protein (wtGFP) from the jellyfish *Aequorea victoria* was used to highlight sensory neurons in the nematode [20], various researchers have produced new and improved versions that are brighter, cover a broad spectral range, and also exhibit enhanced photostability, reduced oligomerization, pH insensitivity, and faster maturation rates. The wtGFP was quickly modified to produce variants emitting in the blue (BFP), cyan (CFP), and yellow (YFP) regions [21]. Further breakthroughs have come with the cloning of novel GFP-like red fluorescent proteins from Anthozoa coral reef species, Discosoma striata, which is named DsRed [22]. DsRed in tandem with GFPs has enhanced the feasibility of multi-color labeling. DsRed has high photostability, red-shifted fluorescence emission, and stability to pH changes. However, DsRed has certain drawbacks, such as obligate oligomerization and slow or incomplete fluorescence maturation. In particular, the obligate tetramerization of DsRed has greatly hindered its use as a genetically encoded fusion tag. Thus, effective approaches for eliminating some of these limitations have been attempted [23,24]. A monomeric mutant, mRFP1, which contains 3 amino acid substitutions within the hydrophobic interface and 10 within the hydrophilic, has been created and can overcome oligomerization [25]. The next generation of monomers derived from Discosoma species, which are called mHoneydew, mBanana, mOrange, d(dimer)Tomato, tdTomato, mTangerine, mStrawberry, and mCherry, has been reported [26]. dTomato, tdTomato, and mCherry are over 10-fold more photostable than mRFP1. mOrange, dTomato, and tdTomato have higher quantum efficiencies. However, the major drawback of tdTomato is its larger size, which may interfere with fusion protein packing in some biopolymers. Further extension of the mFruit spectral class through iterative somatic hypermutation has yielded a far-red fluorescent protein, mPlum, with increased photostability [27]. In addition, a bacteriophytochrome from *Deinococcus radiodurans* was engineered into monomeric infrared-fluorescent proteins (IFPs), which extend the color palette into the near-infrared [28].

The current color palette of FPs is shown in Table 2.3 . The practical FP in ultramarine is Sirius, which has high photo and pH stability [33]. EBFP2 [36] and Azurite [37]

TABLE 2.3 Spectroscopic Properties of Fluorescent Proteins

Protein	Ex [nm]	Em [nm]	ϵ	QY	B	*pKa*	Association State	References
				Ultramarine				
Sirius	355	424	15	0.24	3.6	<3	Monomer	[33]
				Blue				
SBFP2	380	446	34	0.47	16	5.5	Weak dimer[a]	[34]
EBFP	377	446	30	0.15	4.5	6.3	Weak dimer[a]	[35]
EBFP2	383	448	32	0.56	18	5.3	Weak dimer[a]	[36]
Azurite	384	450	22	0.59	13	5	Weak dimer[a]	[37]
mKalama1	385	456	36	0.45	16	5.5	Monomer	[36]
TagBFP	402	457	52	0.63	33	2.7	Monomer	[38]
mBlueberry2	402	467	51	0.48	25	<2.5	Monomer	[36]
				Cyan				
mTurquoise	434	474	30	0.84	25.2	4.5	Monomer	[39]
mTurquoise2	434	474	30	0.93	27.9	3.1	Monomer	[40]
ECFP	433	475	32.5	0.4	13	4.7	Weak dimer[a]	[30]
mCFP	433	475	32.5	0.4	13	4.7	Monomer	[41]
SCFP3A	433	474	30	0.56	17	4.5	Monomer	[42]
Cerulean	433	475	43	0.62	27	4.7	Weak dimer[a]	[43]
mCerulean3	433	475	40	0.87	35		Monomer	[44]
CyPet	435	477	35	0.51	18	5	Weak dimer[a]	[45]
AmCyan1	458	489	44	0.24	11		Tetramer	[22]
MiCy	472	495	27.3	0.9	25	6.6	Dimer	[46]
TagCFP	458	480	37	0.57	21	4.7	Monomer	Evrogen
mTFP1	462	492	64	0.85	54	4.3	Monomer	[47]
				Green				
TagGFP	482	505	58.2	0.59	34	4.7	Monomer	[48]
mAG1	492	505	41.8	0.81	34	6.2	Monomer	[49]
AG	492	505	72.3	0.67	48	<5.0	Tetramer	[49]
TagGFP2	483	506	56.5	0.6	34	5	Monomer	Evrogen
EGFP	488	507	56	0.6	34	6	Weak dimer[a]	[50]
mWasabi	493	509	55	0.8	56	6.5	Monomer	[51]
Emerald	487	509	57.5	0.68	37	6	Weak dimer[a]	[21]
Sapphire	399	511	25	0.64	16	4.9	Weak dimer[a]	[52]
T-Sapphire	399	511	44	0.6	26	4.9	weak dimer[a]	[52]
				Yellow				
TagYFP	508	524	64	0.62	40	5.5	Monomer	Evrogen
mAmetrine1.2	408	525	31	0.59	18	5.8	Monomer	[53]
mAmetrine	406	526	45	0.58	26	6	Monomer	[54]
Topaz	514	527	94.5	0.6	57		Monomer	[55]
EYFP	515	528	80.4	0.61	49	6.9	Weak dimer[a]	[56]
Venus	515	528	92.2	0.57	53	6	Weak dimer[a]	[57]
SYFP2	515	527	101	0.68	67	6	Monomer	[42]
mCitrine	516	529	77	0.76	59	5.7	Monomer	[58]
YPet	517	530	104	0.77	80	5.6	Weak dimer[a]	[45]
TurboYFP	525	538	105	0.53	56	5.9	Dimer	Evrogen

TABLE 2.3 ***(Continued)***

Protein	Ex [nm]	Em [nm]	ϵ	QY	B	*pKa*	Association State	References
				Orange				
mBanana	540	553	6	0.7	4	6.7	Monomer	[26]
mKO	548	559	51.6	0.6	31	5	Monomer	[59]
mOrange	548	562	71	0.69	49	6.5	Monomer	[26]
mKO2	551	565	63.8	0.62	40	5.5	Monomer	[60]
mOrange2	549	565	58	0.6	35	6.5	Monomer	[61]
LSSmOrange	437	572	52	0.45	23.4	5.7	Monomer	[62]
				Red				
TurboRFP	553	574	92	0.67	62	4.4	Dimer	[63]
dTomato	554	581	69	0.69	48	4.7	Dimer	[26]
tdTomato	554	581	138	0.69	95	4.7	Dimer	[26]
DsRed	558	583	75	0.79	59	4.7	Tetramer	[22]
TagRFP	555	584	100	0.48	48	3	Monomer	[63]
TagRFP-T	555	584	81	0.41	33	4.6	Monomer	[61]
mTangerine	568	585	38	0.3	11	5.7	Monomer	[26]
mApple	568	592	75	0.49	37	6.5	Monomer	[61]
mStrawberry	574	596	90	0.29	26	<4.5	Monomer	[26]
TurboFP602	574	602	74.4	0.35	26	4.7	Dimer	Evrogen
mRuby	558	605	112	0.35	39	4.4	Monomer	[64]
LSS-mKate1	463	624	31.2	0.08	2.5	3.2	Monomer	[65]
LSS-mKate2	460	605	26	0.17	4.4	2.7	Monomer	[65]
mRFP1	584	607	50	0.25	13	0.5	Monomer	[25]
J-Red	584	610	44	0.2	9	5	Dimer	[66]
mCherry	587	610	72	0.22	16	<4.5	Monomer	[26]
dKeima-Red	440	620	24.6	0.31	8	6.5	Dimer	[67]
mKeima-Red	440	620	14.4	0.24	3	6.5	Monomer	[67]
				Far Red				
mRaspberry	598	625	86	0.15	13		Monomer	[27]
mKate2	588	633	62.5	0.4	25	5.4	Monomer	[68]
TagFP635	588	635	45	0.33	15	6	Monomer	[69]
mPlum	590	649	41	0.1	4	<4.5	Monomer	[27]
eqFP650	592	650	65	0.07	4.6	5.7	Dimer	[70]
Neptune	600	650	72	0.18	13	5.8	Dimer	[71]
mNeptune	600	650	67	0.2	13	5.4	Monomer	[71]
AQ143	595	655	90	0.04	4		Tetramer	[72]
eqFP670	605	670	70	0.03	2.1	4.5	Dimer	[70]
				Infrared				
IFP1.4	684	708	92	0.07	6.4	4.6	Monomer	[28]
IRFP	690	713	105	0.059	6.2	4.0	Monomer	[73]

Note: Ex and Em denote the absorption and emission peak wavelengths, respectively. Brightness (B $[mM^{-1}cm^{-1}]$) indicates the product of molar extinction coefficient (ϵ $[mM^{-1}cm^{-1}]$) and quantum yield (QY)

[a] Can be made monomeric with A206K mutation.

in monomeric blue FPs are characterized by enhanced photostability and are clearly much better than EBFP. Compared with these blue FPs, the recently reported Tag-BFP possesses superior brightness and high photostability [38]. In monomeric cyan FPs, mTurquoise2 [40], mCerulean3 [44], and TagCFP (Evrogen) are characterized by enhanced brightness and/or maturation rate. Compared with these cyan FPs, the mTFP1 demonstrates higher brightness and photostability [47]. In monomeric green FPs, EGFP has most of desirable characteristics [50,41], and mWasabi is brighter at the expense of some photo- and pH stability [51]. The practical FPs in yellow are TagYFP (Evrogen), Topaz [55], Venus [57], SYFP2 [42], and Citrine [58] because EYFP is characterized by low pH stability and high sensitivity of halide ions [56]. mKO [59] and mOrange [26] are bright variants in monomeric orange FPs. The enhanced version, mKO2, matures faster than mKO but seems to be accompanied by lower pH stability [60]. Although the enhanced variant, mOrange2 matures slower than mOrang, mOrange2 demonstrates higher photostable than mOrange [61]. In monomeric red FPs, mStrawberry [26] and mRuby [64] win in brightness but lose in photostability to mCherry [26]. The brightest monomeric red FP is TagRFP [63], which yields to enhanced TagRFP-T in photostability [61]. The useful monomers in far red FPs include mRaspberry [27], mKate2 [68], mPlum [27], and mNeptune [71]. In particular, mKate2 possesses higher brightness and photostability [68].

Sapphire [52], T-Sapphire [52], mAmetrine [54], mAmetrine1.2 [53], LSSmOrange [62], LSS-mKate2 [65], and mKeima [67] have a large difference between excitation and emission maxima (Stokes Shift). Thus, a combination of these FPs with regular FPs will excite simultaneously multi-color FPs by a single wavelength [67]. The simultaneous excitation by a single wavelength simplifies multi-color imaging [65,67] and fluorescence cross-correlation spectroscopy (FCCS) (see Section 2.2.5 about FCCS) [62,67]. In addition, the FP with a large Stokes shift is applied as a donor of a fluorescence resonance energy transfer (FRET) (see Section 2.2.2 about FRET) pair with a significant spectral overlap of the donor emission spectrum with the acceptor absorption spectrum without excitation crosstalk [53,54,62].

In these GFP-like proteins, internal chromophores are formed without requiring accessory cofactors, external enzymatic catalysis or substrates other than molecular oxygen by spontaneous cyclization and oxidation of three amino acids buried at the heart of the 2.4- by 4-nm beta barrel [29]. This property gives GFP-like proteins many advantages over organic dye probes. These advantages include the generation of the chromophores in live organisms, tissues, or cells, while maintaining their integrity [20], as well as molecular, organelle, or tissue targeting and specificity [30]. GFP-like proteins are also useful for understanding folding pathways and the thermodynamics of intermediates of predominantly β-folded proteins because of their β-can structures [31]. GFP-like proteins are increasingly employed as quantitative, genetically encoded reporters for second messengers, intracellular chemical environments, protein–protein interactions and protein and cell tracking [32,122].

In these approaches, timer FPs, which change fluorescence color with time and are characterized by various maturation rates (from minutes to hours and even days), have been developed [74,75]. Timer FPs are suitable for investigations of processes on a variety of time scales. The first Timer FP was named DsRed-E5 [74]. In DsRed-E5, some of the chromophores within a tetramer mature to the green GFP-like form instead of a red form. And then, their green fluorescence dominates until slower maturing red chromophores, which steal the excitation energy away from the green ones through an intratetrameric FRET [76], appear. Thus, DsRed-E5 produces green fluorescence within several hours after synthesis, but later generates red fluorescence. Therefore, the ratio of red to green fluorescence is proportional to the age of expressed protein and indicates the time since corresponding promoter activation [77–82]. Recently, monomeric mCherry-based timer FPs named fast-FT, medium-FT, and slow-FT, which change fluorescent color from blue to red, have been developed [75]. Because monomeric timer FPs simultaneously track age and localization of proteins of interest in living cells, we can observe the pathways of trafficking of newly produced proteins.

Fluorescent sensors that are based on FRET between a donor FP and an acceptor FP [32] have been also developed to monitor the concentration of Ca^{2+} [83–86], cyclic nucleotides [87–90], glutamate [91], tryptophan [92], specific saccharides [93], GTPase activities [94,95], kinase activities [96–105], activities of proteases [106–115], mechanical stress [116], and membrane potential [117–121]. The FRET-based sensors can detect dynamic conformational changes of protein domains, which are sandwiched between the two FPs, with spatiotemporal resolution because circular permutation of one of the FPs or small changes in linker length alter the FRET response as shown in Fig. 2.3(a) [83]. For example, Cameleon, where the calmodulin-M13 fusion is sandwiched between a BFP and a GFP or between CFP and YFP, is

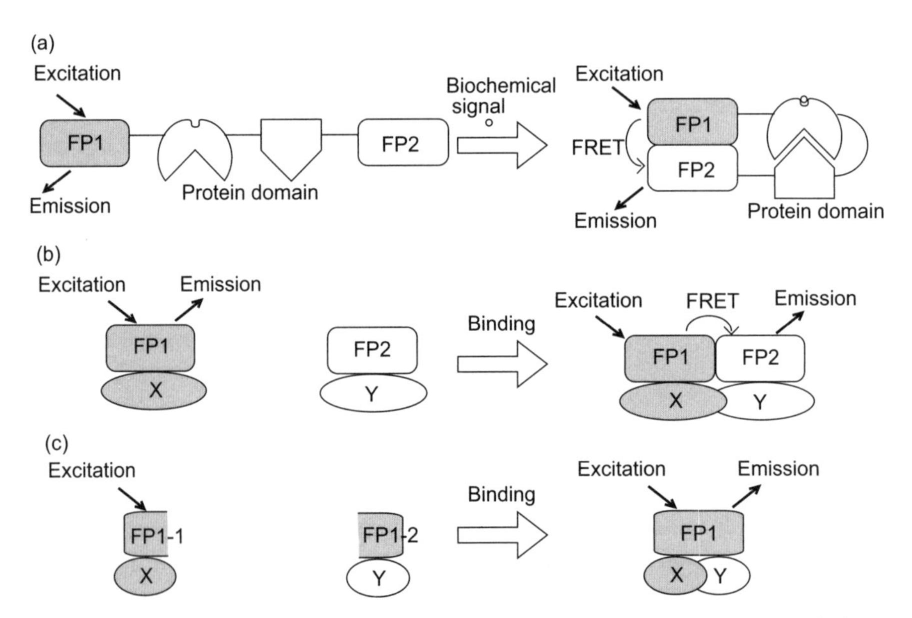

FIGURE 2.3 FRET-based detection of conformational changes (a) and protein–protein interactions (b). (c) Detection of protein–protein interactions using BiFC. FP indicates fluorescent proteins.

an indicator for Ca^{2+} [83]. The protein linking the FPs can be designed to change conformation in response to important biochemical signals. These sensors can be targeted to specific subcellular compartments when fused to an appropriate localization signal. FRET can also detect dynamic protein–protein interaction in live cells of ectopically expressed FP-tagged proteins (X and Y) as shown in Fig. 2.3(b) because binding of X to Y leads to the FRET response.

FPs split at appropriate sites can fold and reconstitute the chromophore when the two halves are fused to interacting partners as shown in Fig. 2.3(c) [123]. This technique is named bimolecular fluorescence complementation (BiFC). BiFC has a high signal-to-background ratio, because it generates new fluorescence rather than modulating existing fluorescence. Multiple protein–protein interactions can be investigated in parallel employing spectrally distinct split FPs [124]. However, BiFC is slow (hours to days) and irreversible, and the geometrical and affinity requirements for the protein–protein interaction have not yet been characterized. BiFC can be used to study gene expression of at least two promotors [125]. Self-assembling fragments of GFP have also been reported [126], in which the two fragments only have to exist in the same compartment to generate fluorescence.

2.2.1.3 Quantum Dots Semiconductor quantum dots (QDs) are nanocrystals that have attracted widespread interest in biology and medicine due to their unique optical and electronic properties [127,128]. QDs typically consists of a spherical core of the semiconductor cadmium selenide (CdSe) surrounded by a zinc sulfide (ZnS) shell, which is in turn surrounded by a hydrophilic polymer surface coating [129]. The ZnS shell has a functional role in not only stabilization of the electronic excited state but also inhibition of release of cytotoxic cadmium from the core [130]. The hydrophilic coating provides water solubility, prevents quenching by water, and allows conjugation to protein-targeting molecules (antibodies, streptavidin, and other targeting groups) [127,128]. The size of QDs conjugated to biomolecules is about 10 to 30 nm. The large size prevents efficient traversal of intact membranes, which restricts their use to permeabilized cells or extracellular or endocytosed proteins. Development of intracellular delivery technique using peptide vectors has been reported by Mattheakis et al. [131]. Although targeting of QDs currently remains challenging, QDs have been developed with outstanding spectroscopic properties, including intense and stable fluorescence for a longer time. QDs have large molar extinction coefficients (10 to 100 times those of small fluorophores and FPs), and good quantum yields. The fluorescence brightness of a single QD has been estimated to be approximately equivalent to 20 rhodamine dye molecules [129]. Because of their predominantly inorganic composition, QDs are resistant to the oxidative photobleaching reactions that affect organic dyes and FPs. The excitation spectra of QDs are essentially continuous and extend from ultraviolet wavelengths up to just below the emission wavelength, whereas the emission spectra are narrow and the emission range increases with increasing the diameter of the semiconductor core. Thus, a single excitation wavelength readily excites QDs of multiple emission maxima simultaneously [132,133]. QD based multiplexed approach have been devoted to the simultaneous identification of many biomarkers.

2.2.1.4 Lanthanide Chelate Aqueous solutions of lanthanide metal ion produce a weak fluorescence. When their ions form a chelate with the appropriate ligands, the chelates absorb light from the near-ultraviolet region, become excited, and emit an extremely strong fluorescence. This is because they give off luminescence based on energy transfer to the central metal ion from the ligand of the chelate. On the one hand, lanthanide chelates have microsecond fluorescence lifetimes that are readily separated from nanosecond autofluorescence background [135–137]. On the other hand, the long decay times of lanthanide chelates result in long image acquisition time in scanning microscopes because the pixel dwell time must be longer than the decay time. Thus, lanthanide chelates will probably be used with wide-field imaging where the exposure time is long compared to its fluorescence lifetime and discrimination from the fast decay of the autofluorescence is still useful. The fluorescence properties of lanthanide chelates are dependent on electron energy transfer levels in an atom rather than those of a molecule, so they are very resistant to photodamage. Because the Stokes shifts of Lanthanide chelates are large, often above 250 nm, they do not experience the concentration quenching (self-quenching) seen in organic fluorescent dyes. The emission spectra of lanthanide chelates are relatively weak and broad. However, time-resolved detection yields excellent signal-to-noise ratio even from these low intensities. Currently, lanthanide chelates have been frequently employed as the oxygen sensor [138].

2.2.1.5 Multi-Color Fluorescence Imaging As mentioned above, the current color palette of fluorophores ranges from ultramarine to near infrared. Multi-color fluorescence imaging using multiple fluorophores with different emission spectra enables us to spectrally discriminate between multiple proteins, organelles, or functions in a single cell. Multi-color fluorescence imaging has been achieved by color separation of fluorescence emission using a set of optical band-pass filters that select different parts of the emission signals for every image channel. In this approach, we assume that the emission spectra of the different fluorophores do not overlap in the selected wavelength ranges. In practice, the emission spectra of the available fluorophores are too broad to choose filter combinations that can eliminate spectral overlap. If fluorescence signals are completely separated without spectral overlap, the significant fluorescence signals are often lost. In order to avoid the loss of fluorescence signals, a complete separation of the fluorescence signals is performed using a linear unmixing technique that is based on a mathematical algorithm.

Let's describe the principle of the linear unmixing technique. The fluorescence spectrum $S(\lambda)$ from samples labeled by N fluorophores is expressed as

$$S(\lambda) = \sum_{\mathrm{j}}^{N} a_{\mathrm{j}} R_{\mathrm{j}}(\lambda), \qquad (2.2.1)$$

where a_j and $R_j(\lambda)$ are the mixing ratio and the measured reference spectrum of the jth fluorophores, respectively. It is a linear algebra problem to solve for the weighting matrix **a**, and the solution is usually obtained with an inverse least-squares procedure that minimizes the difference between the measured and modeled spectrum [141],

$$\chi = \sum_{m=1}^{N_p} \left\{ S(\lambda_m) - \sum_{j=1}^{N} a_j R_j(\lambda_m) \right\}^2, \tag{2.2.2}$$

where the spectrum is measured by a N_p channel detector.

2.2.2 Fluorescence Resonance Energy Transfer Imaging

Fluorescence resonance energy transfer (FRET), which provides information about protein–protein interactions and protein conformal changes, is a nonradiative process in which the excited state energy from one excited fluorophore (donor) is directly transferred to a second fluorophore (acceptor) by means of intermolecular long-range dipole–dipole coupling [142]. FRET imaging is being increasingly used in the biological and biomedical sciences to study a wide variety of cellular and organismal phenomena [143–152]. According to Förster theory, the nonradiative dipole–dipole interaction-induced energy transfer is much slower than the vibrational relaxation. As shown in Fig. 2.4(a), the acceptor transits from the ground state to the excited state simultaneously with return of the donor to the ground state. When a single donor and a single acceptor are separated by a distance r, the rate of resonance energy transfer k_T can be expressed as [153]

$$k_T(r) = \frac{9(\ln 10) Q_D \kappa^2}{128\pi^5 N n^4 \tau_D r^6} \int_0^\infty I_D(\lambda)\varepsilon_A(\lambda)\lambda^4 d\lambda, \tag{2.2.3}$$

where Q_D is the quantum yield of the donor molecule in the absence of the acceptor molecule, n is the refractive index of the medium, N is Avogadro's number, τ_D is the lifetime of the donor molecule in the absence of the acceptor molecule, $I_D(\lambda)$ is the corrected emission spectrum of the donor with the total intensity normalized to unity, $\varepsilon_A(\lambda)$ is the excitation spectrum of the acceptor, and λ is the wavelength. κ is

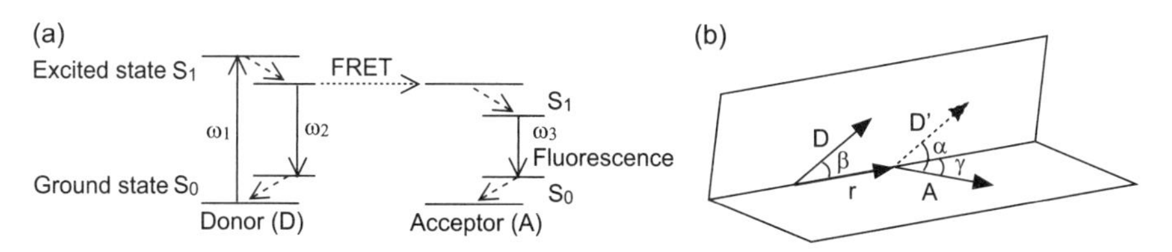

FIGURE 2.4 (a) Energy diagram illustrating the FRET process. (b) The orientation of the emission dipole of the donor (D) and the absorption dipole of the acceptor (A).

the orientation factor that describes the relative orientation of the transition dipole moments of the donor molecule and the acceptor molecule, and is expressed as

$$\kappa = (\cos\alpha - 3\cos\beta\cos\gamma)^2. \tag{2.2.4}$$

Here α is the angle between the transition moments of the donor and the acceptor, and β and γ are the angles between the line joining the centers of the fluorophores and the transition moments of the donor and acceptor, respectively (Fig. 2.4(b)). When both fluorophores are freely rotating and can be considered to be isotropically oriented during the excited state lifetime, κ^2 is often assumed to be 2/3. The efficiency of resonance energy transfer E_{FRET} is described by

$$E_{\mathrm{FRET}} = \frac{k_T}{\tau_D^{-1} + k_T} = \frac{R_0^6}{R_0^6 + r^6}. \tag{2.2.5}$$

Here we have introduced the quantity

$$R_0^6 = \frac{9(\ln 10) Q_D \kappa^2}{128\pi^5 N n^4} \int_0^\infty F_D(\lambda)\varepsilon_A(\lambda)\lambda^4 d\lambda, \tag{2.2.6}$$

where R_0 is called the Förster distance and is defined as the donor-acceptor distance at which the transfer efficiency is 50% [154–156]. Some examples are shown in Table 2.4. A high efficiency of energy transfer requires four conditions that (1) the emission spectrum of the donor has significant overlaps with the absorption spectrum of the acceptor, (2) the two fluorophores are within 1 to 10 nm of each other, (3) the donor emission dipole moment, the accepter absorption dipole, and their separation vectors are in a favorable mutual orientation, and (4) the emission of the donor has a high or equal quantum yield as the acceptor. Because of these characteristics, FRET provides spatial and temporal information about protein–protein interactions and protein conformal changes.

Basically, the FRET sample is double-labeled by donor and acceptor molecules, and the sample is then excited with the donor excitation wavelength. However, FRET is never observed directly. If the above-mentioned four conditions are satisfied, the donor fluorescence is quenched due to the transfer of excitation energy and the acceptor fluorescence is increased. Thus, the FRET signal is indirectly obtained from the fluorescence intensities of the donor and the acceptor or that of one at least. Although donor–acceptor pairs with a significant spectral overlap of the donor emission spectrum with the absorption spectrum of the acceptor give high FRET efficiencies, the emission band of the donor extends into the emission band of the acceptor (emission cross talk), and the absorption band of the acceptor extends into the absorption band of the donor (excitation cross talk). Both of these signals are termed spectral bleed-through (SBT) signal into the acceptor channel. The concentration ratio of donor–acceptor fluorophores present in living cells and the altered localization of fluorophores will also influence the measurement of FRET. Therefore, it is difficult to sense the FRET signal accurately. We discuss the methods for the measurement of FRET in the following sections.

TABLE 2.4 Donor/acceptor pairs

Donor	Acceptor	Förster Distance [nm]	References
Fluorescein	Tetramethylrhodamine	5.6	D. Kosk-Kosicka et al., 1989 [144]
Cy3	Cy5	5	P. I. H. Bastiaens et al., 1996 [145]
Alexa488	Alexa555	7	M. Elangovan et al., 2003 [146]
BFP	GFP	4.1	A. Miyawaki et al., 1997 [83]
BFP	YFP	3.8	R. N. Day et al., 2003 [148]
CFP	YFP	4.9	A. Miyawaki et al., 1997 [83]
CFP	DsRed	4.2	M. G. Erickson et al., 2003 [149]
CFP	mRFP1	3.8	E. Galperin et al., 2004 [150]
GFP	DsRed	4.7	M. G. Erickson et al., 2003 [149]
GFP	mRFP1	4.7	M. Peter et al., 2005 [151]
GFP	mCherry	5.1	L. Albertazzi et al., 2009 [152]
YFP	mRFP1	4.9	E. Galperin et al., 2004 [150]

2.2.2.1 Ratio Method FRET can be determined by ratio imaging [83]. In this approach, only the donor is excited, while the emission of the donor and acceptor are measured simultaneously. A change of the ratio of the fluorescence intensities of the donor and acceptor represents a FRET change. Thus, changes of the fluorescence intensities with the concentration change of fluorophores are corrected. However, no correction of emission and excitation crosstalk is performed to obtain pure donor and acceptor fluorescence intensities. Therefore, the FRET ratio method is usually a qualitative measurement. Nevertheless, this technique is a very sensitive method of detecting FRET changes with milisecond temporal resolution.

2.2.2.2 Correction Method of Emission and Excitation Cross Talk by Three-Cube FRET Imaging Three-cube FRET imaging, in which three images of the interest sample double-labeled by a pair of donor and acceptor (f) are recorded not only with the acceptor channel during donor excitation (AD_{f}) but also with the acceptor channel during acceptor excitation (AA_{f}) and the donor channel during donor excitation (DD_{f}), can be used to correct for the emission and excitation crosstalk in the FRET measurements [157–163]. Assuming that the ratio of cross talk and fluorescence intensity is constant, the emission and excitation cross talk in the measurement of FRET can be corrected by the simple arithmetic processing where the net FRET signal (AD_{nf}) is calculated as follows [157–159]:

$$AD_{\mathrm{nf}} = AD_{\mathrm{f}} - r_{\mathrm{d}} DD_{\mathrm{f}} - r_{\mathrm{a}} AA_{\mathrm{f}}, \qquad (2.2.7)$$

where the emission cross-talk coefficient r_{d} and the excitation cross-talk coefficient r_{a} are expressed by $AD_{\mathrm{d}}/DD_{\mathrm{d}}$ and $AD_{\mathrm{a}}/AA_{\mathrm{a}}$, respectively. AD, DD, and AA are the fluorescence intensity of the acceptor channel during donor excitation, that of the donor channel during donor excitation, and that of the acceptor channel during acceptor excitation, respectively. The indexes f, d, and a denote samples labeled by a pair of donor and acceptor, by the donor, and by the acceptor, respectively. AD_{d} and DD_{d} are the fluorescence intensity of the acceptor channel during donor excitation and that of the donor channel during donor excitation, respectively, which are obtained from two fluorescence images of a sample single-labeled by the donor (d), while AD_{a} and AA_{a} are the fluorescence intensity of the acceptor channel during donor excitation and that of the acceptor channel during acceptor excitation, respectively, which are acquired from two fluorescence images of a sample single-labeled by the acceptor (a).

This algorism is based on the assumption that the ratio of crosstalk and fluorescence intensity is constant. In wide-field FRET microscopy, it is true that the ratio is constant and is independent of the fluorescence intensity level [163]. In confocal microscopy and multiphoton excited fluorescence microscopy; however, the ratio of cross talk and fluorescence intensity is dependent on different fluorescence intensity level [160–163]. Elangovan et al. [160] and Chen et al. [161,162] have developed the intensity range based correction method, in which the net FRET signal is obtained as follows:

$$AD_{\mathrm{nf}} = AD_{\mathrm{f}} - \mathrm{DSBT} - \mathrm{ASBT}, \tag{2.2.8}$$

where

$$\mathrm{DBST} = \sum_{j=1}^{k} \sum_{p=1}^{n} DD_{\mathrm{f}(p)} r_{\mathrm{d}(j)}, \tag{2.2.9}$$

$$r_{\mathrm{d}(j)} = \frac{\sum_{i=1}^{i=m} AD_{\mathrm{d}(i)}}{\sum_{i=1}^{i=m} DD_{\mathrm{d}(i)}}, \tag{2.2.10}$$

$$\mathrm{ABST} = \sum_{j=1}^{k} \sum_{p=1}^{n} AA_{\mathrm{f}(p)} r_{\mathrm{a}(j)}, \tag{2.2.11}$$

$$r_{\mathrm{a}(j)} = \frac{\sum_{i=1}^{i=m} AD_{\mathrm{a}(i)}}{\sum_{i=1}^{i=m} AA_{\mathrm{a}(i)}}. \tag{2.2.12}$$

Here j is the jth range of intensity, $r_{\mathrm{d}(j)}$ and $r_{\mathrm{a}(j)}$ are the donor and acceptor bleed-through ratios for the jth intensity range, m is the number of pixel in DD_{d} and AA_{a} for the jth range, $DD_{\mathrm{d}(i)}$ and $AA_{\mathrm{a}(i)}$ are the intensities of pixel i,n is the number of pixel in DD_{f} and AA_{f} for the jth range, $DD_{\mathrm{f}(p)}$ and $AA_{\mathrm{f}(p)}$ are the intensities of pixel p, k is the number of range, and DSBT and ASBT are the total donor and acceptor SBTs, respectively.

Conventionally, the efficiency of resonance energy transfer is calculated by "ratioing" the donor image in the presence I_{DA} and absence I_{D} of acceptor. With using the algorism as described, we indirectly obtained the I_{D} image by using the FRET image [160]. I_{D} is obtained by adding the F_{nf} to the donor image in the presence of acceptor I_{DA}. Thus, the efficiency of resonance energy transfer E_{FRET} can be written as

$$E_{FRET} = 1 - \frac{I_{DA}}{I_{DA} + AD_{nf}}. \tag{2.2.13}$$

However, we must correct the variation of the detector spectral sensitivity of donor and acceptor channels, and the donor and acceptor quantum yields with the net FRET signal. Therefore, the efficiency of resonance energy transfer EFRET can be acquired from [160]

$$E_{FRET} = 1 - \frac{I_{DA}}{I_{DA} + AD_{nf}\,(g_D\psi_D/g_A\psi_A)\,(q_D/q_A)}, \tag{2.2.14}$$

where g_D and g_A are the detector gains of the donor and acceptor channels, ψ_D and ψ_A are the spectral sensitivities of the donor and acceptor channels, and q_D and q_A are the donor and acceptor quantum yields, respectively.

2.2.2.3 Acceptor Photobleaching Method Photobleaching methods, in which either the donor or the acceptor are selectively photobleached, provide quantitating energy transfer efficiencies in cells [164,165]. In the acceptor photobleaching method, destruction of the acceptor leads to an increase in the fluorescence intensity of the donor. The difference between the quenched donor and the unquenched donor is used to calculate the efficiency of resonance energy transfer. In the experiment, the efficiency of resonance energy transfer EFRET can be obtained directly from

$$E_{FRET} = 1 - \frac{I_{DA}}{I_D}, \tag{2.2.15}$$

where I_{DA} is the fluorescence intensity of the quenched donor in the presence of acceptor and I_D is the fluorescence intensity of the unquenched donor after photobleaching of the acceptor. This approach is influenced by a little emission and excitation crosstalk. Thus, the photobleaching method is usually a qualitative measurement. However, it is generally not suitable for live cell imaging because during the photobleaching time, the cells themselves may move and cause image shifting between images collected before and after photobleaching. In addition, the acceptor photobleaching method does not provide the distinction between FRET and reabsorption where fluorescence emitted from the donor is absorbed by the acceptor or the other molecules in a sample, since the fluorescence intensity of the donor is also increased by inhibiting reabsorption with photobleaching of the acceptor. Because this fluorescence intensity change is independent of the distance between the donor and acceptor, favorable information cannot be obtained. To avoid reabsorption, we recommend the observation of samples with low-concentration fluorophores. Nevertheless, acceptor photobleaching is a simple method for qualitative FRET analysis.

2.2.2.4 Donor Photobleaching Method The donor photobleaching method is based on the measurement of the photobleaching rate of the donor [166,167]. The photobleaching rate depends on the excited state's lifetime, which is shortened by the energy transfer. This reason is discussed below. Photobleaching is mainly induced by photodynamic interactions between excited fluorophores and molecular oxygen (O_2) in a triplet ground state in the environment [168–171]. If the fluorophore has a relatively high quantum yield for intersystem crossing, a significant number of fluorophores may cross from a singlet excited state S^* to a long-lived triplet excited state T^*, which permits these fluorophores to interact with their environment for a much longer time (milliseconds instead of nanoseconds). Interactions between O_2 and fluorophores in the long-lived triplet excited state may change triplet oxygen to singlet oxygen according to $T^*{+}3O_2 \rightarrow S{+}^1O_2$. Singlet oxygen has a longer lifetime than the excited triplet states of the fluorophores. Moreover, several types of oxygen free radicals that oxidize the fluorophores can be created when it decays. A fluorophore in the excited triplet state is also highly reactive and may undergo irreversible chemical reactions with other intracellular organic molecules. Because of these chemical reactions, fluorophores are photobleached. The photobleaching rate of a fluorophore is proportional to the quantum yield for the intersystem crossing, which is proportional to the excited state lifetime. Therefore, the photobleaching rate is proportional to the excited state lifetime. Because FRET shortens the excited state lifetime of the donor, the photobleaching rate is decreased by FRET. In the experiment, the efficiency of resonance energy transfer E_{FRET} can be acquired by

$$E_{\mathrm{FRET}} = 1 - \frac{\tau_{\mathrm{BD}}}{\tau_{\mathrm{BDA}}}, \tag{2.2.16}$$

where τ_{BD} and τ_{BDA} are the time constants of bleaching in the absence and presence of the acceptor, respectively. Here the sample labeled only by the donor needs to be prepared as proper controls. The donor photobleaching method is not influenced by reabsorption of donor fluorescence. However, the acceptor may be directly photobleached by the excitation light for photobleaching of the donor. Then, the breaching rate of the donor cannot be accurately evaluated. Thus, the donor photobleaching method requires the use of the acceptor that has strong resistance to photobleaching.

2.2.2.5 Donor Fluorescence Lifetime Method FRET shortens the fluorescence lifetime of the donor. The efficiency of resonance energy transfer EFRET is obtained by

$$E_{\mathrm{FRET}} = 1 - \frac{\tau_{\mathrm{DA}}}{\tau_{\mathrm{D}}}, \tag{2.2.17}$$

where τ_{D} and τ_{DA} are the fluorescence lifetime in the absence and presence of the acceptor. Here we require the sample to be labeled only by the donor as the proper control. In practice, because the free donor and the donor bound with an acceptor usually coexist in the observation volume of the sample double-labeled by the donor and acceptor, the double-label lifetime data exhibit two different lifetimes. Thus, the lifetime of the donor in the absence of an acceptor can be obtained from the double-label lifetime results. This approach is not influenced by the fluorescence intensity changes, which is induced by the concentration of fluorophores and the altered localization of fluorophores, and by reabsorption of donor fluorescence. Thus, the FRET efficiency is accurately and qualitatively measured. Therefore, this approach allows a most precise estimation of the distance between the donor- and acceptor-labeled proteins in the FRET measurement techniques. The measurement method of the fluorescence lifetime will be discussed in Section 2.2.3.

2.2.3 Fluorescence Lifetime-Resolved Imaging

Fluorescence of fluorophores is not only characterized by the emission spectrum but also by the fluorescence lifetime, which provides information on the molecular microenvironment of a fluorophore. The excited state of a fluorophore is depleted by energy transfer between fluorophores and their local environment and between different fluorophores with factors such as ionic strength, hydrophobicity, oxygen concentration, binding to macromolecules, and the proximity of molecules. Therefore, fluorescence lifetime imaging microscopy (FLIM) can be used for probing separation of fractions of the same fluorophores in different binding states to lipids, proteins, or DNA, and distances on the nanometer scale by FRET [173].

Figure 2.5 shows a simplified version of a Jablonski diagram describing the energy states and relaxation processes of fluorophores. By absorption of a photon of the transition energy between the ground state S_0 and the first excited state S_1, the fluorophore transits into the S_1 state. In condensed phases, after absorption, almost all fluorophores rapidly relax to the lowest vibrational state of the excited state S_1, from which fluorophores return to the ground state S_0 via one of several decay processes. Without interaction with its environment, the fluorophore can be returned from the S_1 state by emitting fluorescence (FL) with a rate of k_{FL}, or by nonradiative process such as internal conversion (IC) and intersystem crossing (ISC) of the absorbed energy

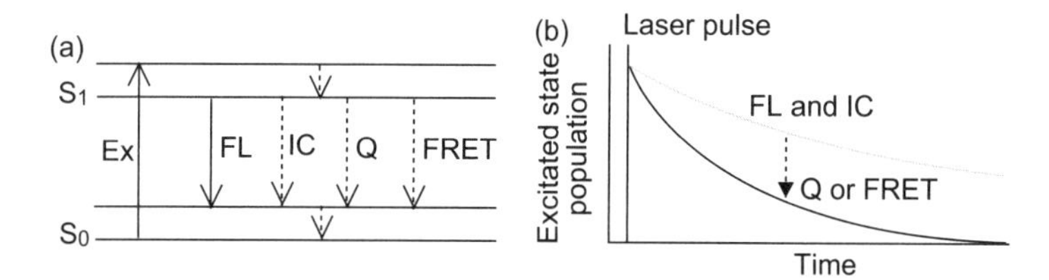

FIGURE 2.5 (a) Fluorophore excitation and decay paths: Ex, excitation; FL, fluorescence; IC, internal conversion; Q, quenching; FRET, fluorescence resonance energy transfer. (b) Fluorescence intensity decay.

internally into heat with a rate of k_{IC}. Then, the decay of the population $N_{S1}(t)$ in the excited state S_1 is expressed by

$$\frac{dN_{S1}(t)}{dt} = -(k_{FL} + k_{IC})N_{S1}(t). \tag{2.2.18}$$

Because the fluorescence decay is proportional to the decay of the population in the excited state S_1, the fluorescence decay function is expressed as a single exponential $I_f(t) = I_{f0}\exp(-t/\tau_0)$. τ_0 is the fluorescence lifetime expressed as $\tau_0 = 1/(k_{FL} + k_{IC})$. The excited state population can be also decayed with a rate of k_Q by interaction with another molecule. Thus, the effective fluorescence lifetime $\tau_{eff} = 1/(k_{FL} + k_{IC} + k_Q)$ becomes shorter than τ_0. This effect is called "fluorescence quenching." The quenching (Q) depends linearly on the concentration of the quencher such as oxygen, halogens, heavy metal ions, and a large number of organic molecules [174]. In practice, the fluorescence lifetime depends on the concentration of heavy metal ions [175,176], the oxygen concentration [177,178], the pH value [179,180], and the binding to proteins [181], DNA [182,183] or lipids [184]. Therefore, the fluorescence lifetime can be used to probe the local environment of fluorophores on the molecular size [185–189]. In the case of different microenvironments or different conformations of the fluorophores, the decay curve may become multi-exponential.

FRET usually results in an extremely efficient quenching of the donor fluorescence and decrease of the donor lifetime. Thus, FLIM is also used to measure the FRET efficiency. Because free donor and the donor bound with an acceptor usually coexist in the observation volume of the sample double-labeled by the donor and acceptor, the double-label lifetime data in FLIM-based FRET exhibits two different lifetimes. Thus, FLIM-based FRET techniques have the benefit that a single-label lifetime image of the donor in the absence of an acceptor can be obtained from the double-label lifetime results [190]. In FLIM-based FRET techniques, therefore, calibration by different cells containing only the donor is not required. Moreover, FLIM is able to discriminate the interacting and non-interacting donor fractions in FRET.

The fluorescence lifetime of the commonly used fluorophores with high quantum efficiencies are typically in the range from 1 to 5 ns. The fluorescence lifetimes of less efficient fluorophores can easily be below 100 ps. The fluorescence lifetime of autofluorescence ranges from 100 to 5 ns [191,192]. Quenching effects can reduce the lifetime down to 50 ps. The lifetime of the donor in FRET is in the order of 100 to 300 ps [190]. Therefore, the time resolution of FLIM should be range from 5 ns to 50 ps.

The measurement techniques of the fluorescence lifetime are classified into time-domain techniques and frequency-domain techniques. In time-domain techniques, the fluorescence intensity decay after excitation with a pulsed light source is obtained as a function of time. The fluorescence intensity decay can be reconstructed in different ways. Most common are time-gated detection [193–196] and time-correlated single photon counting (TCSPC) [198–200]. In frequency-domain techniques, an intensity modulated light source is used for excitation [201–208]. Then, the resultant fluorescence intensity is modulated at the same frequency as the excitation. However, the phase and the modulation depth of the fluorescence differ from those of the excitation. The fluorescence lifetime is obtained from the phase shift and the reduced modulation depth. We next consider these techniques in detail.

2.2.3.1 Time-Domain Techniques One type of time-domain technique is time-gated detection. The principle of time-gated detection is described here. The generated fluorescence signal is expressed by a convolution of the intrinsic fluorescence decay of the sample with the intensity profile of the excitation light $I_{ex}(t)$. For a sample with a j-component multi-exponential decay, the fluorescence signal is expressed by

$$I_{\mathrm{f}}(r, t) = \int_{-\infty}^{t} I_{\mathrm{ex}}(t') \sum_{j} a_j(r) e^{-(t-t')/\tau_j(r)} dt', \tag{2.2.19}$$

where $\tau_j(r)$ and $a_j(r)$ are the lifetime and amplitude of the jth component, respectively. In the case of the use of picosecond and femtosecond pulses as the excitation, $I_{ex}(t)$ can be effectively regarded as a delta function because the fluorescence lifetime ranges from 0.1 to 10 ns. Then, we can write the fluorescence signal as

$$I_{\mathrm{f}}(r, t) \approx \sum_{j} a_j(r) e^{-t/\tau_j(r)}. \tag{2.2.20}$$

The fluorescence signal is measured during a short gate width Δt. For a gate operating at time of t_G, the time-gated detection signal is expressed by

$$I_{\mathrm{Det}}(r, t_G) = \int_{-\infty}^{\infty} G(t' - t_{\mathrm{G}}) I_{\mathrm{f}}(r, t') dt', \tag{2.2.21}$$

where $G(t)$ is the gated detector gain with a gate width of Δt. By using a gated intensified CCD or a streak camera with gate widths as fast as tens of picosecond, the effects of $G(t)$ convolution can be removed [193,194]. Thus, the fluorescence decay is directly obtained.

For rapid lifetime image acquisition, an analytic lifetime determination algorithm can be used. For a sample with a single exponential lifetime, the gated signal is written as

$$I_{\mathrm{Det}}^{(\mathrm{s})}(r, t_G) = \int_{t_G}^{t_G+\Delta t} a e^{-t/\tau(r)} dt = a\tau(r)\left\{e^{-t_{\mathrm{G}}/\tau(r)} - e^{-(t_{\mathrm{G}}+\Delta t)/\tau(r)}\right\}. \tag{2.2.22}$$

Thus, as shown in Fig. 2.6(a), the lifetime can be determined from two gated signals at times t_1 and t_2 by [195]

$$\tau(r) = \frac{t_2 - t_1}{\ln\{I_{\mathrm{Det}}^{(\mathrm{s})}(r, t_1)/I_{\mathrm{Det}}^{(\mathrm{s})}(r, t_2)\}}. \tag{2.2.23}$$

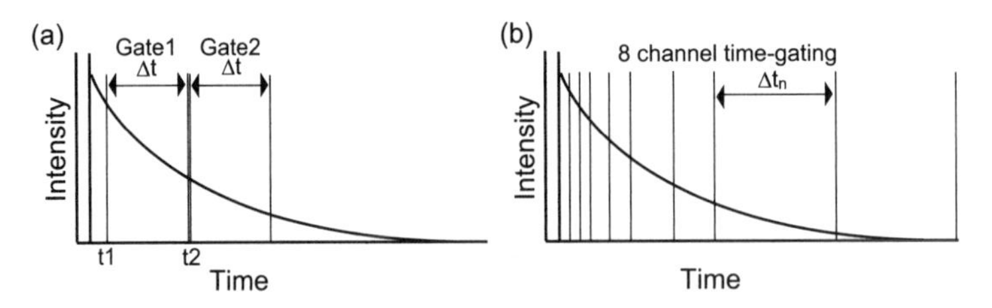

FIGURE 2.6 Time-gating scheme. (a) 2-Channel time-gating. (b) 8-Channel time-gating.

In practice, the two-gate scheme can be sensitive to noise. In addition, since this approach yields only an average fluorescence lifetime, the individual components of a multi-exponential decay cannot be distinguished. Therefore, as illustrated in Fig. 2.6(b), the multiple-gate (4 to 8 time-gate channels) scheme, which enables the measurement of multi-exponential lifetimes over a large lifetime range, is used [196]. Since this technique is realized by opening all the gates sequentially after each and every excitation pulse, the whole decay can be virtually captured without losing any photons. Therefore, high sensitivity is achieved. The sensitivity is further enhanced by employing single photon counting. The lifetime is determined by fitting the data with an exponential decay where a nonlinear least-squares Levenberg–Marquardt fit algorithm is used [173,197].

A more common way in time-domain technique is time-correlated single photon counting (TCSPC). Figure 2.7 shows the concept of TCSPC. In TCSPC, a sample is repetitively excited by a pulsed light source under the condition that the excitation intensity is so low that the probability of the detection of a single photon in each pulse excitation is far less than one. There are signal detection periods without fluorescence photons and the other detection periods contain a single photon. Detection periods

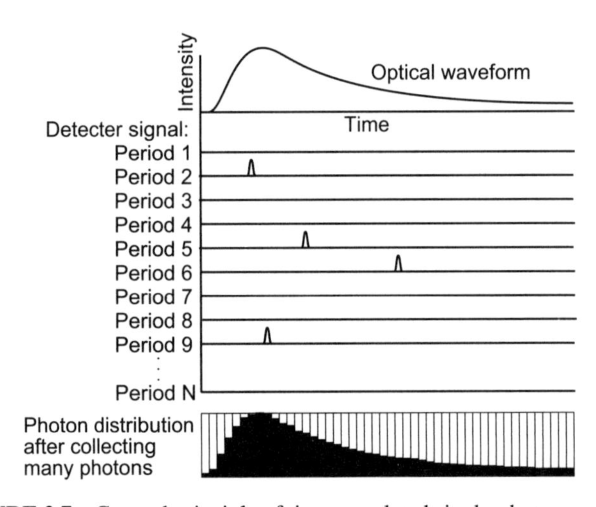

FIGURE 2.7 General principle of time-correlated single-photon counting.

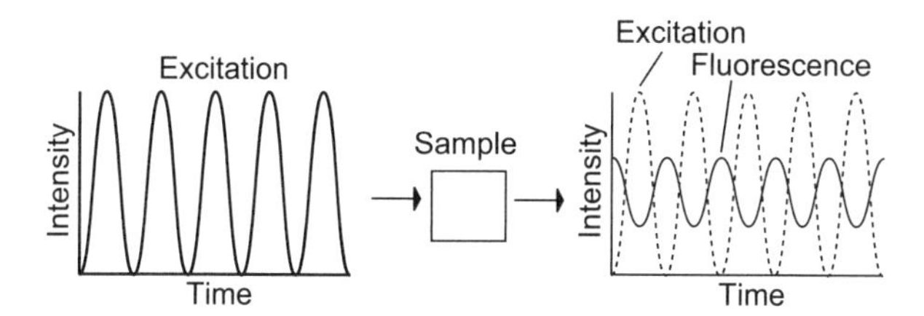

FIGURE 2.8 Modulation technique in frequency-domain measurements.

with more than one photon are very rare. After the excitation at each pulse, the timing of single-photon emission is record by the use of a time-to-amplitude converter (TAC) whose output will be proportional to the time difference between the start and stop pulses. A trigger pulse synchronized with the excitation light pulse is used to start the TAC. The fluorescence emitted by the sample is detected by a detector, sent through a discriminator, and then used to stop the TAC. The TAC output is converted to a digital word by means of an analog-to-digital converter (ADC). This digital word is used as a pointer to an address in a histogram memory, and the events are collected in the memory by adding a "1" to the value at this specific address. Since the exponential fluorescence decay curve is proportional to the probability distribution that a single photon is emitted at time t, the optical waveform of exponential decay can be reconstructed by the histogram after collecting many photons. The dead-time of the TAC electronics is comparatively long, typically 300 to 1000 ns. Therefore, care must be taken that the count rate is sufficiently low to avoid pulse pileup. The TAC alternatively operates in the reversed start–stop scheme in which the TAC is started by the fluorescence signal and stopped by the excitation pulse. The advantage of the reverse operation is that the TAC does not have to be constantly activated and reset in every excitation pulses, since in most excitation pulses no fluorescence photon is even detected. By reducing the excitation intensity to about 1 to 5 detected photons per 100 excitation pulses, the reverse operation mode suffers less from dead-time effects. By operating the TCSPC detection system at too high count rate above 5% of the excitation, the recorded decay curve is distorted [173,200]. In addition, in spectroscopy, excitation frequencies not exceeding 10 MHz are employed to ensure that the fluorescence decay signal from one excitation pulse is not affected by that of the other excitation pulses. Conventional TCSPC equipment has been employed in a confocal laser scanning microscope for fluorescence lifetime imaging. However, this imaging application results in very long acquisition times. More recently, TCSPC plug-in cards for personal computers have been developed that are optimized for imaging applications. These cards have a much lower dead-time than the conventional TCSPC electronics.

In general, the decay curves recorded by TCSPC are fitted to the multi-exponential decay expressed in Eq. (2.2.20). The nonlinear least-squares method is the most general and reliable method for analysis of the recorded decay curves [173]. A nonlinear least-squares analysis starts with a model that is assumed to describe the data. The goal is to test whether the model is consistent with the data and to obtain the parameter values for the model that provide the best match between the measured data and the calculated data using assumed parameter values. The nonlinear least squares method is accomplished by minimizing the goodness-of-fit parameter, which is given by

$$\chi^2 = \sum_{k=1}^{m} \frac{1}{\sigma_k^2} [N(t_k) - N_c(t_k)]^2 = \sum_{k=1}^{m} \frac{[N(t_k) - N_c(t_k)]^2}{N(t_k)}. \tag{2.2.24}$$

Here $N(t_k)$ is the measured count number in time channel k, $N_c(t_k)$ is the calculated count number in channel k from model prediction, and m is the total number of time channels. σ_k is the standard deviation of $N(t_k)$, which is equal to $\sqrt{N(t_k)}$ from Poisson statistics. In fact, it is not convenient to interpret the value of χ^2 because χ^2 depends on the number of time channels. The value of χ^2 will be larger for data sets with more time channels. Thus, we use the value of the reduced χ^2, which is defined by

$$\chi_R^2 = \frac{\chi^2}{m - p} = \frac{\chi^2}{v}. \tag{2.2.25}$$

Here p is the number of floating parameters and $v = m - p$ is the number of degrees of freedom. Because the number of time channels m is usually much larger than the number of floating parameters p, v is nearly equal to m. If only random errors contribute to χ^2, the average of χ^2 per time channel should be about one. Then, χ_R^2 is expected to be approximately unity. If the model does not fit, the individual values of χ^2 and χ_R^2 are both larger than the expected random errors.

The value of χ_R^2 can be used to judge the goodness-of-fit. When the experimental uncertainties σ_k are known, the value of χ_R^2 is expected to be close to unity. If the model does not fit the data, the value of χ_R^2 will be significantly larger than unity. Even though the values of χ_R^2 are used to judge the fit, the first step should be a visual comparison of the data and the fitted function, and then a visual examination of the residuals, which are the differences between the measured data and the fitted function. If the data and the fitted function significantly mismatched, the analysis may be trapped in a local minimum far from the correct parameter values, or the model may be incorrect.

2.2.3.2 Frequency-Domain Techniques In frequency-domain techniques, the excitation intensity is temporally modulated with a modulation depth of M_{ex} at a frequency of f_{M} in the form

$$I_{\text{ex}}(t) = I_0\{1 + M_{\text{ex}} \cos(2\pi f_{\text{M}} t + \varphi_{\text{ex}})\}. \tag{2.2.26}$$

Then, the fluorescence intensity, which is modulated with a modulation depth of M_{f} and a phase shift of φ_{f} at the same frequency as the modulated excitation light, is expressed as

$$I_{\text{f}}(t) = I_{\text{f0}}\{1 + M_{\text{f}} \cos(2\pi f_{\text{M}} t + \varphi_{\text{f}})\}. \tag{2.2.27}$$

The phase shift $\varphi_s = \varphi_{ex} - \varphi_f$ and the change of the modulation depth M_f/M_{ex} are related to the effective fluorescence lifetime τ_{eff} by

$$\tan(\varphi_{ex} - \varphi_f) = 2\pi f_M \tau_{eff}, \tag{2.2.28}$$

$$\frac{M_f}{M_{ex}} = \frac{1}{\sqrt{1 + (2\pi f_M \tau_{eff})^2}}. \tag{2.2.29}$$

In practice, the phase shift and modulation are introduced by electronics and optics. Thus, we use the phase and modulation depth that are measured using a reference sample with a known fluorescence lifetime τ_{ref}. The fluorescence lifetimes of samples are retrieved from the phase φ_{fr} and the modulation depth M_{fr} at the reference sample measurement by employing the following relations:

$$\tau_{eff} = \frac{1}{2\pi f_M} \tan\left\{\varphi_{fr} - \varphi_f + \tan^{-1}(2\pi f_M \tau_{ref})\right\}, \tag{2.2.30}$$

$$\tau_{eff} = \frac{1}{2\pi f_M}\sqrt{\left(\frac{M_{fr}}{M_f}\right)^2 \left\{1 + (2\pi f_M \tau_{ref})^2\right\} - 1}. \tag{2.2.31}$$

The phase and the modulation depth that are obtained by measuring the excitation light are also used to account for the phase shift and modulation introduced by electronics and optics. Therefore, a frequency-domain system generally contains two detectors, one for the sample and the other to serve as the reference.

To resolve the components of multi-exponential decay functions, the phase shifts are measured at different modulation frequencies. In this case, the phase shift and the modulation depth of a sample with a multi-exponential decay in the time domain is expressed in the frequency domain by

$$\varphi_f(f) = Arg\left(\frac{FT_t\left[\sum_j a_j \exp\left(-\frac{t}{\tau_j}\right)\right]}{\sum_j a_j \tau_j}\right) + \varphi_{ex}$$

$$= Arg\left(\frac{\sum_j a_j \tau_j / 1 - i2\pi f \tau_j}{\sum_j a_j \tau_j}\right) + \varphi_{ex}, \tag{2.2.32}$$

$$M_f(f) = \left|\frac{FT_t\left[\sum_j a_j \exp(-t/\tau_j)\right]}{\sum_j a_j \tau_j}\right| = \left|\frac{\sum_j a_j \tau_j / 1 - i2\pi f \tau_j}{\sum_j a_j \tau_j}\right|. \tag{2.2.33}$$

Here $Arg()$ indicates the argument of the complex number and a_j is the fraction of fluorophores whose lifetime is τ_j. The fluorescence lifetime of samples is recovered by fitting the experimental data with Eqs. (2.2.32) and (2.2.33). In general, a Lenvenberg–Marquarbt fit algorithm to minimize the error function χ_R^2 between the data (φ_f

and M_f), and the calculated values (φ_{fc} and M_{fc}) are used. The error function is given by

$$\chi_R^2 = \frac{1}{2m-p}\left[\sum_{k=1}^{m}\frac{1}{\sigma_{k\varphi}^2}[\varphi_f(f_k)-\varphi_{fc}(f_k)]^2+\sum_{k=1}^{m}\frac{1}{\sigma_{kM}^2}[M_f(f_k)-M_{fc}(f_k)]^2\right]. \tag{2.2.34}$$

Here m is the number of frequency, p is the number of floating parameters, $\sigma_{k\varphi}$ is the standard deviation of $\varphi_f(f_k)$, and σ_{kM} is the standard deviation of$M_f(f_k)$.

The phase shift and the change of the modulation depth can be used to determine the fluorescence lifetime. However, phase measurements give a higher accuracy for the fluorescence lifetime than modulation depth measurements. If we assume a strong signal, the phase shift can usually be determined to an accuracy of $\pm 0.2°$. Therefore, the fluorescence lifetime is generally obtained from the phase shift. The theoretical optimal modulation frequencies for a specific fluorescence lifetime in modulation depth measurements f_{MM} and in phase measurements $f_{M\varphi}$ are given by [202]

$$f_{MM} = \frac{\sqrt{2}}{2\pi\tau_{eff}}, \tag{2.2.35}$$

$$f_{M\varphi} = \frac{1}{2\pi\tau_{eff}}. \tag{2.2.36}$$

When moving apart from this optimum, the sensitivity decreases. To account for the fluorescence lifetime range from 5 ns to 50 ps, the modulation frequency should be in the range from 30 MHz to 3 GHz. However, it turns out that the fluorescence lifetime can be obtained with little loss in accuracy at much lower frequencies [203]. Therefore, in practice, the modulation frequencies ranging from 50 MHz to several 100 MHz are used.

Intensity modulation of the excitation light from a xenon lamp, a mercury lamp, or a continuous wave laser can easily be accomplished by using electro-optical modulators (EOMs) or acoustic optical modulators (AOMs). These devices allow modulation over a wide range of modulation frequencies up to about 1 GHz. Alternatively, a pico- or femtosecond laser operating at a repetition rate of 70 to 90 MHz can be used as the modulated light source. The train of output pulses contains harmonic components of the repetition frequency. A frequency-domain system can be synchronized to any of harmonics. In practice, modulation frequencies are limited by the detection system.

On the detection side, phase-sensitive detection requires the detector signal to be multiplied with the modulation signal at the exactly same frequency (homodyne detection) or a different frequency (heterodyne detection) as the excitation light source. In homodyne detection, after filtering out the DC component of the detector signal, the remaining AC component is fed into a mixer and is multiplied by a reference signal at the same frequency f_M as the excitation light source and at a reference phase φ_r, which is given by

$$I_{ref}(\varphi_r) = \cos(2\pi f_M t + \varphi_r). \tag{2.2.37}$$

The result of the multiplication is

$$I_{homo}(\varphi_r) = \frac{I_{f0} M_f}{2}\{\cos(\varphi_f - \varphi_r) + \cos(4\pi f_M t + \varphi_f + \varphi_r)\}. \tag{2.2.38}$$

After filtering out the high-frequency component at $2f_M$, the two DC components are measured at different reference phases of 0° and 90°, which are described by

$$I_{homo}(0^\circ) = \frac{I_{f0} M_f}{2}\cos(\varphi_f), \tag{2.2.39}$$

$$I_{homo}(90^\circ) = \frac{I_{f0} M_f}{2}\sin(\varphi_f). \tag{2.2.40}$$

The modulation depth M_f and the phase φ_f can be calculated from Eqs. (2.2.39) and (2.2.40). Modulation frequencies are limited by the bandwidth of the mixers, the detector, or the modulation bandwidth of the detector. By using a multichannel plate PMT with external mixers, a modulation frequency up to 10 GHz has been achieved [204]. This detection technique can be used to record the fluorescence from several fluorophores simultaneously in several parallel mixer systems. Several lasers that are modulated at different frequencies allow different excitation wavelengths. Fluorescence is split into several wavelength bands and detected by different detectors. The detector signals are mixed with the modulation frequencies of individual lasers in several parallel groups of mixers [205,206].

In heterodyne detection, both an excitation light and fluorescence are typically detected. After the DC component of the detector signal is filtered out, the AC components of the detector signal and the intensity modulation signal at the modulation frequency f_M are mixed with the reference signal at a different frequency f_{OSC} from the excitation light source and at a reference phase φ_r:

$$I_{hetEx}(t) = \frac{I_0 M_{ex}}{2}[\cos\{2\pi(f_M - f_{OSC})t + (\varphi_{ex} - \varphi_r)\} + \cos\{2\pi(f_M + f_{OSC})t + \varphi_{ex} + \varphi_r\}]. \tag{2.2.41}$$

$$I_{hetF}(t) = \frac{I_{f0} M_f}{2}[\cos\{2\pi(f_M - f_{OSC})t + (\varphi_f - \varphi_r)\} + \cos\{2\pi(f_M + f_{OSC})t + \varphi_f + \varphi_r\}]. \tag{2.2.42}$$

The oscillator frequency f_{OSC} is typically chosen to obtain the difference frequency $f_M - f_{OSC}$ in the kHz range. The two signals are directly digitized as a function of time by the analog-to-digital converter and are filtered to extract the low-frequency components at the difference frequency $f_M - f_{OSC}$. Then, the digital filtering algorithms can be applied to determine the phase via fast Fourier transform. The phase shift $\varphi_{ex} - \varphi_f$ is obtained from the phase difference between the fluorescence and the excitation light.

The mixers used in the detector signal channel in homodyne and heterodyne systems can be replaced with gain-modulated detectors such as a PMT and a CCD equipped with an image intensifier [207,208]. In homodyne detection, the gain M_G is modulated at exactly the same frequency (homodyne detection) as the excitation light source and at a phase φ_r. Then, the detector signal is given by

$$\begin{aligned} I_{homoG}(\varphi_r) &= I_{f0}M_G\{1 + \cos(2\pi f_M t + \varphi_r)\} \\ &\quad + \frac{I_{f0}M_f M_G}{2}\{\cos(\varphi_f - \varphi_r) + \cos(4\pi f_M t + \varphi_f + \varphi_r)\}. \end{aligned} \tag{2.2.43}$$

By reducing the frequency response (increasing the integration time) of the detector or preamplifiers, the DC components are separated from the high-frequency components at $2f_M$ and f_M. However, the DC component contains the component that is independent of the phase of the fluorescence signals. To remove the phase-independent DC component, a sequence of images are acquired at different gain phases and are used for calculating the phase of fluorescence signals.

In heterodyne detection using gain-modulated detectors, the gain is modulated at the frequency f_{OSC} that are different from the modulation frequency f_M of the excitation light source. Then, the frequency components at f_M, f_{OSC}, $f_M + f_{OSC}$, $f_M - f_{OSC}$ and zero (DC) may appear at the outputs of the detectors. After filtering out of the high-frequency components (f_M, f_{OSC} and $f_M + f_{OSC}$), the fluorescence and excitation intensities are digitized as a function of time. The phase and amplitude are acquired from the difference frequency component at $f_M - f_{OSC}$ by digital filtering.

2.2.4 Fluorescence Recovery after Photobleaching Imaging

Fluorescence recovery after photobleaching (FRAP) imaging has become a powerful tool for investigating biological systems of cell membrane diffusion and protein binding because it provides transport properties of fluorescently labeled molecules. The FRAP technique was introduced by Peters et al. in 1974 [209–216] and first utilized to measure quantitatively diffusion coefficients by Axelrod et al. in 1976 [210]. FRAP imaging is performed by monitoring the time-evolution of a fluorescence signal after photobleaching spatially localized fluorophores. A focused laser beam with a high intensity is employed to locally bleach fluorophores that were used to label target molecules in a biological sample. After photobleaching the fluorophores, the time-evolution of a fluorescence signal from unbleached fluorophores that diffuse into the bleached region from the unbleached region due to the Brownian motion is measured at the same region as the photobleaching. If the target molecules have no mobility, the fluorescence signal is not recovered after photobleaching. On the other hand, in the case of the target molecules with a low mobility, the fluorescence signal is slowly recovered. On the other hand, in the case of the target molecules with a high mobility, the fluorescence signal is fast recovered. Quantitative information about the nature (diffusion and velocity-driven movement) and kinetics (diffusion coefficients or velocities) of molecular movements are determined by fitting recovery curves with appropriate analytical results or numerical simulations [217,218]. The interpretation of the results depends on the geometry of the sample, the nature of the molecular movements, and the photobleaching conditions used in the model [219].

Solute diffusion is described by the Smoluchowski equation,

$$\frac{\partial C(\mathbf{r},t)}{\partial t} = D\nabla^2 C(\mathbf{r},t), \tag{2.2.44}$$

where D is the diffusion coefficient, and $C(r,t)$ is the space- and time-dependent solute concentration distribution of unbleached fluorophores. Now, we assume an arbitrary axially symmetric initial concentration distribution $C(r,z,0)$ corresponding to the concentration distribution of unbleached fluorophores immediately after photobleaching. Then, Eq. (2.2.44) is written in the cylindrical coordinates as

$$\frac{\partial C(r,z,t)}{\partial t} = D\left(\frac{\partial^2}{\partial r^2} + \frac{1}{r}\frac{\partial}{\partial r} + \frac{\partial^2}{\partial z^2}\right) C(r,z,t). \tag{2.2.45}$$

By first applying a Hankel transform to the radial coordinate r, followed by a Fourier transform to the axial coordinate z, the general solution of Eq. (2.2.45) can be obtained in the form [217]

$$\begin{aligned} C(r,z,t) = {} & \frac{1}{4\sqrt{\pi}(Dt)^{3/2}} \int_{-\infty}^{\infty}\int_{0}^{\infty} C(r',z',0) I_0\left(\frac{rr'}{2Dt}\right) \\ & \times e^{-\{r^2+r'^2+(z-z')^2\}/4Dt} r' dr' dz', \end{aligned} \tag{2.2.46}$$

where I_0 is the modified Bessel function of the first kind. If photobleaching of the fluorophore to a nonfluorescent species is a simple irreversible first-order reaction, the m-photon bleaching of fluorophores is described by a first-order differential equation [218]:

$$\frac{dC(r,z,t)}{dt} = -\frac{1}{m\,(\hbar\omega)^m} q_{\mathrm{bl}}^{(m)} \sigma_{\mathrm{bl}}^{(m)} \left\langle I_{\mathrm{bl}}^m(r,z)\right\rangle C(r,z,t), \tag{2.2.47}$$

where $\sigma_{\mathrm{bl}}^{(m)}$ is the m-photon absorption cross section for the bleaching of the fluorophore, and $q_{\mathrm{bl}}^{(m)}$ is the quantum efficiency for m-photon photobleaching, $\langle I_{\mathrm{bl}}^m(r,z)\rangle$ is the time average of the bleaching intensity raised to the mth power, $\hbar\omega$ is photon energy, and m is the number of photons absorbed in a bleaching event. When the exposure time Δt for photobleaching is much shorter than the diffusion time of the fluorescently labeled molecules, the solution of Eq. (2.2.47) yields the concentration distribution of unbleached fluorophore immediately after illumination by the photobleaching beam:

$$C(r, z, 0) = C_0 \exp\left[-\frac{1}{m(\hbar\omega)^m} q_{\mathrm{bl}}^{(m)} \sigma_{\mathrm{bl}}^{(m)} \left\langle I_{\mathrm{bl}}^m(r, z)\right\rangle \Delta t\right], \quad (2.2.48)$$

where C_0 is the initial equilibrium concentration of fluorophore. Now, let us assume that the effective excitation intensity distribution of for the photobleaching can be modeled using a 3D Gaussian approximation

$$\left\langle I_{\mathrm{bl}}^m(r, z)\right\rangle = \left\langle I_{\mathrm{bl}}^m(0, 0)\right\rangle \exp\left\{-2m\left(\frac{r^2}{W_r^2} + \frac{z^2}{W_z^2}\right)\right\}, \quad (2.2.49)$$

where W_r and W_z are the e^{-2} radial and axial dimensions, respectively, and $\left\langle I_{\mathrm{bl}}^m(0, 0)\right\rangle$ is the time average of the intensity at the center ($r = z = 0$) of the focal volume raised to the mth power. Substituting Eq. (2.2.48) into Eq. (2.2.46) and using the series expansion of the exponential function in Eq. (2.2.48) yields the time-dependent concentration distribution of unbleached fluorophores:

$$C(r, z, t) = C_0 \sum_{n=0}^{\infty} \frac{(-\beta)^n}{n!\left\{1 + 8mnDt/W_r^2\right\}\sqrt{1 + 8mnDt/W_z^2}} \times \exp\left[-\frac{2mnr^2}{W_r^2\left(1 + 8mnDt/W_r^2\right)} - \frac{2mnz^2}{W_z^2\left(1 + 8mnDt/W_z^2\right)}\right], \quad (2.2.50)$$

where

$$\beta = \frac{1}{m(\hbar\omega)^m} q_{\mathrm{bl}}^{(m)} \sigma_{\mathrm{bl}}^{(m)} \left\langle I_{\mathrm{bl}}^m(0, 0)\right\rangle \Delta t \quad (2.2.51)$$

is the bleaching parameter. The time-evolution of the fluorescence signal from unbleached fluorophores after illumination by the photobleaching beam is given by

$$F(t) = \frac{\eta q_{\mathrm{fl}}^{(p)} \sigma_{\mathrm{fl}}^{(p)} N_A}{p(\hbar\omega)^p} \int_{-\infty}^{\infty}\int_0^{\infty} \left\langle I_{\mathrm{fl}}^p(r, z)\right\rangle C(r, z, t) 2\pi r\, dr\, dz, \quad (2.2.52)$$

where $\sigma_{\mathrm{fl}}^{(p)}$ is the p-photon absorption cross section for fluorescence of the fluorophore, and $q_{\mathrm{fl}}^{(p)}$ is the quantum efficiency for p-photon excitation fluorescence, N_A is the Avogadro's number, $\left\langle I_{\mathrm{fl}}^p(r, z)\right\rangle$ is the time average of the pth power of the intensity of the monitoring beam at position r and z, p is the number of photons required to generate a fluorescence photon, and η is the overall efficiency of the detection system. Let us also assume that the effective excitation intensity distribution for the fluorescence can be modeled using a 3D Gaussian approximation:

$$\left\langle I_{\text{fl}}^{p}(r,z)\right\rangle = \left\langle I_{\text{fl}}^{p}(0,0)\right\rangle \exp\left\{-2p\left(\frac{r^2}{W_r^2}+\frac{z^2}{W_z^2}\right)\right\}, \tag{2.2.53}$$

where W_r and W_z are the e^{-2} radial and axial dimensions, respectively, and $\left\langle I_{\text{fl}}^{p}(0,0)\right\rangle$ is the time average of the intensity at the center ($r = z = 0$) of the focal volume raised to the pth power. Inserting Eqs. (2.2.50) and (2.2.53) into Eq. (2.2.52), we obtain the time-evolution of the fluorescence signal from unbleached fluorophores:

$$\begin{aligned} F(t) &= \frac{2\pi\eta q_{\text{fl}}^{(p)}\sigma_{\text{fl}}^{(p)}C_0N_A}{p\,(\hbar\omega)^p}\left\langle I_{\text{fl}}^{p}(0,0)\right\rangle \\ &\quad\times\sum_{n=0}^{\infty}\frac{(-\beta)^n}{n!\left(1+8mnDt/W_r^2\right)\sqrt{1+8mnDt/W_z^2}} \\ &\quad\times\int_0^{\infty} r\exp\left[-\frac{2mn+2p\left(1+8mnDt/W_r^2\right)}{W_r^2\left(1+8mnDt/W_r^2\right)}r^2\right]dr \\ &\quad\times\int_{-\infty}^{\infty}\exp\left[-\frac{2mn+2p\left(1+8mnDt/W_z^2\right)}{W_z^2\left(1+8mnDt/W_z^2\right)}z^2\right]dz \\ &= \sum_{n=0}^{\infty}\frac{F_0p^{3/2}(-\beta)^n}{n!\left(p+mn+8pmnDt/W_r^2\right)\sqrt{p+mn+8pmnDt/W_z^2}}, \end{aligned} \tag{2.2.54}$$

where F_0 is the equilibrium fluorescence signal detected before the photobleaching and is given by

$$F_0 = \frac{\pi\sqrt{\pi}\eta q_{\text{fl}}^{(p)}\sigma_{\text{fl}}^{(p)}C_0N_AW_r^2W_z}{2\sqrt{2}p^2\sqrt{p}\,(\hbar\omega)^p}\left\langle I_{\text{fl}}^{p}(0,0)\right\rangle. \tag{2.2.55}$$

To understand the influence of various experimental parameters on the behavior of the fluorescence recovery curve, we consider the bleach depth, which is defined as $\Delta F = \{F(0) - F_0\}/F_0$, and the initial slope of the fluorescence recovery. The bleach depth is described by

$$\Delta F = \frac{F(0)-F_0}{F_0} = 1-\sum_{n=0}^{\infty}\frac{(-\beta)^n}{n!}\frac{1}{(1+mn/p)^{3/2}}. \tag{2.2.56}$$

For small values of the bleaching parameter and $p = m$, the bleach depth is expressed as

$$\Delta F \approx \frac{\beta}{2^{3/2}}. \tag{2.2.57}$$

From Eq. (2.2.57), we see that the bleaching depth is proportional to the bleaching parameter; consequently, it is proportional to the exposure time for the photobleaching and to the mth power of the bleaching intensity. The initial slope of the recovery curve is given by

$$\left(\frac{\partial F}{\partial t}\right)_{t=0} = -F_0 m \frac{8D}{W_r^2}\left(1+\frac{W_r^2}{2W_z^2}\right)\sum_{n=0}^{\infty}\frac{(-\beta)^n}{n!}\frac{n}{(1+mn/p)^{5/2}}. \tag{2.2.58}$$

According to Eq. (2.2.58), we find that the initial slope is proportional to the diffusion coefficient. For small values of the bleaching parameter and $p = m$, the initial slope is expressed as

$$\left(\frac{\partial F}{\partial t}\right)_{t=0} = F_0 m \frac{8D}{W_r^2}\left(1+\frac{W_r^2}{2W_z^2}\right)\frac{\beta}{2^{5/2}}, \tag{2.2.59}$$

which may prove useful for rapid fitting estimates.

For free diffusion described by Eq. (2.2.44), the diffusion coefficient is constant in space and time. In complex biological systems, this is often not the case. For simple diffusion in three dimensions, the mean square displacement of a diffusion particle increases linearly with time [220]

$$\langle r^2 \rangle = 6Dt. \tag{2.2.60}$$

For anomalous diffusion, the mean square displacement increases with the power law [221]

$$\langle r^2 \rangle = 6\Gamma t^{\alpha}, \tag{2.2.61}$$

where Γ is the transport coefficient, and α is less than 1. The time-evolution of the fluorescence signal from unbleached fluorophores described by Eq. (2.2.54) can be modified to account for anomalous diffusion simply by replacing terms of the form Dt with terms of the form $(Dt)^{\alpha}$ [220,222].

$$F(t) = \sum_{n=0}^{\infty}\frac{F_0 p^{3/2}(-\beta)^n}{n!\left\{p+mn+8pmn(Dt)^{\alpha}/W_r^2\right\}\sqrt{p+mn+8pmn(Dt)^{\alpha}/W_z^2}}. \tag{2.2.62}$$

Then, the effective diffusion coefficient at any time t is equal to $D^{\alpha}t^{\alpha-1}$. To obtain the parameters F_0, β, D, and α, the Levenberg–Marquardt fit is generally performed using Eq. (2.2.62).

The fluorescence recover curve in a compartment diffusionally coupled to a different component is described as [211,213,223]

$$F(t) = F_0\left(1 - \Delta F e^{-t/\tau}\right), \tag{2.2.63}$$

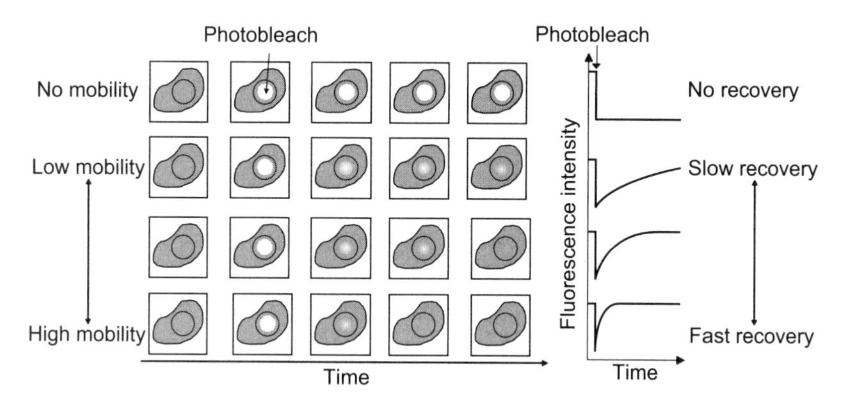

FIGURE 2.9 Fluorescence recovery after photobleaching.

where τ is the time scale of diffusion between the two compartments. The recovery between components provides insight into the diffusion coefficient of the tracer, the characteristic resistivity of the coupling pathways, and the dependence on the geometry of the system.

2.2.5 Fluorescence Correlation Spectroscopy

Fluorescence correlation spectroscopy (FCS) has become a versatile tool for biomolecular studies, including determination of translational and rotational diffusion, concentration and mobility of molecules, chemical kinetics, and binding reactions. In FCS, fluctuations of the fluorescence intensity of a small number of fluorophore-labeled molecules, excited by a focused laser beam, are measured. Owing to the high sensitivity of FCS, detection at the level of single molecules is possible. Fluorescence intensity fluctuations can arise from molecular diffusion in and out of the very small observation volume [224] or from any physical or chemical dynamics that modify the photophysical properties of the fluorophores such as the absorption crosssection and the fluorescence quantum yield [225]. Individual fluctuation events randomly occur. However, on average, the amplitude and time scale of the observed fluctuation dynamics reflect the underlying physical and chemical dynamics of the molecules in the observation volume. Thus, by statically analyzing the observed fluctuations, one can recover significant information about the mobility, dynamics, and interactions of biomolecules. FCS was first applied to the determination of ethidium bromide binding to DNA by Madg et al. in 1972 [226]. In this original work, the fluorescence fluctuations resulted from the change of fluorescence quantum yield by ethidium bromide binding to DNA. Following this pioneering study, FCS has been successfully applied to a variety of the investigations of translational and rotational mobility [227,228], binding of nucleic acids [229] and proteins [230], fluorescence quenching due to reversible protonation [231], electron transfer [232], oxygen [233] and ion concentrations.

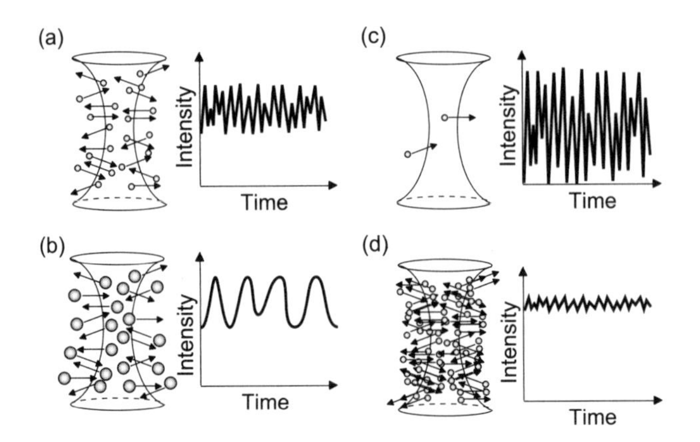

FIGURE 2.10 Fluorescence fluctuations from small molecules (a), large molecules (b), a small number of molecules (c), and a large number of molecules (d).

2.2.5.1 Autocorrelation Spectroscopy In the absence of saturation, the fluorescence signal measured with m-photon excitation is expressed by

$$F(t) = \frac{\eta N_A}{m\,(\hbar\omega)^m} \int_V \left\langle I_{\text{ex}}^m(\mathbf{r}) \right\rangle I_{\text{det}}(\mathbf{r}) q_{\text{fl}}^{(m)}(\mathbf{r}, t) \sigma_{\text{ex}}^{(m)}(\mathbf{r}, t) C(\mathbf{r}, t) dV, \qquad (2.2.64)$$

where η denotes the overall detection efficiency of the optical system, N_A is the Avogadro's number, $\hbar\omega$ is photon energy, $q_{\text{fl}}^{(m)}$ is the fluorescence quantum yield, $\sigma_{\text{ex}}^{(m)}$ is the m-photon excitation cross section, $C(r, t)$ is the local concentration of fluorophore-labeled molecules, $\langle I_{\text{ex}}^m(\mathbf{r})\rangle$ is the time average of the excitation intensity raised to the mth power, $I_{\text{det}}(\mathbf{r})$ is the normalized spatial collection efficiency determined by objective and pinhole transfer properties, and the factor $1/m$ simply reflects that m photons are needed for each excitation event. As shown in Fig. 2.10, the fluorescence fluctuation signals depend on the concentration and size of the fluorophore-labeled molecules. Because the laser excitation conditions are effectively constant on the relevant fluctuation time scales, the fluorescence fluctuation is written as

$$\delta F(t) = \frac{\eta N_A}{m\,(\hbar\omega)^m} \int_V \left\langle I_{\text{ex}}^m(\mathbf{r}) \right\rangle I_{\text{det}}(\mathbf{r}) \delta(q_{\text{fl}}^{(m)}, \sigma_{\text{ex}}^{(m)}, C(\mathbf{r}, t)) dV, \qquad (2.2.65)$$

where $\delta(q_{\text{fl}}^{(m)}, \sigma_{\text{ex}}^{(m)}$, and $C(\mathbf{r}, t))$ indicates the dynamics of the fluorophore on the single-particle level. Here $\delta q_{\text{fl}}^{(m)}$ indicates fluctuation in the fluorescence quantum yield, $\delta\sigma_{\text{ex}}^{(m)}$ fluctuation in the m-photon excitation cross section, and $\delta C(\mathbf{r}, t)$ fluctuation in the local concentration of fluorophore-labeled molecules due to Brownian motion. We can assume that the spatial profile of the observation volume can be modeled using a three-dimensional (3D) Gaussian approximation that is decayed to e^{-2} at W_r in lateral direction and at W_z in axial direction:

$$\left\langle I_{\text{ex}}^{m}(\mathbf{r})\right\rangle I_{\text{det}}(\mathbf{r}) = \left\langle I_{\text{ex}}^{m}(\mathbf{0})\right\rangle \exp\left\{-2\left(\frac{x^2+y^2}{W_r^2}+\frac{z^2}{W_z^2}\right)\right\}, \qquad (2.2.66)$$

where $\langle I_{\text{ex}}^{m}(\mathbf{0})\rangle$ is the time average of the intensity at the center $(x = y = z = 0)$ of the focal volume raised to the mth power. Using Eq. (2.2.66), we can rewrite Eq. (2.2.65) as

$$\delta F(t) = \frac{\eta N_A}{m\,(\hbar\omega)^m}\left\langle I_{\text{ex}}^{m}(\mathbf{0})\right\rangle \int_V S(\mathbf{r})\delta(q_{\text{fl}}^{(m)}, \sigma_{\text{ex}}^{(m)}, C(\mathbf{r}, t))dV, \qquad (2.2.67)$$

where $S(\mathbf{r})$ is described by

$$S(\mathbf{r}) = \exp\left\{-2\left(\frac{x^2+y^2}{W_r^2}+\frac{z^2}{W_z^2}\right)\right\}. \qquad (2.2.68)$$

Since individual fluctuations are random events, statistical analysis such as autocorrelation and cross-correlation is required to obtain the above mentioned information from the measured fluctuation signals. The normalized autocorrelation function, which indicates statistical analysis with respect to its self-similarity after the lag time τ, is defined as

$$G(\tau) = \frac{\langle \delta F(t)\delta F(t+\tau)\rangle}{\langle F(t)\rangle^2}. \qquad (2.2.69)$$

Here the autocorrelation amplitude $G(0)$ is merely the normalized variance of the fluctuating fluorescence signal $\delta F(t)$. Assuming that the fluorophore's fluorescence properties are not changing within the observation time, Eq. (2.2.69) can be rewritten as

$$G(\tau) = \frac{\int_V \int_{V'} S(\mathbf{r})S(\mathbf{r}')\left\langle \delta C(\mathbf{r}, t)\delta C(\mathbf{r}', t+\tau)\right\rangle dV dV'}{\left(\langle C\rangle \int_V S(\mathbf{r})dV\right)^2}. \qquad (2.2.70)$$

Considering only particles that are freely diffusing in three dimensions with the diffusion coefficient D, the so-called number density autocorrelation term is given by [234]

$$\langle \delta C(\mathbf{r}, t)\delta C(\mathbf{r}', t+\tau)\rangle = \langle C\rangle \frac{1}{(4\pi D\tau)^{3/2}} \exp\left\{-\frac{\left|\mathbf{r}-\mathbf{r}'\right|^2}{4D\tau}\right\}. \qquad (2.2.71)$$

Employing Eqs. (2.2.70) and (2.2.71), the autocorrelation function for a free 3D diffusion can be described by [235]

$$G_{3D}(\tau) = \frac{1}{V_{\text{eff}}\,\langle C\rangle}\frac{1}{(1+\tau/\tau_D)\sqrt{1+(W_r/W_z)^2\,\tau/\tau_D}}, \tag{2.2.72}$$

where V_{eff} is the effective observation volume and τ_D is the averaged lateral diffusion time that a molecule stays in the effective observation volume. Here V_{eff} and τ_D are defined as

$$V_{\text{eff}} = \frac{\left(\int_V S(\mathbf{r})dV\right)^2}{\int_V S(\mathbf{r})^2 dV} = \pi^{3/2} W_r^2 W_z, \tag{2.2.73}$$

$$\tau_D = \frac{W_r^2}{4D}. \tag{2.2.74}$$

From Eq. (2.2.72), we see that $G_{3D}(0)$ is inversely proportional to the average number of molecules $\langle N\rangle$,

$$G_{3D}(0) = \frac{1}{V_{\text{eff}}\langle C\rangle} = \frac{1}{\langle N\rangle}. \tag{2.2.75}$$

In cellular applications, we expect not only a free 3D diffusion but also two-dimensional (2D) diffusion on membranes, active, flow-like transport, and anomalous subdiffusion. Considering $W_z = \infty$ in Eq. (2.2.72) gives the autocorrelation function in 2D diffusion on the membranes as [235]

$$G_{2D}(\tau) = \frac{1}{V_{\text{eff}}\langle C\rangle}\frac{1}{(1+\tau/\tau_D)}, \tag{2.2.76}$$

If the particle motion is composed of a diffusion coefficient D and uniform active transport (plug flow) with velocity v, with the y-axis being oriented along the flow direction, the number density autocorrelation term is given by [236]

$$\langle \delta C(\mathbf{r},t)\delta C(\mathbf{r}',t+\tau)\rangle = \langle C\rangle \frac{1}{(4\pi D\tau)^{3/2}} \times \exp\left\{-\frac{(x-x')^2+(y-y'-v\tau)^2+(z-z')^2}{4D\tau}\right\}. \tag{2.2.77}$$

Using Eq. (2.2.77), we can express the autocorrelation function in active, flow-like transport by

$$G_{FD}(\tau) = G_{3D}(\tau)\exp\left\{-\left(\frac{v\tau}{W_r}\right)^2\frac{1}{(1+\tau/\tau_D)}\right\}, \tag{2.2.78}$$

The formalism of anomalous subdiffusion with nonlinear time dependence of the mean square displacement $\langle \Delta r^2 \rangle = \Gamma t^\alpha$, $\alpha < 1$, may be necessary to fully describe situations in and on real cells [237]. The autocorrelation function described by Eq. (2.2.72) can be modified to account for anomalous subdiffusion simply by replacing terms of the form $D\tau$ with terms of the for $(D\tau)^\alpha$:

$$G_{\mathrm{AD}}(\tau) = \frac{1}{V_{\mathrm{eff}}\langle C\rangle}\frac{1}{\{1+(\tau/\tau_{\mathrm{D}})^\alpha\}\sqrt{1+(W_r/W_z)^2(\tau/\tau_{\mathrm{D}})^\alpha}}. \quad (2.2.79)$$

Although it is assumed that the fluorophore's fluorescence properties are not changing within the observation time, the dynamic fluorescence properties for most fluorophore systems are in fact much more complex. For example, the fluorophores may cross from a singlet excited state to a long-lived triplet excited state, in which the fluorophore appears dark. In general, there are physical or chemical dynamics that result in reversible transitions between a fluorescent state B and a nonfluorescent state D in which no fluorescence photons are emitted. Now, let us assume that the nonfluorescent state D is converted from the fluorescent state B with a rate of k_{D} and is returned to the fluorescent state B with a rate of k_{B}. The relaxation time τ_{R} of the reversible transition is expressed by

$$\tau_{\mathrm{R}} = \frac{1}{k_{\mathrm{B}} + k_{\mathrm{D}}}. \quad (2.2.80)$$

If diffusion is much slower that the relaxation time, the dynamics is separated into two factors [238,239]

$$G(\tau) = G_{\mathrm{D}}(\tau)X(\tau), \quad (2.2.81)$$

where $G_{\mathrm{D}}(\tau)$ is the autocorrelation function for normal diffusive motion from Eqs. (2.2.72), (2.2.76), (2.2.78), and (2.2.79), and $X(\tau)$ accounts for the physical or chemical kinetics process. In general, the $X(\tau)$ has the following functional form:

$$X(\tau) = 1 - A + A\exp\left(-\frac{\tau}{\tau_{\mathrm{R}}}\right), \quad (2.2.82)$$

where A is the average fraction of fluorophores in the nonfluorescent state D and is described as

$$A = \frac{k_{\mathrm{D}}}{k_{\mathrm{B}} + k_{\mathrm{D}}}. \quad (2.2.83)$$

It is also possible to normalize this expression by dividing by $(1 - A)$ [240–242]. If the nonfluorescent state D is not completely dark, the relative emission rates q_{B} and q_{D} of the two states have to be taken into account to get the correct expression for A:

$$A = \frac{k_{\mathrm{D}}k_{\mathrm{B}}(q_B - q_D)^2}{(k_{\mathrm{B}} + k_{\mathrm{D}})\left(k_{\mathrm{B}}q_B^2 + k_{\mathrm{D}}q_D^2\right)}. \quad (2.2.84)$$

FCS analysis have been applied to investigate intramolecular reversible quenching owing to electron transfer [232], light-induced blinking revealed by several GFP mutants [242], and, with a slightly modified reaction scheme, protonation [231], and ion binding [239].

There may also be reactions that influence the mobility of the particle in some way. In this case, the autocorrelation function for diffusive motion must be generalized to take into account all different kinds of possible motion, weighted with the relative emission rate q_j [234],

$$G_D(\tau) = \frac{1}{V_{\text{eff}}} \frac{\sum_j q_j \langle C_j \rangle M_j(\tau)}{\left(\sum_j q_j \langle C_j \rangle \right)^2}, \quad (2.2.85)$$

with

$$M_j(\tau) = \frac{1}{(1+\tau/\tau_{D,j})\sqrt{1+(W_r/W_z)^2\,\tau/\tau_{D,j}}} \quad \text{for 3D diffusion,} \quad (2.2.86)$$

$$M_j(\tau) = \frac{1}{(1+\tau/\tau_{D,j})} \quad \text{for 2D membrane diffusion,} \quad (2.2.87)$$

$$M_j(\tau) = \exp\left\{ -\left(\frac{v_j \tau}{W_r} \right)^2 \right\} \quad \text{for active lateral transport with velocity } v_j, \quad (2.2.88)$$

$$M_j(\tau) = \frac{1}{\{1+(\tau/\tau_{D,j})^\alpha\}\sqrt{1+(W_r/W_z)^2(\tau/\tau_{D,j})^\alpha}} \quad \text{for anomalous subdiffusion.} \quad (2.2.89)$$

Information recovery from FCS measurements requires curve fitting of measured correlation data to an appropriate physical model for the underlying fluctuations. A nonlinear least-squares Levenberg–Marquardt algorithm is typically used. The goodness-of-fit is evaluated employing the parameter given by

$$\chi_R^2 = \frac{1}{m-p} \sum_{k=1}^{m} \frac{1}{\sigma_k^2} [G(\tau_k) - G_c(\tau_k)]^2, \quad (2.2.90)$$

where $G(\tau_k)$ is the measured value, $G_c(\tau_k)$ is the calculated value from model prediction, m is the total number of data, p is the number of floating parameters, and σ_k is the standard deviation of the measured value $G(\tau_k)$.

The effective observation volume is obtained by measuring a FCS curve for a monodisperse sample with a known diffusion coefficient. W_r and W_z are acquired by fitting of this FCS curve with the known diffusion coefficient.

By determining the effective observation volume, the average sample concentration $\langle C \rangle$ can be easily calculated from the autocorrelation amplitude $G_{3D}(0)$ in Eq. (2.2.75). It should be noted that the background signal such as stray light, scattered light, and the dark count can distort the correlation amplitude. Since the background signal remain constant, the correlation amplitude can be corrected by

$$G_{3D,c}(0) = G_{3D,m}(0)\left(\frac{\langle F_m \rangle}{\langle F_m \rangle - \langle B \rangle}\right), \qquad (2.2.91)$$

where $G_{3D,c}(0)$ is the corrected correlation amplitude, $G_{3D,m}(0)$ is the measured correlation amplitude, $\langle F_m \rangle$ is the measured fluorescence signal including the background, and $\langle B \rangle$ is the measured background signal.

The diffusion coefficient D is related to the hydrodynamic radius R of the (spherical) particle in solution by the Stokes–Einstein equation

$$D = \frac{k_B T}{6\pi \eta_v R}, \qquad (2.2.92)$$

where η_v is the viscosity of the medium, T is the temperature, and k_B is the Boltzmann constant. Thus, the particle size can be estimated from the diffusion coefficient. We see that the diffusion time of the small molecules is larger than that of the large molecules. Any changes in molecular shape or size that affect the hydrodynamic radius of the particle are reflected in the diffusion coefficient and thus in the average diffusion time. As shown in Fig. 2.11, the autocorrelation curve depends on the size and the number of fluorophore-labeled molecules.

2.2.5.2 *Cross-Correlation Spectroscopy* Next, we show fluorescence cross-correlation spectroscopy (FCCS) in which the cross-correlation function between two fluctuation signals from two different fluorophore-labeled molecules in a common observation volume is measured. Autocorrelation analysis provides information about characteristic repetitive processes of significant duration, while cross-correlation analysis between two fluctuation signals yields information about dynamics and interactions that relate the measured quantities with each other. Although the

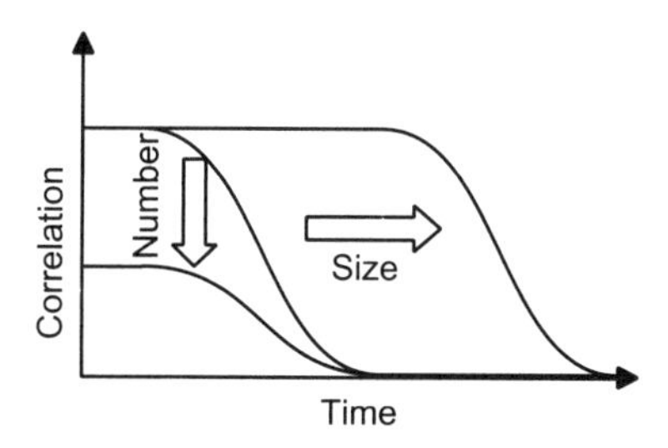

FIGURE 2.11 Influence of the number and the size of molecules to the correlation function.

diffusion coefficient in autocorrelation analysis depends on the hydrodynamic radius, which is proportional to the cubic root of the molecular mass for a spherical particle, a homogeneous increase in mass by a factor of 8 only doubles the diffusion time. On the other hand, it is difficult to detect small changes in molecular shape or size by the molecular interactions. On the other hand, FCCS can be used to detect the synchronous movement of two biomolecules [243].

In general, the two fluorescent species used in FCCS must have sufficiently separated emission spectra and must be simultaneously excited with equal efficiency. The two fluctuation signals must be measured with minimal cross-talk between the detector channels by two distinct detection channels that are significantly distinguished by their emission spectral properties. This is because the cross-correlation between the cross-talk signals reveals the coincidence of 100% even if the cross-correlation between the pure fluctuation signals is zero. In addition, the observation volume for one fluorescent species must overlap with that for the other fluorescent species. Anywhere that the two observation volumes do not overlap, the detection is of only one fluorophore emission, even if two fluorescently labeled molecules coexist. Thus, lack of observation volume overlap results in no cross-correlation signal. By using the simultaneous excitation with a single excitation wavelength, such artifacts can be eliminated [244]. Now, let us consider the fluorophore's fluorescence properties that are not changing within the observation time. The fluctuating signals recorded in the two detection channels are given as

$$\delta F_1(t) = \frac{\eta_1 N_A}{m\,(\hbar\omega)^m} q_{\mathrm{fl},1}^{(m)}, \sigma_{\mathrm{ex},1}^{(m)} \left\langle I_{\mathrm{ex},1}^{m}(\mathbf{0})\right\rangle \int_V S_1(\mathbf{r}) \left\{\delta C_1(\mathbf{r},t) + \delta C_{12}(\mathbf{r},t)\right\} dV, \tag{2.2.93}$$

$$\delta F_2(t) = \frac{\eta_2 N_A}{m\,(\hbar\omega)^m} q_{\mathrm{fl},2}^{(m)}, \sigma_{\mathrm{ex},2}^{(m)} \left\langle I_{\mathrm{ex},2}^{m}(\mathbf{0})\right\rangle \int_V S_2(\mathbf{r}) \left\{\delta C_2(\mathbf{r},t) + \delta C_{12}(\mathbf{r},t)\right\} dV, \tag{2.2.94}$$

where the spatial profile of the observation volume $S_j(\mathbf{r})$ is defined for either fluorescent species, the C_1 and C_2 are the concentrations for the single-color species with mobility-associated terms $M_j(\tau)$, and C_{12} is the concentration for the dual-color species with mobility-associated terms $M_{12}(\tau)$. The normalized cross-correlation function is defend as

$$G_X(\tau) = \frac{\langle \delta F_1(t)\delta F_2(t+\tau)\rangle}{\langle F_1(t)\rangle \langle F_2(t)\rangle}. \tag{2.2.95}$$

Assuming ideal conditions where the two detection channels have the same effective observation volume V_{eff}, fully separable emission spectra of species 1 and 2, and a negligible emission–absorption overlap integral, we can derive the autocorrelation

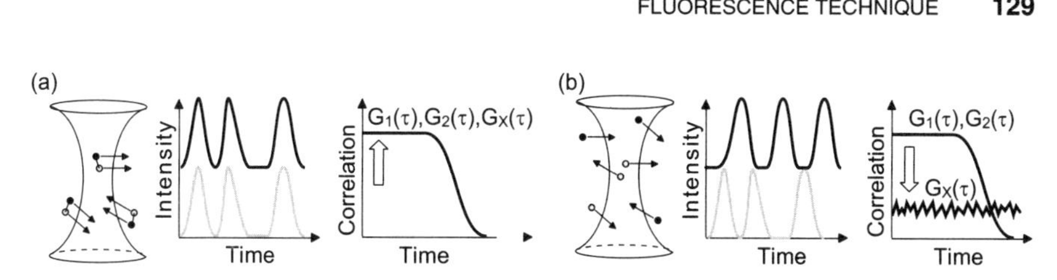

FIGURE 2.12 Fluorescence fluctuation with the interaction between two molecules (a) and without the interaction (b).

and cross-correlation curves in the following form:

$$G_{\mathrm{j}}(\tau) = \frac{\langle C_j \rangle M_j(\tau) + \langle C_{12} \rangle M_{12}(\tau)}{V_{\mathrm{eff}} \left(\langle C_j \rangle + \langle C_{12} \rangle \right)^2}, \tag{2.2.96}$$

$$G_{\mathrm{X}}(\tau) = \frac{\langle C_{12} \rangle M_{12}(\tau)}{V_{\mathrm{eff}} \left(\langle C_1 \rangle + \langle C_{12} \rangle \right) \left(\langle C_2 \rangle + \langle C_{12} \rangle \right)}. \tag{2.2.97}$$

If the efficiency of the interaction between two molecules is 100%, the cross-correlation curve comes close to agreement with the two autocorrelation curves, and the amplitude of the cross-correlation is large as shown in Fig. 2.12. However, in practice, these experimental data may not be obtained owing to the difference in the observation volume. According to Eqs. (2.2.96) and (2.2.97), the absolute amount of dual-color product (interacting molecules) can be easily derived from the autocorrelation amplitudes $G_1(0)$, $G_2(0)$ and cross-correlation amplitude $G_{\mathrm{X}}(0)$:

$$\langle C_{12} \rangle = \frac{G_{\mathrm{X}}(0)}{V_{\mathrm{eff}} G_1(0) G_2(0)}. \tag{2.2.98}$$

However, the absolute amount of interacting molecules often depends on the experimental condition such as the amount of protein expression. In practice, the relative cross-correlation amplitudes, which can be obtained by

$$\frac{G_{\mathrm{X}}(0)}{G_1(0)} = \frac{\langle C_{12} \rangle}{\langle C_2 \rangle + \langle C_{12} \rangle} = \frac{\langle N_{12} \rangle}{\langle N_2 \rangle + \langle N_{12} \rangle}, \tag{2.2.99}$$

$$\frac{G_{\mathrm{X}}(0)}{G_2(0)} = \frac{\langle C_{12} \rangle}{\langle C_1 \rangle + \langle C_{12} \rangle} = \frac{\langle N_{12} \rangle}{\langle N_1 \rangle + \langle N_{12} \rangle}, \tag{2.2.100}$$

are employed. The relative cross-correlation amplitude indicates the fraction of species 1 (2) interacted with species 2 (1) to all species 1 (2). FCCS has been used to detect the interactions between molecular species [245].

REFERENCES

1. J. B. Pawley, *Handbook of Biological Confocal Microscopy*, Springer Science and Business Media, New York, 2006.
2. M. D. Egger and M. Petráň, "New reflected-light microscope for viewing unstained brain and ganglion cells," *Science*, **157**, 305–307 (1967).
3. G. Q. Xiao, T. R. Corle, and G. S. Kino, "Real-time confocal scanning optical microscope," *Appl. Phys. Lett.*, **53**, 716–718 (1988).
4. T. Tanaami, Y. Sugiyama, and K. Mikuriya, "High-speed confocal laser microscopy," *Yokogawa Tech. Rep.*, **9**, 7–10 (1994).
5. A. Ichihara, T. Tanaami, K. Isozaki, Y. Sugiyama, Y. Kosugi, K. Mikuriya, M. Abe, and I. Umeda, "High-speed confocal fluorescent microscopy using a Nipkow scanner with microlenses for 3-d imaging of single fluorescent molecule in real time," *Bioimages*, **4**, 57–62 (1996).
6. I. Johnson, "Fluorescent probes for living cells," *Histochem. J.*, **30**, 123–140 (1998).
7. B. N. G. Giepmans, S. R. Adams, M. H. Ellisman, and R. Y. Tsien, "The fluorescent toolbox for assessing protein location and function," *Science*, **312**, 217–224 (2006).
8. R. P. Haugland, *The Handbook: A Guide to Fluorescent Probes and Labeling Technologies*, 10th ed., Molecular Probes, Eugene, OR, 2005.
9. H. J. Wories, J. H. Koek, G. Lodder, J. Lugtenburg, R. Fokkens, O. Driessen, and G. R. Mohn, "A novel water-soluble fluorescent probe: synthesis, luminescence and biological properties of the sodium salt of the 4-sulfonato-3,3′,5,5′-tetramethyl-2,2′-pyrromethen-1,1′-BF_2 complex," *Recl. Trav. Chim. Pays-Bas*, **104**, 288–291 (1985).
10. H. C. Kang, P. J. Fisher, F. G. Prendergast, and R. P. Haugland, "Bodipy: a novel fluorescein and NBD substitute," *J. Cell Biol.*, **107**, 34a (1988).
11. R. B. Mujumdar, L. A. Ernst, S. R. Mujumdar, C. J. Lewis, and A. S. Waggoner, "Cyanine dye labeling reagents: sulfoindocyanine succinimidyl esters," *Bioconjug. Chem.*, **4**, 105–111 (1993).
12. N. Panchuk-Voloshina, R. P. Haugland, J. Bishop-Stewart, M. K. Bhalgat, P. J. Millard, F. Mao, W.-Y. Leung, and R. P. Haugland, "Alexa dyes, a series of new fluorescent dyes that yield exceptionally bright, photostable conjugates," *J. Histochem. Cytochem.*, **47**, 1179–1188 (1999).
13. S. R. Adams, A. T. Harootunian, Y. J. Buechler, S. S. Taylor, and R. Y. Tsien, "Fluorescence ratio imaging of cyclic AMP in single cells," *Nature*, **349**, 694–697 (1991).
14. G. V. Richieri, R. T. Ogata, and A. M. Kleinfeld, "A fluorescently labeled intestinal fatty acid binding protein," *J. Biol. Chem.*, **267**, 23495–23501 (1992).
15. N. C. Shaner, P. A. Steinbach, and R. Y. Tsien, "A guide to choosing fluorescent proteins," *Nat. Methods*, **2**, 905–909 (2005).
16. N. C. Shaner, G. H. Patterson, and M. W. Davidson, "Advances in fluorescent protein technology," *J. Cell Science*, **120**, 4247–4260 (2007).
17. D. M. Chudakov, S. Lukyanov, and K. A. Lukyanov, "Fluorescent proteins as a toolkit for in vivo imaging," *Trends Biotechnol.*, **23**, 605–613 (2005).
18. D. M. Chudakov, M. V. Matz, S. Lukyanov, and K. A. Lukyanov, "Fluorescent proteins and their applications in imaging living cells and tissues," *Physiol. Rev.*, **90** 1103–1163 (2010).

19. A. Miyawaki, "Proteins on the move: insights gained from fluorescent protein technologies," *Nat. Rev. Mol. Cell Biol.*, **12**, 656–668 (201
20. M. Chalfie, Y. Tu, G. Euskirchen, W. W. Ward, and D. C. Prasher, "Green fluorescent protein as a marker for gene expression," *Science*, **263**, 802–805 (1994).
21. R. Y. Tsien, "The green fluorescent protein," *Annu. Rev. Biochem.*, **67**, 509–544 (1998).
22. M. V. Matz, A. F. Fradkov, Y. A. Labas, A. P. Savitsky, A. G. Zaraisky, M. L. Markelov, and S. A. Lukyanov, "Fluorescent proteins from nonbioluminescent Anthozoa species," *Nat. Biotechnol.*, **17**, 969–973 (1999).
23. A. F. Fradkov, Y. Chen, L. Ding, E. V. Barsova, M. V. Matz, and S. A. Lukyanov, "Novel fluorescent protein from Discosoma coral and its mutants possesses a unique far-red fluorescence," *FEBS Lett.*, **479**, 127–130 (2000).
24. V. V. Verkhusha and K. A. Lukyanov, "The molecular properties and applications of Anthozoa fluorescent proteins and chromoproteins," *Nat. Biotechnol.*, **22**, 289–296 (2004).
25. R. E. Campbell, O. Tour, A. E. Palmer, P. A. Steinbach, G. S. Baird, D. A. Zacharias, and R. Y. Tsien, "A monomeric red fluorescent protein," *Proc. Natl. Acad. Sci. USA*, **99**, 7877–7882 (2002).
26. N. C. Shaner, R. E. Campbell, P. A. Steinbach, B. N. G. Giepmans, A. E. Palmer, and R. Y. Tsien, "Improved monomeric red, orange and yellow fluorescent proteins derived from Discosoma sp. red fluorescent protein," *Nat. Biotechnol.*, **22**, 1567–1572 (2004).
27. L. Wang, W. C. Jackson, P. A. Steinbach, and R. Y. Tsien, "Evolution of new nonantibody proteins via iterative somatic hypermutation," *Proc. Natl. Acad. Sci. USA*, **101**, 16745–16749 (2004).
28. X. Shu, A. Royant, M. Z. Lin, T. A. Aguilera, V. Lev-Ram, P. A. Steinbach, and R. Y. Tsien, "Mammalian expression of infrared fluorescent proteins engineered from a bacterial phytochrome," *Science*, **324**, 804–807 (2009).
29. R. Heim, D. C. Prasher, and R. Y. Tsien, "Wavelength mutations and posttranslational autoxidation of green fluorescent protein," *Proc. Natl. Acad. Sci. USA*, **91**, 12501–12504 (1994).
30. A. B. Cubitt, R. Heim, S. R. Adams, A. E. Boyd, L. A. Gross, and R. Y. Tsien, "Understanding, improving and using green fluorescent proteins," *Trends Biochem. Sci.*, **20**, 448–455 (1995).
31. H. Fukuda, M. Arai, and K. Kuwajima, "Folding of green fluorescent protein and the cycle3 mutant," *Biochem.*, **39**, 12025–12032 (2000).
32. J. Zhang, R. E. Campbell, A. Y. Ting, and R. Y. Tsien, "Creating new fluorescent probes for cell biology," *Nat. Rev. Mol. Cell. Biol.*, **3**, 906–918 (2002).
33. W. Tomosugi, T. Matsuda, T. Tani, T. Nemoto, I. Kotera, K. Saito, K. Horikawa, and T. Nagai, "An ultramarine fluorescent protein with increased photostability and pH insensitivity," *Nat. Methods*, **6**, 351–353 (2009).
34. G. J. Kremers, J. Goedhart, D. J. van den Heuvel, H. C. Gerritsen, and T. W. Gadella Jr, "Improved green and blue fluorescent proteins for expression in bacteria and mammalian cells.,"*Biochemistry*, **46**, 3775–3783 (2007).
35. G. H. Patterson, S. M. Knobel, W. D. Sharif, S. R. Kain, and D.W. Piston, "Use of the green fluorescent protein and its mutants in quantitative fluorescence microscopy," *Biophys. J.*, **73**, 2782–2790 (1997).

36. H. W. Ai, N. C. Shaner, Z. Cheng, R. Y. Tsien, and R. E. Campbell, "Exploration of new chromophore structures leads to the identification of improved blue fluorescent proteins," *Biochemistry.*, **46**, 5904–5910 (2007).
37. M. A. Mena, T. P. Treynor, S. L. Mayo, and P. S. Daugherty, "Blue fluorescent proteins with enhanced brightness and photostability from a structurally targeted library," *Nat. Biotechnol.*, **24**, 1569–1571(2006).
38. O. M. Subach, I. S. Gundorov, M. Yoshimura, F. V. Subach, J. Zhang, D. Gruenwald, E. A. Souslova, D. M. Chudakov, and V. V. Verkhusha, "Conversion of red fluorescent protein into a bright blue probe," *Chem. Biol.*, **15**, 1116–1124 (2008).
39. J. Goedhart, L. van Weeren, M. A. Hink, N. O. Vischer, K. Jalink, and T. W. Gadella Jr, "Bright cyan fluorescent protein variants identified by fluorescence lifetime screening," *Nat. Methods*, **7**, 137–139 (2010).
40. J. Goedhart, D. von Stetten, M. Noirclerc-Savoye, M. Lelimousin, L. Joosen, M. A. Hink, L. van Weeren, T. W. Gadella Jr, and A. Royant, "Structure-guided evolution of cyan fluorescent proteins towards a quantum yield of 93%," *Nat. Commun.*, **3**, 751/1–9 (2012).
41. D. A. Zacharias, J. D. Violin, A. C. Newton, and R. Y. Tsien, "Partitioning of lipid-modified monomeric GFPs into membrane microdomains of live cells," *Science*, **296**, 913–916 (2002).
42. G. J. Kremers, J. Goedhart, E. B. van Munster, and T. W. Gadella Jr. "Cyan and yellow super fluorescent proteins with improved brightness, protein folding, and FRET Förster radius," *Biochemistry*, **45**, 6570–6580 (2006).
43. M. A. Rizzo, G. H. Springer, B. Granada, and D. W. Piston, "An improved cyan fluorescent protein variant useful for FRET," *Nat. Biotechnol.*, **22**, 445–449 (2004).
44. M. L. Markwardt, G. J. Kremers, C. A. Kraft, K. Ray, P. J. Cranfill, K. A. Wilson, R. N. Day, R. M. Wachter, M. W. Davidson, and M. A. Rizzo, "An improved cerulean fluorescent protein with enhanced brightness and reduced reversible photoswitching," *PLoS One*, **6**, e17896 (2011).
45. A. W. Nguyen and P. S. Daugherty, "Evolutionary optimization of fluorescent proteins for intracellular FRET," *Nat. Biotechnol.*, **23**, 355–360 (2005).
46. S. Karasawa, T. Araki, T. Nagai, H. Mizuno, and A. Miyawaki, "Cyan-emitting and orange-emitting fluorescent proteins as a donor/acceptor pair for fluorescence resonance energy transfer," *Biochem. J.*, **381**, 307–312 (2004).
47. H. W. Ai, J. N. Henderson, S. J. Remington, and R. E. Campbell, "Directed evolution of a monomeric, bright and photostable version of Clavularia cyan fluorescent protein: structural characterization and applications in fluorescence imaging," *Biochem. J.*, **400**, 531–540 (2006).
48. N. S. Xia, W. X. Luo, J. Zhang, X. Y. Xie, H. J. Yang, S. W. Li, M. Chen, and M. H. Ng, "Bioluminescence of Aequorea macrodactyla, a common jellyfish species in the East China Sea," *Mar. Biotechnol.*, **4**, 155–162 (2002).
49. S. Karasawa, T. Araki, M. Yamamoto-Hino, and A. Miyawaki, "A green-emitting fluorescent protein from Galaxeidae coral and its monomeric version for use in fluorescent labeling," *J. Biol. Chem.*, **278**, 34167–3471 (2003).
50. R. Heim, A. B. Cubitt, and R. Y. Tsien, "Improved green fluorescence," *Nature*, **373**, 663–664 (1995).

51. H. W. Ai, S. G. Olenych, P. Wong, M. W. Davidson, and R. E. Campbell, "Hue-shifted monomeric variants of Clavularia cyan fluorescent protein: identification of the molecular determinants of color and applications in fluorescence imaging," *BMC Biol.*, **6**, 13 (2008).
52. O. Zapata-Hommer, and O. Griesbeck, "Efficiently folding and circularly permuted variants of the sapphire mutant of GFP," *BMC Biotechnol.*, **3**, 5 (2003).
53. Y. Ding, H.-W. Ai, H. Hoi, and R. E. Campbell, "Förster resonance energy transfer-based biosensors for multiparameter ratiometric imaging of Ca^{2+} dynamics and caspase-3 activity in single cells," *Anal. Chem.*, **83**, 9687–9693 (2011).
54. H. W. Ai, K. L. Hazelwood, M. W. Davidson, and R. E. Campbell, "Fluorescent protein FRET pairs for ratiometric imaging of dual biosensors," *Nat. Methods*, **5**, 401–403 (2008).
55. A. B. Cubitt, L. A. Woollenweber, and R. Heim, "Understanding structurefunctionrelationships in the Aequorea victoria green fluorescent protein," *Methods Cell Biol.*, **58**, 19–30 (1999).
56. M. Ormö, A. B. Cubitt, K. Kallio, L. A. Gross, R. Y. Tsien, and S. J. Remington, "Crystal structure of the Aequorea victoria green fluorescent protein," *Science*, **273**, 1392–1395 (1996).
57. T. Nagai, K. Ibata, E. S. Park, M. Kubota, K. Mikoshiba, and A. Miyawaki, "A variant of yellow fluorescent protein with fast and efficient maturation for cell-biological applications," *Nat. Biotechnol.*, **20**, 87–90 (2002).
58. O. Griesbeck, G. S. Baird, R. E. Campbell, D. A. Zacharias, and R. Y. Tsien, "Reducing the environmental sensitivity of yellow fluorescent protein. Mechanism and applications," *J. Biol. Chem.*, **276**, 29188–29194 (2001).
59. S. Karasawa, T. Araki, T. Nagai, H. Mizuno, and A. Miyawaki, "Cyan-emitting and orange-emitting fluorescent proteins as a donor/acceptor pair for fluorescence resonance energy transfer," *Biochem. J.*, **381**, 307–312 (2004).
60. A. Sakaue-Sawano, H. Kurokawa, T. Morimura, A. Hanyu, H. Hama, H. Osawa, S. Kashiwagi, K. Fukami, T. Miyata, H. Miyoshi, T. Imamura, M. Ogawa, H. Masai, and A. Miyawaki, "Visualizing spatiotemporal dynamics of multicellular cell-cycle progression," *Cell*, **132**, 487–498 (2008).
61. N. C. Shaner, M. Z. Lin, M. R. McKeown, P. A. Steinbach, K. L. Hazelwood, M. W. Davidson, and R. Y. Tsien, "Improving the photostability of bright monomeric orange and red fluorescent proteins," *Nat. Methods*, **5**, 545–551(2008).
62. D. M. Shcherbakova, M. A. Hink, L. Joosen, T. W. Gadella, and V. V. Verkhusha, "An orange fluorescent protein with a large stokes shift for single-excitation multicolor FCCS and FRET imaging," *J. Am. Chem. Soc.*, **134**, 7913–7923 (2012).
63. E. M. Merzlyak, J. Goedhart, D. Shcherbo, M. E. Bulina, A. S. Shcheglov, A. F. Fradkov, A. Gaintzeva, K. A. Lukyanov, S. Lukyanov, T. W. Gadella, and D. M. Chudakov, "Bright monomeric red fluorescent protein with an extended fluorescence lifetime," *Nat. Methods*, **4**, 555–557 (2007).
64. S. Kredel, F. Oswald, K. Nienhaus, K. Deuschle, C. Röcker, M. Wolff1, R. Heilker, G. U. Nienhaus, and J. Wiedenmann, "mRuby, a bright monomeric red fluorescent protein for labeling of subcellular structures," *PLoS ONE*, **4**, e4391 (2009).
65. K. D. Piatkevich, J. Hulit, O. M. Subach, B. Wu, A. Abdulla, J. E. Segall, and V. V. Verkhusha, "Monomeric red fluorescent proteins with a large Stokes shift," *Proc. Natl. Acad. Sci. USA*, **107** 5369–5374 (2010).

66. D. A. Shagin, E. V. Barsova, Y. G. Yanushevich, A. F. Fradkov, K. A. Lukyanov, Y. A. Labas, T. N. Semenova, J. A. Ugalde, A. Meyers, J. M. Nunez, E. A. Widder, S. A. Lukyanov, and M. V. Matz, "GFP-like proteins as ubiquitous metazoan superfamily: evolution of functional features and structural complexity," *Mol. Biol. Evol.*, **21**, 841–850 (2004).
67. T. Kogure, S. Karasawa, T. Araki, K. Saito, M. Kinjo, and A. Miyawaki, "A fluorescent variant of a protein from the stony coral Montipora facilitates dual-color single-laser fluorescence cross-correlation spectroscopy," *Nat. Biotechnol.*, **24**, 577–581 (2006).
68. D. Shcherbo, C. S. Murphy, G. V. Ermakova, E. A. Solovieva, T. V. Chepurnykh, A. S. Shcheglov, V. V. Verkhusha, V. Z. Pletnev, K. L. Hazelwood, P. M. Roche, S. Lukyanov, A. G. Zaraisky, M. W. Davidson, and D. M. Chudakov, "Far-red fluorescent tags for protein imaging in living tissues," *Biochem. J.*, **418**, 567–574 (2009).
69. D. Shcherbo, E. M. Merzlyak, T. V. Chepurnykh, A. F. Fradkov, G. V. Ermakova, E. A. Solovieva, K. A. Lukyanov, E. A. Bogdanova, A. G. Zaraisky, S. Lukyanov, and D. M. Chudakov, "Bright far-red fluorescent protein for whole-body imaging" *Nat. Methods*, **4**, 741–746 (2007).
70. D. Shcherbo, I. I. Shemiakina, A. V. Ryabova, K. E. Luker, B. T. Schmidt, E. A. Souslova, T. V. Gorodnicheva, L. Strukova, K. M. Shidlovskiy, O. V. Britanova, A. G. Zaraisky, K. A. Lukyanov, V. B. Loschenov, G. D. Luker, and D. M. Chudakov, "Near-infrared fluorescent proteins," *Nat. Methods*, **7**, 827–829 (2010).
71. M. Z. Lin, M. R. McKeown, H. L. Ng, T. A. Aguilera, N. C. Shaner, R. E. Campbell, S. R. Adams, L. A. Gross, W. Ma, T. Alber, and R. Y. Tsien, "Autofluorescent proteins with excitation in the optical window for intravital imaging in mammals," *Chem. Biol.*, **16**, 1169–1179 (2009).
72. M. A. Shkrob, Y. G. Yanushevich, D. M. Chudakov, N. G. Gurskaya, Y. A. Labas, S. Y. Poponov, N. N. Mudrik, S. Lukyanov, and K. A. Lukyanov, "Far-red fluorescent proteins evolved from a blue chromoprotein from Actinia equine," *Biochem. J.*, **392**, 649–654 (2005).
73. G. S. Filonov, K. D. Piatkevich, L.-M. Ting, J. Zhang, K. Kim, and V. V. Verkhusha, "Bright and stable near-infrared fluorescent protein for in vivo imaging," *Nat Biotechnol.*, **29** 757–761 (2011).
74. A. Terskikh, A. Fradkov, G. Ermakova, A. Zaraisky, P. Tan, A. V. Kajava, X. Zhao, S. Lukyanov, M. Matz, S. Kim, I. Weissman, and P. Siebert, " 'Fluorescent timer': protein that changes color with time," *Science*, **290**, 1585–1588 (2000).
75. F. V. Subach, O. M. Subach, I. S. Gundorov, K. S. Morozova, K. D. Piatkevich, A. M. Cuervo, and V. V. Verkhusha, "Monomeric fluorescent timers that change color from blue to red report on cellular trafficking," *Nat. Chem. Biol.*, **5**, 118–126 (2009).
76. V. V. Verkhusha, D. M. Chudakov, N. G. Gurskaya, S. Lukyanov, and K. A. Lukyanov, "Common pathway for the red chromophore formation in fluorescent proteins and chromoproteins," *Chem. Biol.*, **11**, 845–854 (2004).
77. R. Mirabella, C. Franken, G. N. van der Krogt, T. Bisseling, and R. Geurts, "Use of the fluorescent timer DsRED-E5 as reporter to monitor dynamics of gene activity in plants," *Plant Physiol.*, **135**, 1879–1887 (2004).
78. R. R. Duncan, J. Greaves, U. K. Wiegand, I. Matskevich, G. Bodammer, D. K. Apps, M. J. Shipston, and R. H. Chow, "Functional and spatial segregation of secretory vesicle pools according to vesicle age," *Nature*, **422**, 176–180 (2003).

79. M. Solimena and H. H. Gerdes, "Secretory granules: and the last shall be first," *Trends Cell Biol.*, **13**, 399–402 (2003).
80. B. A. Kozel, B. J. Rongish, A. Czirok, J. Zach, C. D. Little, E. C. Davis, R. H. Knutsen, J. E. Wagenseil, M. A. Levy, and R. P. Mecham, "Elastic fiber formation: a dynamic view of extracellular matrix assembly using timer reporters," *J. Cell Physiol.*, **207**, 87–96 (2006).
81. A. Czirok, J. Zach, B. A. Kozel, R. P. Mecham, E. C. Davis, and B. J. Rongish, "Elastic fiber macro-assembly is a hierarchical, cell motionmediated process," *J. Cell Physiol.*, **207**, 97–106 (2006).
82. T. Miyatsuka, Z. Li, and M. S. German, "The chronology of islet differentiationrevealed by temporal cell labeling," *Diabetes*, **58**, 1863–1868 (2009).
83. A. Miyawaki, J. Llopis, R. Heim, J. M. McCaffery, J. A. Adams, M. Ikura, and R. Y. Tsien, "Fluorescent indicators for Ca^{2+} based on green fluorescent proteins and calmodulinm," *Nature*, **388**, 882–887 (1997).
84. T. Nagai, S. Yamada, T. Tominaga, M. Ichikawa, and A. Miyawaki, "Expanded dynamic range of fluorescent indicators for Ca2+ by circularly permuted yellow fluorescent proteins," *Proc. Natl. Acad. Sci. USA*, **101**, 10554–10559 (2004)
85. M. Mank, A. F. Santos, S. Direnberger, T. D. Mrsic-Flogel, S. B. Hofer, V. Stein, T. Hendel, D. F. Reiff, C. Levelt, A. Borst, T. Bonhoeffer, M. Hubener, and O. Griesbeck, "A genetically encoded calcium indicator for chronic in vivo two-photon imaging," *Nat. Methods*, **5**, 805–811 (2008).
86. A. E. Palmer, M. Giacomello, T. Kortemme, S. A. Hires, V. Lev-Ram, D. Baker, and R. Y. Tsien, "Ca^{2+} indicators based on computationally redesigned calmodulin-peptide pairs," *Chem. Biol.*, **13**, 521–530 (2006).
87. M. Sato, N. Hida, T. Ozawa, and Y. Umezawa, "Fluorescent indicators for cyclic GMP based on cyclic GMP-dependent protein kinase Ialpha and green fluorescent proteins," *Anal. Chem.*, **72**, 5918–5924 (2000).
88. A. Honda, S. R. Adams, C. L. Sawyer, V. Lev-Ram, and R. Y. Tsien, W. R. Dostmann, "Spatiotemporal dynamics of guanosine 3′,5′-cyclic monophosphate revealed by a genetically encoded, fluorescent indicator," *Proc. Natl. Acad. Sci. USA*, **98**, 2437–2442 (2001).
89. V. O. Nikolaev, S. Gambaryan, and M. J. Lohse, "Fluorescent sensors for rapid monitoring of intracellular cGMP," *Nat. Methods*, **3**, 23–25 (2006).
90. V. O. Nikolaev and M. J. Lohse, "Monitoring of cAMP synthesis and degradation in living cells," *Physiology*, **21**, 86–92 (2006).
91. S. Okumoto, L. L. Looger, K. D. Micheva, R. J. Reimer, S. J. Smith, and W.B. Frommer, "Detection of glutamate release from neurons by genetically encoded surface-displayed FRET nanosensors," *Proc. Natl. Acad. Sci. USA*, **102**, 8740–8745 (2005).
92. T. Kaper, L. L. Looger, H. Takanaga, M. Platten, L. Steinman, and W. B. Frommer, "Nanosensor detection of an immunoregulatory tryptophan influx/kynurenine efflux cycle," *PLoS Biol.*, **5**, e257 (2007).
93. T. Kaper, I. Lager, L. L. Looger, D. Chermak, and W. B. Frommer, "Fluorescence resonance energy transfer sensors for quantitative monitoring of pentose and disaccharide accumulation in bacteria," *Biotechnol. Biofuels*, **1**, 11 (2008).

94. N. Mochizuki, S. Yamashita, K. Kurokawa, Y. Ohba, T. Nagai, A. Miyawaki, and M. Matsuda, "Spatio-temporal images of growth-factor-induced activation of Ras and Rap1," *Nature*, **411**, 1065–1068 (2001).
95. R. E. Itoh, K. Kurokawa, Y. Ohba, H. Yoshizaki, N. Mochizuki, and M. Matsuda, "Activation of rac and cdc42 video imaged by fluorescent resonance energy transfer-based single-molecule probes in the membrane of living cells," *Mol. Cell Biol.*, **22**, 6582–6591 (2002).
96. Y. Nagai, M. Miyazaki, R. Aoki, T. Zama, S. Inouye, K. Hirose, M. Iino, and M. Hagiwara, "A fluorescent indicator for visualizing cAMPinduced phosphorylation in vivo," *Nat. Biotechnol.*, **18**, 313–316 (2000).
97. K. Kurokawa, N. Mochizuki, Y. Ohba, H. Mizuno, A. Miyawaki, and M. Matsuda, "A pair of fluorescent resonance energy transfer-based probes for tyrosine phosphorylation of the CrkII adaptor protein in vivo," *J. Biol. Chem.*, **276**, 31305–31310 (2001).
98. A. Y. Ting, K. H. Kain, R. L. Klemke, and R. Y. Tsien, "Genetically encoded fluorescent reporters of protein tyrosine kinase activities in living cells," *Proc. Natl. Acad. Sci. USA*, **98**, 15003–15008 (2001).
99. J. Zhang, Y. Ma, S. S. Taylor, and R. Y. Tsien, "Genetically encoded reporters of protein kinase A activity reveal impact of substrate tethering," *Proc. Natl. Acad. Sci. USA*, **98**, 14997–15002 (2001).
100. K. Sasaki, M. Sato, and Y. Umezawa, "Fluorescent indicators for Akt/protein kinase B and dynamics of Akt activity visualized in living cells," *J. Biol. Chem.*, **278**, 30945–30951 (2003).
101. J. D. Violin, J. Zhang, R. Y. Tsien, and A.C. Newton, "A genetically encoded fluorescent reporter reveals oscillatory phosphorylation by protein kinase C," *J. Cell. Biol.*, **161**, 899–909 (2003).
102. H. M. Green and J. Alberola-Ila, "Development of ERK activity sensor, an in vitro, FRET-based sensor of extracellular regulated kinase activity," *BMC Chem. Biol.*, **5**, 1 (2005).
103. M. T. Kunkel, Q. Ni, R. Y. Tsien, J. Zhang, and A. C. Newton, "Spatiotemporal dynamics of protein kinase B/Akt signaling revealed by a genetically encoded fluorescent reporter," *J. Biol. Chem.*, **280**, 5581–5587 (2005).
104. M. T. Kunkel, A. Toker, R. Y. Tsien, and A. C. Newton, "Calcium-dependent regulation of protein kinase D revealed by a genetically encoded kinase activity reporter," *J. Biol. Chem.*, **282**, 6733–6742 (2007).
105. C. D. Harvey, A. G. Ehrhardt, C. Cellurale, H. Zhong, R. Yasuda, R. J. Davis, and K. Svoboda, "A genetically encoded fluorescent sensor of ERK activity," *Proc. Natl. Acad. Sci. USA*, **105**, 19264–19269 (2008).
106. N. P. Mahajan, D. C. Harrison-Shostak, J. Michaux, and B. Herman, "Novel mutant green fluorescent protein protease substrates reveal the activation of specific caspases during apoptosis," *Chem. Biol.*, **6**, 401–409 (1999).
107. L. Tyas, V. A. Brophy, A. Pope, A. J. Rivett, and J. M. Tavare, "Rapid caspase-3 activation during apoptosis revealed using fluorescenceresonance energy transfer," *EMBO Rep.*, **1**, 266–270 (2000).
108. K. Q. Luo, V. C. Yu, Y. Pu, and D. C. Chang, "Application of the fluorescence resonance energy transfer method for studying the dynamics of caspase-3 activation during UV-induced apoptosis in living HeLa cells," *Biochem. Biophys. Res. Commun.*, **283**, 1054–1060 (2001).

109. A. G. Harpur, F. S. Wouters, and P. I. Bastiaens, "Imaging FRET between spectrally similar GFP molecules in single cells," *Nat. Biotechnol.*, **19**, 167–169 (2001).
110. T. Nagai and A. Miyawaki, "A high-throughput method for development of FRET-based indicators for proteolysis," *Biochem. Biophys. Res. Commun.*, **319**, 72–77 (2004).
111. S. Karasawa, T. Araki, T. Nagai, H. Mizuno, and A. Miyawaki, "Cyanemitting and orange-emitting fluorescent proteins as a donor/acceptor pair for fluorescence resonance energy transfer," *Biochem. J.*, **381**, 307–312 (2004).
112. X. Wu, J. Simone, D. Hewgill, R. Siegel, P. E. Lipsky, and L. He, "Measurement of two caspase activities simultaneously in living cells by a novel dual FRET fluorescent indicator probe," *Cytometry A*, **69**, 477–486 (2006).
113. Y. Wu, D. Xing, and W. R. Chen, "Single cell FRET imaging for determination of pathway of tumor cell apoptosis induced by photofrin-PDT," *Cell Cycle*, **5**, 729–734 (2006).
114. H. W. Ai, K. L. Hazelwood, M. W. Davidson, and R. E. Campbell, "Fluorescent protein FRET pairs for ratiometric imaging of dual biosensors," *Nat. Methods*, **5**, 401–403 (2008).
115. D. Shcherbo, E. A. Souslova, J. Goedhart, T. V. Chepurnykh, A. Gaintzeva, I. I. Shemyakina, T. W. Gadella, S. Lukyanov, and D. M. Chudakov, "Practical and reliable FRET/FLIM pair of fluorescent proteins," *BMC Biotechnol.*, **9**, 24 (2009).
116. F. Meng, T. M. Suchyna, and F. Sachs, "A fluorescence energy transferbased mechanical stress sensor for specific proteins in situ," *FEBS J.*, **275**, 3072–3087 (2008).
117. M. S. Siegel and E. Y. Isacoff, "A genetically encoded optical probe of membrane voltage," *Neuron*, **19**, 735–741 (1997)
118. Y. Murata, H. Iwasaki, M. Sasaki, K. Inaba, and Y. Okamura, "Phosphoinositide phosphatase activity coupled to an intrinsic voltage sensor," *Nature*, **435**, 1239–1243 (2005).
119. D. Dimitrov, Y. He, H. Mutoh, B. J. Baker, L. Cohen, W. Akemann, and T. Knöpfel, "Engineering and characterization of an enhanced fluorescent protein voltage sensor," *PLoS ONE*, **2**, e440 (2007).
120. H. Tsutsui, S. Karasawa, Y. Okamura, and A. Miyawaki, "Improving membrane voltage measurements using FRET with new fluorescent proteins," *Nat. Methods*, **5**, 683–685 (2008).
121. H. Mutoh, A. Perron, D. Dimitrov, Y. Iwamoto, W. Akemann, D. M. Chudakov, and T. Knopfel, "Spectrally-resolved response properties of the three most advanced FRET based fluorescent protein voltage probes," *PLoS ONE*, **4**, e4555 (2009).
122. J. Lippincott-Schwartz, and G. H. Patterson, "Development and use of fluorescent protein markers in living cells," *Science*, **300**, 87–91 (2003).
123. C.-D. Hu, Y. Chinenov, and T. K. Kerppola, "Visualization of interactions among bZIP and Rel proteins in living cells using bimolecular fluorescence complementation," *Mol. Cell*, **9**, 789–798 (2002).
124. C.-D. Hu and T. K. Kerppola, "Simultaneous visualization of multiple protein interactions in living cells using multicolor fluorescence complementation analysis," *Nat. Biotechnol.*, **21**, 539–545 (2003).
125. S. Zhang, C. Ma, and M. Chalfie, "Combinatorial marking of cells and organelles with reconstituted fluorescent proteins," *Cell*, **119**, 137–144 (2004).
126. S. Cabantous, T. C. Terwilliger, and G. S. Waldo, "Protein tagging and detection with engineered self-assembling fragments of green fluorescent protein," *Nat. Biotechnol.*, **23**, 102–107 (2005).

127. J. K. Jaiswal, S. M. Simon, "Potentials and pitfalls of fluorescent quantum dots for biological imaging," *Trends Cell Biol.*, **14**, 497–504 (2004).

128. X. Michalet, F. F. Pinaud, L. A. Bentolila, J. M. Tsay, S. Doose, J. J. Li, G. Sundaresan, A. M. Wu, S. S. Gambhir and S. Weiss, "Quantum dots for live cells, in vivo imaging, and diagnostics," *Science*, **307**, 538–544 (2005). A. P. Alivisatos, "Semiconductor clusters, nanocrystals, and quantum dots," *Science*, **271**, 933–937 (1996).

129. W. C. Chan and S. Nie, "Quantum dot bioconjugates for ultrasensitive nonisotopic detection," *Science*, **281**, 2016–2018 (1998).

130. A. M. Derfus, W. C. W. Chan, and S. N. Bhatia, "Probing the cytotoxicity of semiconductor quantum dots," *Nano Lett.*, **4**, 11–18 (2004).

131. L. C. Mattheakis, J. M. Dias, Y.-J. Choi, J. Gong, M. P. Bruchez, J. Liu, and E. Wang, "Optical coding of mammalian cells using semiconductor quantum dots," *Anal. Biochem.*, **327**, 200–208 (2004).

132. J. K. Jaiswal, H. Mattoussi, J. M. Mauro, and S. M. Simon1, "Long-term multiple color imaging of live cells using quantum dot bioconjugates," *Nat. Biotechnol.*, **21**, 47–51 (2002)

133. X. Wu, H. Liu, J. Liu, K. N. Haley, J. A. Treadway, J. P. Larson, N. Ge, F. Peale, and M. P. Bruchez, "Immunofluorescent labeling of cancer marker Her2 and other cellular targets with semiconductor quantum dots," *Nat. Biotechnol.*, **21**, 41–46 (2002).

134. C. A. Leatherdale, W. K.Woo, F. V. Mikulec, and M. G. Bawendi, "On the absorption cross section of CdSe nanocrystal quantum dots," *J. Phys. Chem. B*, **106**, 7619–7622 (2002).

135. E. J. Soini, L. J. Pelliniemi, I. A. Hemmilä, V. M. Mukkala, J. J. Kankare, and K. Fröjdman, "Lanthanide chelates as new fluorochrome labels for cytochemistry," *J. Histochem. Cytochem.*, **36**, 1449–1451 (1988).

136. L. Sevéus, M. Väisälä, I. Hemmilä, H. Kojola, G. M. Roomans, and E. Soini, "Use of fluorescent europium chelates as labels in microscopy allows glutaraldehyde fixation and permanent mounting and leads to reduced autofluorescence and good long-term stability," *Microsc. Res. Tech.*, **28**, 149–154 (1994).

137. G. Vereb, E. Jares-Erijman, P. R. Selvin, and T. M. Jovin, "Temporally and spectrally resolved imaging microscopy of lanthanide chelates," *Biophys. J.*, **74**, 2210–2222 (1998).

138. F. N. Castellano and J. R. Lakowicz, "A water-soluble luminescence oxygen sensor," *Photochem. Photobiol.*, **67**, 179–183 (1998).

139. P. R. Selvin, "Principles and biophysical applications of lanthanide-based proves," *Annu. Rev. Biophys. Biomol. Struct.*, **31**, 275–302 (2002).

140. M. E. Dickinson, G. Bearman, S. Tille, R. Lansford, and S. E. Fraser, "Multi-spectral imaging and linear unmixing add a whole new dimension to laser scanning fluorescence microscopy," *BioTechniques*, **31**, 1272–1278 (2001).

141. T. Zimmermann, "Spectral imaging and linear unmixing in light microscopy," *Adv. Biochem. Eng. Biotechnol.*, **95**, 245–265 (2005).

142. T. Förster, "Zwischenmolekulare Energiewanderung und Fluoreszenz," *Ann. Physik*, **437**, 55–75 (1948).

143. A. Miyawaki, "Development of probes for cellular functions using fluorescent proteins and fluorescence resonance energy transfer," *Annu. Rev. Biochem.*, **80**, 357–373 (2011).

144. D. Kosk-Kosicka, T. Bzdega, and A. Wawrynow, "Fluorescence energy transfer studies of purified erythrocyte Ca^{2+}-ATPase Ca^{2+}- regulated activation by oligomerization," *J. Biol. Chem.*, **264**, 19495–19499 (1989).
145. P. I. H. Bastiaens and T. M. Jovin, "Microspectroscopic imaging tracks the intracellular processing of a signal transduction protein: Fluorescent-labeled protein kinase C βI.," *Proc. Natl. Acad. Sci. USA*, **93**, 8407–8412 (1996).
146. M. Elangovan, H. Wallrabe, Y. Chen, R. N. Day, M. Barroso, and A. Periasamy, "Characterization of one- and two-photon excitation fluorescence resonance energy transfer microscopy," *Methods*, **29**, 58–73 (2003).
147. N. P. Mahajan, K. Linder, G. Berry, G. W. Gordon, R. Heim, and B. Herman, "Bcl-2 and Bax interactions in mitochondria probed with green fluorescent protein and fluorescence resonance energy transfer," *Nat. Biotechnol.*, **6**, 547–552 (1998).
148. R. N. Day, T. C. Voss, J. F. Enwright III, C. F. Booker, A. Periasamy, and F. Schaufele, "Imaging the localized protein interactions between Pit-1 and the CCAAT/enhancer binding protein ?' in the living pituitary cell nucleus," *Mol. Endocrinol.*, **17**, 333–345 (2003).
149. M. G. Erickson, D. L. Moon, and D. T. Yue, "DsRed as a potential FRET partner with CFP and GFP," *Biophys. J.*, **85**, 599–611 (2003).
150. E. Galperin, V. V. Verkhusha, and A. Sorkin, "Three-chromophore FRET microscopy to analyze multiprotein interactions in living cells," *Nat. Methods*, **1**, 209–217 (2004).
151. M. Peter, S. M. Ameer-Beg, M. K. Y. Hughes, M. D. Keppler, S. Prag, M. Marsh, B. Vojnovic, and T. Ng, "Multiphoton-FLIM quantification of the EGFP-mRFP1 FRET pair for localization of membrane receptor-kinase interactions," *Biophys. J.*, **88**, 1224–1237 (2005).
152. L. Albertazzi, D. Arosio, L. Marchetti, F. Ricci, and F. Beltram, "Quantitative FRET analysis with the EGFP-mCherry fluorescent protein pair," *Photochem. Photobiol.*, **85**, 287–297 (2009).
153. J. R. Lakowicz, *Principle of Fluorescence Spectroscopy*, 3rd ed., Springer, New York, 2006.
154. G. H. Patterson, D. W. Piston, and B. G. Barisas, "Förster distances between green fluorescent protein pairs," *Anal. Biochem.*, **284**, 438–440 (2000).
155. M. A. Hink, N. V. Visser, J. W. Borst, A. van Hoek, and A. J. W. G. Visser, "Practical use of corrected fluorescence excitation and emission spectra of fluorescent proteins in Förster resonance energy transfer (FRET) studies," *J. Fluoresc.*, **13**, 185–188 (2003).
156. N. Akrap, T. Seidel, and B. G. Barisas, "Förster distances for fluorescence resonant energy transfer between mCherry and other visible fluorescent proteins," *Anal. Biochem.*, **402**, 105–106 (2010).
157. L. Trón, J. Szöllősi, S. Damjanovich, S. H. Helliwell, D. J. Arndt-Jovin, and T. M. Jovin, "Flow cytometric measurement of fluorescence resonance energy transfer on cell surfaces. Quantitative evaluation of the transfer efficiency on a cell-by-cell basis," *Biophys. J.*, **45**, 939–946 (1984).
158. D. C. Youvan, C. M. Silva, E. J. Bylina, W. J. Coleman, M. R. Dilworth, and M. M. Yang, "Calibration of fluorescence resonance energy transfer in microscopy using genetically engineered GFP derivatives on nickel chelating beads," *Biotechnology*, **3**, 1–18 (1997).
159. G. W. Gordon, G. Berry, X. H. Liang, B. Levine, and B. Herman, "Quantitative fluorescence resonance energy transfer measurements using fluorescence microscopy." *Biophys. J.*, **74**, 2702–2713 (1998).

160. M. Elangovan, H. Wallrabe, Y. Chen, R. N. Day, M. Barroso, and A. Periasamy, "Characterization of one- and two-photon excitation fluorescence resonance energy transfer microscopy," *Methods*, **29**, 58–73 (2003).
161. Y. Chen, M. Elangovan, and A. Periasamy, "FRET data analysis—the algorithm" *Molecular Imaging: FRET Microscopy and Spectroscopy*, edited by A. Periasamy and R. N. Day, Oxford University Press, New York, 126–145, 2005.
162. Y. Chen and A. Periasamy, "Intensity range based quantitative FRET data analysis to localize protein molecules in live cell nuclei," *J. Fluoresc.*, **16**, 95–104 (2006).
163. B. R. Masters and P. T. C. So, *Handbook of Biomedical Nonlinear Optical Microscopy*, Oxford University Press, Oxford, UK, 2008.
164. P. H. Bastiaens, I. V. Majoul, P. J. Verveer, H.-D. Soling, and T. M. Jovin, "Imaging the intracellular trafficking and state of the AB5 quaternary structure of cholera toxin," *EMBO J.*, **15**, 4246–4253 (1996).
165. G. Vereb, J. Matkó, and J. Szöllősi, "Cytometry of fluorescence resonance energy transfer," *Methods Cell Biol.*, **75**, 105–152 (2004).
166. T. M. Jovin, and D. J. Arndt-Jovin, "Luminescence digital imaging microscopy," *Ann. Rev. Biophys. Biophys. Chem.*, **18**, 271–308 (1989).
167. P. Nagy, G. Vámosi, A. Bodnár, S. J. Lockett, and János Szöllősi, J. "Intensity-based energy transfer measurements in digital imaging microscopy," *Eur. Biophys. J.*, **27**, 377–389 (1998).
168. V. Kasche and L. Lindqvist, "Reactions between the triplet state of fluorescein and oxygen," *J. Chem. Phys.*, **68**, 817–823 (1964).
169. L. Song, E. J. Hennink, I. T. Young, and H. J. Tanke, "Photobleaching kinetics of fluorescein in quantitative fluorescence microscopy," *Biophys. J.*, **68**, 2588–2600 (1995).
170. L. Song, C. A. Varma, J. W. Verhoeven, and H. J. Tanke, "Influence of the triplet excited state on the photobleaching kinetics of fluorescein in microscopy," *Biophys. J.*, **70**, 2959–2968 (1996).
171. C. Eggeling, J. Widengren, R. Rigler, and C. A. M. Seidel, "Photobleaching of fluorescent dyes under conditions used for single-molecule detection: evidence of two-step photolysis," *Anal. Chem.*, **70**, 2651–2659 (1998).
172. M. Elangovan, R. N. Day, and A. Periasamy, "Nanosecond fluorescence resonance energy transfer-fluorescence lifetime imaging microscopy to localize the protein interactions in a single living cell," *J. Microsc.*, **205**, 3–14 (2002).
173. J. R. Lakowicz, H. Szmacinski, K. Nowaczyk, K. W. Berndt, and M. Johnson, "Fluorescence lifetime imaging," *Anal. Biochem.*, **202**, 316–330 (1992).
174. J. R. Lakowicz, *Principle of Fluorescence Spectroscopy*, Plenum Press, New York, 1999.
175. H. Szmacinski, J. R. Lakowicz, and M. Johnson, "Fluorescence lifetime imaging microscopy: homodyne technique using high-speed gated image intensifier," *Methods Enzymol.*, **240**, 723–748 (1994).
176. J. R. Lakowicz and H. Szmacinski, "Fluorescence lifetime-based sensing of pH, Ca^{2+}, K^+ and glucose," *Sens. Actuator Chem.*, **11**, 133–143(1993).
177. H. C. Gerritsen, R. Sanders, A. Draaijer, C. Ince, and Y. K. Levine, "Fluorescence lifetime imaging of oxygen in living cells," *J. Fluoresc.*, **7**, 11–15 (1997).
178. D. Sud, W. Zhong, D. G. Beer, and M.-A. Mycek, "Time-resolved optical imaging provides a molecular snapshot of altered metabolic function in living human cancer cell models," *Opt. Express*, **14**, 4412–4426 (2006).

179. R. Sanders, A. Draaijer, H. C. Gerritsen, P. M. Houpt, and Y. K. Levine, "Quantitative pH imaging in cells using confocal fluorescence lifetime imaging microscopy," *Anal. Biochem.*, **227**, 302–308 (1995).
180. H. J. Lin, P. Herman, and J. R. Lakowicz, "Fluorescence lifetime-resolved pH imaging of living cells," *Cytometry A*, **52**, 77–89 (2003).
181. J. R. Lakowicz, H. Szmacinski, K. Nowaczyk, and M. L. Johnson, "Fluorescence lifetime imaging of free and protein-bound NADH," *Proc. Natl. Acad. Sci. USA*, **89**, 1271–1275 (1992).
182. J.-P. Knemeyer, N. Marmé, and M. Sauer, "Probes for detection of specific DNA sequences at the single-molecule level," *Anal. Chem.*, **72**, 3717–3724 (2000).
183. M. A. M. J. van Zandvoort, C. J. de Grauw, H. C. Gerritsen, J. L. V. Broers, M. G. A. oude Egbrink, F. C. S. Ramaekers, and D. W. Slaaf, "Discrimination of DNA and RNA in cells by a vital fluorescent probe: Lifetime imaging of SYTO13 in healthy and apoptotic cells," *Cytometry*, **47**, 226–235 (2002).
184. M. Stöckl, A. P. Plazzo, T. Korte, and A. Herrmann, "Detection of lipid domains in model and cell membranes by fluorescence lifetime imaging microscopy of fluorescent lipid analogues," *J. Biol. Chem.*, **283**, 30828–30837 (2008).
185. J. B. Pawley, *Handbook of Biological Confocal Microscopy*, Springer Science and Business Media, New York, 2006.
186. J. E. Whitaker, R. P. Haugland, and F. G. Prendergast, "Spectral and photophysical studies of benzo[c]xanthene dyes: dual emission pH sensors," *Anal. Biochem.*, **194**, 330–344 (1991).
187. A. Gafni and L. Brand, "Excited state proton transfer reactions of acridine studied by nanosecond fluorometry," *Chem. Phys. Lett.*, **58**, 346–350 (1978).
188. K. Clays, M. D. Giambattista, A. Persoons, Y. Engelborghs, "A fluorescence lifetime study of virginiamycin S using multifrequency phase fluorometry," *Biochemistry*, **30**, 7271–7276 (1991).
189. S. M. Keating and T. G. Wensel, "Nanosecond fluorescence microscopy: emission kinetics of fura-2 in single cells," *Biophys. J.*, **59**, 186–202 (1991).
190. M. Tramier, I. Gautier, T. Piolot, S. Ravalet, K. Kemnitz, J. Coppey, C. Durieux, V. Mignotte and M. Coppey-Moisan, "Picosecond-hetero-FRET microscopy to probe protein-protein interactions in live cells," *Biophys. J.*, **83**, 3570–3577 (2002).
191. K. König, P. T. So, W. W. Mantulin, B. J. Tromberg, and E. Gratton, "Two-photon excited lifetime imaging of autofluorescence in cells during UVA and NIR photostress," *J Microsc.*, **183**, 197–204 (1996).
192. K. König and I. Riemann, "High-resolution multiphoton tomography of human skin with subcellular spatial resolution and picosecond time resolution," *J. Biomed. Opt.*, **8**, 432–439 (2003).
193. K. Dowling, S. C. W. Hyde, J. C. Dainty, P. M. W. French, and J. D. Hares, "2-D fluorescence lifetime imaging using a time-gated image intensifier," *Opt. Commun.*, **135**, 27–31(1997).
194. R. V. Krishnan, H. Saitoh, H. Terada, V. E. Centonze, and B. Herman, "Development of multiphoton fluorescence lifetime imaging microscopy (FLIM) system using a streak camera," *Rev. Sci. Instrum.*, **74**, 2714–2721 (2003).
195. R. M. Ballew and J. N. Demas, "An error analysis of the rapid lifetime determination method for the evaluation of single exponential decays," *Anal. Chem.*, **61**, 30–33 (1989).

196. C. J. de Grauw and H. C. Gerrintsen, "Multiple time-gate module for fluorescence lifetime imaging," *Appl. Spectrosc.*, **55**, 670–678 (2001).
197. J. Bradshaw, P. D. Marsh, K. M. Schilling, and D. Cummins, "A modified chemostat system to study the ecology of oral biofilms," *J. Appl. Bacteriol.*, **80**, 124–130 (1996).
198. L. M. Bollinger and G. E. Thomas, "Measurement of the time dependence of scintillation intensity by a delayed-coincidence method," *Rev. Sci. Instrum.*, **32**, 1044–1050(1961).
199. G. A. Morton, "Photon counting," *Appl. Opt.*, **7**, 1–10 (1968).
200. D. V. O'Connor and D. Phillips, *Time-Correlated Single Photon Counting*, Academic Press, London, 1984.
201. B. Chance, M. Cope, E. Gratton, N. Ramanujam, and B. Tromberg, "Phase measurement of light absorption and scatter in human tissue," *Rev. Sci. Instrum.*, **69**, 3457–3481 (1998).
202. A. D. Elder, J. H. Frank, J. Swartling, X. Dai, and C. F. Kaminski, "Calibration of a wide-field frequency-domain fluorescence lifetime microscopy system using light emitting diodes as light sources," *J. Microsc.*, **224**, 166–180 (2006).
203. J. Philip and K. Carlsson, "Theoretical investigation of the signal-to-noise ratio in fluorescence lifetime imaging," *J. Opt. Sos. Am. A*, **20**, 368–379 (2003).
204. G. Laczko, I. Gryczynski, Z. Gryczynski, W. Wiczk, H. Malak, and J. R. Lakowicz, "A 10-GHz frequency-domain fluorometer," *Rev. Sci. Instrum.*, **61**, 2331–2337 (1990).
205. K. Carlsson and A. Liljeborg, "Confocal fluorescence microscopy using spectral and lifetime information to simultaneously record four fluorophores with high channel separation," *J. Microsc.*, **185**, 37–46 (1997).
206. K. Carlsson and A. Liljeborg, "Simultaneous confocal lifetime imaging of multiple fluorophores using the intensity-modulated multiple-wavelength scanning (IMS) technique," *J. Microsc.*, **191**, 119–127 (1998).
207. E. B. Van Munster and T. W. J. Gadella Jr, "φ FLIM: a new method to avoid aliasing in frequency-domain fluorescence lifetime imaging microscopy," *J. Microsc.*, **213**, 29–38 (2004).
208. A. C. Mitchell, J. E. Wall, J. G. Murray, and C. G. Morgan, "Direct modulation of the effective sensitivity of a CCD detector: a new approach to time-resolved fluorescence imaging," *J. Microsc.*, **206**, 225–232 (2002).
209. R. Peters, J. Peters, K. Tews, and W. Bahr, "Microfluorimetric study of translational diffusion of proteins in erythrocyte membranes," *Biochim. Biophys. Acta.*, **367**, 282–294 (1974).
210. D. Axelrod, D. E. Koppel, J. Schlessinger, E. Elson, and W. W. Webb, "Mobility measurement by analysis of fluorescence photobleaching recovery kinetics," *Biophys. J.*, **16**, 1055–1069 (1976).
211. A. Majewska, E. Brown, J. Ross, and R. Yuste, "Mechanisms of calcium decay kinetics in hippocampal spines: role of spine calcium pumps and calcium diffusion through the spine neck in biochemical compartmentalization," *J. Neurosci.*, **20**, 1722–1734 (2000).
212. A. Majewska, A. Tashiro, and R. Yuste, "Regulation of spine calcium dynamics by rapid spine motility," *J. Neurosci.*, **20**, 8262–8268 (2000).
213. K. Svoboda, D. W. Tank, and W. Denk, "Direct measurement of coupling between dendritic spines and shafts," *Science*, **272**, 716–719 (1996).
214. J. Lippincott-Schwartz, E. Snapp, and A. Kenworthy, "Studying protein dynamics in living cells," *Nat. Rev. Mol. Cell Biol.*, **2**, 444–456 (2001).

215. R. D. Phair and T. Misteli, "Kinetic modelling approaches to in vivo imaging," *Nat. Rev. Mol. Cell Biol.*, **2**, 898–907 (2001).

216. A. B. Houtsmuller and W. Vermeulen, "Macromolecular dynamics in living cell nuclei revealed by fluorescence redistribution after photobleaching," *Histochem. Cell. Biol.*, **115**, 13–21 (2001).

217. K. Braeckmans, L. Peeters, N. N. Sanders, S. C. De Smedt, and J. Demeester, "Three-dimensional fluorescence recovery after photobleaching with the confocal scanning laser microscope," *Biophys. J.*, **85**, 2240–2252 (2003).

218. E. B. Brown, E. S. Wu, W. Zipfel, and W. W. Webb, "Measurement of molecular diffusion in solution by multiphoton fluorescence photobleaching recovery," *Biophys. J.*, **77**, 2837–2849 (1999).

219. I. F. Sbalzarini, A. Mezzacasa, A. Helenius, and P. Koumoutsakos, "Effects of organelle shape on fluorescence recovery after photobleaching," *Biophys. J.*, **89** 1482–1492 (2005).

220. N. Periasamy and A.S. Verkman, "Analysis of fluorophore diffusion by continuous distributions of diffusion coefficients: application to photobleaching measurements of multicomponent and anomalous diffusion," *Biophys. J.*, **75**, 557–567 (1998).

221. J-P. Bouchaud, and A. Georges, "Anomalous diffusion in disordered media: statistical mechanisms, models and physical applications," *Phys. Rep.*, **195**, 127–293, (1990).

222. T. J. Feder, I. Brust-Mascher, J. P. Slattery, B. Baird, and W. W. Webb, "Constrained diffusion or immobile fraction on cell surfaces: a new interpretation," *Biophys. J.*, **70**, 2767–2773 (1996).

223. B. R. Masters and P. T. C. So, *Handbook of Biomedical Nonlinear Optical Microscopy*, Oxford University Press, Oxford, UK, 2008.

224. P. F. Fahey, L. S. Barak, E. L. Elson, D. E. Koppel, D. E. Wolf, and W. W. Webb, "Lateral diffusion in planar lipid bilayers," *Science*, **195**, 305–306 (1977).

225. D. Magde, "Chemical kinetics and fluorescence correlation spectroscopy," *Quart. Rev. Biophys.*, **9**, 35–47 (1976).

226. D. Magde, E. Elson, and W. W. Webb, "Thermodynamic fluctuations in a reacting system-Measurement by fluorescence correlation spectroscopy," *Phys. Rev. Lett.*, **29**, 705–708 (1972).

227. M. Ehrenberg and R. Rigler, "Rotational Brownian motion and fluorescence intensify fluctuations," *Chem. Phys.*, **4**, 390–401 (1974).

228. S. R. Aragón and R. Pecora, "Fluorescence correlation spectroscopy as a probe of molecular dynamics," *J. Chem. Phys.*, **64**, 1791–1803 (1976).

229. M. Kinjo and R. Rigler, "Ultrasensitive hybridization analysis using fluorescence correlation spectroscopy," *Nucleic Acids Res.*, **23**, 1795–1799 (1995).

230. B. Rauer, E. Neumann, J. Widengren, and R. Rigler, "Fluorescence correlation spectrometry of the interaction kinetics of tetramethylrhodamin alpha-bungarotoxin with Torpedo californica acetylcholine receptor," *Biophys. Chem.*, **58**, 3–12 (1996).

231. U. Haupts, S. Maiti, P. Schwille, and W. W. Webb, "Dynamics of fluorescence fluctuations in green fluorescent protein observed by fluorescence correlation spectroscopy," *Proc. Natl. Acad. Sci. USA*, **95**, 13573–13578 (1998).

232. L. Edman, Ü. Mets, and R. Rigler, "Conformational transitions monitored for single molecules in solution," *Proc. Natl. Acad. Sci. USA*, **93**, 6710–6715 (1996).

233. C. G. Hübner, A. Renn, I. Renge, and U. P. Wild, "Direct observation of the triplet lifetime quenching of single dye molecules by molecular oxygen," *J. Chem. Phys.*, **115**, 9619–9622 (2001).

234. P. Schwille, "Fluorescence correlation spectroscopy and its potential for intracellular applications," *Cell Biochem. Biophys.*, **34**, 383–408 (2001)

235. P. Schwille, U. Haupts, S. Maiti, and W. W. Webb, "Molecular dynamics in living cells observed by fluorescence correlation spectroscopy with one- and two-photon excitation," *Biophys. J.*, **77**, 2251–2265 (1999).

236. R. H. Köhler, P. Schwille, W. W. Webb, and M. R. Hanson, "Active protein transport through plastid tubules: velocity quantified by fluorescence correlation spectroscopy," *J. Cell Sci.*, **113**, 3921–3930 (2000).

237. D. S. Banks and C. Fradin, "Anomalous diffusion of proteins due to molecular crowding," *Biophys. J.*, **89**, 2960–2971 (2005).

238. A. G. Palmer and N. L. Thompson, "Theory of sample translation in fluorescence correlation spectroscopy," *Biophys. J.*, **51**, 339–343 (1987).

239. J. Widengren and R. Rigler, "Fluorescence correlation spectroscopy as a tool to investigate chemical reactions in solutions and on cell surfaces," *Cell Mol. Biol.*, **44**, 857–879 (1998).

240. J. Widengren, U. Mets, and R. Rigler, "Fluorescence correlation spectroscopy of triplet states in solution: a theoretical and experimental study," *J. Phys. Chem.*, **99**, 13368–13379 (1995).

241. F. Malvezzi-Campeggi, M. Jahnz, K. G. Heinze, P. Dittrich, and P. Schwille, "Light-induced flickering of DsRed provides evidence for distinct and interconvertible fluorescent," *Biophys. J.*, **81**, 1776–1785 (2001).

242. P. Schwille, S. Kummer, A. A. Heikal, W. E. Moerner, and W. W. Webb, "Fluorescence correlation spectroscopy reveals fast optical excitation driven intramolecular dynamics of yellow fluorescent proteins," *Proc. Natl. Acad. Sci. USA*, **97**, 151–156 (2000).

243. M. Eigen and R. Rigler, "Sorting single molecules: application to diagnostics and evolutionary biotechnology," *Proc. Natl. Acad. Sci. USA*, **91**, 5740–5747 (1994).

244. L. C. Hwang and T. Wohland, "Dual-color fluorescence cross-correlation spectroscopy using single laser wavelength excitation," *Chem. Phys. Chem.*, **5**, 549–551 (2004).

245. P. Schwille, J. Bieschke, and F. Oehlenschläger, "Kinetic investigations by fluorescence correlation spectroscopy: the analytical and diagnostic potential of diffusion studies," *Biophys. Chem.*, **66**, 211–228 (1997).

CHAPTER 3

NONLINEAR OPTICAL SPECTROSCOPY (NLOS)

Nonlinear optical microscopy has become a powerful tool for investigating biological phenomena because of its advantages over conventional optical microscopy. To obtain high image contrast in nonlinear optical microscopy, we need to know some spectroscopic information about various molecules of interest. There are three techniques for obtaining spectroscopic information of various molecules by nonlinear optical spectroscopy. One technique is the laser-wavelength scanning method using a narrowband laser pulse. Another technique is the multiplex method employing a combination of a broadband laser pulse and a narrowband laser pulse. The last technique is the nonlinear Fourier transform method using broadband pulses. In this chapter, we describe these spectroscopic techniques.

3.1 LASER-WAVELENGTH SCANNING METHOD

In the laser-wavelength scanning method, the central wavelength of a narrowband laser pulse can be tuned to various excitation wavelengths and the wavelength dependence of signal intensity is then obtained sequentially. Multiphoton absorption cross sections can be directly measured by detecting the loss of the excitation beam. Until recently, the direct measurement of multiphoton absorption cross sections required a high-peak-power pulse laser system such as an amplified Ti:sapphire laser operating at a repetition rate of less than a few hundred kilohertz. By using Z-scan techniques [1] and modulation transfer techniques [2], the direct measurement of multiphoton

Functional Imaging by Controlled Nonlinear Optical Phenomena, First Edition.
Keisuke Isobe, Wataru Watanabe and Kazuyoshi Itoh.

absorption cross sections has been measured with low-peak-power pulses from Ti:sapphire oscillators. Multiphoton absorption cross sections can be also obtained by the fluorescence technique. Because the fluorescence technique provides very high detection sensitivity, multiphoton absorption cross sections can be easily measured with low-peak-power pulses. In this section, we focus on the fluorescence technique.

Now we consider the measurements of the n-photon excitation spectra of fluorophores by recording multiphoton excited fluorescence using the laser-wavelength scanning method. Since the fluorescence technique allows very high sensitivity, the m-photon excitation spectra can be measured in dilute solution. Let us assume no stimulated emission and no self-quenching, the number of fluorescence photons collected per unit time is expressed as [3]

$$F^{(m)}(t) = \frac{1}{m\,(\hbar\omega)^m}\eta q^{(m)} N_A \int_V \sigma^{(m)}(m\omega) C(\mathbf{r}, t) I^m(\mathbf{r}, t) d\mathbf{r}, \tag{3.1.1}$$

where η denotes the overall detection efficiency of the optical system, $q^{(m)}$ is the fluorescence quantum efficiency, N_A is the Avogadro's number, $\hbar\omega$ is photon energy, $\sigma^{(m)}(m\omega)$ is the m-photon excitation cross-section at a frequency of ω, $C(r, t)$ is the fluorophore concentration, V is the illuminated volume, $I(r, t)$ is the excitation intensity, and the factor $1/m$ simply reflects the fact that one fluorescence photon is emitted after absorption of m excitation photons in each excitation event. If time and space dependence of the excitation intensity can be separated, the excitation intensity can be rewritten by

$$I(\mathbf{r}, t) = I_{\text{focal}}(t) S(\mathbf{r}), \tag{3.1.2}$$

where $S(\mathbf{r})$ denotes a unitless spatial distribution functions and $I_{\text{focal}}(t)$ indicates the excitation intensity at the center of the focal volume. If we assume the absence of ground state depletion and photobleaching, the concentration is constant in a solution. Then, Eq. (3.1.1) can be described as

$$F^{(m)}(t) = \frac{1}{m\,(\hbar\omega)^m}\eta q^{(m)} C_0 N_A \sigma^{(m)}(n\omega) I^m_{\text{focal}}(t) \int_V S^m(\mathbf{r}) d\mathbf{r}. \tag{3.1.3}$$

In practice, the time-averaged fluorescence photon flux $\langle F^{(m)}(t)\rangle$ is measured. The time average of the excitation intensity at the center of the focal volume raised to the nth power $\langle I^m_{\text{focal}}(t)\rangle$ depends on the temporal coherence. Considering an excitation pulse with a pulse duration of Δt (FWHM) and a repetition of f, we can express $\langle I^m_{\text{focal}}(t)\rangle$ by

$$\langle I^m_{\text{focal}}(t)\rangle = g^{(m)} \frac{\langle I_{\text{focal}}(t)\rangle^m}{(\Delta t f)^{m-1}}. \tag{3.1.4}$$

Here we introduced the parameter $g^{(m)}$ in the following form:

$$g^{(m)} = \frac{\Delta t^{m-1} \int_{-1/(2f)}^{1/(2f)} I_{\text{focal}}^m(t)dt}{\left[\int_{-1/(2f)}^{1/(2f)} I_{\text{focal}}(t)dt\right]^m}. \tag{3.1.5}$$

For pulses with a Gaussian temporal profile, $g^{(2)}$ and $g^{(3)}$ are 0.66 and 0.51, respectively, whereas for pulses with a hyperbolic-secant square temporal profile, $g^{(2)}$ and $g^{(3)}$ are 0.59 and 0.41, respectively [3]. According to eqs. (3.1.3) and (3.1.4), the time-averaged fluorescence photon flux $\langle F^{(m)}(t)\rangle$ can be described as

$$\left\langle F^{(m)}(t)\right\rangle = \frac{1}{m\,(\hbar\omega)^m}\eta q^{(m)} C_0 N_A \sigma^{(m)}(m\omega) g^{(m)} \frac{\langle I_{\text{focal}}(t)\rangle^m}{(\Delta t f)^{m-1}} \int_V S^m(\mathbf{r})d\mathbf{r}. \tag{3.1.6}$$

By assuming that the incident beam is the Gaussiam beam as shown in Eq. (1.1.61), the expression for $S(\mathbf{r})$ and $I_{\text{focal}}(t)$ can be obtained by

$$S(\mathbf{r}) = \frac{w_c^2}{W_c^2(z)} e^{-2(x^2+y^2)/W_c^2(z)}, \tag{3.1.7}$$

$$I_{\text{focal}}(t) = \frac{2P(t)}{\pi w_c^2}, \tag{3.1.8}$$

where $P(t)$ is the incident power. If the sample thickness is much greater than the Rayleigh length, we can readily obtain [3]

$$\int_{V\to\infty} S^m(\mathbf{r})d\mathbf{r} = \frac{(2m-5)!!}{2m(2m-4)!!} \frac{n\pi^3 w_c^4}{\lambda_c}, \tag{3.1.9}$$

where !! denotes a double factorial:

$$2m!! = 2\cdot 4\cdot 6\cdots 2m,$$
$$(2m-1)!! = 1\cdot 3\cdot 5\cdots(2m-1),$$
$$1!! = 0!! = 1.$$

To determine the m-photon excitation cross section $\sigma^{(m)}(m\omega)$, we need to characterize the pulse duration, the concentration, the overall detection efficiency of the optical system, and the spatial distribution of the excitation light. This problem can be solved by using a reference sample whose absolute m-photon excitation cross section is known. Then, the m-photon excitation spectrum of the sample can be obtained by

$$\sigma_{\text{s}}^{(m)}(m\omega) = \frac{C_{\text{r}} q_{\text{r}}^{(m)} \langle P_{\text{r}}(t)\rangle^m \left\langle F_{\text{s}}^{(m)}(t)\right\rangle}{C_{\text{s}} q_{\text{s}}^{(m)} \langle P_{\text{s}}(t)\rangle^m \left\langle F_{\text{r}}^{(m)}(t)\right\rangle} \sigma_{\text{r}}^{(m)}(m\omega), \tag{3.1.10}$$

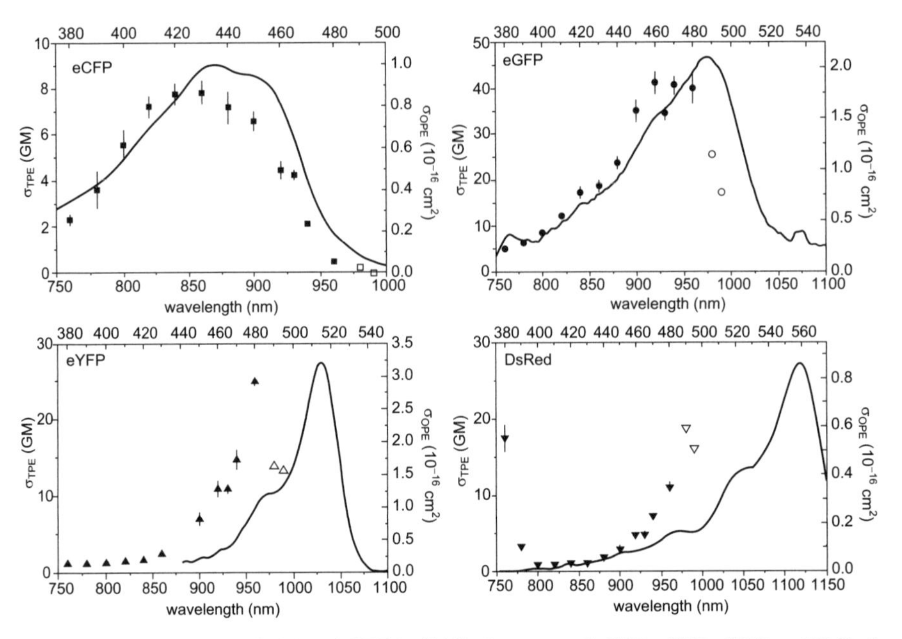

FIGURE 3.1 TPE (symbols) and OPE (solid line) spectra of eCFP, eGFP, eYFP, and DsRed. 1 GM is 10^{-50} cm^4 s photon^{-1}. Reprinted with permission from [10]. Copyright 2001 Elsevier.

where the subscript s and r indicate the parameters for the test sample and the reference sample, respectively. Here $\langle P(t) \rangle$ is the time-averaged excitation power. Because the pulse duration and the spatial distribution of the excitation light depend on the central wavelength of the narrowband pulse, their calibration or the fluorescence signals from the reference sample must be measured at each central wavelength. In addition, to confirm that the fluorescence signals are generated with the m-photon excitation, we must investigate the fluorescence intensity dependence on the excitation intensity at each central wavelength.

The two-photon excitation (TPE) spectra of various organic dyes [4–9], fluorescent proteins [10–15], and quantum dot [16] have been obtained from two-photon excited fluorescence (TPEF) measured by the laser-wavelength scanning method. Figures 3.1 and 3.2 show TPE and one-photon excitation (OPE) spectra of various fluorescent proteins [10, 14]. We see that the TPE peak wavelengths appear blue shifted and never red shifted relative to twice the OPE peak wavelengths. There are several reasons for this behavior. One reason is the selection rules. The selection rules for TPE are different from those for OPE. In centrosymmetric molecules, a TPE transition is allowed only between two states that have the same parity, while OPE transition is allowed only between an initial state and final state with opposite parity. Thus, TPE spectra is different from OPE spectra. In molecules having no center of symmetry, TPE spectra are similar to OPE spectra because parity restrictions may be relaxed. The second explanation is that some higher excited singlet states are reached with greater probability by TPE than by OPE [18, 19]. Other factors may affect the TPE spectra.

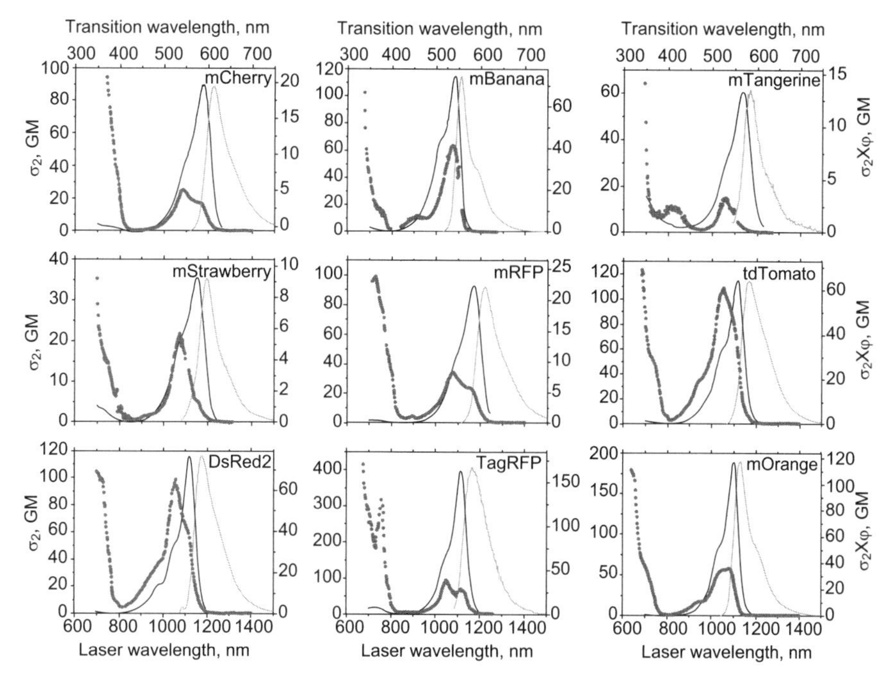

FIGURE 3.2 TPE (symbols), OPE (black solid line), and emission (blue solid line) spectra of mCherry, mBanana, mTangerine, mStrawberry, mRFP, td Tomato, DsRed2, TagRFP, and mOrange. The scale on the right represents two-photon brightness. 1 GM is 10^{-50} cm^4 s $photon^{-1}$. Reprinted with permission from [14]. Copyright 2009 American Chemical Society. (*See insert for color representation of the figure.*)

For example, vibronic coupling may alter the inversion symmetry and partially allow the parity forbidden bands [20].

3.2 MULTIPLEX SPECTROSCOPY

In the laser-wavelength scanning method, it takes several seconds or more to tune the wavelength of the mode-locked femtosecond lasers that are currently available. In nonlinear optical spectroscopy (NLOS) using multiplex techniques, a combination of a broadband pulse and a narrowband pulse is employed as excitation light sources. By simultaneously exciting molecules to various excited states, broadband spectroscopic information can be acquired without tuning the central wavelength of a tunable narrowband laser. Thus, the acquisition time is several hundred milliseconds or less. The measurable spectral bandwidth is limited by the spectral bandwidth of the broadband pulse, while the spectral resolution is determined by the spectral bandwidth of the narrowband pulse. The multiplex techniques have been applied to coherent anti-Stokes Raman scattering (CARS) [21–24], stimulated parametric emission (SPE), sum-frequency generation (SFG) [25], two-photon absorption (TPA)

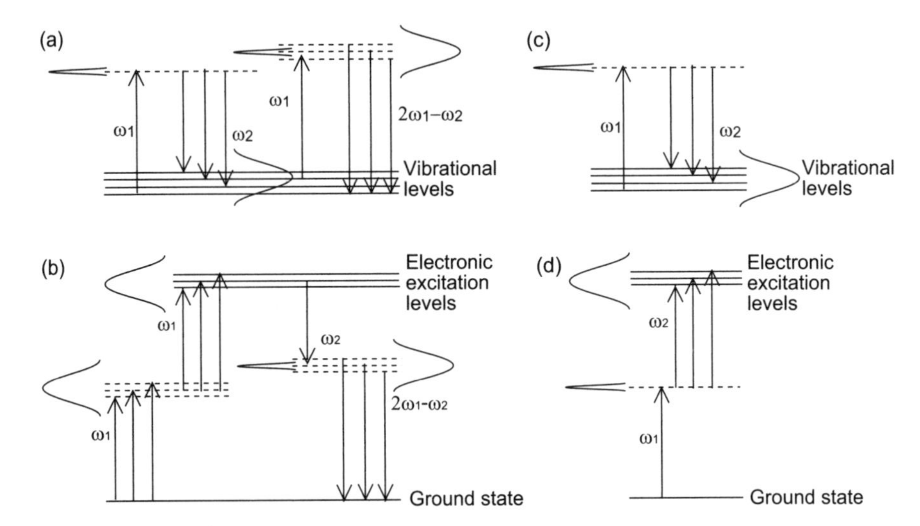

FIGURE 3.3 Energy level diagram of the CARS (a), SPE (b), SRS (c), and TPA (d) processes with a multiplex excitation scheme.

[26, 27], and stimulated Raman scattering (SRS) [28]. In this section, we discuss the multiplex techniques in CARS, SPE, SRS, and TPA spectroscopies.

3.2.1 Principle of Multiplex Nonlinear Optical Spectroscopy

3.2.1.1 Multiplex Coherent Anti-Stokes Raman Scattering (CARS) Spectroscopy We show the principle of multiplex CARS spectroscopy. The energy level diagram of the CARS process with a multiplex excitation scheme is shown in Fig. 3.3(a). A broadband light source is used as a Stokes pulse at a frequency of ω_2 for simultaneously addressing a broad spectral range of various vibrational resonances. A narrowband light source is employed as a pump pulse at a frequency of ω_1 for spectrally resolving various vibrational resonances.

The FWM signal is given by

$$I_{\text{FWM}}(\omega) \propto \left| \tilde{P}^{(3)}_{\text{FWM}}(\omega) \right|^2 = \left| \int P^{(3)}(t) \exp(i\omega t) dt \right|^2, \tag{3.2.1}$$

where $P^{(3)}(t)$ is the third-order polarization of the FWM process. According to (1.1.163), the polarization of the Raman resonant FWM process at a frequency of $2\omega_1 - \omega_2$, $P^{(3)}_{\text{RR}}(t)$, is expressed by

$$P^{(3)}_{\text{RR}}(t) = \varepsilon_0 E_1(t) \int \chi^{(\text{RR})}(t - t_1) E_1(t_1) E_2^*(t_1) dt_1, \tag{3.2.2}$$

where $\chi^{(\text{RR})}$ the Raman resonance contribution of the third-order susceptibility, and E_1 and E_2 are electric fields of a pump pulse and a Stokes pulse, respectively. Here

* indicates the complex conjugate. The Fourier transform of Eq. (3.2.2) is described by

$$\tilde{P}^{(3)}_{\mathrm{RR}}(\omega) = \varepsilon_0 \int \tilde{\chi}^{(\mathrm{RR})}(\Omega_1)\tilde{E}^{(2\#)}_{12}(\Omega_1)\tilde{E}_1(\omega - \Omega_1)d\Omega_1, \tag{3.2.3}$$

where $\tilde{E}^{(2\#)}_{12}(\omega)$ is the difference-frequency (DF) spectrum and is defined as

$$\tilde{E}^{(2\#)}_{12}(\omega) = \int \tilde{E}_1(\Omega_1)\tilde{E}^*_2(\Omega_1 - \omega)d\Omega_1. \tag{3.2.4}$$

In multiplex CARS spectroscopy, the spectral bandwidth of the pump pulse is much narrower than that of the Stokes pulse. Thus, $\tilde{E}_1(\omega)$ in Eq. (3.2.4) can be approximated by $A_1\delta(\omega - \omega_1)$. Then, the DF spectrum can be rewritten by

$$\tilde{E}^{(2\#)}_{12}(\omega) \approx A_1\tilde{E}^*_2(\omega_1 - \omega). \tag{3.2.5}$$

Assuming that the spectral bandwidth of the pump pulse is also much narrower than the Raman resonance bandwidth, we can than approximate $\tilde{E}_1(\omega)$ in Eq. (3.2.3) by $A_1\delta(\omega - \omega_1)$. Then, $\tilde{P}^{(3)}_{\mathrm{RR}}(\omega)$ is expressed by

$$\tilde{P}^{(3)}_{\mathrm{RR}}(\omega) \approx \varepsilon_0 A_1^2\tilde{E}^*_2(2\omega_1 - \omega)\tilde{\chi}^{(\mathrm{RR})}(\omega - \omega_1). \tag{3.2.6}$$

According to Eq. (1.1.163), the polarization of the nonresonant FWM (NFWM) process at a frequency of $2\omega_1 - \omega_2$ in time domain, $P^{(3)}_{\mathrm{NR}}(t)$, is given by

$$P^{(3)}_{\mathrm{NR}}(t) = \varepsilon_0\chi^{(\mathrm{NR})}E_1^2(t)E_2(t), \tag{3.2.7}$$

where $\chi^{(\mathrm{NR})}$ the nonresonant contribution of the third-order susceptibility. In the frequency domain, the polarization in the NFWM process is expressed by

$$\tilde{P}^{(3)}_{\mathrm{NR}}(\omega) = \varepsilon_0\chi^{(\mathrm{NR})} \int \tilde{E}^{(2)}_{11}(\Omega_1)\tilde{E}^*_2(\Omega_1 - \omega)d\Omega_1, \tag{3.2.8}$$

where $E^{(2)}_{12}(\omega)$ is the second harmonic (SH) spectrum of the pump pulse and is defined as

$$\tilde{E}^{(2)}_{12}(\omega) = \int \tilde{E}_1(\Omega_1)\tilde{E}_2(\omega - \Omega_1)d\Omega_1. \tag{3.2.9}$$

Because in the multiplex CARS scheme, $\tilde{E}_1(\omega)$ in Eq. (3.2.8) can be approximated by $A_1\delta(\omega - \omega_1)$, we obtain

$$\tilde{P}^{(3)}_{\mathrm{NR}}(\omega) \approx \varepsilon_0\chi^{(\mathrm{NR})}A_1^2\tilde{E}^*_2(2\omega_1 - \omega). \tag{3.2.10}$$

According to Eq. (1.1.163), the polarization of the two-photon electronic resonant FWM process at a frequency of $2\omega_1 - \omega_2$ in time domain $P_{\rm ER}^{(3)}(t)$, and that in frequency domain $\tilde{P}_{\rm NR}^{(3)}(\omega)$, are given by

$$P_{\rm ER}^{(3)}(t) = \varepsilon_0 E_2^*(t) \int \chi^{\rm (ER)}(t - t_1) E_1^2(t_1) dt_1, \tag{3.2.11}$$

$$\tilde{P}_{\rm ER}^{(3)}(\omega) = \varepsilon_0 \int \tilde{\chi}^{\rm (ER)}(\Omega_1) \tilde{E}_{11}^{(2)}(\Omega_1) \tilde{E}_2^*(\Omega_1 - \omega) d\Omega_1, \tag{3.2.12}$$

where $\chi^{\rm (ER)}$ is the electronic resonance contribution of the third-order susceptibility. In the multiplex CARS technique, $\tilde{P}_{\rm ER}^{(3)}(\omega)$ can be approximated by

$$\tilde{P}_{\rm ER}^{(3)}(\omega) \approx \varepsilon_0 \tilde{\chi}^{\rm (ER)}(2\omega_1) A_1^2 \tilde{E}_2^*(2\omega_1 - \omega). \tag{3.2.13}$$

From Eqs. (3.2.6), (3.2.10), and (3.2.13), the polarization in the FWM process at a frequency of $2\omega_1 - \omega_2$ is written as

$$\begin{aligned} \tilde{P}_{\rm FWM}^{(3)}(\omega) &= \tilde{P}_{\rm RR}^{(3)}(\omega) + \tilde{P}_{\rm ER}^{(3)}(\omega) + \tilde{P}_{\rm NR}^{(3)}(\omega) \\ &\approx \left\{ \tilde{\chi}^{\rm (RR)}(\omega - \omega_1) + \tilde{\chi}^{\rm (ER)}(2\omega_1) + \chi^{\rm (NR)} \right\} \\ &\quad \times \varepsilon_0 A_1^2 \tilde{E}_2^*(2\omega_1 - \omega). \end{aligned} \tag{3.2.14}$$

Substituting Eq. (3.2.14) into Eq. (3.2.1), we can write the FWM spectrum in the multiplex CARS scheme as

$$I_{\rm FWM}(\omega) \propto \left(\varepsilon_0 |A_1|^2\right)^2 \left|\tilde{E}_2(2\omega_1 - \omega)\right|^2 \left|\tilde{\chi}^{\rm (RR)}(\omega - \omega_1) + \tilde{\chi}^{\rm (ER)}(2\omega_1) + \chi^{\rm (NR)}\right|^2. \tag{3.2.15}$$

According to Eq. (3.2.15), we find that the spectral feature of the FWM spectrum is proportional to the product of the spectrum of the Stokes pulse and the Raman spectrum, which reflects the vibrational energy levels of samples. Note that the measurable Raman bandwidth is equal to the spectral bandwidth of the Stokes pulse.

3.2.1.2 Multiplex Stimulated Parametric Emission (SPE) Spectroscopy

Next, we consider multiplex SPE spectroscopy. Figure 3.3(b) shows the energy level diagram of the SPE process with a multiplex excitation scheme. A broadband light source is used as the pump pulses at a frequency of ω_1 for simultaneously exciting various electronic resonances. A narrowband light source is employed as the probe pulse at a frequency of ω_2 for spectrally resolving various electronic resonances. The spectral features of the resultant SPE signals reflect the electronic excitation energy levels of samples. The spectral resolution of the SPE spectrum is proportional to the spectral bandwidth of the probe pulse. For the electronic resonance bandwidth, 100-fs laser pulses with a bandwidth of 100 cm^{-1} are appropriate for the probe pulse.

In multiplex SPE spectroscopy, the spectral bandwidth of the probe pulse is much narrower than that of the pump pulse and the electronic resonance bandwidth. Thus, $\tilde{E}_2(\omega)$ in Eqs. (3.2.8) and (3.2.12) can be approximated by $A_2\delta(\omega - \omega_2)$. Then, $\tilde{P}^{(3)}_{\mathrm{NR}}(\omega)$ and $\tilde{P}^{(3)}_{\mathrm{ER}}(\omega)$ are described as

$$\tilde{P}^{(3)}_{\mathrm{NR}}(\omega) \approx \varepsilon_0 \chi^{(\mathrm{NR})} A_2^* \tilde{E}^{(2)}_{11}(\omega + \omega_2), \tag{3.2.16}$$

$$\tilde{P}^{(3)}_{\mathrm{ER}}(\omega) \approx \varepsilon_0 \tilde{\chi}^{(\mathrm{ER})}(\omega + \omega_2) A_2^* \tilde{E}^{(2)}_{11}(\omega + \omega_2). \tag{3.2.17}$$

$\tilde{E}_2(\omega)$ in Eq. (3.2.4) can be also approximated by $A_2\delta(\omega - \omega_2)$. Thus, the DF spectrum is expressed by

$$\tilde{E}^{(2\#)}_{12}(\omega) \approx A_2^* \tilde{E}_1(\omega + \omega_2). \tag{3.2.18}$$

Inserting Eq. (3.2.18) into Eq. (3.2.3), $\tilde{P}^{(3)}_{\mathrm{RR}}(\omega)$ is obtained by

$$\tilde{P}^{(3)}_{\mathrm{RR}}(\omega) \approx \varepsilon_0 A_2^* \int \tilde{\chi}^{(\mathrm{RR})}(\Omega_1) \tilde{E}_1(\Omega_1 + \omega_2) \tilde{E}_1(\omega - \Omega_1) d\Omega_1. \tag{3.2.19}$$

In the multiplex SPE technique, the spectral bandwidth of the probe pulse is much broader than the Raman resonance bandwidth. Thus, $\tilde{\chi}^{(\mathrm{RR})}(\Omega_1)$ in Eq. (3.2.19) can be approximated by $\chi^{(\mathrm{RR})}\delta(\omega - \omega_{\mathrm{RR}})$. Then, $\tilde{P}^{(3)}_{\mathrm{RR}}(\omega)$ can be rewritten as

$$\tilde{P}^{(3)}_{\mathrm{RR}}(\omega) \approx \varepsilon_0 \chi^{(\mathrm{RR})} A_2^* \tilde{E}_1(\omega_{\mathrm{RR}} + \omega_2) \tilde{E}_1(\omega - \omega_{\mathrm{RR}}). \tag{3.2.20}$$

From Eqs. (3.2.17), (3.2.18), and (3.2.20), the polarization in the FWM process at a frequency of $2\omega_1 - \omega_2$ is written by

$$\begin{aligned}\tilde{P}^{(3)}_{\mathrm{FWM}}(\omega) &= \tilde{P}^{(3)}_{\mathrm{RR}}(\omega) + \tilde{P}^{(3)}_{\mathrm{ER}}(\omega) + \tilde{P}^{(3)}_{\mathrm{NR}}(\omega) \\ &\approx \left\{\chi^{(\mathrm{NR})} + \tilde{\chi}^{(\mathrm{ER})}(\omega + \omega_2)\right\} \varepsilon_0 A_2^* \tilde{E}^{(2)}_{11}(\omega + \omega_2) \\ &\quad + \varepsilon_0 \chi^{(\mathrm{RR})} A_2^* \tilde{E}_1(\omega_{\mathrm{RR}} + \omega_2) \tilde{E}_1(\omega - \omega_{\mathrm{RR}}),\end{aligned} \tag{3.2.21}$$

Substituting Eq. (3.2.21) into Eq. (3.2.1), we can write the FWM spectrum in the multiplex SPE scheme as

$$\begin{aligned}I_{\mathrm{FWM}}(\omega) \propto\ & \left|\chi^{(\mathrm{NR})} + \tilde{\chi}^{(\mathrm{ER})}(\omega + \omega_2)\right|^2 |A_2|^2 \left|\tilde{E}^{(2)}_{11}(\omega + \omega_2)\right|^2 \\ & + \left|\chi^{(\mathrm{RR})}\right|^2 |A_2|^2 \left|\tilde{E}_1(\omega_{\mathrm{RR}} + \omega_2)\right|^2 \left|\tilde{E}_1(\omega - \omega_{\mathrm{RR}})\right|^2 \\ & + 2\mathrm{Re}\left\{\left\{\chi^{(\mathrm{NR})} + \tilde{\chi}^{(\mathrm{ER})}(\omega + \omega_2)\right\} \chi^{(\mathrm{RR})*}\right. \\ & \left. \times |A_2|^2 \, \tilde{E}^{(2)}_{11}(\omega + \omega_2) \tilde{E}_1^*(\omega_{\mathrm{RR}} + \omega_2) \tilde{E}_1^*(\omega - \omega_{\mathrm{RR}})\right\},\end{aligned} \tag{3.2.22}$$

From Eq. (3.2.22), we find that the spectral feature reflecting the electronic excitation energy levels of samples appears in the FWM spectrum obtained by multiplex SPE spectroscopy, while which reflects the vibrational energy levels of samples is not included. Note that the measurable electronic resonance bandwidth depends on the spectral bandwidth of the pump pulse.

3.2.1.3 Multiplex Stimulated Raman Scattering (SRS) Spectroscopy

We describe the principle of multiplex SRS spectroscopy. The energy level diagram of the SRS process with a multiplex excitation scheme is shown in Fig. 3.3(c). A narrowband light source and a broadband light source are employed either as a pump pulse at a frequency of ω_1 and a Stokes pulse at a frequency of ω_2, respectively, or as a Stokes pulse at a frequency of ω_2 and a pump pulse at a frequency of ω_1, respectively. First, we show multiplex SRS spectroscopy using a narrowband pump pulse and a broadband Stokes pulse. According to Eq. (1.1.242), the polarization $P^{(3)}_{\mathrm{SRS2}}(t)$ in the SRS process at a frequency of ω_2 is given by

$$P^{(3)}_{\mathrm{SRS2}}(t) = \varepsilon_0 E_1(t) \int \chi^{(\mathrm{RR})*}(t - t_1) E_1^*(t_1) E_2(t_1) dt_1. \tag{3.2.23}$$

The Fourier transform of Eq. (3.2.23) is described by

$$\tilde{P}^{(3)}_{\mathrm{SRS2}}(\omega) = \varepsilon_0 \int \tilde{\chi}^{(\mathrm{RR})*}(\Omega_1 - \omega) \tilde{E}^{(2\#)*}_{12}(\Omega_1 - \omega) \tilde{E}_1(\Omega_1) d\Omega_1. \tag{3.2.24}$$

The spectral bandwidth of the pump pulse is much narrower than that of the Stokes pulse and the Raman resonance bandwidth. Thus, $\tilde{E}_1(\omega)$ in Eq. (3.2.24) can be approximated by $A_1\omega(\omega - \omega_1)$. Then, $\tilde{P}^{(3)}_{\mathrm{SRS2}}(\omega)$ can be rewritten as

$$\tilde{P}^{(3)}_{\mathrm{SRS2}}(\omega) \approx \varepsilon_0 \left|A_1\right|^2 \tilde{\chi}^{(\mathrm{RR})*}(\omega_1 - \omega) \tilde{E}_2(\omega). \tag{3.2.25}$$

Assuming that the interaction length L is much shorter, and the loss and the dispersion can be neglected, we use Eqs. (1.1.37) and (3.2.25), to express the electric field of the Stokes pulse after SRS by

$$\begin{aligned} \tilde{E}^{(\mathrm{SRS})}_2(\omega) &= \tilde{E}_2(\omega) + i \frac{\omega L}{2\tilde{n}^{(1)}(\omega_2)c} \tilde{P}^{(3)}_{\mathrm{SRS2}}(\omega) \\ &\approx \left\{ 1 + i \frac{\omega L}{2\tilde{n}^{(1)}(\omega_2)c} \left|A_1\right|^2 \varepsilon_0 \tilde{\chi}^{(\mathrm{RR})*}(\omega_1 - \omega) \right\} \tilde{E}_2(\omega). \end{aligned} \tag{3.2.26}$$

Using Eq. (3.2.26), we then express the spectrum of the Stokes pulse after SRS by

$$\tilde{I}^{(\mathrm{SRS})}_2(\omega) \approx \left| 1 + i \frac{\omega L}{2\tilde{n}^{(1)}(\omega_2)c} \left|A_1\right|^2 \varepsilon_0 \tilde{\chi}^{(\mathrm{RR})*}(\omega_1 - \omega) \right|^2 \tilde{I}_2(\omega). \tag{3.2.27}$$

Assuming that the Stokes intensity change, which is undergone by SRS, is much lower than its initial intensity, we can approximate the spectrum of the Stokes pulse by

$$\tilde{I}_2^{(\mathrm{SRS})}(\omega) \approx \left[1 - i\frac{\omega L}{2\tilde{n}^{(1)}(\omega_2)c}\,|A_1|^2 \right.$$
$$\left. \times\, \varepsilon_0\left\{\tilde{\chi}^{(\mathrm{RR})}(\omega_1-\omega) - \tilde{\chi}^{(\mathrm{RR})*}(\omega_1-\omega)\right\}\right]\tilde{I}_2(\omega)$$
$$= \left[1 + \frac{\omega L}{\tilde{n}^{(1)}(\omega_2)c}\,|A_1|^2\,\varepsilon_0 \mathrm{Im}\left\{\tilde{\chi}^{(\mathrm{RR})}(\omega_1-\omega)\right\}\right]\tilde{I}_2(\omega). \quad (3.2.28)$$

Here $\mathrm{Im}\{h\}$ denotes the imaginary part of the complex $h = \mathrm{Re}\{h\} + i\mathrm{Im}\{h\}$. Equation (3.2.28) indicates that the spectral feature of the Stokes pulse reflects the vibrational energy levels of samples. We also find that the measurable Raman resonance bandwidth is proportional to the spectral bandwidth of the Stokes pulse.

Next, we describe using a broadband pump pulse and a narrowband Stokes pulse. The spectral bandwidth of the Stokes pulse is much narrower than that of the pump pulse and the Raman resonance bandwidth. According to Eq. (1.1.242), the polarization $P_{\mathrm{SRS1}}^{(3)}(t)$ in the SRS process at a frequency of ω_1 is written as

$$P_{\mathrm{SRS1}}^{(3)}(t) = \varepsilon_0 E_2(t)\int \chi^{(\mathrm{RR})}(t-t_1)E_2^*(t_1)E_1(t_1)dt_1. \quad (3.2.29)$$

The Fourier transform of Eq. (3.2.29) is written as

$$\tilde{P}_{\mathrm{SRS1}}^{(3)}(\omega) = \varepsilon_0 \int \tilde{\chi}^{(\mathrm{RR})}(\omega-\Omega_1)\tilde{E}_{12}^{(2\#)}(\omega-\Omega_1)\tilde{E}_2(\Omega_1)d\Omega_1. \quad (3.2.30)$$

The spectral bandwidth of the Stokes pulse is much narrower than that of the pump pulse and the Raman resonance bandwidth. Thus, $\tilde{E}_2(\omega)$ in Eq. (3.2.30) can be approximated by $A_2\delta(\omega-\omega_2)$. Then, $\tilde{P}_{\mathrm{SRS1}}^{(3)}(\omega)$ takes the form

$$\tilde{P}_{\mathrm{SRS1}}^{(3)}(\omega) \approx \varepsilon_0\,|A_2|^2\,\tilde{\chi}^{(\mathrm{RR})}(\omega-\omega_2)\tilde{E}_1(\omega). \quad (3.2.31)$$

According to Eqs. (1.1.37) and (3.2.31), the electric field of the pump pulse after SRS is described by

$$\tilde{E}_1^{(\mathrm{SRS})}(\omega) = \tilde{E}_1(\omega) + i\frac{\omega L}{2\tilde{n}^{(1)}(\omega_1)c}\tilde{P}_{\mathrm{SRS1}}^{(3)}(\omega)$$
$$\approx \left\{1 + i\frac{\omega L}{2\tilde{n}^{(1)}(\omega_1)c}\,|A_2|^2\,\varepsilon_0\tilde{\chi}^{(\mathrm{RR})}(\omega-\omega_2)\right\}\tilde{E}_1(\omega). \quad (3.2.32)$$

Employing Eq. (3.2.32), we can express the spectrum of the pump pulse after SRS by

$$\tilde{I}_1^{(\mathrm{SRS})}(\omega) \approx \left|1 + i\frac{\omega L}{2\tilde{n}^{(1)}(\omega_1)c}\,|A_2|^2\,\varepsilon_0\tilde{\chi}^{(\mathrm{RR})}(\omega-\omega_2)\right|^2\tilde{I}_1(\omega). \quad (3.2.33)$$

If the pump intensity change is much lower than its initial intensity, the spectrum of the pump pulse can be approximated by

$$\begin{aligned}\tilde{I}_1^{(\mathrm{SRS})}(\omega) &\approx \left[1 + i\frac{\omega L}{2\tilde{n}^{(1)}(\omega_1)c}\,|A_2|^2\right.\\&\left.\times\ \varepsilon_0\left\{\tilde{\chi}^{(\mathrm{RR})}(\omega-\omega_2) - \tilde{\chi}^{(\mathrm{RR})*}(\omega-\omega_2)\right\}\right]\tilde{I}_1(\omega)\\&= \left[1 - \frac{\omega L}{\tilde{n}^{(1)}(\omega_1)c}\,|A_2|^2\,\varepsilon_0\mathrm{Im}\left\{\tilde{\chi}^{(\mathrm{RR})}(\omega-\omega_2)\right\}\right]\tilde{I}_1(\omega). \quad (3.2.34)\end{aligned}$$

From Eq. (3.2.34), it is clear that the spectral feature of the pump pulse reflects the vibrational energy levels of samples, and that the measurable Raman resonance bandwidth is proportional to the spectral bandwidth of the pump pulse.

3.2.1.4 Multiplex Two-Photon Absorption (TPA) Spectroscopy

Now we analyze multiplex TPA spectroscopy. Figure 3.3(d) shows the energy level diagram of the TPA process with a multiplex excitation scheme. In multiplex TPA spectroscopy, TPA is typically induced by the combination of two-color pump pulses. A narrowband light source and a broadband light source are used as one pump pulses at a frequency of ω_1 and the other pulse at a frequency of ω_2.

From Eq. (1.1.213), the polarization $P_{\mathrm{TPA}}^{(3)}(t)$ in the TPA process at a frequency of ω_2 can be written as

$$P_{\mathrm{TPA}}^{(3)}(t) = \varepsilon_0 E_1^*(t_1)\int \chi^{(\mathrm{ER})}(t-t_1)E_1(t_1)E_2(t_1)dt_1. \quad (3.2.35)$$

The polarization in frequency domain is described by

$$\tilde{P}_{\mathrm{TPA}}^{(3)}(\omega) = \varepsilon_0\int \tilde{\chi}^{(\mathrm{ER})}(\Omega_1)\tilde{E}_{12}^{(2)}(\Omega_1)\tilde{E}_1^*(\Omega_1-\omega)d\Omega_1. \quad (3.2.36)$$

In the multiplex TPA technique, the spectral bandwidth of the pump pulse at ω_1 is much narrower than that of the pump pulse at ω_2 and the electronic resonance bandwidth. Thus, $\tilde{E}_1(\omega)$ in Eq. (3.2.36) can be approximated by $A_1\delta(\omega-\omega_1)$. Then, $\tilde{P}_{\mathrm{TPA}}^{(3)}(\omega)$ can be approximated by

$$\tilde{P}_{\mathrm{TPA}}^{(3)}(\omega) \approx \varepsilon_0\,|A_1|^2\,\tilde{\chi}^{(\mathrm{ER})}(\omega+\omega_1)\tilde{E}_2(\omega). \quad (3.2.37)$$

Let us assume that the interaction length L is much shorter, and that the loss and the dispersion can be neglected. Then, from Eqs. (1.1.37) and (3.2.37), the electric field of the pump pulse at ω_2 after TPA can be written as

$$\begin{aligned}\tilde{E}_2^{(\mathrm{TPA})}(\omega) &= \tilde{E}_2(\omega) + i\frac{\omega L}{2\tilde{n}^{(1)}(\omega_2)c}\tilde{P}_{\mathrm{TPA}}^{(3)}(\omega)\\&\approx \left\{1 + i\frac{\omega L}{2\tilde{n}^{(1)}(\omega_2)c}\,|A_1|^2\,\varepsilon_0\tilde{\chi}^{(\mathrm{ER})}(\omega_1+\omega)\right\}\tilde{E}_2(\omega). \quad (3.2.38)\end{aligned}$$

From Eq. (3.2.38), the spectrum of the pump pulse at ω_2 after TPA is expressed by

$$\tilde{I}_2^{(\mathrm{TPA})}(\omega) \approx \left| 1 + i \frac{\omega L}{2\tilde{n}^{(1)}(\omega_2)c} \left|A_1\right|^2 \varepsilon_0 \tilde{\chi}^{(\mathrm{ER})}(\omega_1 + \omega) \right|^2 \tilde{I}_2(\omega). \qquad (3.2.39)$$

If we assume that the intensity change of the pump pulse at ω_2, which is experienced by TPA, is much lower than its initial intensity, the spectrum of the pump pulse at ω_2 can be approximated by

$$\begin{aligned} \tilde{I}_2^{(\mathrm{TPA})}(\omega) &\approx \left[1 + i \frac{\omega L}{2\tilde{n}^{(1)}(\omega_2)c} \left|A_1\right|^2 \right. \\ &\quad \left. \times\, \varepsilon_0 \left\{ \tilde{\chi}^{(\mathrm{ER})}(\omega_1 - \omega) - \tilde{\chi}^{(\mathrm{ER})*}(\omega_1 - \omega) \right\} \right] \tilde{I}_2(\omega) \\ &= \left[1 - \frac{\omega L}{\tilde{n}^{(1)}(\omega_2)c} \left|A_1\right|^2 \varepsilon_0 \mathrm{Im}\left\{ \tilde{\chi}^{(\mathrm{ER})}(\omega_1 - \omega) \right\} \right] \tilde{I}_2(\omega). \qquad (3.2.40) \end{aligned}$$

Here $\mathrm{Im}\{h\}$ denotes the imaginary part of the complex $h = \mathrm{Re}\{h\} + i\mathrm{Im}\{h\}$. As is evident from Eq. (3.2.40), the spectral feature of the pump pulse at ω_2 reflects the electronic excitation energy levels of samples. We also find that the measurable electronic resonance bandwidth is proportional to the spectral bandwidth of the pump pulse at ω_2.

3.2.2 Experimental Setup for Multiplex-NLOS

Figure 3.4 shows a schematic of typical multiplex-NLOS. The excitation light sources of multiplex-NLOS, a broadband pulse laser and a narrowband pulse laser, are synchronized as shown in Fig. 3.4(a); the combination of a spectral broadening system and an original narrowband laser are employed as shown in Fig. 3.4(b). In the spectral broadening system, output from a narrowband pulse laser is split into two pulses by a beam splitter (BS) or by a dichroic mirror (DM). To broaden the spectral bandwidth of one pulse, a supercontinuum (SC) is generated in a highly nonlinear material such as a photonic crystal fiber (PCF) [29]. Then, the SC is shaped by passing through an optical filter (F1), such as a band-pass filter, a long-pass filter, and a short-pass filter. The dispersion of the optics is precompensated through a prechirper such as a prism pair or a chirp mirror pair. The narrowband pulse and the broadband pulse can be overlapped in time by using an optical delay line and in space by adjusting the angle of a DM. The two pulses are focused into a sample using a microscope objective lens (OB1). The generated signals are collected with the second objective (OB2) in the forward direction. After passing through a second optical filter (F2), the signals are measured by a spectrometer.

3.2.3 Measurement of Spectroscopic Information by Multiplex-NLOS

By using a combination of a narrowband pump laser at a pulse duration of 10 ps with a bandwidth of 1.5 cm^{-1} and a broadband Stokes laser at a pulse duration of 80 fs with a bandwidth of 280 cm^{-1}, Müller and Schine have demonstrated multiplex

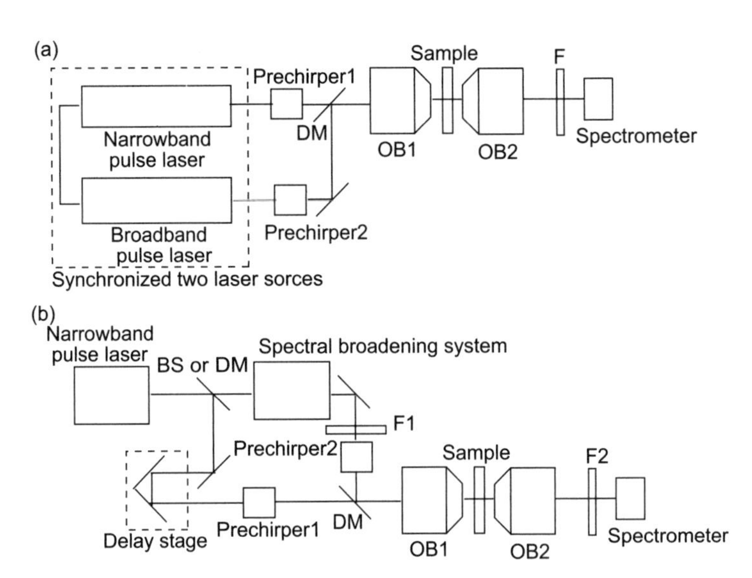

FIGURE 3.4 Experimental setup for multiplex-NLOS using two synchronized lasers (a), and using a combination of a wavelength converter and an original narrowband laser (b): BS, beam splitter, OB, objective; DM, dichroic mirror; F, filter.

CARS spectroscopy with a spectral resolution of 5 cm^{-1} [22]. Figure 3.5(a) and (b) shows multiplex CARS spectra of a 1,2-distearoyl-*sn*-glycero-3-phosphocholine (DSPC; 18:0) and a 1,2-dioleoyl-*sn*-glycero-3-phosphocholine (DOPC; 18:1, 9-cis) multi-lamellar vesicle, respectively. The CARS spectrum was collected over spectral band of approximately 300 cm^{-1}. Figure 3.5(c) and (d) displays the CARS spectra of DSPC and DOPC, respectively, after dividing out the influence of the Stokes spectral-intensity profile from the CARS spectra shown in Fig. 3.5(a) and (b). This is accomplished by fitting the experimental data to the theoretical CARS signal. These results indicate that the mixing of the nonresonant and resonant contributions is responsible for the typical dispersive shape of the CARS spectra.

In order to achieve more measurable bandwidth, it is necessary to use an ultrabroadband Stokes pulse at a pulse duration of 5~10 fs. However, it is difficult to synchronize an ultrabroadband pulse laser at a pulse duration of 5~10 fs with a narrowband pulse laser at a pulse duration of 1 to 10 ps because of the relatively large timing jitter between them. Thus, an SC generated with a narrowband pump pulse in the nonlinear fiber is used as a broadband Stokes pulse. By applying SCs generated with the picosecond pulses in the nonlinear fiber to the broadband Stokes pulses, Kee et al. [30] and Kano et al. [31] have developed ultrabroadband (>2500 cm^{-1}) multiplex CARS spectroscopy with a spectral resolution of 10 cm^{-1}.

3.3 FOURIER-TRANSFORM SPECTROSCOPY

In the laser-wavelength scanning method and in the multiplex method, spectral resolution depends on the spectral width of the excitation light. Thus, there is a trade-off

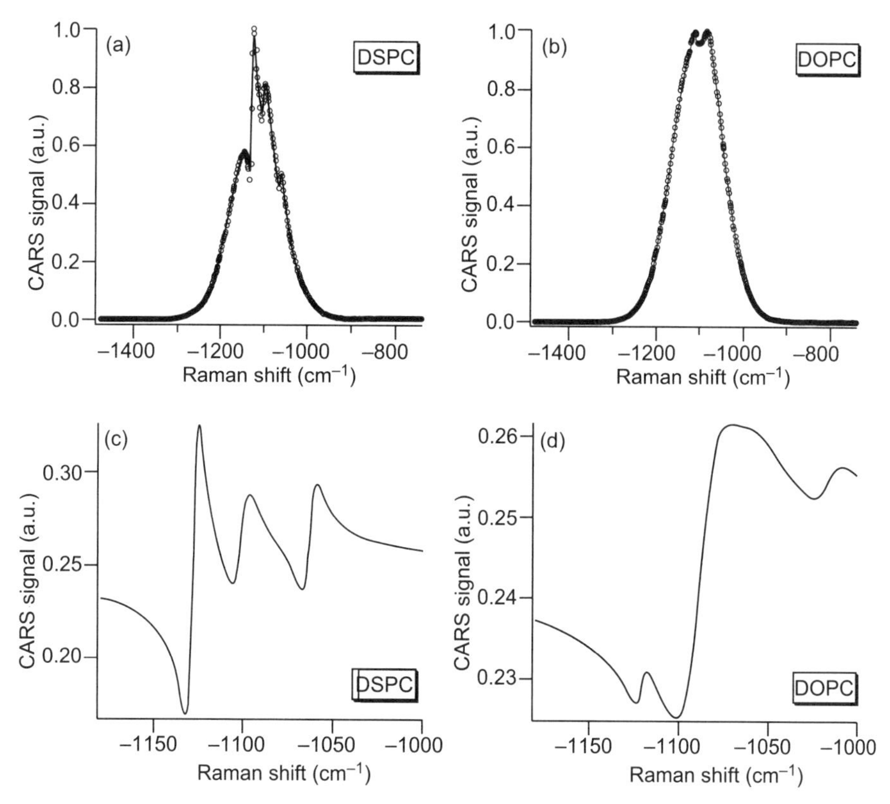

FIGURE 3.5 (a, b) CARS spectra of a DSPC and a DOPC multi-lamellar vesicle at room temperature. The solid line is the fit of the data, including a Gaussian Stokes spectral-intensity profile. (c, d) CARS spectra of the DSPC (c) and DOPC (d) vesicle after eliminating the Stokes spectral-intensity profile. Reprinted with permission from [22]. Copyright 2002 American Chemical Society.

between spectral resolution and the excitation efficiency of nonlinear optical processes. The reason for this is as follows: pulse duration is inversely proportional to the spectral bandwidth, and excitation efficiency increases with the shortening pulse duration of the excitation light. Therefore, the excitation efficiency is low if the spectral resolution is high. Fourier-transform (FT) nonlinear optical spectroscopy (NLOS) is based on the measurement of an interferometric autocorrelation (IAC) signal by broadband pulses [32]. The spectrum is obtained by the Fourier transformation of an IAC signal. Thus, the spectral resolution is not determined by the spectral bandwidth of the excitation light source, but by the inverse of the maximum delay time between excitation pulses. Therefore, a high spectral resolution can be achieved, even though 5-fs ultrabroadband pulses are used. In the case of the use of broadband pulses, it is advantageous that the pulses are tightly focused with a high numerical-aperture (NA) objective lens because the phase-matching condition in a coherent nonlinear process is markedly relaxed owing to the wide cone of wave vectors and the short interaction length. With tight focusing, excitation efficiency is also enhanced by increasing intensity. Because broadband spectroscopic information can be obtained from a single measurement, the acquisition time is several seconds or less. Since FT coherent anti-Stokes Raman scattering (CARS) spectroscopy was first reported by Felker et al. [33], FT-NLOS has been applied to CARS [34–37], stimulated parametric emission (SPE) [37], two-photon excited fluorescence (TPEF) [37–40], and sum-frequency generation (SFG) [41,42].

3.3.1 Principle of FT-NLOS

Now we show the principle of FTNLOS spectroscopy [37]. We analyze the IAC signal and show that the response function of a sample can be derived from the IAC signal. The nth-order IAC signal $S^{(n)}(\tau)$ is written as

$$S^{(n)}(\tau) = \int_{-\infty}^{\infty} \left| P^{(n)}(t, \tau) \right|^2 dt, \tag{3.3.1}$$

where τ is the delay time between broadband pulses and $P^{(n)}(t, \tau)$ is the nonlinear polarization described by

$$\begin{aligned} P^{(n)}(t, \tau) = & \int_{-\infty}^{t} dt_1 \int_{-\infty}^{t} dt_2 \cdots \int_{-\infty}^{t} dt_n R^{(n)}(t - t_1, t_1 - t_2, \cdots, t_{n-1} - t_n) \\ & \times E(t_1, \tau) E(t_2, \tau) \cdots E(t_n, \tau). \end{aligned} \tag{3.3.2}$$

Here $E(t, \tau)$ is the interferometric electric field and $R^{(n)}(t - t_1, t_1 - t_2, \ldots, t_{n-1} - t_n)$ is the response function of a sample. To simplify our discussion, let us assume that a nonlinear process is due to a single-resonance contribution. Under a single-resonance condition, one of the real atomic or molecular levels is nearly coincident with one of the virtual levels of the indicated process. When the m-photon transition is nearly resonant, the time-delayed transition involving a real level can be separated from the instantaneous transition between virtual levels, and $R(n)(t - t_1, t_1 - t_2, \ldots, t_{n-1} - t_n)$ is described as

$$\begin{aligned} & R^{(n)}(t - t_1, t_1 - t_2, \cdots, t_{n-1} - t_n) \\ & \quad = \delta(t - t_1) \cdots \delta(t - t_{m-1}) R(t - t_m) \delta(t_m - t_{m+1}) \cdots \delta(t_m - t_n). \end{aligned} \tag{3.3.3}$$

In a nonresonance contribution, $R(t - t_m)$ is equal to $\delta(t - t_m)$, and all transitions are instantaneous. In FT-NLOS, the power spectrum $\tilde{S}^{(n)}(\Omega)$ is calculated to obtain the spectral information, that is, the Fourier transform of $R(t - t_m)$. $\tilde{S}^{(n)}(\Omega)$ is obtained by the Fourier transform of the IAC signal $S^{(n)}(\tau)$ expressed as

$$\tilde{S}^{(n)}(\Omega) = \int_{-\infty}^{\infty} S^{(n)}(\tau) \exp(i\Omega\tau) d\tau. \tag{3.3.4}$$

3.3.1.1 Second-Order IAC Signal We present that two-photon excitation (TPE) spectra in second-harmonic generation (SHG), and TPEF processes can be obtained from the second-order IAC signal. In TPEF and SHG processes, the signal intensity is proportional to the square of incident intensity. If two-photon transition is nearly resonant, the response function is given by $R^{(2)}(t-t_1, t_1-t_2) = R^{(\mathrm{TP})}(t-t_1)\delta(t_1-t_2)$. In using this response function, we obtain the nonlinear polarizations $P^{(\mathrm{TP})}(t,\tau)$ in TPEF and SHG:

$$P^{(\mathrm{TP})}(t,\tau) = \int_{-\infty}^{\infty} R^{(\mathrm{TP})}(t-t_1)E(t_1,\tau)^2 dt_1, \tag{3.3.5}$$

where we represent the interferometric field $E(t,\tau)$ as

$$E(t,\tau) = \left[A(t) + A(t-\tau)\exp(i\omega_0\tau)\right]\exp(-i\omega_0 t). \tag{3.3.6}$$

Here we define the electric fields as $E(t) = A(t)\exp(-i\omega_0 t)$, where $A(t)$ is a slowly varying amplitude and ω_0 is the central frequency of the electric field. The specific condition, under which the concept of envelope equations can be applicable to the single cycle regime [43], is satisfied as follows: We insert Eq. (3.3.6) into Eq. (3.3.5) to obtain

$$\begin{aligned} P^{(\mathrm{TP})}(t,\tau) = H_1(t) &+ 2H_2(t,\tau)\exp(i\omega_0\tau) \\ &+ H_3(t,\tau)\exp(i2\omega_0\tau), \end{aligned} \tag{3.3.7}$$

where we define $H_1(t)$, $H2(t,\tau)$, and $H_3(t,\tau)$ as

$$H_1(t) = \int R^{(TP)}(t-t_1)A(t_1)^2\exp(-i2\omega_0 t_1)dt_1, \tag{3.3.8}$$

$$H_2(t,\tau) = \int R^{(TP)}(t-t_1)A(t_1)A(t_1-\tau)\exp(-i2\omega_0 t_1)dt_1, \tag{3.3.9}$$

$$H_3(t,\tau) = \int R^{(TP)}(t-t_1)A(t_1-\tau)^2\exp(-i2\omega_0 t_1)dt_1. \tag{3.3.10}$$

Substituting Eq. (3.3.7) into Eq. (3.3.1), we write the second-order IAC signal $S^{(2)}(\tau)$ as

$$\begin{aligned} S^{(2)}(\tau) = \int dt \Big[& |H_1(t)|^2 + 4|H_2(t,\tau)|^2 + |H_3(t,\tau)|^2 \\ & +2\left\{H_1(t)H_2^*(t,\tau) + H_2(t)H_3^*(t,\tau)\right\}\exp(-i\omega_0\tau) \\ & +2\left\{H_1^*(t)H_2(t,\tau) + H_2^*(t)H_3(t,\tau)\right\}\exp(i\omega_0\tau) \\ & +H_1(t)H_3^*(t,\tau)\exp(-i2\omega_0\tau) \\ & +H_1^*(t)H_3(t,\tau)\exp(i2\omega_0\tau)\Big]. \end{aligned} \tag{3.3.11}$$

We find that the second-order IAC signal is composed of fringe components at frequencies of around 0, ω_0, and $2\omega_0$, and the fringe contrast $S^{(2)}(0)/S^{(2)}(\infty)$ is 8.

We show that the response function can be obtained from the fringe component at a frequency of around $2\omega_0$. The power spectrum at a frequency of around $2\omega_0$ is then given by

$$\begin{aligned}
\tilde{S}^{(2)}_{2\omega_0}(\Omega) &= \int\int H_1(t) H_3^*(t,\tau) \exp(-i2\omega_0\tau) dt \exp(i\Omega\tau) d\tau \\
&= \int\int\int dt dt_1 d\tau \, H_1(t) \exp(i\Omega t) \\
&\quad \times R^{(\mathrm{TP})*}(t-t_1) \exp\{-i\Omega(t-t_1)\} \\
&\quad \times A^*(t_1-\tau)^2 \exp\{-i(\Omega-2\omega_0)(t_1-\tau)\} \\
&= \tilde{H}_1(\Omega) \tilde{R}^{(\mathrm{TP})*}(\Omega) \tilde{A}^{(2)*}(\Omega-2\omega_0) \\
&= \left|\tilde{R}^{(\mathrm{TP})}(\Omega)\right|^2 \left|\tilde{A}^{(2)}(\Omega-2\omega_0)\right|^2,
\end{aligned} \tag{3.3.12}$$

where $\tilde{A}^{(2)}(\Omega)$ and $\tilde{H}_1(\Omega)$ take the forms

$$\tilde{A}^{(2)}(\Omega) = \int \tilde{A}(\Omega_1)\tilde{A}(\Omega-\Omega_1) d\Omega_1, \tag{3.3.13}$$

and

$$\tilde{H}_1(\Omega) = R^{(\mathrm{TP})}(\Omega) A^{(2)}(\Omega-2\omega_0), \tag{3.3.14}$$

respectively. $|\tilde{A}^{(2)}(\Omega-2\omega_0)|^2$ is the second-harmonic (SH) power spectrum as determined by the properties of the excitation pulses. The SH power spectrum can be acquired by the measurement of $\tilde{S}^{(2)}_{2\omega_0}(\Omega)$ from a reference sample whose response is instantaneous, and $\tilde{R}^{(\mathrm{TP})}(\Omega)$ can be assumed to be constant. By dividing the value of $\tilde{S}^{(2)}_{2\omega_0}(\Omega)$ from a test sample by that from the reference sample, we can obtain $\tilde{R}^{(\mathrm{TP})}(\Omega)$, which is called the TPE spectrum. To obtain the absolute TPE spectra, we also need to measure $\tilde{S}^{(2)}_{2\omega_0}(\Omega)$ from a standard sample whose absolute TPE spectrum is known. The known absolute TPE spectrum of the standard sample is expressed by

$$\left|\tilde{R}^{(\mathrm{TP})}_{\mathrm{st}}(\Omega)\right|^2 = a \frac{\tilde{S}^{(2)}_{2\omega_0,\mathrm{st}}(\Omega)}{\tilde{S}^{(2)}_{2\omega_0,\mathrm{r}}(\Omega)}, \tag{3.3.15}$$

where a is the calibration constant, and the subscript "st" and "r" indicate the parameters for the standard sample and the reference sample, respectively. By using the calibration constant a, we obtain the absolute TPE spectrum of the test sample from

$$\left|\tilde{R}^{(\mathrm{TP})}_{\mathrm{s}}(\Omega)\right|^2 = a \frac{\tilde{S}^{(2)}_{2\omega_0,\mathrm{s}}(\Omega)}{\tilde{S}^{(2)}_{2\omega_0,\mathrm{r}}(\Omega)} = \frac{\tilde{S}^{(2)}_{2\omega_0,\mathrm{s}}(\Omega)}{\tilde{S}^{(2)}_{2\omega_0,\mathrm{st}}(\Omega)} \left|\tilde{R}^{(\mathrm{TP})}_{\mathrm{st}}(\Omega)\right|^2, \tag{3.3.16}$$

where the subscript "s" denotes the parameter for the test sample. Note that in the theoretical derivation, there is no need to take care of a difference between dispersions of the test sample, the standard sample, and the reference sample. This is because the dispersion has a negligible effect in the broadband pulse due to the short interaction length resulting from tight focusing.

In practice, there are three important requirements in obtaining the ideal reference spectrum. First, the wavelength dependence of the nonlinear response should be as weak as possible over the full spectral range of the broadband light source. Second, the nonlinear signal needs to be large enough to obtain the spectrum with high signal-to-noise ratio (SNR). Third, other nonlinear phenomena competing with the SH conversion, such as four-wave mixing (FWM), should not occur. A β-BBO crystal is one of the most widely used for SHG media because of its large nonlinear optical constant and the wide wavelength range over which the phase-matching condition is satisfied. However, even when using a thin β-BBO crystal cut on the proper phase-matching angle at the central wavelength, the bandwidth of the SH spectrum is still limited by the phase matching. In particular, the confocal parameter of the tightly focused beam is $< 1\ \mu$m, which is much less than the thickness of crystals currently commercially available. In this case the nonuniform spatial and spectral distributions of the SH light cause the reference SH spectrum to deviate from the ideal one. For a simple implementation, in semiconductors the two-photon-induced free-carrier current such as GaAsP or GaP photodiodes has been used to measure the second-order IAC signal of ultrashort laser pulses [44]. However, the bandwidth is narrower than that of the typical β-BBO SHG, and the one-photon absorption on the long-wavelength side overlaps with the two-photon absorption on the short-wavelength side. The SH signal generated in the region near the surface of a quartz crystal cut on 90° is appropriate for the reference spectrum measurement [40]. Although the SH signal is weak due to the non–phase-matched process, it is less sensitive to wavelength. In addition, the unfavorable FWM process can be weak enough not to disturb the precise measurement by controlling the incident laser power.

From the Fourier spectra of the IAC signals, we can also confirm that the fringe component at a frequency of around $2\omega_0$ is proportional to the square of the excitation intensity. The fringe component at a frequency of around $n\omega_0$ implies that the signal intensity is proportional to the $m(\geq n)$th power of the excitation intensity. Thus, if the fringe component at a frequency of around $3\omega_0$ does not exist in the IAC signal, we can see that the fringe component at a frequency of around $2\omega_0$ is proportional to the square of the excitation intensity. Therefore, the measurement of the signal intensity dependence on the excitation intensity is not required in FT-NLOS.

A second-order IAC measurement can be also applied to obtain not only the SH power spectrum but also the deference-frequency (DF) power spectrum, both of which are determined by the properties of the excitation pulses. If we assume that the response of a reference sample is instantaneous, $R^{(\mathrm{TP})}(t) = \delta(t)$ and $\left|\tilde{R}^{(\mathrm{TP})}(\Omega)\right|^2$ is constant, the power spectrum at a frequency of around 0 is given by

$$\begin{aligned}
\tilde{S}_0^{(2)}(\Omega) &= \int\int \left[|H_1(t)|^2 + 4\,|H_2(t,\tau)|^2 + 4\,|H_3(t,\tau)|^2\right]dt\,\exp(i\Omega\tau)d\tau \\
&= 2\,|A(t)|^2\,\delta(\Omega) + 4\int\int |A(t)|^2\,|A(t-\tau)|^2 dt\,\exp(i\Omega\tau)d\tau \\
&= 2\,|A(t)|^2\,\delta(\Omega) + 4\left|\int \tilde{A}(\omega)\tilde{A}^*(\omega-\Omega)d\omega\right|^2. \qquad (3.3.17)
\end{aligned}$$

The second term in Eq. (3.3.17) indicates the DF spectrum. Thus, we can measure the DF spectrum without an IR spectrometer, which is used for directly measuring the DFG generated from a nonlinear crystal.

3.3.1.2 Third-Order IAC Signal Consider the four-wave mixing (FWM) process in which a photon with a frequency of $2\omega_1 - \omega_2$ is generated by the interaction between the medium and photons with frequencies of ω_1 and ω_2. In two-photon resonance FWM, electronic resonance and Raman resonance exist as shown in Chapter 1.3. The frequency of the FWM signal does not depend on the contributions of Raman resonance and electronic resonance. To discriminate the signal under the electronic resonance from that under the Raman resonance, it has been necessary to measure the dependences of signal intensity on sum frequency ($2\omega_1$) and difference frequency ($\omega_1 - \omega_2$). We show that the origin of the FWM signal can be distinguished by taking advantage of the FWM power spectrum obtained by FT-NLOS.

We analyze the FWM power spectrum obtained by the Fourier transform of the third-order IAC signal by CARS, which is FWM under the Raman resonance. The polarization in FWM under the Raman resonance is described by

$$P^{(\mathrm{Raman})}(t,\tau) = E_1(t,\tau)\int R^{(\mathrm{Raman})}(t-t_1)E_1(t_1,\tau)E_2^*(t_1,\tau)dt_1, \quad (3.3.18)$$

where $E_{\mathrm{j}}(t,\tau) = \left[A_{\mathrm{j}}(t) + A_{\mathrm{j}}(t-\tau)\exp(i\omega_{\mathrm{j}}\tau)\right]\exp(-i\omega_{\mathrm{j}}t)$, (j $= 1, 2$). The absolute square of $P^{(Raman)}(t,\tau)$ is expressed as

$$\begin{aligned}
\left|P^{(\mathrm{Raman})}(t,\tau)\right|^2 &= |E_1(t,\tau)|^2\left|\int R^{(\mathrm{Raman})}(t-t_1)E_1(t_1,\tau)E_2^*(t_1,\tau)dt_1\right|^2 \\
&= B_1(t,\tau) \\
&\quad + B_2(t,\tau)\exp(-i\omega_1\tau) + B_2^*(t,\tau)\exp(i\omega_1\tau) \\
&\quad + B_3(t,\tau)\exp(-i\omega_2\tau) + B_3^*(t,\tau)\exp(i\omega_2\tau) \\
&\quad + B_4(t,\tau)\exp\{-i(\omega_1-\omega_2)\tau\} \\
&\quad + B_4^*(t,\tau)\exp\{i(\omega_1-\omega_2)\tau\} \\
&\quad + B_5(t,\tau)\exp\{-i(\omega_1+\omega_2)\tau\} \\
&\quad + B_5^*(t,\tau)\exp\{i(\omega_1+\omega_2)\tau\} \\
&\quad + B_6(t,\tau)\exp(-i2\omega_1\tau) + B_6^*(t,\tau)\exp(i2\omega_1\tau) \\
&\quad + B_7(t,\tau)\exp\{-i(2\omega_1-\omega_2)\tau\} \\
&\quad + B_7^*(t,\tau)\exp\{i(2\omega_1-\omega_2)\tau\} \\
&\quad + B_8(t,\tau)\exp\{-i(2\omega_1+\omega_2)\tau\} \\
&\quad + B_8^*(t,\tau)\exp\{i(2\omega_1+\omega_2)\tau\}, \qquad (3.3.19)
\end{aligned}$$

where we define $B_1(t,\tau)$, $B_2(t,\tau)$, $B_3(t,\tau)$, $B_4(t,\tau)$, $B_5(t,\tau)$, $B_6(t,\tau)$, $B_7(t,\tau)$, and $B_8(t,\tau)$ as

$$\begin{aligned}
B_1(t,\tau) = & \left[\left[|A_1(t)|^2 + |A_1(t-\tau)|^2\right]\right. \\
& \times \left[|G_1(t)|^2 + |G_2(t,\tau)|^2 + |G_3(t,\tau)|^2 + |G_4(t,\tau)|^2\right] \\
& + A_1^*(t)A_1(t-\tau)\left[G_1(t)G_3^*(t,\tau) + G_2^*(t,\tau)G_4(t,\tau)\right] \\
& \left. + A_1(t)A_1^*(t-\tau)\left[G_1^*(t)G_3(t,\tau) + G_2(t,\tau)G_4^*(t,\tau)\right]\right], \qquad (3.3.20)
\end{aligned}$$

$$\begin{aligned}
B_2(t,\tau) = & \left[\left[|A_1(t)|^2 + |A_1(t-\tau)|^2\right]\right. \\
& \times \left[G_1(t)G_3^*(t,\tau) + G_2^*(t,\tau)G_4(t,\tau)\right] \\
& + A_1(t)A_1^*(t-\tau) \\
& \left. \times \left[|G_1(t)|^2 + |G_2(t,\tau)|^2 + |G_3(t,\tau)|^2 + |G_4(t,\tau)|^2\right]\right], \qquad (3.3.21)
\end{aligned}$$

$$\begin{aligned}
B_3(t,\tau) = & \left[\left[|A_1(t)|^2 + |A_1(t-\tau)|^2\right]\right. \\
& \times \left[G_1^*(t)G_4(t,\tau) + G_2(t,\tau)G_3^*(t,\tau)\right] \\
& + A_1(t)A_1^*(t-\tau)G_1^*(t)G_2(t,\tau) \\
& \left. + A_1^*(t)A_1(t-\tau)G_3^*(t,\tau)G_4(t,\tau)\right], \qquad (3.3.22)
\end{aligned}$$

$$\begin{aligned}
B_4(t,\tau) = & \left[\left[|A_1(t)|^2 + |A_1(t-\tau)|^2\right]G_1(t)G_2^*(t,\tau)\right. \\
& \left. + A_1(t)A_1^*(t-\tau)\left[G_1(t)G_4^*(t,\tau) + G_2^*(t,\tau)G_3(t,\tau)\right]\right], \qquad (3.3.23)
\end{aligned}$$

$$\begin{aligned}
B_5(t,\tau) = & \left[\left[|A_1(t)|^2 + |A_1(t-\tau)|^2\right]G_3^*(t,\tau)G_4(t,\tau)\right. \\
& \left. + A_1(t)A_1^*(t-\tau)\left[G_1^*(t)G_4(t,\tau) + G_2(t,\tau)G_3^*(t,\tau)\right]\right], \qquad (3.3.24)
\end{aligned}$$

$$B_6(t,\tau) = A_1(t)A_1^*(t-\tau)\left[G_1(t)G_3^*(t,\tau) + G_2^*(t,\tau)G_4(t,\tau)\right], \qquad (3.3.25)$$

$$B_7(t,\tau) = A_1(t)A_1^*(t-\tau)G_1(t)G_2^*(t,\tau), \qquad (3.3.26)$$

$$B_8(t,\tau) = A_1(t)A_1^*(t-\tau)G_3^*(t,\tau)G_4(t,\tau). \qquad (3.3.27)$$

Here we define $G_1(t)$, $G_2(t,\tau)$, $G_3(t,\tau)$, and $G_4(t,\tau)$ as

$$G_1(t) = \int R^{(\mathrm{Raman})}(t-t_1)A_1(t_1)A_2^*(t_1)\exp\{-i(\omega_1-\omega_2)t_1\}dt_1, \qquad (3.3.28)$$

$$\begin{aligned}
G_2(t,\tau) = & \int R^{(\mathrm{Raman})}(t-t_1)A_1(t_1-\tau)A_2^*(t_1-\tau) \\
& \times \exp\{-i(\omega_1-\omega_2)t_1\}\,dt_1, \qquad (3.3.29)
\end{aligned}$$

$$\begin{aligned}
G_3(t,\tau) = & \int R^{(\mathrm{Raman})}(t-t_1)A_1(t_1-\tau)A_2^*(t_1) \\
& \times \exp\{-i(\omega_1-\omega_2)t_1\}\,dt_1, \qquad (3.3.30)
\end{aligned}$$

$$\begin{aligned}
G_4(t,\tau) = & \int R^{(\mathrm{Raman})}(t-t_1)A_1(t_1)A_2^*(t_1-\tau) \\
& \times \exp\{-i(\omega_1-\omega_2)t_1\}\,dt_1. \qquad (3.3.31)
\end{aligned}$$

From Eq. (3.3.19), we find that the FWM-IAC signal under the Raman resonance includes fringe components at frequencies of around 0, ω_1, ω_2, $\omega_1-\omega_2$, $\omega_1+\omega_2$,

$2\omega_1$, $2\omega_1 - \omega_2$, and $2\omega_1 + \omega_2$, and that the fringe contrast $S^{(\text{Raman})}(0)/S^{(\text{Raman})}(\infty)$ is 32.

If the delay time τ is longer than the pulse duration but is shorter than the Raman coherence lifetime, $G_1(t)$ and $G_2(t,\tau)$ are not equal to zero, whereas $G_3(t,\tau)$, $G_4(t,\tau)$, and $A_1(t)A_1^*(t-\tau)$ are equal to zero. Then, from Eqs. (3.3.1) and (3.3.19), we obtain

$$\begin{aligned} S^{(\text{dRaman})}(\tau) = \int dt \Big[&|A_1(t)|^2 \, |G_1(t)|^2 + |A_1(t-\tau)|^2 \, |G_2(t,\tau)|^2 \\ &+ \left[|A_1(t)|^2 + |A_1(t-\tau)|^2\right] \\ &\times G_1(t) G_2^*(t,\tau) \exp\{-i(\omega_1 - \omega_2)\tau\} \\ &+ \left[|A_1(t)|^2 + |A_1(t-\tau)|^2\right] \\ &\times \, G_1^*(t) G_2(t,\tau) \exp\{i(\omega_1 - \omega_2)\tau\}\Big]. \end{aligned} \tag{3.3.32}$$

Equation (3.3.32) indicates that when no Raman coherence is induced, no FWM signal is generated by the interaction between two pulses that are not temporally overlapped. Even though the nonresonant FWM signal and electronic resonant FWM signal are mixed with the Raman resonant FWM signal, the FWM-IAC signal shown in Eq. (3.3.32) can be obtained in a delay time longer than the pulse duration and the electronic coherence lifetime. Note that in Eq. (3.3.32), the FWM-IAC signal under the Raman resonance condition is composed of only fringe components at frequencies of around 0 and $\omega_1 - \omega_2$. The FWM power spectrum at a frequency of around $\omega_1 - \omega_2$ is obtained by

$$\begin{aligned} \tilde{S}^{(\text{dRaman})}_{(\omega_1-\omega_2)}(\Omega) = \; & \tilde{R}^{(\text{Raman})}(\Omega) \tilde{A}_{12}^{(2\#)}(\Omega - \omega_1 + \omega_2) \\ & \times \left[\int \tilde{A}_{11}^{(2\#)}(\Omega - \Omega_1) \tilde{R}^{(\text{Raman})}(\Omega_1) \tilde{A}_{12}^{(2\#)}(\Omega_1 - \omega_1 + \omega_2) d\Omega_1 \right]^* \\ & + c.c. \end{aligned} \tag{3.3.33}$$

where $\tilde{A}_{12}^{(2\#)}(\omega)$ is given by

$$\tilde{A}_{\text{ij}}^{(2\#)}(\omega) = \int \tilde{A}_{\text{i}}(\Omega_1) \tilde{A}_{\text{j}}^*(\Omega_1 - \omega) d\Omega_1, \tag{3.3.34}$$

Since the Raman bandwidth is typically on the order of 10 cm^{-1}, the bandwidth of $\tilde{A}_{12}^{(2\#)}(\omega)$ in a sub-10-fs pulse is much broader than the Raman bandwidth. Therefore, we can regard $\tilde{R}^{(\text{Raman})}(\Omega)$ as $\delta(\Omega - \Omega_0)$ and simplify Eq. (3.3.33) to

$$\begin{aligned} \tilde{S}^{(\text{dRaman})}_{(\omega_1-\omega_2)}(\Omega) \approx \; & \delta(\Omega - \Omega_0) \tilde{A}_{11}^{(2\#)}(\Omega - \Omega_0) \left| \tilde{A}_{12}^{(2\#)}(\Omega_0 - \omega_1 + \omega_2) \right|^2 \\ & + c.c. \end{aligned} \tag{3.3.35}$$

Equation (3.3.35) indicates the Fourier spectrum of the response function. By measuring the FWM-IAC signal in a delay time range longer than the pulse duration but

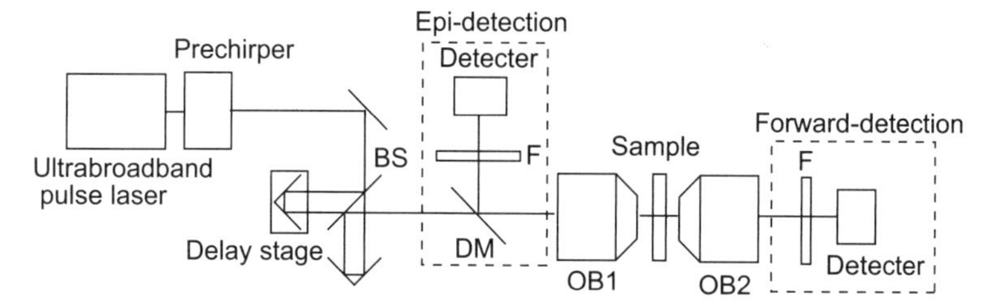

FIGURE 3.6 Fourier-transform nonlinear optical spectroscope.

shorter than the Raman coherence time, we can obtain the response function of a Raman resonance sample. The response function of an electronic resonance sample can be acquired from the FWM power spectrum at a frequency of around $2\omega_1$, which is obtained by the Fourier transform of the third-order IAC signal by FWM under the two-photon electronic resonance, SPE [37]. Note that the origin of the FWM signal can be discriminated by focusing on the FWM power spectrum at frequencies of around $2\omega_1$ or $\omega_1 - \omega_2$.

3.3.2 Experimental Setup for FT-NLOS

A schematic of the typical experimental setup for FT-NLOS is shown in Fig. 3.6. As an excitation light source, an ultrabroadband pulse laser is employed. In order to compensate for dispersion of all the optical components, the laser pulse is passed through a prechirper such as a prism pair and/or a pulse shaper with a spatial light modulator. The prechirped pulse is then launched into a Michelson interferometer. The output pulse is focused into a sample or a reference sample by an objective lens (OB). The resultant coherent signals and incoherent signals are detected with point detectors such as photomultiplier tubes (PMTs) and photodiodes (PDs) in the forward and backward directions, respectively. For example, TPEF signals are obtained with epi-detection, while SHG and CARS signals are typically acquired with forward detection. IAC signals are obtained by scanning a delay stage in the interferometer.

3.3.3 Measurement of Broadband Excitation Spectra by FT-NLOS

Isobe et al. [37] and Hashimoto et al. [40] have reported FT-TPEF spectroscopy employing the 5-fs broadband pulse whose spectrum is shown in Fig. 3.7(a). Figure 3.7(b) shows the second-order IAC signal for the SH signal generated in the region near the surface of a quartz crystal cut on 90°. Because the SH signal was generated in a non–phase-matched configuration, it is less sensitive to the wavelength. Thus, we selected the SH signal generated in the region near the surface of a quartz crystal cut on 90° for the measurement of the reference spectrum without the wavelength dependence. As shown in Fig. 3.7(c), frequency components of 0, ω_0, and $2\omega_0$ exist in the Fourier transform of the SHG-IAC signal. Figure 3.7(d) shows the SH power spectrum obtained by the Fourier transform of the SHG-IAC signal at a frequency

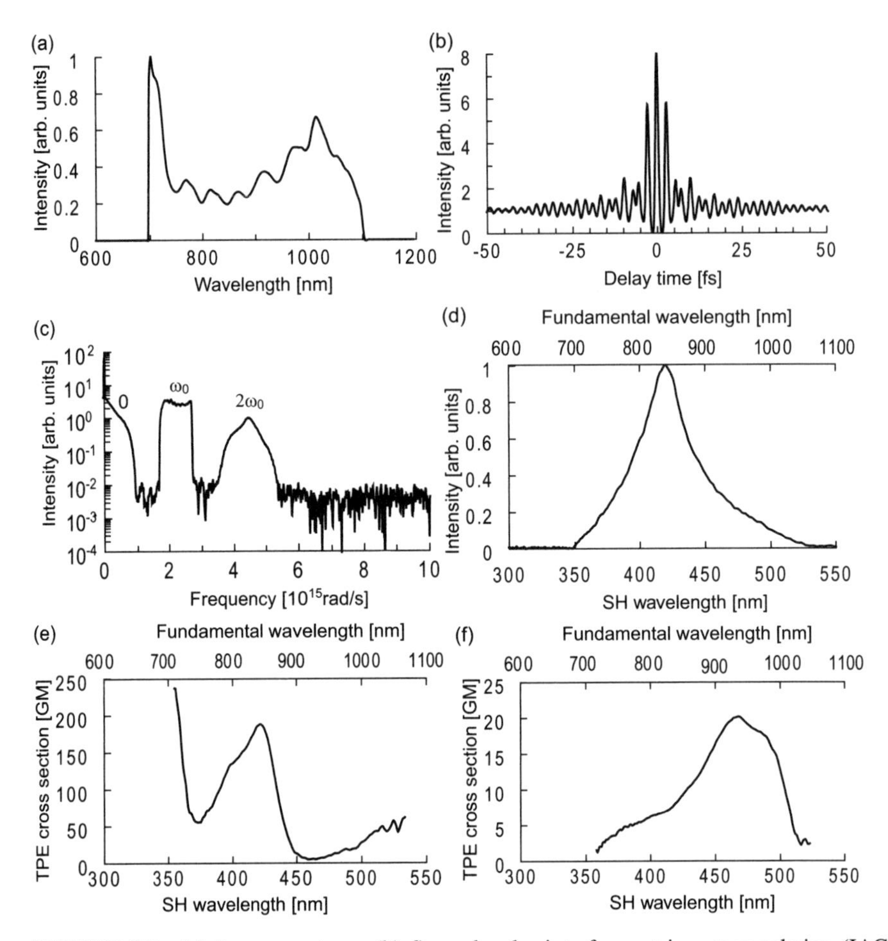

FIGURE 3.7 (a) Laser spectrum. (b) Second-order interferometric autocorrelation (IAC) signal by SHG from generated near the surface of a quartz crystal cut on 90°. (c) SH power spectrum of obtained by the Fourier transform of the IAC signal shown in (b). (d) SH power spectrum at a frequency of around $2\omega_0$ shown in (c). (e, f) Absolute TPE spectra for rhodamine B (e) and EGFP (f).

of around $2\omega_0$. Figure 3.7(e) and (f) show the absolute TPE spectra of rhodamine B and green fluorescent protein (GFP), respectively, as were obtained by dividing their Fourier spectra by the SH power spectrum from the quartz crystal. The absolute TPE spectra are calibrated against that of rhodamine B [9]. Note that the measurable spectral bandwidth covers the wavelength tuning range of a narrowband Ti:sapphire laser.

Isobe et al. demonstrated FT-CARS spectroscopy employing a 5-fs broadband pulse [36,37]. Figure 3.8(a) shows the third-order IAC signals by CARS from acetone. The Fourier spectrum, which is obtained by the FT of the IAC signal from acetone in the delay time range from −266.5 to 266.5 fs is shown in Fig. 3.8(b). By dividing

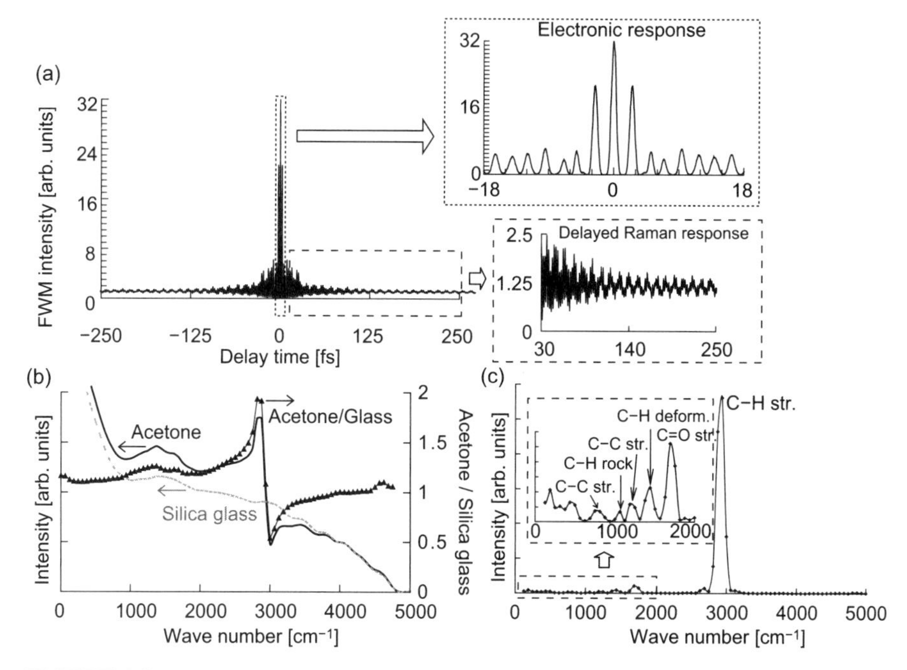

FIGURE 3.8 (a) Third-order IAC signal by Raman resonant FWM from acetone. (b, c) Fourier spectra of the IAC signals in the delay-time range from −266.5 to 266.5 fs (b) and in the delay-time range over 30 fs (c).

the Fourier spectrum from acetone by that from silica glass, the dependence of the Fourier spectrum from acetone on the spectral amplitude and phase of the excitation pulse is calibrated. Although the nonresonant CARS signals are included in the CARS spectrum, the CARS spectrum with a bandwidth of about 5000 cm^{-1} can be measured simultaneously. Figure 3.8(c) shows the Fourier spectrum of the IAC signal in the delay-time range from over 30 fs. It should be emphasized that a broadband CARS spectrum without any nonresonant CARS signals can be simultaneously obtained.

REFERENCES

1. M. Sheik-Bahae, A. A. Said, T. H. Wei, D. J. Hagan, and E. W. Van Stryland, "Sensitive measurement of optical nonlinearities using a single beam," *IEEE J. Quantum Electron.*, **QE-26**, 760–769 (1990).
2. P. Tian and W. S. Warren, "Ultrafast measurement of two-photon absorption by loss modulation," *Opt. Lett.*, **27**, 1634–1636 (2002).
3. B. R. Masters and P. T. C. So, *Handbook of Biomedical Nonlinear Optical Microscopy*, Oxford University Press, Oxford, UK, 2008.
4. C. Xu and W. W. Webb, "Measurement of two-photon excitation cross sections of molecular fluorophores with data from 690 to 1050 nm," *J. Opt. Soc. Am. B*, **13**, 481–491 (1996).

5. C. Xu, W. Zipfel, J. B. Shear, R. M. Williams, and W. W. Webb, "Multiphoton fluorescence excitation: new spectral windows for biological nonlinear microscopy," *Proc. Natl. Acad. Sci. USA*, **93**, 10763–10768 (1996).
6. F. Bestvater, E. Spiess, G. Stobrawa, M. Hacker, T. Feurer, T. Porwol, U. Berchner-Pfannschmidt, C. Wotzlaw, and H. Acker, "Two-photon fluorescence absorption and emission spectra of dyes relevant for cell imaging," *J. Microsc.*, **208**, 108–115 (2002).
7. M. A. Albota, C. Xu, and W. W. Webb, "Two-photon fluorescence excitation cross sections of biomolecular probes from 690 to 960 nm," *Appl. Opt.*, **37**, 7352–7356 (1998).
8. J. Lukomska, I. Gryczynski, J. Malicka, S. Makowiec, J. R. Lakowicz, and Z. Gryczynski, "Two-photon induced fluorescence of Cy5-DNA in buffer solution and on silver island films," *Biochem. Biophys. Res. Commun.*, **328**, 78–84 (2005).
9. N. S. Makarov, M. Drobizhev, and A. Rebane, "Two-photon absorption standards in the 550-1600 nm excitation wavelength range," *Opt. Express*, **16**, 4029–4047 (2008).
10. G. A. Blab, P. H. M. Lommerse, L. Cognet, G. S. Harms, and T. Schmidt, "Two-photon excitation action cross-sections of the autofluorescent proteins," *Chem. Phys. Lett.*, **350**, 71–77 (2001).
11. W. R. Zipfel, R. M. Williams, and W. W. Webb, "Nonlinear magic: multiphoton microscopy in the bioscience," *Nat. Biotechnol.*, **21**, 1369–1377 (2003).
12. E. Spiess, F. Bestvater, A. Heckel-Pompey, K. Toth, M. Hacker, G. Stobrawa, T. Feurer, C. Wotzlaw, U. Berchner-Pfannschmidt, T. Porwol, and H. Acker, "Two-photon excitation and emission spectra of the green fluorescent protein variants ECFP, EGFP and EYFP," *J. Microsc.*, **217**, 200–204 (2005).
13. P. S. Tsai, B. Friedman, A. I. Ifarraguerri, B. D. Thompson, V. Lev-Ram, C. B. Schaffer, Q. Xiong, R. Y. Tsien, J. A. Squier, and D. Kleinfeld, "All-optical histology using ultrashort laser pulses," *Neuron*, **39**, 27–41 (2003).
14. M. Drobizhev, S. Tillo, N. S. Makarov, T. E. Hughes, and A. Rebane, "Absolute two-photon absorption spectra and two-photon brightness of orange and red fluorescent proteins," *J. Phys. Chem. B*, **113**, 855–859 (2009).
15. M. Drobizhev, N. S. Makarov S. E. Tillo T. E. Hughes, and A. Rebane, "Two-photon absorption properties of fluorescent proteins," *Nat. Methods*, **8**, 393–399 (2011).
16. D. R. Larson, W. R. Zipfel, R. M. Williams, S. W. Clark, M. P. Bruchez, F. W. Wise, and W. W. Webb, "Water-soluble quantum dots for multiphoton fluorescence imaging in vivo," *Science*, **300**, 1434–1436 (2003).
17. B. Dick and G. Hohlneicher, "Importance of initial and final states as intermediate states in two-photon spectroscopy of polar molecules," *J. Chem. Phys.*, **76**, 5755–5760 (1982).
18. D. J. Bradley, M. H. R. Hutchinson, T. M. H. Koetser, G. H. C. New, and M. S. Petty, "Interaction of picosecond laser pulses with organic molecules I: two-photon fluorescence quenching and singlet states excitation in rhodamine dyes," *Proc. R. Soc. London Ser. A*, **328**, 97–121 (1972).
19. J. P. Hermann and J. Ducuing, "Dispersion of the two-photon cross-section in rhodamine dyes," *Opt. Commun.*, **6**, 101–105 (1972).
20. L. Goodman and R. P. Pava, "Two-photon spectra of aromatic molecules," *Acc. Chem. Res.*, **17**, 250–257 (1984).
21. J. Tretzel and F. W. Schneider, "Resonance multiplex cars of fluorescing acridines," *Chem. Phys. Lett.*, **59**, 514–518 (1978).

22. M. Müller and J. M. Schins, "Imaging the thermodynamic state of lipid membranes with multiplex CARS microscopy," *J. Phys. Chem. B*, **106**, 3715–3723 (2002).
23. J.-X. Cheng, A. Volkmer, L. D. Book, and X. S. Xie, "Multiplex coherent anti-Stokes Raman scattering microspectroscopy and study of lipid vesicles," *J. Phys. Chem. B*, **106**, 8493–8498 (2002).
24. G. W. H. Wurpel, J. M. Schins, and M. Müller, "Chemical specificity in three-dimensional imaging with multiplex coherent anti-Stokes Raman scattering microscopy," *Opt. Lett.*, **27**, 1093–1095 (2002).
25. L. J. Richter, T. P. Petralli-Mallow, and J. C. Stephenson, "Vibrationally resolved sum-frequency generation with broad-bandwidth infrared pulses," *Opt. Lett.*, **23**, 1594–1596 (1998).
26. S. Yamaguchi and T. Tahara, "Two-photon absorption spectrum of all-trans retinal," *Chem. Phys. Lett.*, **376**, 237–243 (2003).
27. H. Hosoi, S. Yamaguchi, H. Mizuno, A. Miyawaki, and T. Tahara, "Hidden electronic excited state of enhanced green fluorescent protein," *J. Phys. Chem. B*, **112**, 2761–2763 (2008).
28. D. W. McCamant, P. Kukura, and R. A. Mathies, "Femtosecond broadband stimulated Raman: a new approach for high-performance vibrational spectroscopy," *Appl. Spectrosc.*, **57**, 1317–1323 (2003).
29. J. K. Ranka, R. S. Windeler, and A. J. Stentz, "Visible continuum generation in airsilica microstructure optical fibers with anomalous dispersion at 800 nm," *Opt. Lett.*, **25**, 25–27 (2000).
30. T. W. Kee and M. T. Cicerone, "Simple approach to one-laser, broadband coherent anti-Stokes Raman scattering microscopy," *Opt. Lett.*, **29**, 2701–2703 (2004).
31. H. Kano and H. Hamaguchi, "Ultrabroadband (>2500 cm^{-1}) multiplex coherent anti-Stokes Raman scattering microspectroscopy using a supercontinuum generated from a photonic crystal fiber," *Appl. Phys. Lett.*, **86**, 121113/1–3 (2005).
32. K. Isobe, A. Suda, M. Tanaka, H. Hashimoto, F. Kannari, H. Kawano, H. Mizuno, A. Miyawaki, and K. Midorikawa, "Nonlinear optical microscopy and spectroscopy employing octave spanning pulses," *IEEE J. Sel. Top. Quant. Electron.*, **16**, 767–780 (2010).
33. P. M. Felker and G. V. Hartland, "Fourier transform coherent Raman spectroscopy," *Chem. Phys. Lett.*, **134**, 503–506 (1987).
34. G. V. Hartland and P. M. Felker, "High spectral resolution in coherent Raman scattering using broad-band, nanosecond-pulsed sources and nonlinear interferometry," *J. Phys. Chem.*, **91**, 5527–5531 (1987).
35. J. P. Ogilvie, E. Beaurepaire, A. Alexandrou, and M. Joffre, "Fourier-transform coherent anti-Stokes Raman scattering microscopy," *Opt. Lett.*, **31**, 480–482 (2006).
36. K. Isobe, A. Suda, M. Tanaka, H. Hashimoto, F. Kannari, H. Kawano, H. Mizuno, A. Miyawaki, and K. Midorikawa, "Single-pulse coherent anti-Stokes Raman scattering microscopy employing an octave spanning pulse," *Opt. Express*, **17**, 11259–11266 (2009).
37. K. Isobe, A. Suda, M. Tanaka, F. Kannari, H. Kawano, H. Mizuno, A. Miyawaki, and K. Midorikawa, "Fourier transform spectroscopy combined with 5-fs broadband pulse for multispectral nonlinear microscopy," *Phys. Rev. A*, **77**, 063832/1–13 (2008).
38. M. Bellini, A. Bartoli, and T. W. Hänsch, "Two-photon Fourier spectroscopy with femtosecond light pulses," *Opt. Lett.*, **22**, 540–542 (1997).

39. J. P. Ogilvie, Kevin J. Kubarych, A. Alexandrou, and M. Joffre, "Fourier transform measurement of two-photon excitation spectra: applications to microscopy and optimal control," *Opt. Lett.*, **30**, 911–913 (2005).
40. H. Hashimoto, K. Isobe, A. Suda, F. Kannari, H. Kawano, H. Mizuno, A. Miyawaki, and K. Midorikawa, "Measurement of two-photon excitation spectra of fluorescent proteins with nonlinear Fourier-transform spectroscopy," *Appl. Opt.*, **49**, 3323–3329 (2010).
41. J. A. McGuire, W. Beck, X. Wei, and Y. R. Shen, "Fourier-transform sum-frequency surface vibrational spectroscopy with femtosecond pulses," *Opt. Lett.*, **24**, 1877–1879 (1999).
42. J. A. McGuire and Y. R. Shen, "Signal and noise in Fourier-transform sum-frequency surface vibrational spectroscopy with femtosecond lasers," *J. Opt. Soc. Am. B*, **23**, 363–369 (2006).
43. T. Brabec and F. Krausz, "Nonlinear optical pulse propagation in the single-cycle regime," *Phys. Rev. Lett.*, **78**, 3282–3285 (1997).
44. J. K. Ranka, A. L. Gaeta, A. Baltuska, M. S. Pshenichnikov, and D. A. Wiersma, "Autocorrelation measurement of 6 fs pulses based on the two-photon-induced photocurrent in a GaAsP photodiode," *Opt. Lett.*, **22**, 1344–1366 (1997).
45. K. Isobe, H. Hashimoto, A. Suda, F. Kannari, H. Kawano, H. Mizuno, A. Miyawaki, and K. Midorikawa, "Measurement of two-photon excitation spectrum used to photoconvert a fluorescent protein (Kaede) by nonlinear Fourier-transform spectroscopy," *Biomed. Opt. Express*, **1**, 687–693 (2010).

CHAPTER 4

NONLINEAR OPTICAL MICROSCOPY

4.1 INTRODUCTION TO NONLINEAR OPTICAL MICROSCOPY

4.1.1 Various Nonlinear Optical Microscopies

A series of nonlinear optical techniques, such as second-harmonic generation (SHG) [1,2], third-harmonic generation (THG) [3,4], nonresonant four-wave mixing (NFWM) [5], stimulated parametric emission (SPE) [6], coherent anti-Stokes Raman scattering (CARS) [7,8], two-photon absorption (TPA) [9,10], two-photon excited fluorescence (TPEF) [11–13], and stimulated Raman scattering (SRS) [14–16] have been incorporated within the field of high spatial resolution microscopy as shown in Fig. 4.1. Nonlinear optical microscopy (NLOM) offers a number of advantages over linear optical microscopy. The nonlinear dependence of the signal intensity on the intensity of tightly focused excitation light provides inherent optical sectioning capability, along with the reduction of detrimental out-of-focus interactions such as photo-induced damage and photobleaching [1–16]. TPEF microscopy is typically used for imaging particular molecular species in samples that have been labeled by fluorophores [11–13]. Other nonlinear microscopies permit the imaging of label-free samples. SPE [6] and TPA [9,10] microscopies, which utilize two-photon electronic resonance, allow imaging of absorbing molecules. CARS [7,8] and SRS [14–16] microscopies, which employ two-photon vibrational resonance, provide image contrasts that are derived from vibrations of the chemical compositions and thermodynamic states of the sample. Since SHG is allowed only for noncentrosymmetric media, SHG microscopy has been used for imaging oriented and organized structures [1,2].

Functional Imaging by Controlled Nonlinear Optical Phenomena, First Edition.
Keisuke Isobe, Wataru Watanabe and Kazuyoshi Itoh.

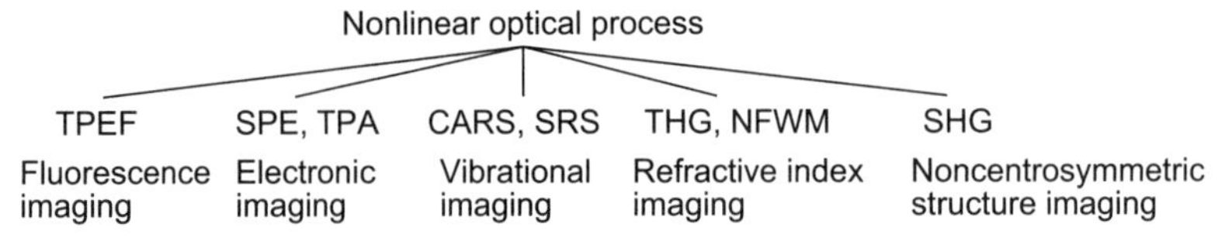

FIGURE 4.1 Various nonlinear optical microscopies.

THG [3,4] and nonresonant FWM [5] microscopies can produce images based on a refractive index.

4.1.2 Basic Architecture of a Nonlinear Optical Microscope

Figure 4.2(a) shows a schematic of typical nonlinear optical microscope. An ultrashort pulse laser is used as an excitation light source because of its high peak intensity. In the case of the requirement of multicolor excitations, multiple pulse lasers as shown in Fig. 4.2(b), a combination of wavelength converters and original laser pulses as shown in Fig. 4.2(c), or a broadband pulse laser are employed. Moreover, in combined excitation by different color pulses, it is necessary that the multiple lasers be synchronized, or the wavelength-converted pulses and the original pulses be temporally overlapped with a delay stage. To compensate for the dispersion of a laser scanning microscope, the laser output pulses are passed through prechirpers before sent into the microscope. In all types of nonlinear optical microscopes, a specimen is generally illuminated with a point light source produced by focusing

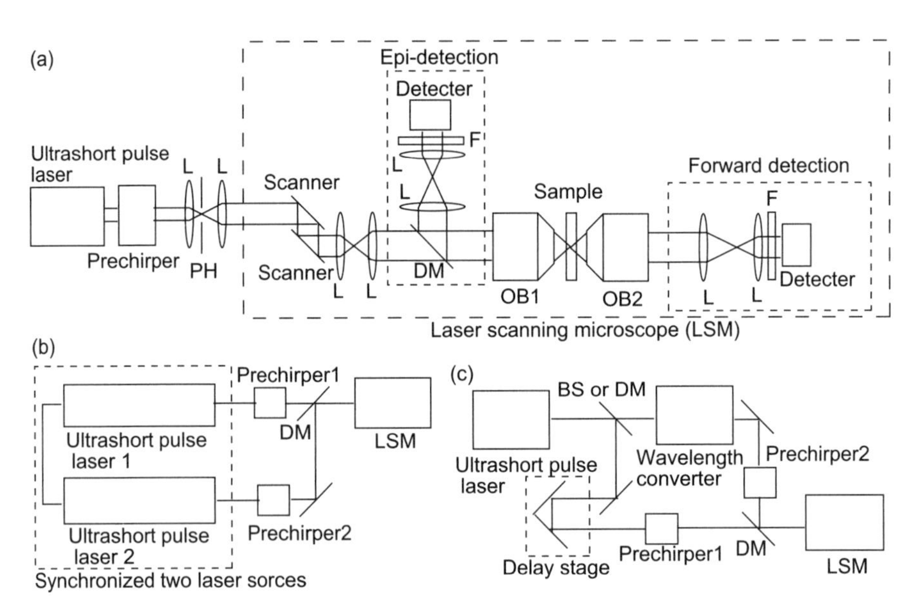

FIGURE 4.2 (a) Typical nonlinear optical microscope. (b) Nonlinear optical microscope using a synchronized two-laser system. (c) Nonlinear optical microscope employing the combination of a wavelength converter and the original laser pulse.

laser beams with a high numerical-aperture (NA) objective lens (OB) to obtain the sectioning capability derived from the nonlinear dependence of the signal intensity on the excitation intensities and to minimize the contribution of out-of-focus signals. Thus, the nonlinear optical microscopes require sample scanning or beam scanning configurations to form images. The generated nonlinear signals are collected by the objective lens and are detected in the forward and backward directions.

4.1.2.1 Scanning Methods Pivoting collimated beams at the back focal plane of the objective lens with galvanometer-controlled mirrors is the most common beam scanning method (see Section 2.1.1). For two-dimensional (2D) scanning in the lateral direction, the reflection at two orthogonally pivoting mirrors is used, and their angles are controlled with high precision by galvanometric actuators. In the axial direction, the objective lens or the sample is typically scanned with motorized scanning stages. The beam scanning methods employing resonant galvonometric scanners [17], rotating polygonal mirrors [18], and acousto-optic deflectors (AODs) [19] can be applied to video-rate imaging. However, the exposure time per pixel is reduced. Several recent technical advances have increased frame rates without reducing the exposure time by employing multi-point scanning methods using microlens arrays [20–22], cascaded-beam splitter arrays [23,24], diffractive optical elements (DOEs) [25], or temporal focusing techniques [26,27].

4.1.2.2 Dispersion Management The net group delay dispersion (GDD) of the nonlinear optical microscope including beam expanders, dichroic mirrors, scan optics, and an objective lens can typically be on the order of 5000 fs^2. This GDD results in expanding of the pulse duration. For a 100-fs pulse at 800 nm, the pulse duration at the focal point is broadened to 171 fs. In the case of a 10-fs pulse at 800 nm, the pulse duration is broadened to 1384 fs. Such broadened pulse duration significantly decreases the efficiency of any nonlinear optical processes. Thus, the dispersion of the microscope is precompensated through a prechirper such as a prism pair or a chirp mirror pair. Pulses as short as 5 fs have been generated at the focus of high numerical-aperture objectives by compensating for the dispersion [28]. When the pulse duration of a bandwidth-limited pulse is longer than 200 fs, it is usually not necessary to precompensate for the dispersion of the microscope. This is because the pulse duration of such long pulse is a little broadened by the dispersion, as we mentioned earlier.

4.1.2.3 Signal Detection The generated signals by the coherent processes are commonly detected in the forward direction, while those by the incoherent processes are generally detected in the backward direction. The nonlinear signals can be detected immediately after the collection objective because beam scanners are not used for descanning the nonlinear signal as is required for confocal microscopy (see Section 2.1.2). This advantage in nonlinear microscopy produces more efficient detection than confocal microscopy. Point detectors such as photomultiplier tubes (PMTs) and avalanche photodiodes (APDs) are used in the single-point scanning methods, whereas imaging detectors such as charged coupled devices (CCDs) and complementary metal oxide semiconductors (CMOSs) are employed in the multi-point scanning methods. In the point scanning methods, large area PMTs are the most appropriate detectors because they provide the efficient detection of photons that are scattered in thick samples, as long as the scattered photons are collected by the objective lens. APDs generally have higher quantum efficiencies (QEs) than PMTs. However, the active area is typically very small (less than 1 mm in diameter). Therefore, the efficient detection of the scattered photons is low.

4.1.3 Spatial Resolution

The spatial resolution in nonlinear optical microscopy is evaluated with the interaction volume. By approximating the interaction volume as a three-dimensional Gaussian volume, when the signal intensity is proportional to the jth power of the excitation intensity, the full widths at half maximum (FWHMs) of the interaction volume in the lateral (r_0) and axial (z_0) directions can be approximated by [13]

$$r_0 \approx \begin{cases} \dfrac{2\sqrt{\ln 2} \times 0.320\lambda}{\sqrt{j}NA}, & NA \leq 0.7 \\ \dfrac{2\sqrt{\ln 2} \times 0.325\lambda}{\sqrt{j}NA^{0.91}}, & NA > 0.7 \end{cases}, \tag{4.1.1}$$

$$z_0 \approx \frac{2\sqrt{\ln 2} \times 0.532\lambda}{\sqrt{j}\left(n - \sqrt{n^2 - NA^2}\right)}, \tag{4.1.2}$$

where λ is the excitation wavelength, NA is the effective numerical aperture, and n is the refractive index.

4.2 FLUORESCENCE IMAGING

4.2.1 Basic Two-Photon Excited Fluorescence (TPEF) Microscopy

In the two-photon absorption process, two photons whose sum frequency matches the electronic transition frequency are simultaneously absorbed. For fluorescent molecules, the fluorescence is emitted after the TPA process. This process is called two-photon excited fluorescence (TPEF). Since TPEF was first applied to microscopy by Denk et al. in 1990 [11], its advantages over confocal microscopy have been clearly demonstrated [29–34]. The advantages include lower phototoxicity due to using near-infrared (NIR) wavelengths, restriction of photobleaching and photodynamic damage to the vicinity of the focal region, decreased background fluorescence, and deeper penetration into thick specimen than confocal microscopy. Because of these advantages, TPEF microscopies combined with various fluorescence techniques have been used for investigating biological phenomena. An oscillator providing 100-fs pulses enabled Svoboda et al. [29,30] and Helmchen et al. [31] to achieve deep-imaging exceeding 500 μm within thick, strongly scattering samples. Theer et al. employed a regenerative amplifier and demonstrated that it is possible to image deeper (up to 1 mm) in the neocortex [32]. Kobat et al. achieved in vivo imaging of adult mouse brains an imaging depth of 1 mm by using a optical parametric oscillator at a central wavelength of 1280 nm [33]. Squirrell et al. achieved the long-term imaging of hamster embryos over a period of 24 h without compromising cell viability [34]. Confocal imaging for 8 hours of the same specimen inhibited the development of the embryos.

4.2.1.1 Photobleaching in Multiphoton Excited Fluorescence Microscopy In multiphoton excited fluorescence (MPEF) microscopy, the region of fluorescence excitation is limited to a single-focal plane, with virtually no out-of-focus photobleaching. Photobleaching is mainly induced by photodynamic interactions among fluorophores in the first-excited triplet state and molecular oxygen (O_2) in the triplet ground-state environment [35–38]. However, the large photon flux necessary to facilitate multiphoton excitation can open up new photobleaching channels due to the possibility of sequential excitation into the higher excited electronic states than the first-excited singlet or triplet states [39] and can lead to higher order photobleaching within the focal volume [40–43]. Patterson and Piston showed that log–log plots of excitation power versus the photobleaching rate for one-photon excited fluorescence (OPEF) microscopy increased with a slope of 1, while the photobleaching rate for TPEF microscopy increased with a slope $\geq$3 [40]. Kalies et al. showed the excitation wavelength dependence of the photobleaching order, in which the photobleaching orders of EGFP were $\sim$3 from 720 up to 800 nm and $\sim$4 from 860 nm to 960 nm, even though the MPEF orders were $\sim$2 over the whole wavelength range [43]. In Hoechst 33342, the photobleaching orders were $\sim$2 from 720 up to 900 nm and $\sim$3 at 950 nm, and the MPEF orders of that were $\sim$2 from 720 up to 800 nm and increased to values within 2 to 3 between 850 and 950 nm [43]. Donnert et al. demonstrated that a relaxation of the triplet state, which was achieved by reducing the repetition rate but maintaining the peak intensity of the excitation pulse, causes a photobleaching reduction and a signal increase in TPEF microscopy [44]. This is because the molecules trapped in the first excited triplet state, which are susceptible to bleaching upon the absorption of an additional photon, can relax back to ground state before the next pulse arrives. In contrast, Ji et al. demonstrated that a reduction of the peak intensity of the excitation pulse due to the increasing repetition rate caused a photobleaching reduction in TPEF microscopy [45]. The reason is that a reduction of the peak intensity prevents an excitation to the higher excited states with a single pulse. Therefore, both decreasing and increasing the repetition rate reduce photobleaching by acting on different photochemical pathways.

4.2.1.2 Three-Dimensional Image Formation We describe an imaging property of TPEF microscopy using the three-dimensional (3D) intensity point spread function (IPSF) and 3D optical transfer function (OTF). The 3D IPSF is an image of an ideal point object and gives insight into the imaging behavior in practical space, while the 3D OTF describes the imaging characteristics in the Fourier space. We first consider the 3D IPSF. Now, we assume that the optical imaging system has one identical circular objective lens, that the excitation and fluorescence wavelengths are same wavelength λ in one-photon excitation, and that the excitation and fluorescence wavelengths in vacuum are 2λ and λ in two-photon excitation, respectively. Using normalized optical coordinates (v, u), we can express the 3D IPSF for one-photon excited fluorescence (OPEF) microscopy by the paraxial form of normalized IPSF for a diffraction-limited lens [46] as

$$I_{\text{OPEF}}(v, u) = \left| 2 \int_0^1 J_0(v\rho) \exp\left(\frac{iu\rho^2}{2} \right) \rho d\rho \right|^2, \tag{4.2.1}$$

where J_0 is a Bessel function of the first kind of order zero. Here the normalized optical coordinates v and u are radial and axial optical coordinates and are given by the following formula:

$$v = \frac{2\pi r n \sin\theta}{\lambda}, \tag{4.2.2}$$

$$u = \frac{8\pi z n \sin^2(\theta/2)}{\lambda}, \tag{4.2.3}$$

where r and z are the distance from the focal point along the focal plane and the optical axis, respectively, n is refractive index of the immersion medium and $n \sin\theta$ is the numerical aperture (NA) of the objective lens. The paraxial approximation is adequate for $\sin\alpha < 0.7$ [47]. Since the excitation probability of the two-photon excitation is proportional to the square of the incident intensity, the IPSF for TPEF microscopy can be expressed as

$$I_{\text{TPEF}}(v, u) = I_{\text{OPEF}}^2 \left(\frac{v}{2}, \frac{u}{2} \right). \tag{4.2.4}$$

Figure 4.3 shows the 3D IPSFs for OPEF and TPEF microscopies. Unlike the 3D IPSF for OPEF microscopy, most of the light is confined to a single small spot for IPSF for TPEF microscopy. However, the resolution in TPEF microscopy is worse than that in OPEF microscopy by almost a factor of $\sqrt{2}$. This is because the excitation wavelength in TPEF microscopy is twice as long as that in OPEF microscopy and the TPEF intensity is proportional to the square of the incident intensity. Because only

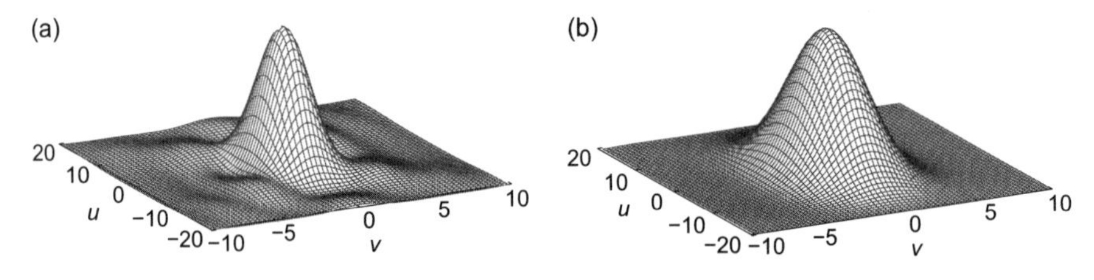

FIGURE 4.3 Three-dimensional IPSFs for OPEF microscopy (a) and TPEF microscopy (b).

a point object is considered, the depth discrimination capability cannot be evaluated from the 3D IPSF.

In order to evaluate the optical depth discrimination capability, we use the 3D OTF. The 3D OTF for an optical imaging system can be calculated with the 3D Fourier transform of the 3D IPSF of the system [49]. Consequently, the 3D OTFs for OPEF and TPEF microscopies can be expressed as [50]

$$\tilde{I}_{\mathrm{OPEF}}(l,s) = FT_3[I(u,v)] = \frac{1}{|l|}\mathrm{Re}\left\{\left[1-\left(\frac{|s|}{l}+\frac{l}{2}\right)^2\right]^{1/2}\right\}, \quad (4.2.5)$$

$$\tilde{I}_{\mathrm{TPEF}}(l,s) = \tilde{I}_{\mathrm{OPEF}}(2l,2s) \otimes_3 \tilde{I}_{\mathrm{OPEF}}(2l,2s), \quad (4.2.6)$$

where FT_3, $\otimes_3$ and Re{} denote the 3D Fourier transform, the 3D convolution operation, and the real part of its argument, respectively. Here l and s are the radial and axial spatial frequencies normalized by

$$\frac{n\sin\theta}{\lambda} \quad (4.2.7)$$

and

$$\frac{4n\sin^2(\theta/2)}{\lambda}. \quad (4.2.8)$$

After considering the cylindrical symmetry of the microscopic system, we can expand Eq. (4.2.6) as [51]

$$\tilde{I}_{\mathrm{TPEF}}(l,s) = \int\int\int_V \frac{1}{l_1^2}\mathrm{Re}\left\{\left[l_1^2-\left(\left|s'-2s\right|+\frac{l_1^2}{2}\right)^2\right]^{1/2}\right\} \times \frac{1}{l_2^2}\mathrm{Re}\left\{\left[l_2^2-\left(\left|s'\right|+\frac{l_2^2}{2}\right)^2\right]^{1/2}\right\} dp\,dq\,ds', \quad (4.2.9)$$

where V denotes an overlapped volume by the two functions Re{} in the integrand of Eq. (4.2.9) and

$$l_1 = \sqrt{(p-2l)^2+q^2}, \quad (4.2.10)$$

$$l_2 = \sqrt{p^2+q^2}, \quad (4.2.11)$$

where variables p and q represent the spatial frequencies along two orthogonal transverse directions, respectively. Figure 4.4 shows the OTFs for OPEF and TPEF microscopies. OPEF microscopy is missing a section of its cone of spatial frequencies, while TPEF microscopy shows complete 3D image formation. The TPEF microscopy

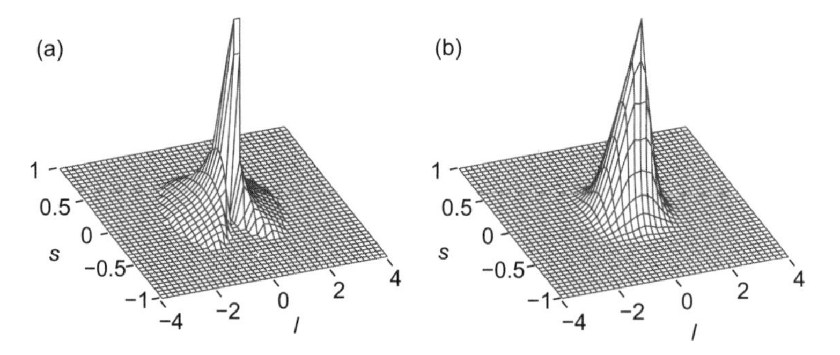

FIGURE 4.4 Three-dimensional OTFs for OPEF microscopy (a) and TPEF microscopy (b).

has an optical sectioning property that circumvents use of the pinhole. This optical sectioning property enables the detection of scattered fluorescence photons in a sample with high scattering.

4.2.1.3 Reduction of Detrimental Out-of-Focus Interactions Nonlinear dependence of fluorescence intensity on the intensity of tightly focused pulses enables not only inherent optical sectioning but also a reduction of detrimental out-of-focus interactions such as photo-induced damage and photobleaching.

Let us consider the difference of the interaction volumes between one-photon excitation (OPE) and two-photon excitation (TPE) by focused laser pulses under the paraxial approximation. The electric field in the frequency domain just before the objective lens can be expressed by spatially transform-limited Gaussian beams in the form

$$\tilde{E}_{\mathrm{LB}}(x_0, y_0, \omega) = \exp^{-\left(\frac{x_0^2}{S_x^2}+\frac{y_0^2}{S_y^2}\right)} \exp^{-\frac{(\omega-\omega_0)^2}{\Omega^2}}, \tag{4.2.12}$$

where S_x and S_y are the beam radii in the x and y directions, respectively, ω_0 is the central angular frequency of the pulse, and Ω is related to the spectral bandwidth $\Delta\omega$ at FWHM:

$$\Delta\omega = \sqrt{2\ln 2}\Omega. \tag{4.2.13}$$

After adding the quadratic phase $\exp\left\{-\frac{ik}{2f}\left(x_0^2+y_0^2\right)\right\}$ along x and y directions due to the objective lens with a focal length of f, the field in the frequency domain becomes

$$\tilde{E}_{\mathrm{LA}}(x_0, y_0, \omega) = e^{-\left(\frac{x_0^2}{S_x^2}+\frac{y_0^2}{S_y^2}\right)} e^{-\frac{(\omega-\omega_0)^2}{\Omega^2}} e^{-\frac{ik}{2f}\left(x_0^2+y_0^2\right)}, \tag{4.2.14}$$

where k is the wave number. By propagating the field a distance f from the objective lens with Fresnel diffraction, we can write the field in the frequency domain at the focal plane of the objective lens as [52]

$$\begin{aligned}
\tilde{E}_{\text{focal}}(x_1, y_1, \omega) &= \frac{\exp(ikf)}{i\lambda f} \int_{-\infty}^{\infty} \int_{-\infty}^{\infty} \tilde{E}_{\text{LA}}(x_0, y_0, \omega) \\
&\quad \times \exp\left[\frac{ik}{2f}\left\{(x_1 - x_0)^2 + (y_1 - y_0)^2\right\}\right] dx_0 dy_0 \\
&= \frac{\pi S_x S_y}{i\lambda f} e^{ikf} e^{-\frac{k^2}{4f^2}(S_x^2 x_1^2 + S_y^2 y_1^2)} e^{\frac{ik}{2f}(x_1^2 + y_1^2)} \\
&\quad \times e^{-\frac{(\omega-\omega_0)^2}{\Omega^2}}, \qquad (4.2.15)
\end{aligned}$$

where we used $\int_{-\infty}^{\infty} \exp(-ax^2)dx = \sqrt{\pi/a}$. The field in the time domain at the focal plane of the objective lens is obtained by inverse Fourier, transforming Eq. (4.2.15) along the frequency coordinate:

$$\begin{aligned}
E_{\text{focal}}(x_1, y_1, t) &= \int_{-\infty}^{\infty} \tilde{E}_{\text{focal}}(x_1, y_1, \omega) e^{-i\omega t} d\omega \\
&= \frac{\sqrt{\pi}\Omega k_0 S_x S_y}{i2f} e^{ik_0 f} e^{-\frac{k_0^2}{4f^2}(S_x^2 x_1^2 + S_y^2 y_1^2)} e^{\frac{ik_0}{2f}(x_1^2 + y_1^2)} \\
&\quad \times e^{-\frac{\Omega^2}{4}t^2} e^{-i\omega_0 t}. \qquad (4.2.16)
\end{aligned}$$

Here, to simplify calculations, we assume that the wavevector k for each frequency is approximately k_0, the wave vector of the central wavelength of the pulse. By propagating the field a distance z from the focal plane with Fresnel diffraction, we can describe the field near the focal plane by

$$\begin{aligned}
E(x, y, z, t) &= \frac{\exp(ik_0 z)}{i\lambda_0 z} \int_{-\infty}^{\infty} \int_{-\infty}^{\infty} E_{\text{focal}}(x_1, y_1, t) \\
&\quad \times \exp\left[\frac{ik_0}{2z}\left\{(x - x_1)^2 + (y - y_1)^2\right\}\right] dx_1 dy_1 \\
&= -\frac{\sqrt{\pi}\Omega k_0 S_x S_y f}{\sqrt{\left\{k_0 S_x^2 z - i2f(f+z)\right\}\left\{k_0 S_y^2 z - i2f(f+z)\right\}}} \\
&\quad \times \exp^{-\left[\frac{k_0 f^2\left\{k_0 S_x^2 z + i2f(f+z)\right\}}{z\left\{k_0 S_x^2 z + 4f^2(f+z)^2\right\}} x^2 + \frac{k_0 f^2\left\{k_0 S_y^2 z + i2f(f+z)\right\}}{z\left\{k_0 S_y^2 z + 4f^2(f+z)\right\}} y^2\right]} \\
&\quad \times \exp^{-\frac{\Omega^2}{4}t^2} \exp^{ik_0(f+z)} \exp^{-i\omega_0 t} \exp^{\frac{ik_0}{2z}(x^2+y^2)}. \qquad (4.2.17)
\end{aligned}$$

From Eq. (4.2.17), the instantaneous intensity is

$$\begin{aligned}
I(x, y, z, t) &= |E(x, y, z, t)|^2 \\
&= \frac{\pi\Omega^2 k_0^2 S_x^2 S_y^2 f^2}{\sqrt{\left\{k_0^2 S_x^4 z^2 + 4f^2(f+z)^2\right\}\left\{k_0^2 S_y^4 z^2 + 4f^2(f+z)^2\right\}}} e^{-\frac{\Omega^2}{2}t^2} \\
&\quad \times e^{-\left\{\frac{2k_0^2 f^2 S_x^2}{k_0^2 S_x^4 z^2 + 4f^2(f+z)^2} x^2 + \frac{2k_0^2 f^2 S_y^2}{k_0^2 S_y^4 z^2 + 4f^2(f+z)^2} y^2\right\}}. \qquad (4.2.18)
\end{aligned}$$

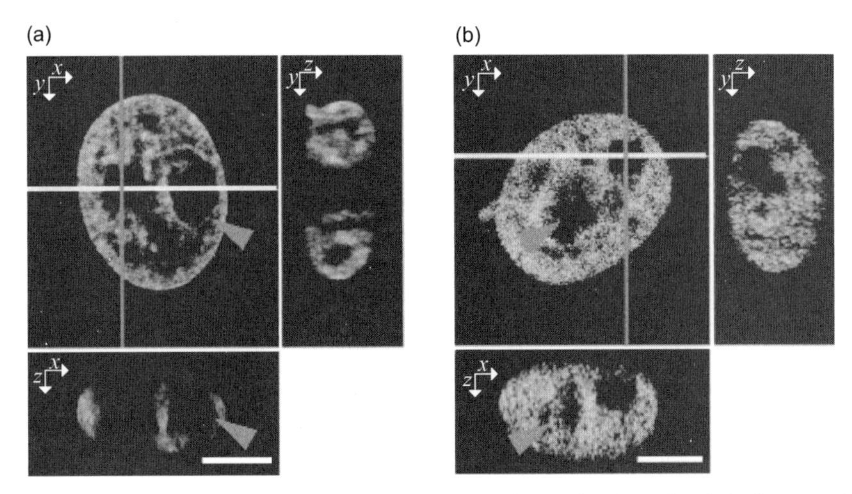

FIGURE 4.5 Photobleaching induced by OPE (a) and TPE (b). Arrowheads indicate nucleolus. Scale bars is 5 μm. Reprinted with permission from [53]. Copyright 2007 Elsevier.

According to Eq. (4.2.18), the axial responses of OPEF and TPEF microscopies take the forms

$$R_{\mathrm{OPE}}(z) = \int_{-\infty}^{\infty}\int_{-\infty}^{\infty}\int_{-\infty}^{\infty} I(x, y, z, t)dtdxdy = \sqrt{\frac{\pi}{2}}\pi^2\Omega S_x S_y, \tag{4.2.19}$$

$$R_{\mathrm{TPE}}(z) = \int_{-\infty}^{\infty}\int_{-\infty}^{\infty}\int_{-\infty}^{\infty} I^2(x, y, z, t)dtdxdy = \frac{\pi^3\sqrt{\pi}\Omega^3 k_0^2 S_x^3 S_y^3 f^2}{4\sqrt{\left\{k_0^2 S_x^4 z^2 + 4f^2(f+z)^2\right\}\left\{k_0^2 S_y^4 z^2 + 4f^2(f+z)^2\right\}}}. \tag{4.2.20}$$

As is evident from Eqs. (4.2.19) and (4.2.20), the interaction volume by TPE is confined only at focal region, that by OPE is not confined along the axial direction.

As shown in Fig. 4.5, the photobleaching region induced by TPE is confined to a region close to the focal plane, while that by OPE is extended to out-of-focus regions [53]. The region of photoinduced damage is also confined to the vicinity of the focal point. Thus, the reduction of detrimental out-of-focus interactions allows the long-term imaging.

4.2.1.4 Achievable Imaging Depth TPEF microscopy provides deeper imaging capability over confocal fluorescence microscopy. In TPEF the near-infrared (NIR) laser wavelengths to the spectral range have 650 of 1300 nm, which is known

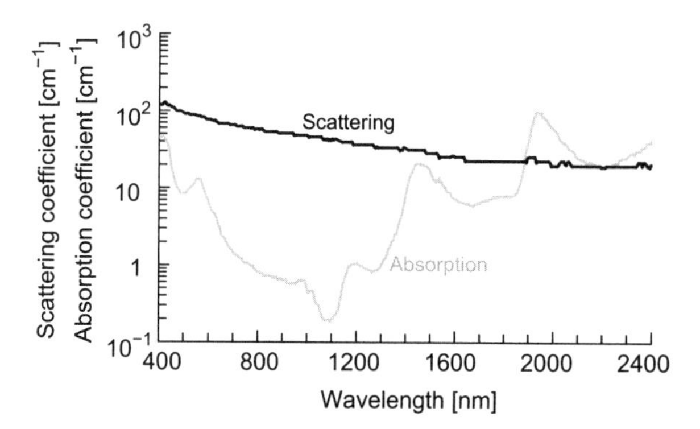

FIGURE 4.6 Absorption and scattering coefficients of porcine liver. Reprinted with permission from [54]. Copyright 2001 John Wiley & Sons, Inc.

as the optical window or therapeutic window of cells and tissues. This spectral range falls into the window for maximum optical transparency in biological systems owing to the lack of efficient endogenous absorbers [12] as shown in Fig. 4.6 [54]. In this spectral range, the major absorbing components are water [55], haemoglobin [56], and melanin [57,58]. At the shorter wavelength end, the window is bound by the absorption of haemoglobin [56]. At the longer wavelength end of the window, the penetration of light is limited by the absorption properties of water [55]. Within the optical window, scattering is dominant over absorption [54].

As the penetration depth z increases, the collected signal of m-photon excited fluorescence in the focal region decreases rapidly because of the scattering and absorption losses of excitation power in the sample with an exponential behavior well known to be

$$I_{\mathrm{MPEF}}^{(m)}(z) = I_0 \exp\{-(m\alpha_{\mathrm{ex}} + \alpha_{\mathrm{em}})\, z\}, \tag{4.2.21}$$

where α_{ex} and α_{em} are the intensity attenuation coefficients for excitation wavelengths and emission wavelengths, respectively. Here we assume that the sample surface is located at $z = 0$. If one compensates for the loss of excitation intensity at the focus by increasing the power at the sample surface, the collected signal in the focal region at z_{in} can be maintained:

$$I_{\mathrm{MPEF}}^{(m)}(z_{\mathrm{in}}) = I_0 P_{\mathrm{inc}}(z_{\mathrm{in}}) \exp\{-(m\alpha_{\mathrm{ex}} + \alpha_{\mathrm{em}})\, z_{\mathrm{in}}\} = I_0, \tag{4.2.22}$$

where $P_{\mathrm{inc}}(z_{\mathrm{in}}) = \exp\{(m\alpha_{\mathrm{ex}} + \alpha_{\mathrm{em}})\, z_{\mathrm{in}}\}$. However, the background signal, which includes nonlinear signals generated in out-of-focus regions at z_{out},

$$I_{\mathrm{MPEF}}^{(m)}(z_{\mathrm{out}}) = I_0 \exp\{(m\alpha_{\mathrm{ex}} + \alpha_{\mathrm{em}})\,(z_{\mathrm{in}} - z_{\mathrm{out}})\}, \tag{4.2.23}$$

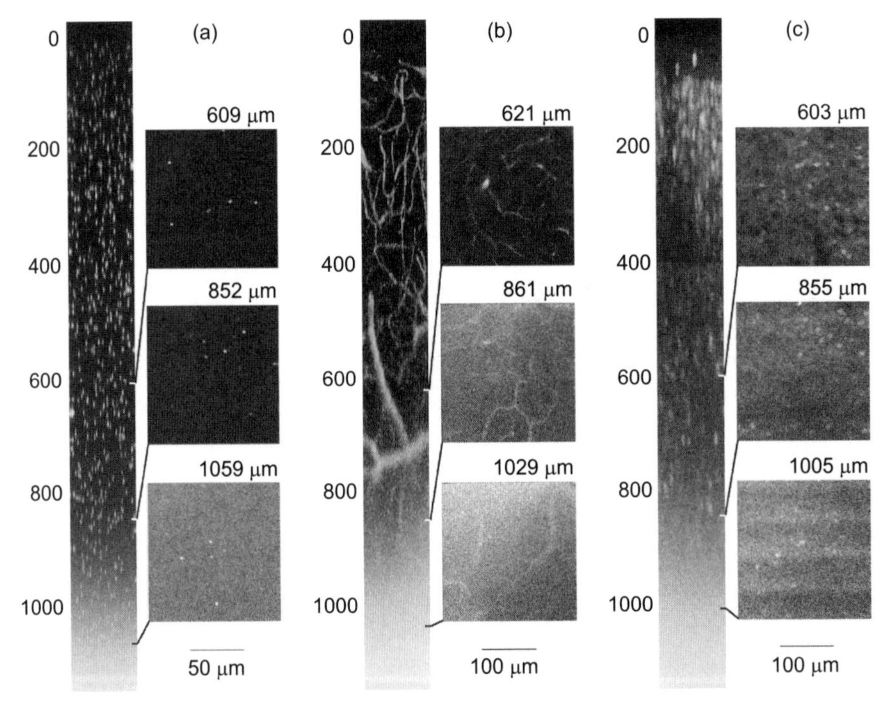

FIGURE 4.7 Volume images of (a) a tissue phantom, (b) stained cerebral blood vessels, and (c) fluorescent-protein-labeled neurons. Shown are *xz* projections and single *xy* planes at three depths (*z* positions) for each sample. Reprinted with permission from [32]. Copyright 2003 Optical Society of America.

increases and eventually becomes comparable to the signal intensity in the focal region as shown in Fig. 4.7. Thus, the large out-of-focus signals limit the obtainable imaging depth [59].

4.2.2 Application of TPEF Microscopy

4.2.2.1 Multi-Color Imaging Multi-color imaging by confocal OPEF microscopy is generally performed using several CW lasers. Multi-color imaging by TPEF microscopy can be also achieved by using several narrowband femtosecond lasers. However, femtosecond lasers are much more expensive than CW lasers. Thus, multi-color TPEF microscopy using a single femtosecond laser has been attempted. A simple approach using a single femtosecond laser is to excite various fluorophores with different TPE spectra by tuning the central wavelength of the narrowband femtosecond laser to the appropriate excitation wavelengths. Multi-spectral fluorescence images are then be obtained sequentially. However, some preparation time is required to tune the laser wavelengths. Thus, the wavelength tuning method is not practical. To overcome this limitation, two other methods that provide simultaneous excitation of fluorophores with a single femtosecond pulse have been developed [60–63]. One uses a pair of fluorophores whose TPE spectra are overlapped enough but whose emission spectra are significantly different [60,61,64]. The first-excited singlet state of a blue fluorescent molecule and the higher excited singlet state of a red fluorescent molecule are simultaneously accessed with TPE using a single wavelength. Tillo et al. demonstrated dual-color imaging with the simultaneous excitation at 780 nm of the lowest energy electronic transition of a blue fluorescent protein mKalama1 and a higher energy electronic transition of a red fluorescent protein tagRFP [64]. TPE into the higher excited state of the red fluorescent protein can often be enhanced by one-photon electronic resonance through a real intermediate level [65]. However, it is reported that fluorescence from DsRed with TPE into the higher excited state at 750 nm bleaches more rapidly than that into the first-excited state at 950 nm owing to three-photon absorption at 750 nm [41]. In addition, multiphoton excitation enhanced by one-photon resonance may also increase photodamage. Thus, we should take care of photobleaching and photodamage.

These effects can be avoided by utilizing the large differences in Stokes shifts between the fluorophores [61,67,68]. Kawano et al. propose using a pair of fluorescent proteins, EGFP and mKeima, in dual-color imaging [61]. The first-excited singlet state of the two fluorescent proteins can be simultaneously accessed with TPE using a single wavelength, since mKeima has a much larger Stokes shift than EGFP [66]. Thus, the TPEF signals that are simultaneously generated from the two fluorescent proteins can be identified. This technique has been extended to pairs of ECFP and mKeima [67], ECFP and LSS-mKate1 (LSS-mKate2) [68], and EGFP and LSS-mKate1 (LSS-mKate2) [68]. If there are large differences in concentration among the various fluorophores, we must adjust their TPEF intensities to the same level for multi-color imaging. In confocal microscopy, by independently adjusting the incident power of each laser, the fluorescence signal ratios among the various fluorophores can be easily controlled. In simultaneous TPE with a single wavelength, however, it is difficult to regulate the TPE signal ratios among the fluorophores with significantly overlapped TPE spectra. This is because even if the laser wavelength is tuned, the TPEF signals still behave the same due to the overlapped TPE spectra.

The other approach is the use of a broadband pulse whose spectrum covers over the different TPE spectra of various fluorophores [62,63,69]. In this approach, the various fluorophores can be simultaneously and efficiently excited with optimal wavelengths within the spectral bandwidth of the single broadband pulse. Isobe et al. demonstrated multi-color TPEF imaging using a broadband supercontinuum [62]. Figure 4.8 shows three-color TPEF images simultaneously taken with a broadband supercontinuum. The sample was a fixed bovine pulmonary artery endothelial (BPAE) cell stained with three different fluorescent dyes: blue-fluorescent 4′,6-diamidino-2-phenylindole (DAPI) to stain the nuclei; green-fluorescent BODIPY FL phallacidin to stain F-actin; and red-fluorescent MitoTracker Red CMXRos to stain the mitochondria [62]. It should be noted that the three-color TPEF images were acquired simultaneously without tuning the laser wavelength.

In simultaneous excitation using a broadband pulse, the TPE signal ratios among the various fluorophores with different TPE spectra can be easily controlled by

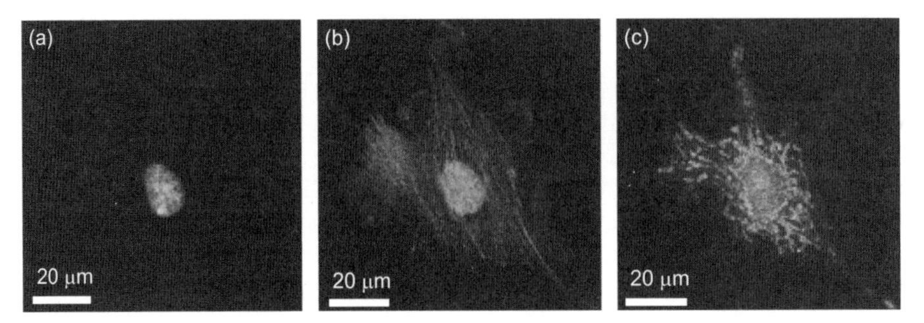

FIGURE 4.8 TPEF images of a fixed bovine pulmonary artery endothelial (BPAE) cell acquired through band-pass filters (a) for DAPI used to stain the nuclei (465–495 nm), (b) for BODIPY FL used to stain F-actin (510–550 nm), and (c) for MitoTracker Red used to stain the mitochondria (560–600 nm). Reprinted with permission from [62]. Copyright 2005 The Japan Society of Applied Physics.

modulating the spectral phase of the broadband pulse [63]. Therefore, even if there are large differences in concentration among the various fluorophores, we can adjust the TPEF intensities to the same level for multi-color imaging. In Section 5.2.1, we will describe the way of controlling the TPEF intensities by spectral phase modulation.

4.2.2.2 FRET and FLIM Fluorescence resonance energy transfer (FRET) provides spatial and temporal information about protein–protein interactions and protein conformal changes in living cells and tissues as described in Section 2.2.2. In FRET measurements, the interest sample double-labeled by donor and acceptor molecules is excited with the donor excitation wavelength. Ratio imaging, in which the fluorescence intensity changes of the donor and acceptor are simultaneously measured, enables us to obtain a FRET change from a change of their ratio with a high sensitivity [70]. In addition, most FRET sensors have been optimized for ratio-imaging [70–73]. Fan et al. achieved two-photon ratio-imaging of yellow cameleon, which is a calcium indicator [71], by using video-rate TPEF microscopy with a resonant galvanometer mirror at 7.875 kHz [74]. Two-photon ratio-imaging has also been used to image actin polymerization [75] and Ca^{2+}/calmodulin-dependent kinase II (CaMKII) activation [76] in single dendritic spines.

In general, the FRET signal is contaminated by donor fluorescence into the acceptor channel and by the excitation of the acceptor by the donor excitation wavelength. In particular, two-photon FRET signals are easily affected by the excitation of the acceptor with the donor excitation wavelength because the TPE spectra of the donor and acceptor are more overlapped compared with their OPE spectra. In three-cube FRET imaging an intensity range based correction method can be used to correct for these spectral bleed-through signals [77,78]. In two-photon FRET, however, the image used to correct for the acceptor spectral bleed-through signals is taken with the acceptor channel during acceptor-excitation of the interest sample double-labeled by donor and acceptor, so it may be contaminated by the excitation of the donor with the acceptor excitation wavelength due to the spectral overlap [78]. Although these bleed-through signals can be corrected [78], there are no corrections for the spectral bleed-through signals in FRET imaging based on the fluorescence lifetime measurements. This is because the fluorescence lifetime is independent of the intensity of the fluorophores.

Fluorescence lifetime imaging microscopy (FLIM) can be used for probing separation of fractions of the same fluorophores in different binding states to lipids, proteins, or DNA, and distances on the nanometer scale by FRET as shown in Section 2.2.3. The fluorescence lifetime is obtained by recording the fluorescence intensity decay after excitation as a function of time, or by extracting the phase and the modulation depth of the fluorescence induced by an intensity modulated light. In FLIM, FRET efficiency can be independently determined, since FRET is not affected by fluorophore concentrations, and this allows measurement of dynamic events at very high temporal resolutions (ns). Another advantage is that only the lifetime of the donor fluorophore is measured. Thus, the acceptor fluorophore can have inefficient emission (even might be a quencher). Therefore, FRET sensors for ratio-imaging and fluorescence lifetime imaging require different types of optimization [79]. Because ratio-imaging employs both donor and acceptor signals, a bright donor and acceptor such as an ECFP-EYFP pair [70] are required for FRET sensors in ratio-imaging. In contrast, FLIM uses only the donor signal, and thus, the brightness of the acceptor is not important, although a high extinction coefficient of the acceptor is required for high FRET efficiency. However, fluorophores with mono-exponential decay are preferable for FLIM so that the binding fraction can be measured. Recently, FRET sensors have been optimized for fluorescence lifetime imaging of Ras [79] and CaMKII [80] activity. ECFP is not an optimal donor for FLIM because of its multiexponential fluorescence lifetime decay. Instead of ECFP, monomeric EGFP (mEGFP) has been used, because it is bright and has a mono-exponential fluorescence lifetime decay [79]. FRas-F, which is the Ras sensor optimized for FLIM, consists of H-Ras tagged with mEGFP and the Ras-binding domain of Raf (RBD) tagged with two mRFPs [79]. Recently, Ganesan et al. have developed resonance energy-accepting chromoprotein (REACh) (nonradiative) variant as a new possibility for an acceptor in a FRET sensor [81]. However, REACh also has the problems of low folding efficiency and environmental sensitivity. Murakoshi et al. improved the fluorescent properties of REACh by introducing several rational mutations to improve solubility, environmental insensitivity, and folding (super-REACh or sREACh) [82]. Green-Camuiα, in which CFP and YFP in Camuiα [76] are replaced by the pair of mEGFP and sREACh, is the CaMKII sensor optimized for FLIM [80].

For FLIM in the time domain, the fluorescence intensity decay after excitation with a pulsed light source is obtained as a function of time. The excitation efficiency of TPE can be enhanced by the use of an ultrashort pulse with a high peak power. The repetition rate of a typical mode-locked Ti:sapphire oscillator used as an excitation light source in TPEF microscopy is 80 MHz. The repetition rate corresponds to an interpulse interval of 12 ns, which is long enough to measure the fluorescence decay of typical fluorescent proteins (2–4 ns) after each pulse. Thus, the Ti:sapphire oscillator is also an appropriate excitation light source in FLIM. The procedure of combining TPEF microscopy using the Ti:sapphire oscillator with FLIM is straightforward.

The reduction of out-of-focus signals by TPE has become particularly important in combination with FLIM. In order to obtain clean lifetime results, mixing the decay functions of out-of-focus regions must be avoided. The axial crosstalk can be sufficiently suppressed by TPE. Recently, the two-photon FLIM-FRET based on time-correlated single photon counting (TCSPC), which can be used for fluorescence lifetime imaging in a time domain with a high signal-to-noise ratio, was applied to the imaging of Ras activation [83], actin polymerization [82], extracellular signal-regulated kinase (ERK) [84], and CaMKII activation [80] in single dendritic spines of brain slices.

4.2.2.3 FRAP Fluorescence recovery after photobleaching (FRAP) or fluorescence photobleaching recovery (FPR) can be used to measure transport properties of fluorescent-labeled molecules as described in Section 2.2.4. FRAP measurements are achieved by monitoring the evolution of a fluorescence signal after a spatially localized population of fluorophores is bleached by light. In one-photon FRAP, the photobleaching occurs along the whole optical axis. Thus, the bleached area for one-photon FRAP experiments is confined to a two-dimensional geometry using thin samples such as membranes [85]. By combining OPE photobleaching using a high NA objective lens with confocal detection, Blonk et al. developed FRAP that was used to determine three-dimensional diffusion of fluorophores in cells [86]. However, the extended double cone of bleached fluorophore created in this technique could not be mathematically defined in a simple enough form to derive fitting functions for the determination of diffusion coefficients [87]. The measured bleaching profiles used as the initial condition for finite-element simulations of FRAP-diffusion experiments showed that in the confocal regime, distortion of the bleaching pattern can cause the diffusion coefficient to be seriously underestimated [88]. This problem was later solved by using photobleaching with multiphoton excitation [87,89,90], which provides accurate estimation of the kinetic parameter [88]. Piston et al. first demonstrated FRAP with TPE photobleaching [89]. The theory used in one-photon FRAP analysis, which was originally developed by Axelrod et al. [85], was extended to multiphoton FRAP by Brown et al. [87]. Fluorescence excitation and photobleaching can be confined to subfemtoliter volumes by using multiphoton excitation with high NA objectives, resulting in significant increase of the potential for sampling different regions of the same cell, point by point. The localization of the photobleaching volume provided by multiphoton excitation makes the quantitative analysis of the FRAP experiments more accurate and allows a reduction of the errors made in the estimation of the diffusion coefficient [88]. In Section 2.2.4, we described FRAP theory in terms of multiphoton excitation. Multiphoton FRAP can be used at different positions throughout the cell to create a three-dimensional diffusion map and the three-dimensionally allows measurement of the diffusion coefficients with significant penetration depth in thick, intact tissues.

Two-photon FRAP has been extensively used in neurobiology to determine the coupling of dendritic spines to their parent dendrite [90–98]. Spine had been difficult to investigate in live tissue because of their small size. However, as TPE photobleaching can be confined to subfemtoliter volumes, two-photon FRAP can be used to examine the kinetics of protein movement within dendritic spines. Two-photon FRAP in fact proved that the diffusional coupling between spines and their parent dendrites occurs [90–95]. Because current propagation in neurons is mediated by electrodiffusion, diffusion time constants can be used to estimate spine neck resistance. In order to estimate the electrical resistance of spine necks, Svoboda et al. measured the rate of diffusion of fluorescein dextrans from the dendrite after photobleaching of fluorescein dextran in the spine head [90]. Majewska et al. showed that the rate of diffusion between the spine and the dendrite depends on the length of the spine neck and that the longer-necked spine acts as a greater barrier to intercompartmental diffusion [91]. Majewska et al. investigated the effect of rapid changes in spine morphology on spine calcium dynamics and showed that the elongation or retraction of the spine neck during spine motility alters the diffusional coupling between the spine and dendrite and significantly changes calcium decay kinetics in spines [92]. Two-photon FRAP has been used to characterize the behavior of Ca^{2+} indicators employed to measure calcium signaling in spines in order to estimate the behavior of calcium in the absence of exogenous fluorophore [93–95]. Two-photon FRAP has also been used to characterize the diffusion of endgenous calcium buffers between spines and dendrites [96,97].

4.2.2.4 FCS Fluorescence correlation spectroscopy (FCS) can provide quantitative information about mobility, concentration, interactions, chemical kinetics, and physical dynamics of biomolecules within living cells and tissues as shown in Section 2.2.5. FCS measurements are performed by recording fluctuations of the fluorescence intensity in a very small observation volume, which is created by focusing laser beams into submicron beam waists in the radial dimensions. However, OPE is not confined along the axial direction. This three-dimensional confinement problem has been successfully solved by confocal detection, which makes possible measurements of diffusion in three-dimensional samples [99–100]. An alternative method to solve the three-dimensional confinement problem is to use TPE. By determining the diffusion coefficient of latex beads in the cytoplasm of mouse fibroblast cells, FCS measurements with TPE was first demonstrated by Berland et al. in 1994 [102]. Two-photon FCS provides a number of advantages such as higher signal-to-noise ratios, reduced photobleaching, and improved penetration in turbid media over confocal one-photon FCS [103]. In FCS measurements in cells with thick walls; such as plant cells, effects of noncorrelating background due to scattering of fluorescence signal are quite severe. Thus, for all applications where scattering is present, like plant cells or deeper tissue layers, two-photon FCS is highly recommended [103]. The superior penetration depth of two-photon FCS was exploited by Alexandrakis et al. in a study to determine transport parameters in melanomas, which is a malignant tumor of melanocytes [104]. Larson et al. examined the mobility and multimerization of the polyprotein Gag, which is a key player in the assembly of retroviruses in the cytosol of live cells, and found that most Gag molecules exist in the cytoplasm as part of large complexes [105].

Fluorescence cross-correlation spectroscopy (FCCS) is a powerful tool for probing interactions of different fluorescently labeled molecules in biological samples as described in Section 2.2.5. In FCCS measurement, the two fluctuation signals from the sample double-labeled by two-color fluorophores must be simultaneously measured with minimal cross talk between the detector channels by two distinct detection channels. Thus, the two fluorescent species used in FCCS must have sufficiently separated emission spectra and must be simultaneously excited with equal efficiency. As described earlier (multi-color imaging), TPE provides the simultaneous excitation of multi-color fluorophores with a single wavelength. Thus, two-photon FCCS has been simplified by using the simultaneous excitation with a single wavelength [106–111]. Because this technique also provides a complete observation volume overlap, artifacts due to lack of observation volume overlap can be eliminated. Two-photon FCCS was used to develop an in vitro protease assay [108] and study the complex binding stoichiometry of calmodulin (CaM), which is a key transducer of intracellular Ca^{2+} signaling, and Ca^{2+}/CaM-dependent protein kinase II (CaMKII) in vivo [109,110]. Collini et al., applied two-photon FCCS to DNA glycosylase assays of human cell extracts and proposed this method to be an efficient tool for screening of DNA repair activity of a large number of samples [112]. By employing two-photon FCCS, Merkle et al., analyzed DNA synapsis and end joining in solution using purified proteins in order to investigate repair mechanisms for DNA double-strand breaks [113].

4.2.3 High-Speed Imaging Using TPEF Microscopy

In this section, we describe multifocal multiphoton microscopy and multiphoton microscopy using temporal focusing both of which provides real-time imaging capability.

4.2.3.1 Multifocal Multiphoton Microscopy The scanning speed of a single excitation beam can be increased by employing resonant galvonometric scanners [17], rotating polygonal mirrors [18], and acousto-optic deflectors (AODs) [19]. However, because the exposure time per pixel is reduced, the fluorescence signals are reduced. If one compensates for the reduction of the fluorescence signals by increasing the incident power, photodamage and photobleaching may occur. Thus, high-speed imaging without reducing the exposure time is desired. Such imaging can be achieved by scanning with an array of foci rather than with a single beam. Multifocal microscopy has been traditionally accomplished by Nipow-disk confocal microscopy with an array of pinholes [114,115]. Multifocal multiphoton microscopy (MMM) that incorporates a microlens array was first reported by Buist et al. [20] and Bewrsdorf et al. [21] in 1998.

A typical setup of multifocal multiphoton microscope is shown in Fig. 4.9. The microlenses are arranged in a hexagonal pattern such that the illuminating beam is split into small beams, referred to as beamlets. The microlenses are illuminated by an expanded laser beam, and then, an array of foci of the beamlets is created at the prefocusing plane (PFP). After the intermediate optics, the beamlets are directed into a conventional fluorescence microscope. The role of the intermediate optics is to ensure that the array of foci is imaged into the focal plane of the lens and that each beamlet is parallel at, and over-illuminates, the objective entrance pupil [116]. The objective lens then produces a pattern of high-resolution foci at the sample. The focal plane of the objective lens is imaged onto imaging detectors such as CCD and

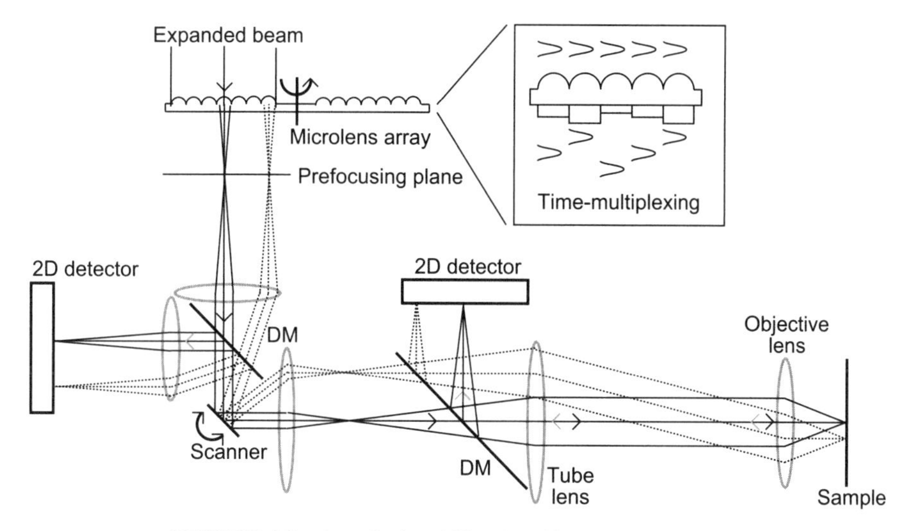

FIGURE 4.9 A typical multifocal multiphoton microscope.

CMOS image sensors. The image is obtained by rotating the disk containing the microlens arrays [21] or by rapidly moving the foci in a Lissajous pattern with an xy galvanometric mirror [20]. In MMM, the image acquisition time is cut down by the number of foci. However, the spacing between the foci has to be chosen carefully. On the one hand, if the foci are too close to one another and light pulses of neighboring foci are simultaneously focused, the axial resolution is degraded by cross talk between foci due to out-of-focus interference. On the other hand, spreading the foci too far apart reduces the efficiency of excitation and increases the image acquisition time. To resolve this trade-off between axial resolution and parallelization, Egner et al. introduced time delays between the different foci that are larger than the pulse duration of the laser pulses (time-multiplexing) as shown in Fig. 4.9 [22]. As a result, the foci can be positioned arbitrarily close to one another without degrading axial resolution. In spinning disk MMM, rotating the disk by 360° renders as many complete lateral scans as segments on the disk, typically 5 to 12. The disk can be rotated at more than 100 Hz, resulting in scanning rates of more than 1000 frames/s [116]. In these techniques, the microlenses illuminated by the expanded laser beam generate the beamlets. However, because the microlenses in the outer part of the laser beam of the Gaussian profile are illuminated by less intense light than those in the center of the laser beam, the intensity of the foci decreases toward the rim [116]. Thus, the excitation efficiency at the edge of the field of view is dramatically reduced due to the nonlinearity of multiphoton excited fluorescence (MPEF). By converting the Gaussian profile to the flat-top profile with beam-shaping techniques [117] or by extracting the center of the Gaussian beam, the homogeneity across the field of view can be balanced.

Fittinghoff et al. [23] and Nielsen et al. [24] developed another type of MMM that uses a cascaded-beamsplitter array to produce temporally decorrelated beamlets.

The image is acquired by scanning the foci with an xy galvanometric mirror [23] or by scanning of the sample with a stage [24]. In either case, the beam profile is maintained for each beamlet. By selecting the ratio between reflected and transmitted light carefully, we can obtain equal beamlet intensities. In addition, by handling the tilting of the beam splitters with particular care, we can achieve equidistant foci with each beamlet centered on the objective entrance pupil [24]. These techniques provide a homogeneously illuminated field of view, and the multiplexing time minimizes cross talk between foci. The disadvantage of these techniques is that their configurations are more complex than those of microlens arrays. Let us consider the spatial resolution of MMM. In MMM, the detector integrates the emission light during the two-dimensional scanning of excitation foci on the sample plane. Therefore, the excitation point spread function (PSF) $H_{\mathrm{exM}}(x, y, z)$ of the focal pattern is integrated over the focal plane of the objective lens, spreading out the PSF laterally. Thus, the effective excitation profile of MMM using m-photon excited fluorescence is proportional to the axial response [22,116]

$$R_{\mathrm{exM}}^{(m)}(z) = \int_{-\infty}^{\infty}\int_{-\infty}^{\infty} H_{\mathrm{exM}}^{(m)}(x, y, z)dxdy. \tag{4.2.24}$$

where

$$H_{\mathrm{exM}}^{(m)}(x, y, z) = [h_{\mathrm{exS}}(x, y, z) \otimes_3 G(x, y, z)]^{2m}. \tag{4.2.25}$$

Here $h_{\mathrm{exS}}(x, y, z)$ describes the (single-focus) amplitude PSF of the illumination; $\otimes_3$ denotes the 3D convolution operation. The grating function $G(x, y, z)$ is the sum of several Dirac's delta functions in the focal plane, one for each focus:

$$G(x, y, z) = \sum_{n=1}^{N} \delta(x - x_n)\delta(y - y_n)\delta(z), \tag{4.2.26}$$

where N is the number of the foci. In the limiting case of a sufficiently sparse grating, $x_n - x_{n-1} \gg \lambda_{\mathrm{ex}}\sqrt{N}$, $y_n - y_{n-1} \gg \lambda_{\mathrm{ex}}\sqrt{N}$, the interference effects between the foci are negligible, and Eq. (4.2.24) can be rewritten as

$$\begin{aligned} R_{\mathrm{exM}}^{(m)}(z) &= \int_{-\infty}^{\infty}\int_{-\infty}^{\infty} \left[h_{\mathrm{exS}}^{2m}(x, y, z) \otimes_3 G(x, y, z)\right] dxdy \\ &= N \int_{-\infty}^{\infty}\int_{-\infty}^{\infty} h_{\mathrm{exS}}^{2m}(x, y, z)dxdy. \end{aligned} \tag{4.2.27}$$

Reducing the distance between the lens foci increases the interference between the focal fields in the planes away from the focal plane. However, by introducing time-multiplexing where time delays between the different foci are longer than the pulse duration of the laser pulses, the interference effects between the foci can be removed [22,116]. The effective PSF of MMM is given by the product of the axial response for excitation and the detection PSF $H_{\text{det}}(x, y, z)$, which is determined not by the longer excitation wavelength but by the shorter fluorescence wavelength [22,116]:

$$PSF_{\text{M}}^{(m)}(x, y, z) = R_{\text{exM}}^{(m)}(z) H_{\text{det}}(x, y, z). \qquad (4.2.28)$$

From Eq. (4.2.28), we see that the lateral resolution of MMM is equal to that of conventional wide-field fluorescence microscopy. Thus, the lateral resolution of MMM is higher than that of conventional single-point-scanning MPEF microscopy. However, MMM is sensitive to variation of the emission PSF due to photon scattering, while single-point-scanning MPEF microscopy is not sensitive to variation of the emission PSF. In MMM, the emission photons scattered in tissues will not arrive at the correct pixel of the imaging detector, resulting in a decrease of the signals and in an increase of background noise. As a consequence, it is difficult to apply MMM to deep imaging in tissues.

To overcome the imaging depth limitation of MMM, Kim et al. developed a new MMM using multi-anode photomultiplier tubes (MAPMTs) with a small number of pixels and with a large effective detection area for a given field of view, which provides more efficient collection of the emission photons scattered in tissues [118]. The detection area for each pixel is large enough to collect the scattered emission photons into the correct pixels. Therefore, MAPMT-based MMM provides imaging depth comparable to single-point-scanning MPEF microscopy. Figure 4.10 shows TPEF images of GFP expressing neurons in the ex vivo mouse brain acquired with CCD-based MMM and MAPMT-based MMM at different depth locations [118]. In CCD-based MMM, the background noise dramatically increases with the penetration depth. In contrast, the background noise of MAPMT-based MMM is low enough at 75-μm depth. We see that the MAPMT provides the suppression of the background noise of MMM.

4.2.3.2 Multiphoton Excited Fluorescence Microscopy Using Temporal Focusing Oron et al. [26] and Zhu et al. [27] developed in 2005 a new type of parallelized MPEF microscopy whose optical sectioning capability is obtained by temporal focusing instead of spatial focusing. The image is acquired by employing a single beam, but line or plane excitation is used. The axial resolution of conventional MPEF microscopy using line excitation is lower than that using point excitation. However, temporal focusing provides line-scanning MPEF microscopy whose axial resolution is equal to the axial resolution of conventional line-scanning MPEF microscopy, and wide-field MPEF microscopy whose axial resolution is equal to the axial resolution of conventional point-scanning MPEF microscopy [119].

Figure 4.11(a) shows a schematic of a typical MPEF microscope using temporal focusing [26,27]. The frequency components of the excitation pulse are diverged by the grating and then they are collimated by the collimating lens. In this process, the geometric dispersion caused by the grating is automatically canceled just after the collimating lens. Thus, the rainbow beam after collimating is chirped in space but is chirp-free in time. Temporal focusing is achieved by focusing this rainbow beam with the objective lens because the spatial chirp is compensated only at the

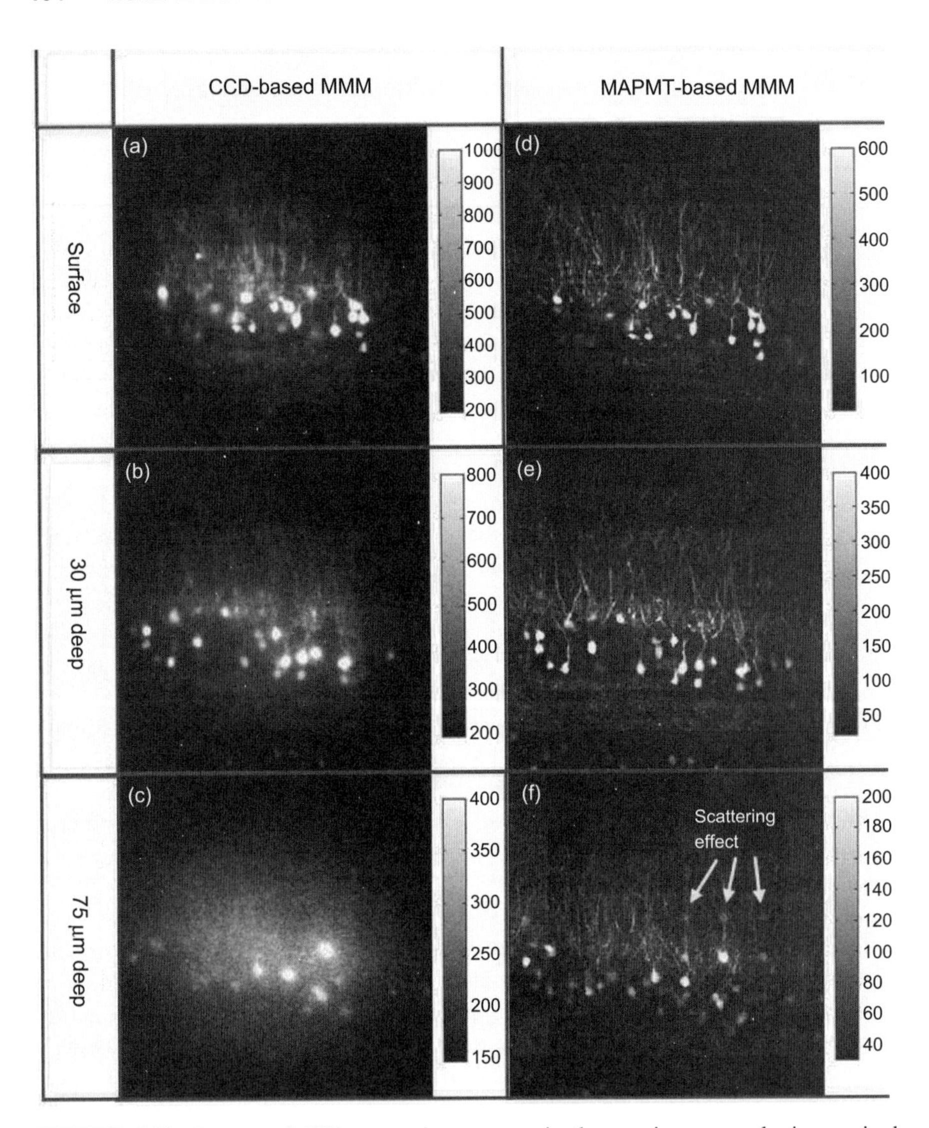

FIGURE 4.10 Images of GFP expressing neurons in the ex vivo mouse brain acquired with CCD-based MMM (a–c) and MAPMT-based MMM (d–f) at different depth locations. Reprinted with permission from [118]. Copyright 2007 Optical Society of America.

focal point (all frequency components are spatially overlapped only within the focal region of the objective lens), and thus the pulse duration is shortest only at the focal plane. Because wide-field illumination or line illumination is used in MPEF microscopy's temporal focusing, the emitted fluorescence is imaged onto imaging detectors such as CCD and CMOS image sensors. This process is described by the detection PSF $H_{det}(x, y, z)$, which depends not on the longer excitation wavelength

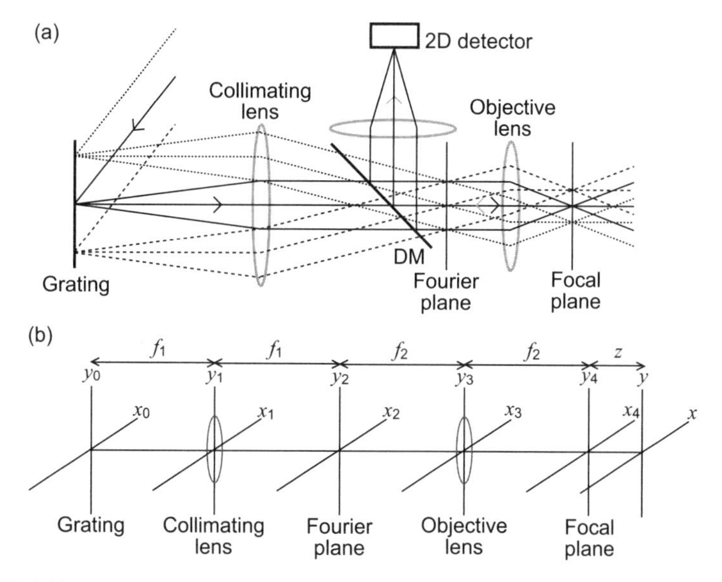

FIGURE 4.11 (a) A typical setup for multiphoton excited fluorescence microscopy using temporal focusing. (b) A setup for theoretical analysis of temporal focusing.

but on the shorter fluorescence wavelength. The excitation profile of m-photon excited fluorescence microscopy using temporal focusing is given by

$$H_{\mathrm{exTF}}^{(m)}(x, y, z) = \int_{-\infty}^{\infty} I_{\mathrm{TF}}^{m}(x, y, z, t)dt, \tag{4.2.29}$$

where $I_{\mathrm{TF}}(x, y, z, t)$ is the instantaneous intensity of the temporal focusing pulse. In line-scanning MPEF microscopy using temporal focusing, the detector integrates the emission light during the one-dimensional scanning of line excitation on the sample plane [120]. In addition, provided that wide-field illumination and integrated line-illumination do not include high spatial frequency components, the excitation profile is written by the axial response:

$$R_{\mathrm{TF}}^{(m)}(z) = \int_{-\infty}^{\infty} \int_{-\infty}^{\infty} \int_{-\infty}^{\infty} I_{\mathrm{TF}}^{m}(x, y, z, t)dtdxdy, \tag{4.2.30}$$

Thus, the effective PSF of MPEF microscopy using temporal focusing is written as

$$PSF_{\mathrm{TF}}(x, y, z) = R_{\mathrm{TF}}^{(m)}(z)H_{\mathrm{det}}(x, y, z). \tag{4.2.31}$$

Therefore, the lateral resolution of MPEF microscopy using temporal focusing is equal to that of conventional wide-field OPEF microscopy, which is higher than that of conventional single-point-scanning MPEF microscopy (see Section 4.2.1).

Next, we theoretically analyze temporal focusing by using a typical setup shown in Fig. 4.11(b). The angle of incidence and the diffraction angle are related through the grating equation

$$d\,(\sin\theta_i + \sin\theta_d) = m\lambda, \tag{4.2.32}$$

where d is the grating constant, θ_i is the angle of incidence, θ_d is the diffraction angle at a frequency ω (wavelength λ). Now, we consider the first-order diffraction and assume that the grating does not affect the pulse spectrum. Then, the grating can be characterized by a complex optical transfer function

$$H(x, y, \omega) = \exp\,(ik\sin\theta_d x) = \exp\left\{i\left(\frac{2\pi}{d} - \frac{\omega}{c}\sin\theta_i\right)x\right\}, \tag{4.2.33}$$

where we have used Eq. (4.2.32). We assume that the electric field of the excitation pulse in the frequency domain just before the grating can be expressed by spatially transform-limited Gaussian beams in the form

$$\tilde{E}_{\mathrm{Gin}}\,(x_0, y_0, \omega) = \tilde{A}\,(\omega)\,e^{-\left\{\frac{x_0^2}{s_x^2}+\frac{y_0^2}{s_y^2}\right\}}, \tag{4.2.34}$$

where the complex spectral amplitude $\tilde{A}(\omega)$ is expressed by

$$\tilde{A}\,(\omega) = \exp^{-\frac{(\omega-\omega_0)^2}{\Omega_g^2}}\exp^{i\frac{\beta}{2}(\omega-\omega_0)^2} = \exp^{-\left(\frac{2-i\beta\Omega_g^2}{2\Omega_g^2}\right)(\omega-\omega_0)^2}. \tag{4.2.35}$$

Here ω_0 is the central angular frequency of the pulse, β is the group delay dispersion (GDD), and Ω_g is related to the spectral bandwidth (FWHM) of the pulse, $\Delta\omega$, by the following equation:

$$\Delta\omega = \sqrt{2\ln 2}\Omega_g. \tag{4.2.36}$$

The complex temporal amplitude $A(t)$ is obtained by Fourier inversion of Eq. (4.2.35) along the frequency coordinate:

$$A\,(t) = \int_{-\infty}^{\infty}\exp^{-\left(\frac{2-i\beta\Omega_g^2}{2\Omega_g^2}\right)(\omega-\omega_0)^2}\exp^{-i\omega t}\,d\omega = \sqrt{\frac{2\pi\Omega_g^2}{2-i\beta\Omega_g^2}}\exp^{-i\omega_0 t-\frac{\Omega_g^2}{2\left(2-i\beta\Omega_g^2\right)}t^2}, \tag{4.2.37}$$

where we used $\int_{-\infty}^{\infty}\exp(-ax^2)dx = \sqrt{\pi/a}$. The pulse duration (FWHM) of the pulse is expressed by

$$\Delta\tau = \frac{2}{\Omega_g}\sqrt{\frac{\left(4+\beta^2\Omega_g^4\right)\ln 2}{2}} = \frac{\sqrt{16\,(\ln 2)^2 + \beta^2\Delta\omega^2}}{\Delta\omega}. \tag{4.2.38}$$

From Eqs. (4.2.33) and (4.2.34), the electric field in the frequency domain just after the grating is written as

$$\begin{aligned}\tilde{E}_{\text{Gout}}(x_0, y_0, \omega) &= \tilde{E}_{\text{Gin}}(x_0, y_0, \omega)\, H(x_0, y_0, \omega)\\ &= \exp^{-\left\{\frac{x_0^2}{s_x^2}+\frac{y_0^2}{s_y^2}\right\}+i\frac{2\pi}{d}x_0}\, \tilde{A}(\omega) \exp^{-i\frac{x_0\omega\sin\theta_i}{c}}, \end{aligned} \tag{4.2.39}$$

The field in the time domain just after the grating can be obtained by Fourier inversion of Eq. (4.2.39) along the frequency coordinate:

$$E_{\text{Gout}}(x_0, y_0, t) = \exp^{-\left\{\frac{x_0^2}{s_x^2}+\frac{y_0^2}{s_y^2}\right\}+i\frac{2\pi}{d}x_0}\, A\left(t + \frac{x_0 \sin\theta_i}{c}\right), \tag{4.2.40}$$

By propagating the field a distance f_1 from the grating with Fresnel diffraction, the field in the frequency domain just before the collimating lens can be expressed as [52]

$$\begin{aligned}\tilde{E}_{\text{L1in}}(x_1, y_1, \omega) = \frac{\exp(ikf_1)}{i\lambda f_1} \int_{-\infty}^{\infty}\int_{-\infty}^{\infty} \tilde{E}_{\text{Gout}}(x_0, y_0, \omega)\\ \times \exp\left[\frac{ik}{2f_1}\left\{(x_1 - x_0)^2 + (y_1 - y_0)^2\right\}\right] dx_0 dy_0,\end{aligned} \tag{4.2.41}$$

where we used the paraxial approximation. After adding the quadratic phase $\exp\left\{\frac{-ik}{2f_1}\left(x_1^2 + y_1^2\right)\right\}$ along x_1 and y_1 directions due to the collimating lens with a focal length of f_1, the field in the frequency domain becomes

$$\begin{aligned}\tilde{E}_{\text{L1out}}(x_1, y_1, \omega) = \frac{\exp(ikf_1)}{i\lambda f_1} \int_{-\infty}^{\infty}\int_{-\infty}^{\infty} \tilde{E}_{\text{Gout}}(x_0, y_0, \omega)\\ \times \exp\left\{\frac{ik}{2f_1}\left(x_0^2 + y_0^2\right)\right\} \exp\left\{-\frac{ik}{f_1}(x_0 x_1 + y_0 y_1)\right\} dx_0 dy_0.\end{aligned} \tag{4.2.42}$$

By propagating the field a distance f_1 from the collimating lens with Fresnel diffraction, we can express the field in the frequency domain at the Fourier plane as

$$\begin{aligned}\tilde{E}_{\text{FT}}(x_2, y_2, \omega) = \frac{\exp(2ikf_1)}{i\lambda f_1} \int_{-\infty}^{\infty}\int_{-\infty}^{\infty} \tilde{E}_{\text{Gout}}(x_0, y_0, \omega)\\ \times \exp\left\{-\frac{ik}{f_1}(x_0 x_2 + y_0 y_2)\right\} dx_0 dy_0.\end{aligned} \tag{4.2.43}$$

The field in the frequency domain at the focal plane of the objective lens with a focal length of f_2 is written as

$$\begin{aligned}
\tilde{E}_{\text{focal}}(x_4, y_4, \omega) &= \frac{\exp(2ikf_2)}{i\lambda f_2} \int_{-\infty}^{\infty} \int_{-\infty}^{\infty} \tilde{E}_{\text{FT}}(x_2, y_2, \omega) \\
&\quad \times \exp\left\{-\frac{ik}{f_2}(x_2 x_4 + y_2 y_4)\right\} dx_2 dy_2 \\
&= -\frac{f_1}{f_2} \exp\{2ik(f_1 + f_2)\} \tilde{E}_{\text{Gout}}\left(-\frac{f_1}{f_2}x_4, -\frac{f_1}{f_2}y_4, \omega\right) \\
&= M \exp\{2ik(f_1 + f_2)\} \tilde{E}_{\text{Gout}}(Mx_4, My_4, \omega),
\end{aligned} \tag{4.2.44}$$

where

$$M = -\frac{f_1}{f_2}. \tag{4.2.45}$$

Equation (4.2.44) indicates that the grating is imaged at the focal plane of the objective lens. The field in the time domain at the focal plane of the objective lens is obtained by Fourier inversion of Eq. (4.2.44) along the frequency coordinate:

$$\begin{aligned}
E_{\text{focal}}(x_4, y_4, t) &= M \exp\{2ik_0(f_1 + f_2)\} \\
&\quad \times \exp^{-\left\{\frac{(Mx_4)^2}{S_x^2} + \frac{(My_4)^2}{S_y^2}\right\} + i\frac{2\pi}{d}Mx_4} A\left(t + \frac{Mx_4 \sin\theta_i}{c}\right).
\end{aligned} \tag{4.2.46}$$

Here, to simplify calculations, we assumed that the wavevector k for each frequency is approximately k_0, the wavevector of the center wavelength of the pulse. By propagating the field a distance z from the focal plane with Fresnel diffraction, we can write the field at near the focal plane as

$$\begin{aligned}
E(x, y, z, t) &= \frac{\exp(ik_0 z)}{i\lambda_0 z} \int_{-\infty}^{\infty} \int_{-\infty}^{\infty} E_{\text{focal}}(x_4, y_4, t) \\
&\quad \times \exp\left[\frac{ik_0}{2z}\left\{(x - x_4)^2 + (y - y_4)^2\right\}\right] dx_4 dy_4 \\
&= \frac{MS_y\sqrt{k_0}\exp[ik_0\{2(f_1 + f_2) + z\}]}{\sqrt{k_0 S_y^2 + i2zM^2}} \\
&\quad \times \exp\left[\frac{ik_0}{2z}(x^2 + y^2)\right] \exp\left[-\frac{k_0^2 S_y^2}{2z(2zM^2 - ik_0 S_y^2)} y^2\right] X(x, z, t),
\end{aligned} \tag{4.2.47}$$

Here $X(x, z, t)$ is expressed by

$$X(x,z,t) = \sqrt{\frac{2\pi}{\lambda_0 z\left(\beta\Omega_g^2 + 2i\right)}}\Omega_g \exp^{-\left\{\frac{\Omega_g^2}{2\left(2-i\beta\Omega_g^2\right)}t^2 + i\omega_0 t\right\}} \int_{-\infty}^{\infty} \exp^{-\left(\frac{M^2}{S_x^2} - \frac{ik_0}{2z} + \frac{\Omega_g^2 M^2 \sin^2\theta_i}{2c^2\left(2-i\beta\Omega_g^2\right)}\right)x_4^2}$$
$$\times \exp^{\left[i\left\{\frac{k_0}{z}x - M\left(\frac{2\pi}{d} - \frac{\omega_0 \sin\theta_i}{c}\right)\right\} + \frac{\Omega_g^2 M \sin\theta_i}{c\left(2-i\beta\Omega_g^2\right)}t\right]x_4} dx_4$$
$$= \sqrt{\frac{2\pi k_0}{p(z)}}\Omega_g S_x \exp^{-\frac{\Omega_g^2 u(z)}{2W(z)}t^2 - \frac{v(z)}{W(z)}t - i\omega_0 t} \exp^{-\frac{k_0^2 S_x^2 q(z)}{2zW(z)}\left\{x - \frac{M}{k_0}\left(\frac{2\pi}{d} - \frac{\omega_0 \sin\theta_i}{c}\right)z\right\}^2}, \quad (4.2.48)$$

where

$$p(z) = 2\left(M^2\beta\Omega_g^2 z + k_0 S_x^2\right) + i\left\{\left(S_x^2\Omega_g^2 \sin^2\frac{\theta_i}{c^2+4}\right)M^2 z - k_0 S_x^2 \beta\Omega_g^2\right\}, \quad (4.2.49)$$

$$q(z) = 2\left(S_x^2\Omega_g^2 \sin^2\frac{\theta_i}{c^2} + 4 + \beta^2\Omega_g^4\right)M^2 z + i\left\{k_0 S_x^2\left(4 + \beta^2\Omega_g^4\right) - \beta S_x^2 \Omega_g^4 M^2 z \sin^2\frac{\theta_i}{c^2}\right\}, \quad (4.2.50)$$

$$u(z) = 2\left\{\left(S_x^2\Omega_g^2 \sin^2\frac{\theta_i}{c^2} + 4\right)M^4 z^2 + k_0^2 S_x^4\right\} - i\left\{k_0 S_x^4 \Omega_g^2 M^2 z \sin^2\frac{\theta_i}{c^2} - \left(k_0^2 S_x^4 + 4M^4 z^2\right)\beta\Omega_g^2\right\}, \quad (4.2.51)$$

$$v(z) = k_0 S_x^2\left\{x - \frac{M}{k_0}\left(\frac{2\pi}{d} - \frac{\omega_0 \sin\theta_i}{c}\right)z\right\}\frac{\Omega_g^2 M \sin\theta_i}{c} \times \left[2\left(M^2\beta\Omega_g^2 z + k_0 S_x^2\right) - i\left\{\left(S_x^2\Omega_g^2 \sin^2\frac{\theta_i}{c^2} + 4\right)M^2 z - k_0 S_x^2\beta\Omega_g^2\right\}\right], \quad (4.2.52)$$

$$W(z) = \left\{\left(S_x^2\Omega_g^2 \sin^2\frac{\theta_i}{c^2} + 4\right)M^2 z - k_0 S_x^2 \beta\Omega_g^2\right\}^2 + 4\left(M^2\beta\Omega_g^2 z + k_0 S_x^2\right)^2. \quad (4.2.53)$$

From Eq. (4.2.47), the instantaneous intensity is written as

$$I_{TF}(x,y,z,t) = |E(x,y,z,t)|^2 = \frac{2\pi\Omega_g^2 M^2 S_x^2 S_y^2 k_0^2}{\sqrt{W(z)\left(4z^2M^4 + k_0^2 S_y^4\right)}} \exp^{-\frac{2M^2 k_0^2 S_y^2}{4z^2M^4 + k_0^2 S_y^4}y^2}$$
$$\times \exp^{-\frac{2k_0^2 M^2 S_x^2}{B(z)}\left\{x - \frac{M}{k_0}\left(\frac{2\pi}{d} - \frac{\omega_0 \sin\theta_i}{c}\right)z\right\}^2}$$
$$\times \exp^{-\frac{2\Omega_g^2 B(z)}{W(z)}\{t + T(x,z)\}^2}, \quad (4.2.54)$$

where

$$B(z) = \left(S_x^2\Omega_g^2 \sin^2\theta_i/c^2 + 4\right) M^4 z^2 + k_0^2 S_x^4, \tag{4.2.55}$$

$$T(x, z) = k_0 S_x^2 \frac{M \sin\theta_i}{c} \left(M^2 \beta \Omega_g^2 z + k_0 S_x^2\right) \times \frac{\left\{x - \frac{M}{k_0}\left(\frac{2\pi}{d} - \frac{\omega_0 \sin\theta_i}{c}\right) z\right\}}{B(z)}. \tag{4.2.56}$$

Equation (4.2.54) indicates that the illumination area at the focal plane ($z = 0$) is expressed by that at the grating plane divided by M. Thus, line or plane excitation is used in the temporal focusing technique. In order to set the temporal focal plane to be parallel to the xy plane, according to Eq. (4.2.54), we see that the following condition must be satisfied:

$$\sin\theta_i = \frac{2\pi c}{d\omega_0} = \frac{\lambda_0}{d}. \tag{4.2.57}$$

When this condition is satisfied, from Eq. (4.2.32), we find that the diffraction angle at the central wavelength of the pulse is zero. According to Eq. (4.2.54), the pulse duration is expressed by

$$\tau(z) = \frac{1}{\Omega_g} \sqrt{\frac{2W(z) \ln 2}{B(z)}}. \tag{4.2.58}$$

From Eq. (4.2.53), $W(z)$ is rewritten by

$$\begin{aligned} W(z) &= 4B(z) \\ &\quad + \left\{\left(S_x^2\Omega_g^2 \sin^2\frac{\theta_i}{c^2} + 4\right) S_x^2\Omega_g^2 \sin^2\frac{\theta_i}{c^2} + 4\beta^2\Omega_g^4\right\} \\ &\quad \times \left\{M^2 z - \frac{k_0 S_x^2 \beta \Omega_g^2 \left(S_x^2\Omega_g^2 \sin^2\theta_i/c^2\right)}{S_x^2\Omega_g^2 \sin^2\theta_i/c^2 \left(S_x^2\Omega_g^2 \sin^2\theta_i/c^2 + 4\right) + 4\beta^2\Omega_g^4}\right\}^2 \\ &\quad + \frac{4\left(S_x^2\Omega_g^2 \sin^2\theta_i/c^2 + \beta^2\Omega_g^4\right) k_0^2 S_x^4 \beta^2 \Omega_g^4}{S_x^2\Omega_g^2 \sin^2\theta_i/c^2 \left(S_x^2\Omega_g^2 \sin^2\theta_i/c^2 + 4\right) + 4\beta^2\Omega_g^4} \qquad (4.2.59) \\ &\approx 4B(z) + \alpha \left\{M^2 z - \frac{k_0 S_x^2 \beta \Omega_g^2}{\left(S_x^2\Omega_g^2 \sin^2\theta_i/c^2 + 4\right)}\right\}^2 \\ &\quad + O(\beta), \end{aligned} \tag{4.2.60}$$

where

$$\alpha = \left(S_x^2\Omega_g^2 \sin^2\frac{\theta_i}{c^2} + 4\right) S_x^2\Omega_g^2 \sin^2\frac{\theta_i}{c^2}, \tag{4.2.61}$$

$$O(\beta) = \frac{4k_0^2 S_x^4 \beta^2 \Omega_g^4}{S_x^2\Omega_g^2 \sin^2\theta_i/c^2 + 4}. \tag{4.2.62}$$

Here we assumed that $\beta\Omega_g$ is much smaller than $S_x \sin\theta_i/c$. By substituting Eq. (4.2.60) into Eq. (4.2.58), we can rewrite the pulse duration as

$$\tau(z) \approx \frac{\sqrt{2\ln 2}}{\Omega_g}\sqrt{4 + \frac{\alpha\left\{M^2 z - \frac{k_0 S_x^2 \beta\Omega_g^2}{S_x^2\Omega_g^2\sin^2\theta_i/c^2+4}\right\}^2 + O(\beta)}{B(z)}}. \tag{4.2.63}$$

As is evident from Eq. (4.2.63), we see that the pulse duration depends on the distance from the focal plane along the axial direction. The m-photon excited fluorescence signal during temporal focusing is expressed by

$$\begin{aligned} F_{\mathrm{TF}}^{(m)}(x,y,z) &= \int_{-\infty}^{\infty} I_{\mathrm{TF}}^{m}(x,y,z,t)dt \\ &= \frac{\sqrt{\pi}\left(2\pi\Omega_g^2 M^2 S_x^2 S_y^2 k_0^2\right)^m}{\Omega_g\left\{W^{(m-1)/2}(z)\left(4z^2M^4 + k_0^2 S_y^4\right)^{m/2}\right\}} \exp^{-\frac{2mk_0^2 S_y^2 M^2}{4z^2M^4+k_0^2S_y^4}y^2} \\ &\times \frac{1}{\sqrt{2mB(z)}} \exp^{-\frac{2mk_0^2M^2S_x^2}{B(z)}\left\{x - \frac{M}{k_0}\left(\frac{2\pi}{d} - \frac{\omega_0\sin\theta_i}{c}\right)z\right\}^2}. \end{aligned} \tag{4.2.64}$$

From Eq. (4.2.64), the axial response of the m-photon excited fluorescence signal by using temporal focusing is described by

$$\begin{aligned} R_{\mathrm{TF}}^{(m)}(z) &= \int_{-\infty}^{\infty}\int_{-\infty}^{\infty} F^{(m)}(x,y,z)dxdy \\ &= \frac{\pi^{5/2}\Omega_g S_x S_y\left(\sqrt{2\pi}\Omega_g M S_x S_y k_0\right)^{2(m-1)}}{m\sqrt{2m}\left\{W(z)\left(4z^2M^4 + k_0^2 S_y^4\right)\right\}^{(m-1)/2}}. \end{aligned} \tag{4.2.65}$$

Here $W(z)$ and $(4z^2M^4 + k_0^2 S_y^4)$ are the response functions characterizing temporal focusing and spatial focusing, respectively. From Eq. (4.2.53), $W(z)$ is rewritten as

$$\begin{aligned} W(z) &= \left\{\left(S_x^2\Omega_g^2\sin^2\theta_i/c^2 + 4\right)^2 + 4\beta^2\Omega_g^4\right\} \\ &\times \left\{M^2 z - \frac{k_0\beta S_x^4\Omega_g^4\sin^2\theta_i/c^2}{\left(S_x^2\Omega_g^2\sin^2\theta_i/c^2 + 4\right)^2 + 4\beta^2\Omega_g^4}\right\}^2 \\ &+ 4k_0^2 S_x^4\left[1 + \frac{2\left(S_x^2\Omega_g^2\sin^2\theta_i/c^2 + 2\right)\beta^2\Omega_g^4 + \beta^4\Omega_g^8}{\left\{\left(S_x^2\Omega_g^2\sin^2\theta_i/c^2 + 4\right)^2 + 4\beta^2\Omega_g^4\right\}}\right] \\ &\approx \left(S_x^2\Omega_g^2\sin^2\theta_i/c^2 + 4\right)^2\left\{M^2 z - \frac{k_0\beta S_x^4\Omega_g^4\sin^2\theta_i/c^2}{\left(S_x^2\Omega_g^2\sin^2\theta_i/c^2 + 4\right)^2}\right\}^2 \\ &+ 4k_0^2 S_x^4\left[1 + \frac{2\left(S_x^2\Omega_g^2\sin^2\theta_i/c^2 + 2\right)\beta^2\Omega_g^4}{\left(S_x^2\Omega_g^2\sin^2\theta_i/c^2 + 4\right)^2}\right]. \end{aligned} \tag{4.2.66}$$

Here we assume that $\beta\Omega_g$ is much smaller than $S_x \sin\theta_i/c$. Furthermore, we assume that S_x, Ω_g and $\sin\theta_i = \lambda_0/d$ are the order of 1 mm, 10^{13} rad/s, and 0.5, respectively. Then, $W(z)$ and $\tau(z)$ are approximated as

$$W(z) \approx \left(S_x^2\Omega_g^2 \sin^2\theta_i/c^2\right)^2 \left\{M^2 z - \frac{k_0\beta c^2}{\sin^2\theta_i}\right\}^2 + 4k_0^2 S_x^4 \left[1 + \frac{2\beta^2\Omega_g^2 c^2}{S_x^2 \sin^2\theta_i}\right], \tag{4.2.67}$$

$$\tau(z) \approx \frac{\sqrt{2\ln 2}}{\Omega_g}\sqrt{4 + \frac{\alpha\left\{M^2 z - \frac{k_0\beta c^2}{\sin^2\theta_i}\right\}^2 + O(\beta)}{B(z)}}. \tag{4.2.68}$$

From Eqs. (4.2.65), (4.2.67), and (4.2.68), we see that the temporal focal plane has a linear dependence on the GDD:

$$\Delta z = \frac{k_0\beta c^2}{M^2 \sin^2\theta_i} = \frac{k_0\beta c^2 d^2}{M^2\lambda_0^2}. \tag{4.2.69}$$

Thus, the temporal focal plane can be scanned along the axial direction by controlling the GDD in the excitation beam path. According to Eqs. (4.2.65) and (4.2.67), when $|\Omega_g \sin\theta_i/c|$ is much larger than $1/S_y$, the axial response is determined only by temporal focusing characterization $W(z)$. In this case, wide-field illumination in MPEF microscopy whose axial resolution is equal to that of conventional line-scanning MPEF microscopy can be achieved by temporal focusing [119], and the axial scanning of the sample or the objective lens can be replaced with the tuning of the GDD [121]. However, as is evident from Eqs. (4.2.65) and (4.2.67), the signal intensity in the temporal focusing plane ($z = \Delta z$) decreases with the increasing axial scanning distance the GDD tuning. Let us consider the maximum axial scan range. We define the maximum applicable GDD where $R_{\mathrm{TF}}(\Delta z)$ is equal to half of $R_{\mathrm{TF}}(0)$ under the condition that the GDD β is 0:

$$\beta_{\max} = \frac{S_x \sin\theta_i}{\Omega_g c}\sqrt{\frac{4^{m-1}-1}{2}} = \frac{S_x\lambda_0}{\Omega_g c d}\sqrt{\frac{4^{m-1}-1}{2}}. \tag{4.2.70}$$

Insertion of Eq. (4.2.70) into (4.2.69) leads to the maximum axial scan range:

$$\Delta z_{\max} = \frac{k_0 S_x c}{M^2\Omega_g \sin\theta_i}\sqrt{\frac{4^{m-1}-1}{2}} = \frac{k_0 S_x c d}{M^2\Omega_g\lambda_0}\sqrt{\frac{4^{m-1}-1}{2}}. \tag{4.2.71}$$

Straub et al. have achieved remote axial scanning with a speed of 200 Hz by using the piezo-bimorph mirror [122]. Because the axial scan speed was limited by the mechanical resonance of the piezo bimorph, which depends on the size of the piezo-bimorph mirror, the axial scan speed will be increased by reducing its size. By using temporal focusing of amplified pulses (>1 μJ/pulse), Therrien et al. have achieved video-rate imaging over total areas as wide as 4800 μm^2 [123].

In contrast, according to Eq. (4.2.65), when $1/S_y$ is comparable to $|\Omega_g \sin\theta_i/c|$, the axial response is determined not only by the temporal focusing characterization $W(z)$ but also by the spatial focusing characterization $(4z^2M^4 + k_0^2S_y^4)$. The shift of the temporal focal plane by tuning the GDD results in the broadening of the excitation volume in the axial direction [121]. Therefore, temporal focusing and spatial focusing must be synchronized, and the axial scanning cannot be achieved by tuning the GDD. If we compensate for the GDD of the microscopic system, line illumination in MPEF microscopy can be achieved where the axial resolution is equal to that of conventional point-scanning MPEF microscopy [119]. Tal et al. acquired video-rate images by line-scanning TPEF microscopy using temporal focusing [120].

Figures 4.12(a) and (b) show the measured interferometric autocorrelation (IAC) signals at the focal plane of the water immersion objective lens with a NA of 1.1, and at 9 μm away from focal plane, respectively, as were obtained by temporal focusing using optical parametric oscillator pulses at a central wavelength of 1.27 μm. From these IAC signals, the pulse durations were estimated to be 145 fs at the focal plane and 1.64 ps at 9 μm away from focal plane. The measured pulse duration and the TPEF signals are plotted as a function of the axial position in Fig. 4.12(c) and (d), respectively. From these results, we see that temporal focusing provides optical cross-sectional capability.

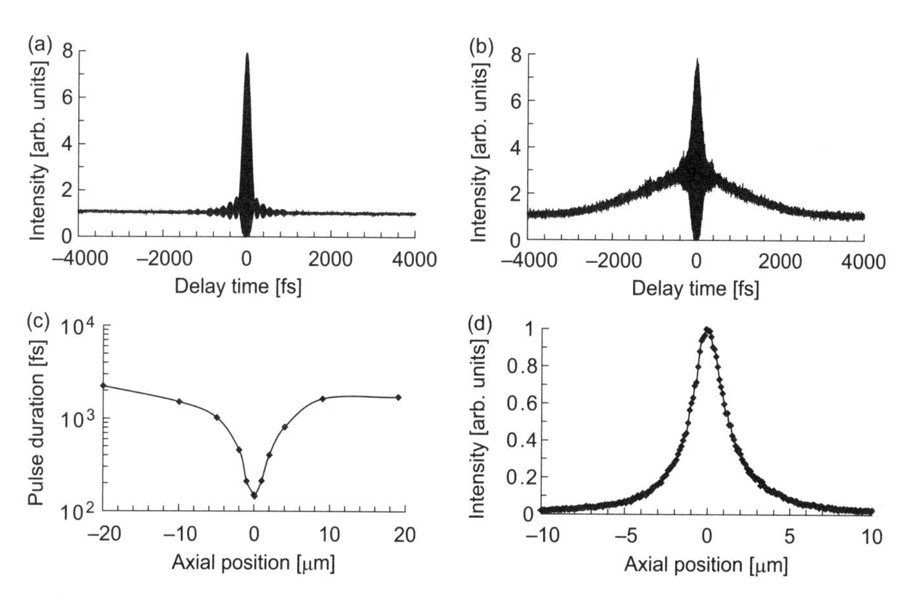

FIGURE 4.12 (a, b) Interferometric autocorrelation signals at the focal plane (a) at 9 μm away from focal plane (b). (c, d) Pulse duration (c) and TPEF signal (d) as a function of axial position.

4.3 ELECTRONIC RESONANCE IMAGING

4.3.1 Stimulated Parametric Emission Microscopy

Stimulated parametric emission (SPE) is an electronic-resonant four-wave mixing (FWM) process where SPE light at a frequency of $\omega_1 + \omega_2 - \omega_3$ is generated by the interaction between a sample and the lights, including the two pump lights at the frequency of ω_1 and ω_2 and the probe light at ω_3. The pump light at ω_1 can be also used as the second pump light at $\omega_2 (\omega_1 = \omega_2)$. The SPE process was first applied to microscopy by Isobe et al. [6]. By tuning the sum frequency $(\omega_1 + \omega_2)$ to specific electronic resonances within a sample, the SPE signal is enhanced. Thus, SPE microscopy allows imaging of absorbing molecules including fluorescent chromophores and nonfluorescent chromophores [6,124–129]. In the nonfluorescent chromophores in biological samples, beta-carotene, oxy-hemoglobin, deoxy-hemoglobin, melanin, and cytochromes are included. SPE signals from fluorophores persist in the presence of photobleaching phenomena [126].

The FWM signal may be composed of the nonresonant contribution $\chi^{(NR)}$, electronic resonance contribution $\chi^{(ER)}$ (SPE), and Raman resonant contribution $\chi^{(RR)}$; coherent anti-Stokes Raman scattering (CARS). Then, according to Eqs. (1.1.175) and (1.1.186), the FWM signal is proportional to the square modules of the total third-order susceptibility:

$$I_{\mathrm{FWM}}(\omega_1, \omega_2, \omega_3) \propto \left| \chi^{(\mathrm{NR})} + \chi^{(\mathrm{ER})}(\omega_1 + \omega_2) + \chi^{(\mathrm{RR})}(\omega_1 - \omega_3) \right|^2 I_1 I_2 I_3, \quad (4.3.1)$$

where

$$\chi^{(\mathrm{NR})} = (6 - 2f_{\mathrm{E}} - 2f_{\mathrm{R}})\chi^{(3)}, \quad (4.3.2)$$

$$\chi^{(\mathrm{ER})}(\omega_1 + \omega_2) = f_{\mathrm{E}}\chi^{(3)} \frac{\omega_{\mathrm{ER}}^2 + \Gamma_{\mathrm{ER}}^2}{\omega_{\mathrm{ER}}} \frac{1}{\omega_{\mathrm{ER}} - (\omega_1 + \omega_2) - i\Gamma_{\mathrm{ER}}}, \quad (4.3.3)$$

$$\chi^{(\mathrm{RR})}(\omega_1 - \omega_3) = f_{\mathrm{R}}\chi^{(3)} \frac{\omega_{\mathrm{RR}}^2 + \Gamma_{\mathrm{RR}}^2}{\omega_{\mathrm{RR}}} \frac{1}{\omega_{\mathrm{RR}} - (\omega_1 - \omega_3) - i\Gamma_{\mathrm{RR}}}. \quad (4.3.4)$$

Here ω_{ER} and ω_{RR} are the electronic resonant frequency and the Raman resonant frequency, and Γ_{ER} and Γ_{RR} is the electronic resonant bandwidth and the Raman resonant bandwidth (half width at half maximum), respectively. From Eq. (4.3.1), we see that the CARS and nonresonant FWM signals become significant background signals in SPE microscopy. The electronic resonant bandwidth Γ_{ER} (>100 cm^{-1}) is much broader than the the Raman resonant bandwidth Γ_{RR} (10 cm^{-1}). Thus, if the bandwidths of the pump pulses are much broader than the Raman resonant bandwidth, the SPE intensity is much larger than the CARS intensity. Of course, by detuning the difference frequency $\omega_1 - \omega_3$ from the Raman resonance frequencies, the CARS signals can be also removed. Figure 4.13(a) shows an SPE image of an anthracene powder in acetone. The signal is generated by the intrapulse FWM effect induced by electric field components with frequencies of ω_1, ω_2, and ω_3 in a single

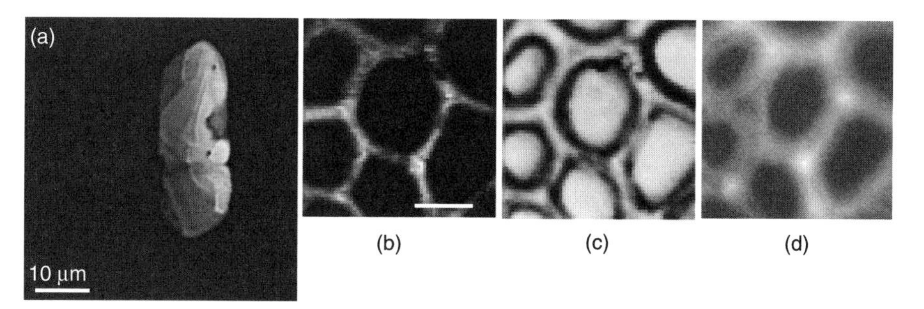

FIGURE 4.13 (a) SPE image of an anthracene powder in acetone. (b–d) Images of vascular bundles in the leaf of *Camellia sinensis* (b) SPE image. (c) Transmission image. (d) Autofluorescence image. Scale bar indicates 10 μm. Reprinted with permission from [6]. Copyright 2006 Optical Society of America.

broadband pulse with a pulse duration of 5 fs. Although the CARS signals based on CH (2900 cm^{-1}) stretching mode from acetone are included in Fig. 4.13(a), the SPE signal from two-photon resonance at 340 nm in the anthracene powder is 12 times higher than the CARS signal from acetone due to the broad bandwidth of the excitation pulse. Isobe et al. have applied SPE microscopy to visualize the cell wall. Figure 4.13(b) and (c) shows, respectively, an SPE image and a transmission image of vascular bundles in the 10-μm cross section of a leaf of *Camellia sinensis* with pump pulses at 800 nm and probe pulses at 1150 nm. Figure 4.13(d) shows a wide-field autofluorescence image of cell wall with the excitation at 436 nm (FWHM: 10 nm). These results indicate that SPE microscopy can detect the cell wall.

Although the SPE-to-CARS ratio can be enhanced by increasing the bandwidth of the excitation pulses, the SPE-to-nonresonant FWM ratio remains. In addition, a high spectral resolution in SPE signals cannot be expected because of the broad bandwidth of the electronic resonance. Thus, the discrimination of materials and molecules requires the detection of several independent spectral bands of SPE signals and the signal-processing techniques. The SPE signals are also frequently accompanied by significant two-photon excited fluorescence (TPEF) owing to the two-photon excitation of the broad spectral range of various electronic resonances. These problems have been solved by various techniques. The nonresonant FWM signals can be suppressed by polarization detection [126,127]. Multiplex SPE (M-SPE) microspectroscopy allows simultaneous multi-spectral SPE imaging and the identification of the molecules. The SPE signals can be separated from the TPEF signals by using interferometric detection [124]. These techniques are shown in the following sections.

4.3.1.1 Polarization Detection The nonresonant background signal can be suppressed by taking advantage of the differences in the polarization orientation properties of the nonresonant and electronic-resonant portions of the third-order polarization [126,127]. For isotropic media, $\chi^{(3)}_{ijkl}$ consists of 21 elements, where i, j, k, l indicate the Cartesian polarization directions of the signal (ω_4), pump1 (ω_1),

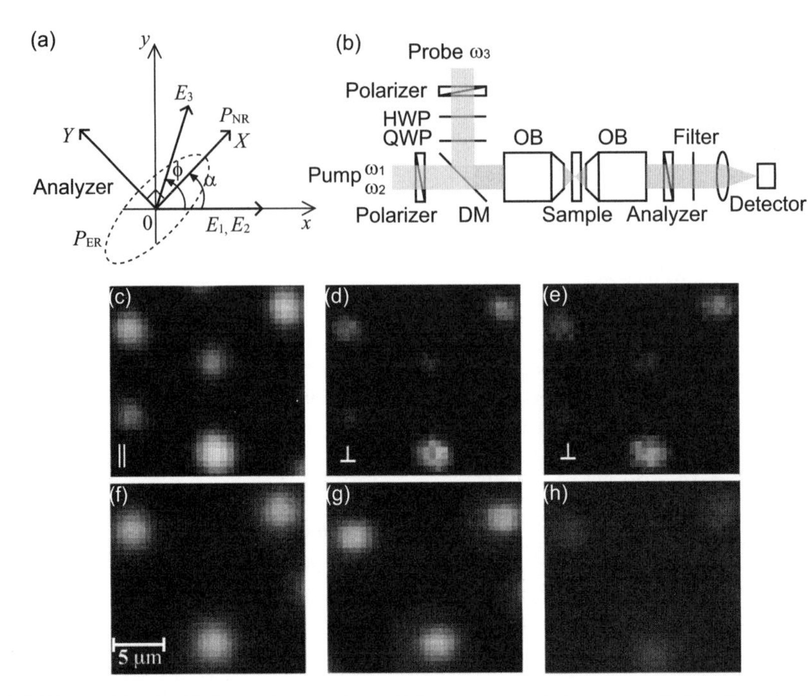

FIGURE 4.14 (a) Polarization vectors of the pump and the probe fields, the electronic resonant and nonresonant SPE signals, and the analyzer polarizer. (b) Schematic of the typical P-CARS microscope: HWP, half-wave plate; QWP, quarter-wave plate; DM, dichroic mirror; OB, objective lens. (c–h) SPE and TPEF images of the sample containing fluorescent beads and polystyrene beads. (c–e) SPE images obtained by the ordinary detection (c) and by the polarization detection (d, e) in alphabetical order. (f–h) The TPEF images obtained in alphabetical order. Reprinted with permission from [126]. Copyright 2009 Optical Society of America.

pump2 (ω_2), and probe (ω_3) fields, respectively. According to Eq. (1.1.120), only three elements are independent, and their sum is given by

$$\chi^{(3)}_{xxxx} = \chi^{(3)}_{xxyy} + \chi^{(3)}_{xyxy} + \chi^{(3)}_{xyyx}. \tag{4.3.5}$$

The two pump beams are linearly polarized along the x-axis, and the probe beam is linearly polarized along a direction at an angle of ϕ relative to the x-axis, as shown in Fig. 4.14(a). We assume that $\omega_1 + \omega_2$ is resonant with an electronic transition. The x and y components of the nonlinear polarization at frequency $\omega_4 = \omega_1 + \omega_2 - \omega_3$ is given by

$$P_x^{(\mathrm{SPE})} = \varepsilon_0 \left(\chi^{(\mathrm{NR})}_{xxxx} + \chi^{(\mathrm{ER})}_{xxxx}\right) E_1 E_2 E_3^* \cos\phi, \tag{4.3.6}$$

$$P_y^{(\mathrm{SPE})} = \varepsilon_0 \left(\chi^{(\mathrm{NR})}_{yxxy} + \chi^{(\mathrm{ER})}_{yxxy}\right) E_1 E_2 E_3^* \sin\phi, \tag{4.3.7}$$

where $\chi^{(\mathrm{NR})}$ and $\chi^{(\mathrm{ER})}$ are nonresonant and Raman resonant components of third-order susceptibility, respectively. Because the nonresonant component $\chi^{(\mathrm{NR})}$ is a real quantity that is independent of frequency, the direction of the nonlinear polarization arising from the nonresonant term $\chi^{(\mathrm{NR})}$ makes an angle

$$\alpha = \tan^{-1}\left[\frac{\chi_{yxxy}^{(NR)}}{\chi_{xxxx}^{(NR)}}\tan\phi\right], \tag{4.3.8}$$

with the x-axis. Let X and Y be the axes parallel and perpendicular to the nonresonant part of the nonlinear polarization. The nonlinear polarization components along X and Y are written by

$$P_X^{(\mathrm{SPE})} = \varepsilon_0\left[\chi_{xxxx}^{(NR)}\frac{\cos\phi}{\cos\alpha} + \chi_{xxxx}^{(\mathrm{ER})}\cos\phi\cos\alpha\,(1+\rho_{\mathrm{ER}}\tan\phi\tan\alpha)\right]E_1E_2E_3^*, \tag{4.3.9}$$

$$P_Y^{(\mathrm{SPE})} = -\varepsilon_0\chi_{xxxx}^{(\mathrm{ER})}\cos\phi\sin\alpha\left(1-\frac{\rho_{\mathrm{ER}}\tan\phi}{\tan\alpha}\right)E_1E_2E_3^*, \tag{4.3.10}$$

where $\rho_{\mathrm{ER}} = \chi_{yxxy}^{(\mathrm{ER})}/\chi_{xxxx}^{(\mathrm{ER})}$ is the depolarization ratio of the resonant SPE field. In transparent media, the nonresonant susceptibility follows Kleinman symmetry $\chi_{xxxx}^{(\mathrm{NR})} = 3\chi_{yxxy}^{(\mathrm{NR})}$. Then, we obtain

$$3\tan\alpha = \tan\phi, \tag{4.3.11}$$

and Eqs. (4.3.9) and (4.3.10) can be rewritten by

$$P_X^{(\mathrm{SPE})} = \varepsilon_0\left[\chi_{xxxx}^{(\mathrm{NR})}\frac{\cos\phi}{\cos\alpha} + \chi_{xxxx}^{(\mathrm{ER})}\cos\phi\cos\alpha\,(1+3\rho_{\mathrm{ER}})\right]E_1E_2E_3^*, \tag{4.3.12}$$

$$P_Y^{(\mathrm{SPE})} = -\varepsilon_0\chi_{xxxx}^{(\mathrm{ER})}\cos\phi\sin\alpha\,(1-3\rho_{\mathrm{ER}})\,E_1E_2E_3^*. \tag{4.3.13}$$

Because the resonant component $\chi^{(\mathrm{ER})}$ is complex, we notice immediately from Eqs. (4.3.12) and (4.3.13) that the SPE signal is elliptically polarized. Thus, by using an analyzer along Y to detect the SPE signal, we can effectively increase the resonant-to-nonresonant signal ratio. Theoretically, the signal from $P_Y^{(\mathrm{SPE})}$ is free of nonresonant background. In practice, there exists a residual background because of the birefringence of the optics in the beam path and the scrambling of polarization at the tight focus. For a weak resonance with $|\chi^{(\mathrm{ER})}| \ll |\chi^{(\mathrm{NR})}|$, the $\chi^{(\mathrm{ER})}$ term in Eq. (4.3.12) can be neglected. Then, we can write

$$\left(\frac{P_Y^{(\mathrm{SPE})}}{P_X^{(\mathrm{SPE})}}\right)^2 \cong \left\{\frac{(1-3\rho_{\mathrm{ER}})\,\chi_{xxxx}^{(\mathrm{ER})}\sin 2\alpha}{2\chi_{xxxx}^{(\mathrm{NR})}}\right\}^2. \tag{4.3.14}$$

According to Eq. (4.3.14), it is clear that the maximum electronic resonant-to-nonresonant signal ratio is obtained for $\alpha = 45^\circ$. From Eq. (4.3.11), the optimal value for the angle ϕ is then 71.6°. The schematic of typical polarization-SPE (P-SPE) microscope is shown in Fig. 4.14(b). Both the pump and the probe beams are linearly polarized. The polarization direction of the probe beam is adjustable with a half-wave plate. The angle ϕ is set at 71.6°. A quarter-wave plate is used to compensate for the birefringence in the probe field that is induced by the dichroic mirror. The polarization direction of the detection beam is set at 135° by an analyzer.

Liu et al. have successfully demonstrated nonresonant background suppression by using the P-SPE technique [126]. Figures 4.14(c–h) shows, the SPE and TPEF images of the samples containing fluorescent (electronic resonant) beads and polystyrene (nonresonant) beads. As shown in Fig. 4.14(c), the SPE image obtained by the ordinary detection shows all beads because both electronic resonant and nonresonant signals are detected. The strong SPE signals from the fluorescent beads and the weak SPE signals from the polystyrene beads are measured by the polarization detection as illustrated in Fig. 4.14(d). Note that the TPEF signals are readily photobleached, while the resonant SPE signals are persistent in the presence of photobleaching.

4.3.1.2 *Multiplex SPE Microspectroscopy* Recall the energy level diagram of the SPE process with a multiples excitation scheme given in Section 3.2. A broadband light source is used as the pump pulses for simultaneously exciting various electronic resonances. A narrowband light source is employed as the probe pulse for spectrally resolving various electronic resonances. The spectral features of the resultant SPE signals reflect the electronic excitation energy levels of samples. The spectral resolution of the SPE spectrum is proportional to the spectral bandwidth of the probe pulse. Because the electronic resonance bandwidth is typically on the order of 100 cm^{-1}, 100-fs laser pulses with a bandwidth of 100 cm^{-1} are adequate for the probe pulse. Since a high spectral resolution in SPE spectra cannot be executed, to discriminate among the molecules, the normalized spectral correlation method is applied. The value C_{SP} of the normalized spectral correlation is defined as

$$C_{\mathrm{SP}} = \frac{\frac{1}{N}\sum_{j=0}^{N} S_{\mathrm{s}}(\lambda_j)S_{\mathrm{r}}(\lambda_j)}{\sqrt{\frac{1}{N}\sum_{j=0}^{N} S_{\mathrm{s}}^2(\lambda_j)}\sqrt{\frac{1}{N}\sum_{j=0}^{N} S_{\mathrm{r}}^2(\lambda_j)}}, \qquad (4.3.15)$$

where $S_{\mathrm{s}}(\lambda_j)$ and $S_{\mathrm{r}}(\lambda_j)$ represent the SPE spectrum from the sample and reference, respectively. The value of C_{SP} provides a quantitative comparison between the sample and reference spectra regardless of the signal intensity. Although information about signal intensity is lost due to the use of a normalized SPE spectrum, normalized spectral correlation analysis allows us to identify regions of specific molecules by way of the similarity of the SPE spectrum.

Isobe et al. demonstrated multiplex SPE microspectroscopy using the combination of a broadband pump with pulses ranging from 750 to 850 nm and a narrowband probe pulse at 1060 nm. Figures 4.15(a), (b), and (c) shows, the TPEF and SPE images of mouse kidney cells stained with hematoxylin and eosin, which are constructed

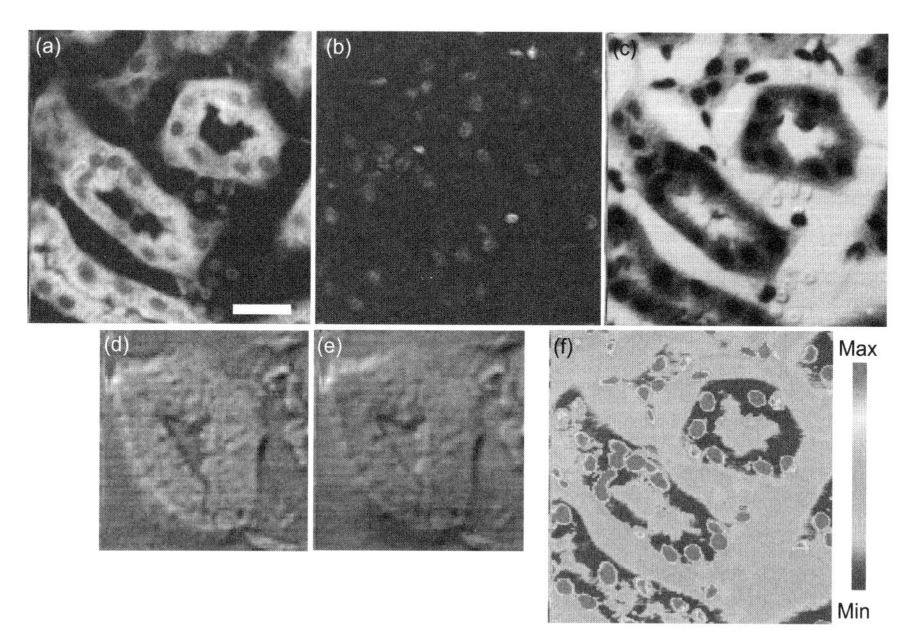

FIGURE 4.15 (a–c) Multi-spectral images of mouse kidney cells stained with hematoxylin and eosin in the spectral ranges of 530 to 590 nm (a), 590 to 620 nm (b), and 650 to 680 nm (c). (d, e) Multi-spectral images of unstained mouse kidney cells in the spectral ranges of 590–620 nm (d) and 650 to 680 nm (e). (f) The reconstructed image after the normalized spectral correlation process using the SPE spectrum from the nucleus as the reference spectrum. Scale bar is 20 μm. (*See insert for color representation of the figure.*)

from the integration in spectral ranges of 530 to 590 nm, 590 to 620 nm, and 650 to 680 nm, respectively. Hematoxylin, which is a popular absorbing dye used to stain nuclei, has a broad absorption band ranging from the ultraviolet region to the visible region. Eosin, which has an absorption band in the visible region at a peak wavelength of 516 nm, is a red fluorescent dye used to stain cellular cytoplasms. The signal ranging from 530 to 590 nm is two-photon excited fluorescence (TPEF) from eosin, which is generated by the probe pulse. Figure 4.15(d) and (e) presents FWM images of unstained mouse kidney cells, which are constructed from the integration in spectral bands of 590–620 nm, and 650–680 nm, respectively. Although the frequency difference between the pump and probe pulses are ranged from 2700 to 3700 cm^{-1} and CH (2900 cm^{-1}) and OH (3200–3800 cm^{-1}) stretching modes correspond to the SPE wavelengths of 665 nm (CH) and 630–586 nm (OH), the wavelength dependence based on Raman resonance does not appear in Fig. 4.15(d) and (e) because of the broad bandwidth of pump pulse. These results indicate that the enhancement by two-photon and three-photon electronic resonances in hematoxylin leads to the image contrast seen in Fig. 4.15(b), while linear absorption of the generated SPE fields in hematoxylin and eosin results in that in Fig. 4.15(c). Figure 4.15(f) shows the image of mouse kidney cells stained with hematoxylin and eosin, which was reconstructed after the normalized spectral correlation process using the SPE spectrum from the nucleus as the reference spectrum. Note that the normalized spectral correlation provides identification of the nuclei.

4.3.1.3 Interferometric Detection By taking advantage of the difference between coherences of the SPE and TPEF processes, the SPE signals can be extracted by interferometric detection where a small SPE field from samples is mixed with a strong local oscillator (LO) field [124]. It is also known that this technique allows the shot-noise limited detection. In the spectral interference method, an SPE-field $E_{\mathrm{SPE}}^{(\mathrm{sample})}(\omega)$ from the sample interferes with an LO field $E_{\mathrm{local}}(\omega)$, which is generated from a nonresonant sample or is spectrally selected from a broadband excitation light source. Then, the interference signal including the TPEF signal $I_{\mathrm{TPEF}}(\omega)$ from the sample is given by

$$S_{\mathrm{i}}(\omega) = \left|E_{\mathrm{SPE}}^{(\mathrm{sample})}(\omega)\right|^2 + |E_{\mathrm{local}}(\omega)|^2 + I_{\mathrm{TPEF}}(\omega) \tag{4.3.16}$$

$$+2\left|E_{\mathrm{SPE}}^{(\mathrm{sample})}(\omega)\right| |E_{\mathrm{local}}(\omega)| \cos\phi_{\mathrm{i}}(\omega), \tag{4.3.17}$$

where $\phi_{\mathrm{i}}(\omega)$ is introduced by

$$\phi_{\mathrm{i}}(\omega) = \phi_{\mathrm{sample}}^{(\mathrm{ER})}(\omega) + \omega\tau + \phi_0(\omega). \tag{4.3.18}$$

Here $\phi_{\mathrm{sample}}^{(\mathrm{ER})}(\omega)$ is the phase of the SPE signal from the sample, which represents the phase of $\chi^{(\mathrm{ER})}$, $\phi_0(\omega)$ is the relative phase delay due to the optical components, and τ is the temporal delay between the two pulses of the SPE signal from the sample and the LO signals. If the phase $\phi_{\mathrm{sample}}^{(\mathrm{ER})}(\omega)$ can be obtained, the real and imaginary parts of $\chi^{(\mathrm{ER})}$, $\phi_0(\omega)$ can be easily obtained. The total phase information $\phi_{\mathrm{i}}(\omega)$ is extracted from the spectral interferogram. According to Eq. (4.3.17), a fringe pattern is produced in the spectral range of SPE signals. Thus, the TPEF signals are not included in the signals extracted through the Fourier filtering. The interferometric technique in the time domain also makes it possible to isolate SPE signals from the TPEF signals.

Isobe et al. demonstrated the separation of SPE signals from TPEF signals [124]. Figure 4.16(a) shows the emission spectrum obtained by ordinary detection from a mixed dye solution (Coumarin 120 and Coumarin 153). The SPE spectrum (570–640 nm) is overlapped and almost obscured by the TPEF spectrum (480–640 nm) from Coumarin 153. Figure 4.16(b) shows the spectral interferogram with the LO field. Note the fringe pattern that appears in the spectral region of the SPE signal due to the added LO field. As shown in Fig. 4.16(c), Fourier filtering through the interferometric detection enables the TPEF signals to be removed without losing SPE signals. Isobe et al. applied interferometric imaging to mouse kidney cells whose nuclei and cellular cytoplasms were stained with hematoxylin and eosin, respectively. Figure 4.16(d) shows the SPE image obtained with ordinary detection. Since the TPEF signals from the eosin are mixed with the SPE signals from the hematoxylin, the nuclei and cellular cytoplasms cannot be identified. Figure 4.16(e) and (f) shows the SPE and TPEF images that are simultaneously obtained with the

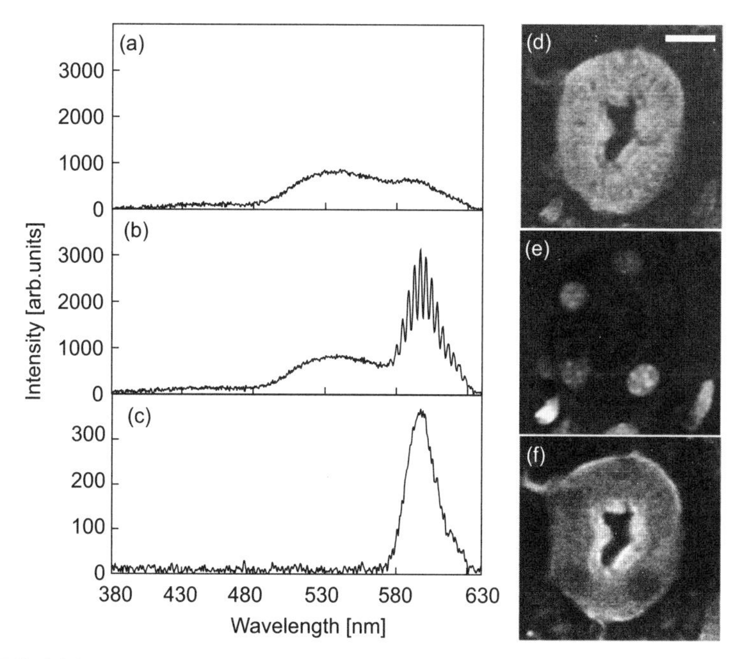

FIGURE 4.16 (a–c) Emission spectra from the mixed dye solution (Coumarin 120 and Coumarin 153). (a) Emission spectrum in the ordinary detection. (b) Spectral interferogram. (c) Extracted SPE spectrum by interferometric detection. Reprinted with permission from [124]. Copyright 2006 Optical Society of America. (d–f) Simultaneously obtained images of mouse kidney cells stained with hematoxylin and eosin. (d) Ordinary SPE image including the TPEF signal. (e) SPE image extracted by the interferometric detection. (f) TPEF image. Scale bar indicates 10 μm.

interferometric detection. It should be noted that the interferometric detection allows simultaneous SPE and TPEF imaging without cross talk.

4.3.2 Two-Photon Absorption Microscopy

The two-photon absorption (TPA) process can be induced by two photons whose sum frequency is tuned to the electronic transition frequency. The excitation probability of the TPA process is proportional to the product of the intensities of the two pump pulses. This nonlinearity provides optical sectioning capability. The TPA cross section is related to the imaginary part of the electronic resonant contribution in the third-order nonlinear susceptibility, and the TPA signals reflect the energy level structure, transition dipole moments, and relaxation time constants. Thus, TPA microscopy allows label-free imaging of nonfluorescent chromophores [9,10], which is similar to SPE microscopy. The SPE signals include a high nonresonant FWM background, where as TPA microscopy is free of the nonresonant background.

In the case of the use of an amplifier laser system, which produces high-peak intensity pulses, the significant intensity loss of the excitation pulses in TPA process

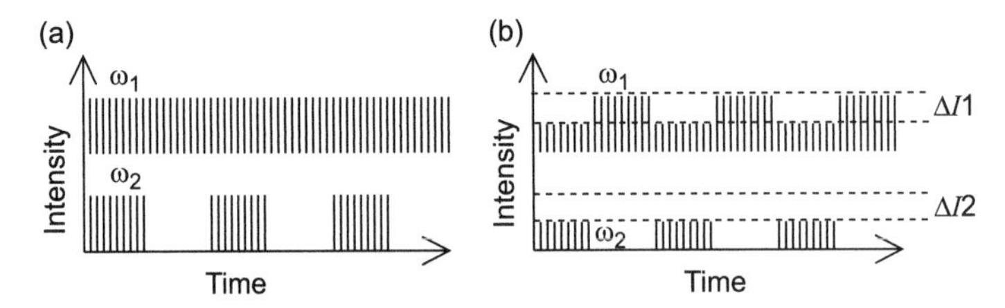

FIGURE 4.17 (a) Intensity modulation of the excitation pulses. (b) Modulation transfer based on loss experienced by TPA.

can be observed with a detector having moderate sensitivity. Although the amplified laser pulses produce a large TPA loss signal, it is not suitable for bioimaging because the excessive peak power can cause sample damage, and the low repetition rate limits the image acquisition speed. If an oscillator laser system is employed as an excitation light source, the TPA loss signal is much lower than the excitation intensity. The small TPEF signal can be separated from the strong excitation light by means of spectral filtering and can be detected without background light. However, TPEF cannot be used for imaging nonfluorescent samples. In contrast the weak TPA loss signal cannot be spectrally discriminated from the excitation light. Thus, the small TPA signals are typically buried in the laser $1/f$ noise. This problem has been solved by the modulation transfer technique, which was first proposed by Tian et al. [130]. In the modulation transfer technique, the nonlinear optical process is induced by the combination of two (color) pulses at ω_1 and ω_2. As shown in Fig. 4.17, the intensity of one of the two pulses is modulated at a high frequency f and the modulation transfer to the other pulse is measured with the lock-in detection at a frequency of f. With the increasing modulation frequency f, the laser $1/f$ noise is reduced. If the wavelength of one of the two pulses corresponds to that of the other pulse, the intensities of one and the other of the two pulses are modulated at a high frequency f_1 and f_2, respectively, the TPA loss can be detected with the lock-in detection at a frequency of $f_1 - f_2$ or $f_1 + f_2$. The modulation transfer technique can be applied to the measurement of the loss and gain experienced by various nonlinear optical processes. However, the nonlinear signals generated by other nonlinear optical, processes, including cross-phase modulation, may be mixed in the modulation transfer technique. Thus, the modulation transfer technique must be carefully employed.

Now we discuss the signal-to-noise ratio (SNR) of TPA microscopy. Let us assume low incident intensities and/or a short length of the nonlinear medium. Then, from Eqs. (1.1.231) and (1.1.232), the intensity changes of the pump pulses at ω_1 and ω_2 by TPA are described as

$$\Delta I_1 = -\frac{\omega_1 L}{\varepsilon_0 \tilde{n}^{(1)}(\omega_1)\tilde{n}^{(1)}(\omega_2)c^2}\mathrm{Im}\{\tilde{\chi}^{(\mathrm{ER})}(\omega_1+\omega_2)\}I_1 I_2, \tag{4.3.19}$$

$$\Delta I_2 = -\frac{\omega_2 L}{\varepsilon_0 \tilde{n}^{(1)}(\omega_1)\tilde{n}^{(1)}(\omega_2)c^2}\mathrm{Im}\{\tilde{\chi}^{(\mathrm{ER})}(\omega_1+\omega_2)\}I_1 I_2, \tag{4.3.20}$$

where

$$\chi^{(ER)}(\omega_1+\omega_2) = f_E \chi^{(3)} \frac{\omega_{ER}^2+\Gamma_{ER}^2}{\omega_{ER}} \frac{1}{\omega_{ER}-(\omega_1+\omega_2)-i\Gamma_{ER}}. \quad (4.3.21)$$

Here $\mathrm{Im}\{h\}$ denotes the imaginary part of the complex $h = \mathrm{Re}\{h\} + i\mathrm{Im}\{h\}$, and I_1 and I_2 are the intensities of the pump pulses at ω_1 and ω_2, respectively. The SNR of TPA microscopy is given by

$$SNR = \frac{\Delta I_1}{a(f)I_1 + \sqrt{I_1}}. \quad (4.3.22)$$

Here the first term in denominator of Eq. (4.3.22) denotes the laser intensity noise of the pump light at ω_1, and the second term is the shot noise of the pump light at ω_1. When the pump intensity at ω_2 is modulated at a high frequency and the pump intensity at ω_1 is detected through the lock-in detection at f, the shot-noise-limit sensitivity can be achieved. Then, the SNR of TPA microscopy becomes

$$SNR = \frac{\mathrm{Im}\{\tilde{\chi}^{(ER)}(\omega_1+\omega_2)\}\,\omega_1 L\sqrt{I_1}I_2}{\varepsilon_0 \tilde{n}^{(1)}(\omega_1)\tilde{n}^{(1)}(\omega_2)c^2}, \quad (4.3.23)$$

Warren et al. first demonstrated the use of the modulation transfer technique for TPA microscopy [9]. They successfully achieved TPA imaging of human melanoma. TPA microscopy provides contrast mechanisms for nonfluorescent chromophores including beta-carotene, oxy-hemoglobin, deoxy-hemoglobin,melanin, and cytochromes.

Pump-probe microscopy using the modulation transfer technique was also reported for imaging based on excited-state absorption (ESA) as shown in Fig. 4.18(a) [10]. Since the excited lifetimes of those nonfluorescent chromophores are less than 1 ps, the pulse duration of the pump (ω_1) and probe (ω_2) pulses have to be less than a few hundred femtoseconds in order for the transient excited states to be investigated effectively. The probe pulse is delayed by a few hundred femtoseconds with respect to the pump pulse to permit the molecule enough time to relax vibrationally on the electronic excited state [10]. Compared with TPA via an intermediate virtual state, ESA can be significantly enhanced by the one-photon resonance between a real intermediate electronic state and the pump pulse. Fu et al. demonstrated ESA imaging based on differences in excited-state decay dynamics between oxy-hemoglobin and deoxy-hemoglobin in a single layer of mouse red blood cells fixed with methanol on a glass slide as shown in Fig. 4.18(b) [10]. Individual red blood cells with endogenous hemoglobin contrast are successfully obtained at micrometer resolution in the image.

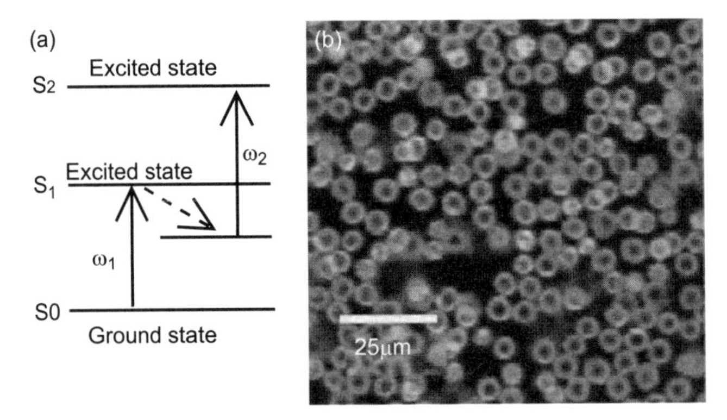

FIGURE 4.18 (a) Energy diagram of the excited-state absorption process. (b) Two-color TPA image of mouse red blood cells with endogenous hemoglobin using a 775 nm pump and 650 nm probe, and with a delay time of 100 fs. Reprinted with permission from [10]. Copyright 2007 Optical Society of America.

4.4 VIBRATIONAL IMAGING

4.4.1 Coherent Anti-Stokes Raman Scattering Microscopy

In spontaneous Raman scattering, a pump light at frequency ω_p illuminates the samples, and signals are generated at the Stokes frequency $\omega_S = \omega_p - \omega_{RR}$ and anti-Stokes frequency $\omega_{as} = \omega_p + \omega_{RR}$, where ω_{RR} is the Raman transition frequency. Coherent anti-Stokes Raman scattering (CARS) is a Raman resonant FWM process in which the CARS light at a frequency of $\omega_p + \omega_{pr} - \omega_s$ is generated by the interaction between a sample and the lights, including the pump light at a frequency of ω_p, the Stokes light at ω_s, and the probe light at ω_{pr}. By tuning the frequency difference ($\omega_p - \omega_s$) to specific vibrational resonances (ω_{RR}) within a sample, the CARS signal is enhanced [133–136]. Duncan et al. were the first to apply the CARS process to microscopy in 1982 [7]. Due to technical difficulties, there were no further developments until 1999, when CARS microscopy was revived by Zumbusch et al. [8]. Hashimoto et al. developed CARS microscopy in fingerprint region in 2000 [132]. Since then, CARS microscopy has been applied to imaging living cells with various vibrational modes because of the development of laser technology. CARS microscopy provides image-contrasts that are derived from vibrations of the chemical compositions and thermodynamic states of the sample. The vibrational contrast in bio-imaging is based on phosphate stretch vibration for deoxyribonucleic acid (DNA) [133], amide I vibration for protein [134], OH stretching vibration for water [135], and the CH group of stretching vibrations for lipids [8,136].

Figure 4.19(a) shows spontaneous Raman spectra for the cytoskeleton, endoplasmic reticulum, mitochondria, Golgi apparatus, and nucleus of LN-18 cells in phosphate-buffered saline (PBS) [137]. The broad band observed at ~1100 cm^{-1} originates from the glass substrate. The Raman spectra include H_2O solvent (1600 cm^{-1} and 3000–3700 cm^{-1}) and aliphatic C–H stretch modes of viral protein and DNA

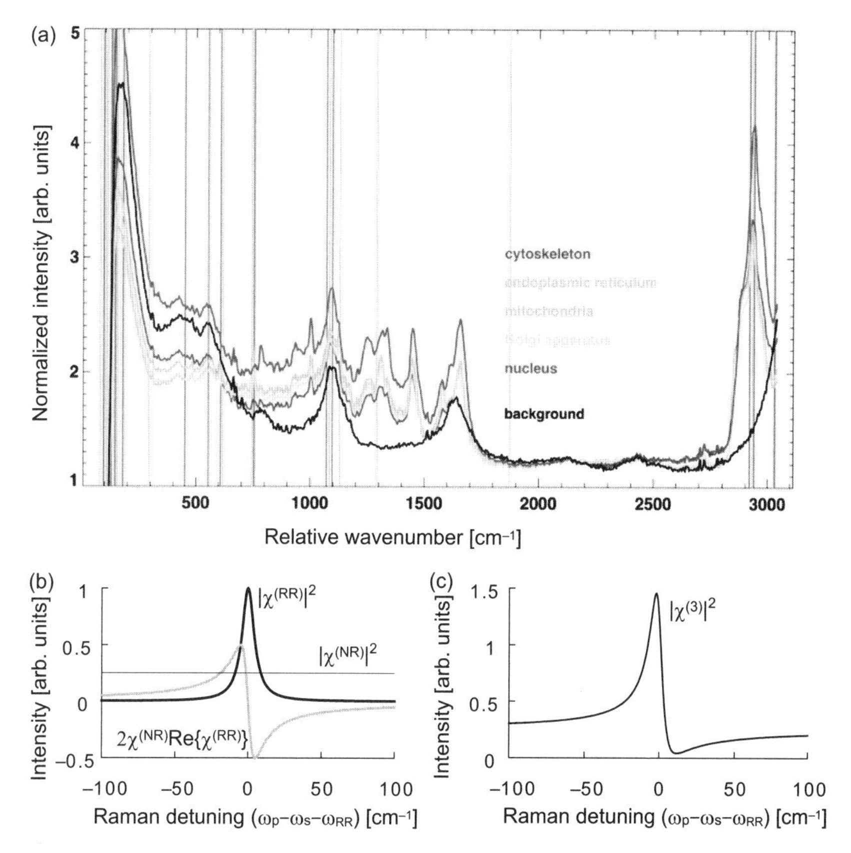

FIGURE 4.19 (a) Spontaneous Raman spectra for cytoskeleton, endoplasmic reticulum, mitochondria, Golgi apparatus, and nucleus of LN-18 cells in phosphate-buffered saline (PBS). Reprinted with permission from [137]. Copyright 2012 Elsevier. (b) Raman resonant contribution, nonresonant contribution, and mixing term of the nonresonant and resonant contributions to the CARS spectrum. (c) Typical CARS spectrum. (*See insert for color representation of the figure.*)

(2900 cm^{-1}). Most structurally informative Raman bands appear in the 600 to 1800 cm^{-1} range. Thomas reported representative Raman markers of dG (681 cm^{-1}), DNA backbone (830 cm^{-1} and 1093 cm^{-1}), Trp (878 cm^{-1}), Phe (1003 cm^{-1}), dA and dG (1490 cm^{-1} and 1579 cm^{-1}), Tyr and Trp (1618 cm^{-1}), and protein amide I (1666 cm^{-1}) in a spontaneous Raman spectrum of P22 virus in H_2O buffer [138]. Spontaneous Raman microscopy requires high laser powers and long integration times per pixel in images owing to its low sensitivity. Thus, spontaneous Raman microscopy has been limited in its application to the study of living biological systems.

The sensitivity of CARS microscopy is much higher than that of spontaneous Raman microscopy. However, in CARS microscopy, the resonant CARS signal is accompanied by significant nonresonant FWM signals from the electronic contribution

[136]. Without electronic resonance, according to Eq. (1.1.175), the total third-order susceptibility is

$$\tilde{\chi}^{(3)}(\omega) = \chi^{(\mathrm{NR})} + \tilde{\chi}^{(\mathrm{RR})}(\omega), \tag{4.4.1}$$

where $\chi^{(\mathrm{NR})}$ and $\tilde{\chi}^{(\mathrm{RR})}$ are nonresonant and Raman resonant components of a third-order susceptibility, and are given by

$$\chi^{(\mathrm{NR})} = (6 - 2f_{\mathrm{R}})\chi^{(3)}, \tag{4.4.2}$$

$$\tilde{\chi}^{(\mathrm{RR})}(\omega_{\mathrm{p}} - \omega_{\mathrm{s}}) = f_{\mathrm{R}}\chi^{(3)}\frac{\omega_{\mathrm{RR}}^2 + \Gamma_{\mathrm{RR}}^2}{\omega_{\mathrm{RR}}}\frac{1}{\omega_{\mathrm{RR}} - (\omega_{\mathrm{p}} - \omega_{\mathrm{s}}) - i\Gamma_{\mathrm{RR}}}. \tag{4.4.3}$$

Here ω_{RR} is the Raman resonant frequency, and Γ_{RR} is the Raman resonant bandwidth (half width at half maximum). The total CARS intensity, including the nonresonant FWM signal, is proportional to the square modules of the total third-order susceptibility

$$\begin{aligned} I_{\mathrm{CARS}}(\omega) &\propto \left|\chi^{(\mathrm{NR})} + \tilde{\chi}^{(\mathrm{RR})}(\omega)\right|^2 \\ &= \left|\chi^{(\mathrm{NR})}\right|^2 + \left|\tilde{\chi}^{(\mathrm{RR})}(\omega)\right|^2 + 2\chi^{(\mathrm{NR})}\mathrm{Re}\left\{\tilde{\chi}^{(\mathrm{RR})}(\omega)\right\}. \end{aligned} \tag{4.4.4}$$

The CARS spectrum is composed of the nonresonant contribution, the resonant contribution, and a mixing term of the nonresonant and resonant contributions as shown in Fig. 4.19(b) and 4.19(c). The CARS spectrum exhibits a peak shift, dispersive shape and nonresonant background. In contrast, the spontaneous Raman signal is proportional to the imaginary part of $\tilde{\chi}^{(\mathrm{RR})}$:

$$I_{\mathrm{Raman}}(\omega) \propto \mathrm{Im}\left\{\tilde{\chi}^{(\mathrm{RR})}(\omega)\right\}. \tag{4.4.5}$$

For weak vibrational resonances and a low concentration of Raman active molecules in the biological samples, the nonresonant background may overwhelm the Raman resonant information. Thus, suppression of the nonresonant FWM signals is essential for bio-imaging. To increase the resonant-to-nonresonant signal ratio, two synchronized picosecond pulse lasers operating at a bandwidth of 10 cm^{-1} are often employed as an excitation light source [139] because Raman bandwidth is typically on the order of 10 cm^{-1}. In addition, the nonresonant FWM signals can be suppressed by various detection techniques, including epi-detection [140–142], polarization detection [142,134,143,144], interferometric detection [145,146], and time-resolved detection [147–151]. Multiplex CARS technique makes it possible to separate the resonant and nonresonant CARS signals by analyzing the spectral profile of the CARS signal [152–157].

4.4.1.1 Epi-Detection Pulses should be preferably tightly focused with a high numerical-aperture (NA) objective lens because the phase-matching condition in the FWM process is markedly relaxed owing to the wide cone of wave vectors and the small interaction length. When the object size is increased, the CARS signals add up in phase along the optical axis in the forward direction because of the small wave vector mismatch Δk_F, and they are reduced in the backward by destructive interference because of the large wave vector mismatch Δk_B. This constructive interference for large objects results in strong highly directional signals in the forward direction. For objects smaller than a quarter wavelength, since destructive interference in the backward direction is incomplete, the CARS waves are radiated in both the forward and the backward directions. Thus, forward-detected CARS (F-CARS) microscopy provides image contrast of large objects whereas epi-detected CARS (E-CARS) microscopy gives image contrast of small objects. In practice, the object size is not the sole criterion for measuring the E-CARS signals because small objects in biological samples are surrounded with an aqueous medium. The signals from the surrounding aqueous medium are commonly a nonresonant contribution $\chi^{(NR)}_{\text{medium}}$, while those from the small objects are both the Raman resonant $\chi^{(RR)}_{\text{target}}$ contribution and the nonresonant $\chi^{(NR)}_{\text{target}}$ contribution. If the nonlinear susceptibility of the small objects $\chi^{(RR)}_{\text{target}} + \chi^{(NR)}_{\text{target}}$ is different from that of its surrounding medium $\chi^{(NR)}_{\text{medium}}$, the destructive interference in the backward direction is incomplete, and the E-CARS signals can be detected. Thus, E-CARS microscopy enables the nonresonant contribution from the surrounding aqueous medium to be suppressed. However, the nonresonant contribution of small objects cannot be limited in E-CARS microscopy. E-CARS microscopy can be used for vibrational imaging with strong resonant CARS signals, for example, imaging of live cells based on the C–H stretching vibration because of the high density of the C–H modes.

Cheng et al. reported a far-field CARS radiation field pattern [140]. The signals are calculated with the tightly focused (NA = 1.4) incident beams copropagating along the $+z$-axis and polarized along the x-axis. The radiation field is polarized along the x-axis. Figure 4.20(a) shows a far-field CARS radiation pattern from spherical scatterers centered at the focus with different diameters [142]. When diameter D of the scatterer is much smaller than the pump wavelength λ_p, the CARS waves are radiated in both the forward and backward directions. In contrast, with increasing sample size, the CARS radiation is confined in a small cone angle in the forward direction. Figure 4.20(b) shows the far-field CARS radiation pattern from scatterers centered at the focus with different shapes (a rod with a diameter of 0.2 λ_p and an axial length of 2.0 λ_p, a sphere with a diameter of 0.78 λ_p, and a disk with a diameter of 0.89 λ_p and a thickness of 0.1 λ_p) [142]. Although the three samples have the same volume, their CARS radiation patterns and intensities are quite different.

Cheng et al. [140] and Volkmer et al. [141] demonstrated the suppression of the nonresonant background by E-CARS microscopy. Figure 4.20(c) and (d) shows F-CARS and E-CARS images of a NIH 3T3 cell in metaphase, respectively [133]. The forward detection allows CARS imaging with low excitation power. However, the F-CARS signals from the cellular components are overwhelmed by the large nonresonant background due to water. The E-CARS greatly improves the image contrast by efficiently suppressing the nonresonant background thus providing high contrast for the vesicles surrounding the chromosomes.

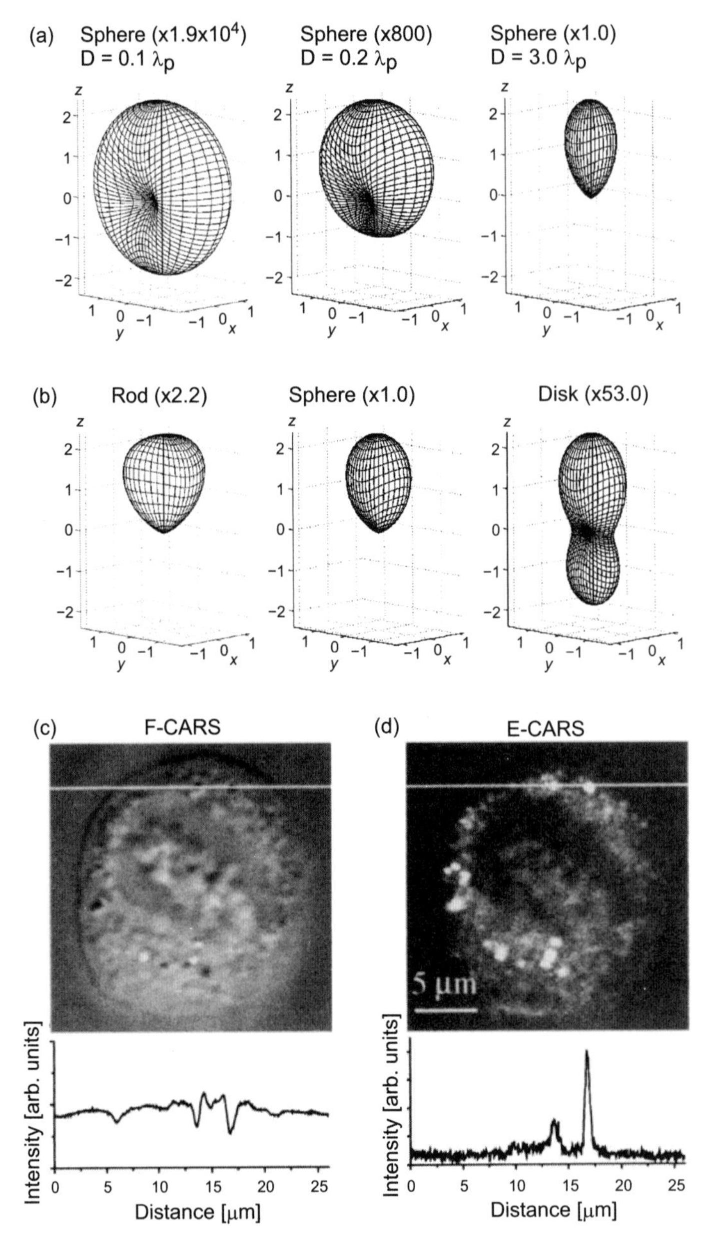

FIGURE 4.20 (a, b) Far-field CARS radiation pattern from spherical scatterers centered at the focus with different diameters (a) and from scatterers centered at focus with the same volume but different shapes (b). Reprinted with permission from [142]. Copyright 2004 American Chemical Society. (c, d) F-CARS (c) and E-CARS (d) images of a NIH 3T3 cell in metaphase with $\omega_p - \omega_s$ tuned to 2873 cm^{-1}. Reprinted with permission from [133]. Copyright 2002 Elsevier.

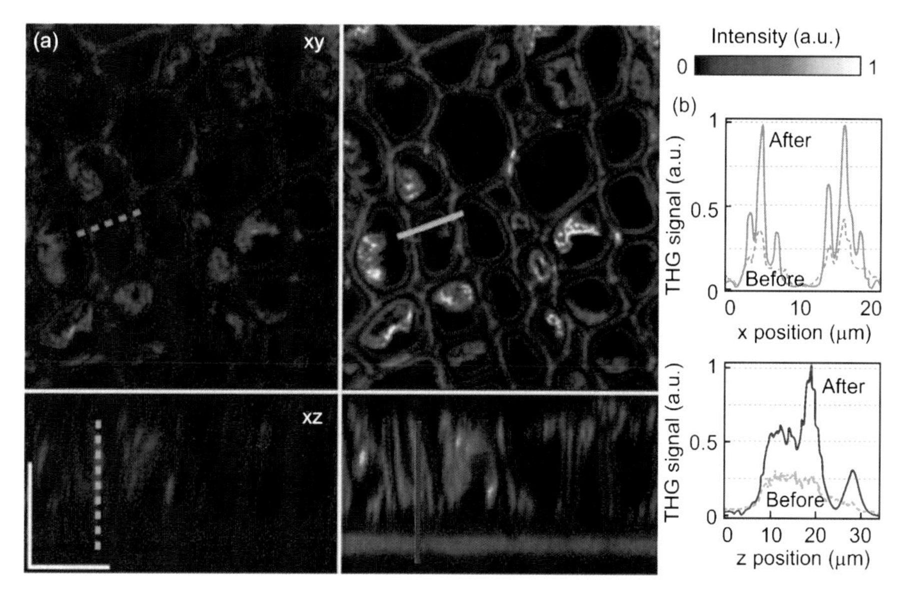

FIGURE 1.23 (a) THG images of an elderberry stem slice (left) before and (right) after aberration correction on the middle plane of the z stack. Scale bars, 20 μm. (b) Profiles along the lines in (a). Reprinted with permission from [112]. Copyright 2009 Optical Society of America.

Functional Imaging by Controlled Nonlinear Optical Phenomena, First Edition.
Keisuke Isobe, Wataru Watanabe and Kazuyoshi Itoh.

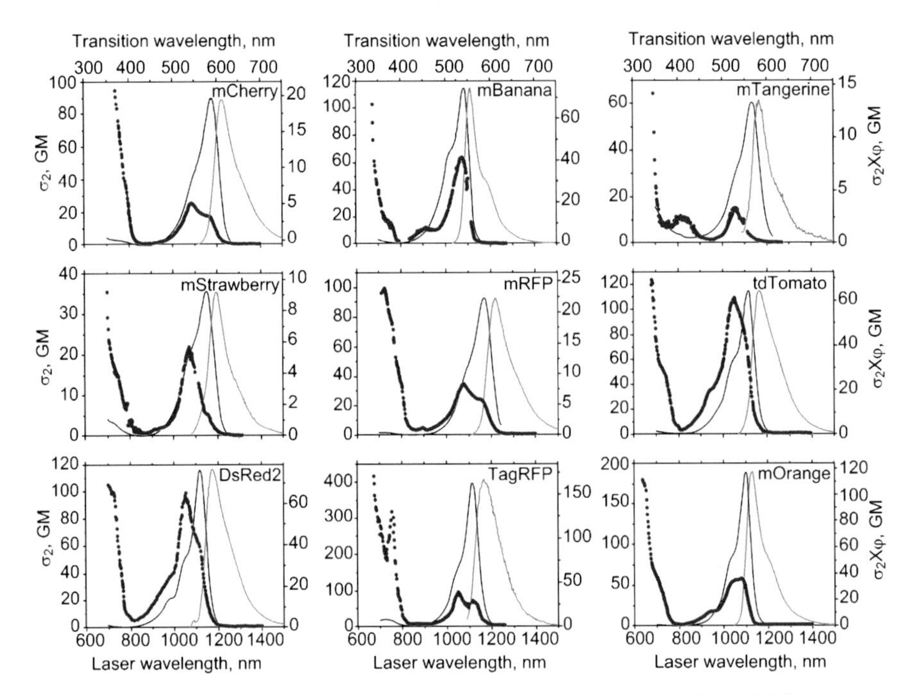

FIGURE 3.2 TPE (symbols), OPE (black solid line), and emission (blue solid line) spectra of mCherry, mBanana, mTangerine, mStrawberry, mRFP, td Tomato, DsRed2, TagRFP, and mOrange. The scale on the right represents two-photon brightness. 1 GM is 10^{-50} cm^4 s $photon^{-1}$. Reprinted with permission from [14]. Copyright 2009 American Chemical Society.

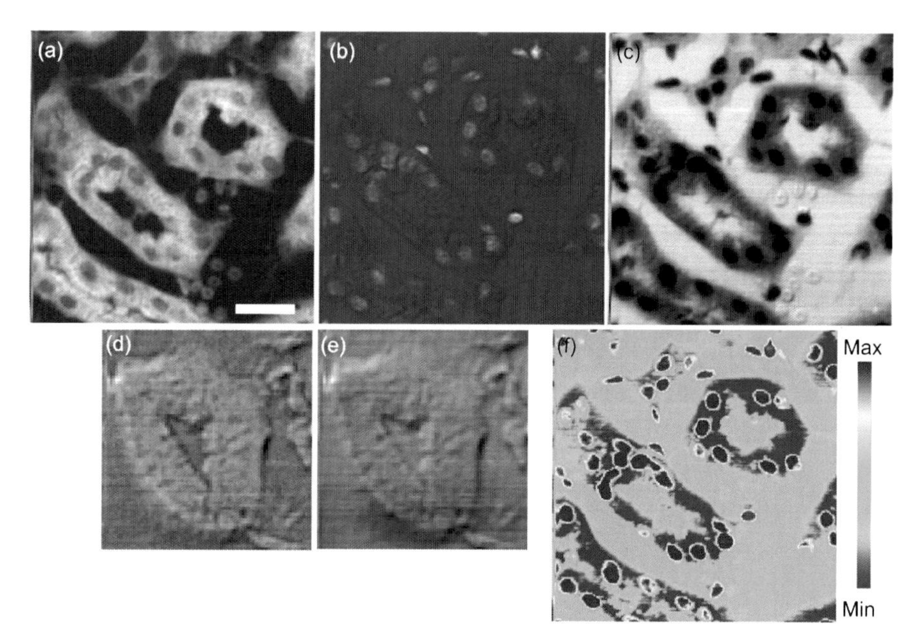

FIGURE 4.15 (a–c) Multi-spectral images of mouse kidney cells stained with hematoxylin and eosin in the spectral ranges of 530 to 590 nm (a), 590 to 620 nm (b), and 650 to 680 nm (c). (d, e) Multi-spectral images of unstained mouse kidney cells in the spectral ranges of 590–620 nm (d) and 650–680 nm (e). (f) The reconstructed image after the normalized spectral correlation process using the SPE spectrum from the nucleus as the reference spectrum. Scale bar is 20 μm.

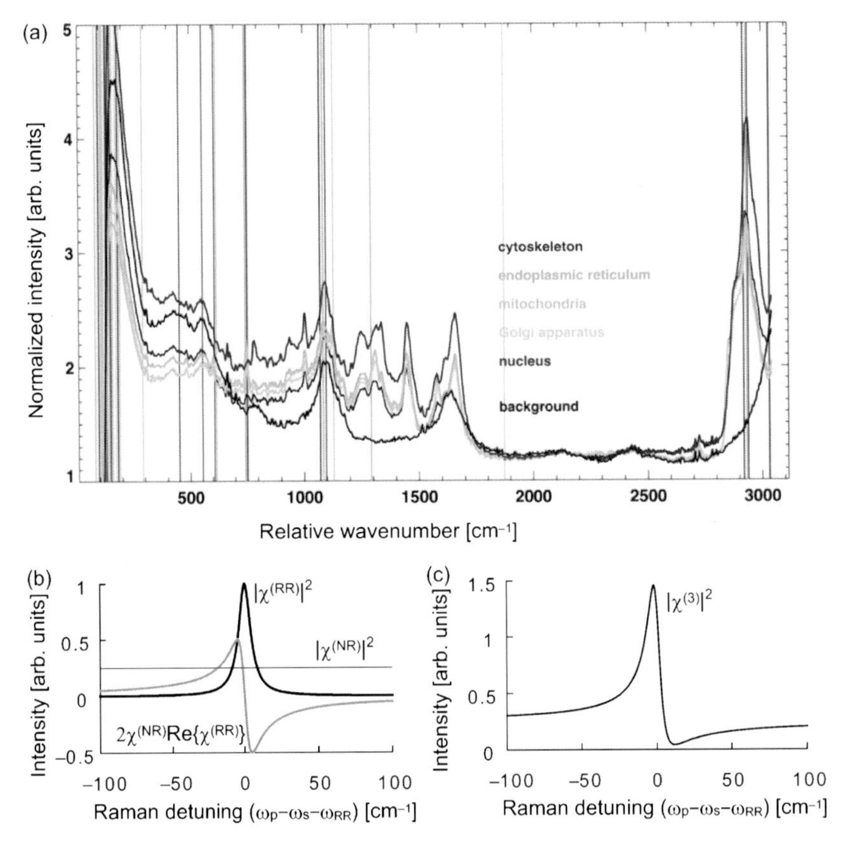

FIGURE 4.19 (a) Spontaneous Raman spectra for cytoskeleton, endoplasmic reticulum, mitochondria, Golgi apparatus, and nucleus of LN-18 cells in phosphate-buffered saline (PBS). Reprinted with permission from [137]. Copyright 2012 Elsevier. (b) Raman resonant contribution, nonresonant contribution, and mixing term of the nonresonant and resonant contributions to the CARS spectrum. (c) Typical CARS spectrum.

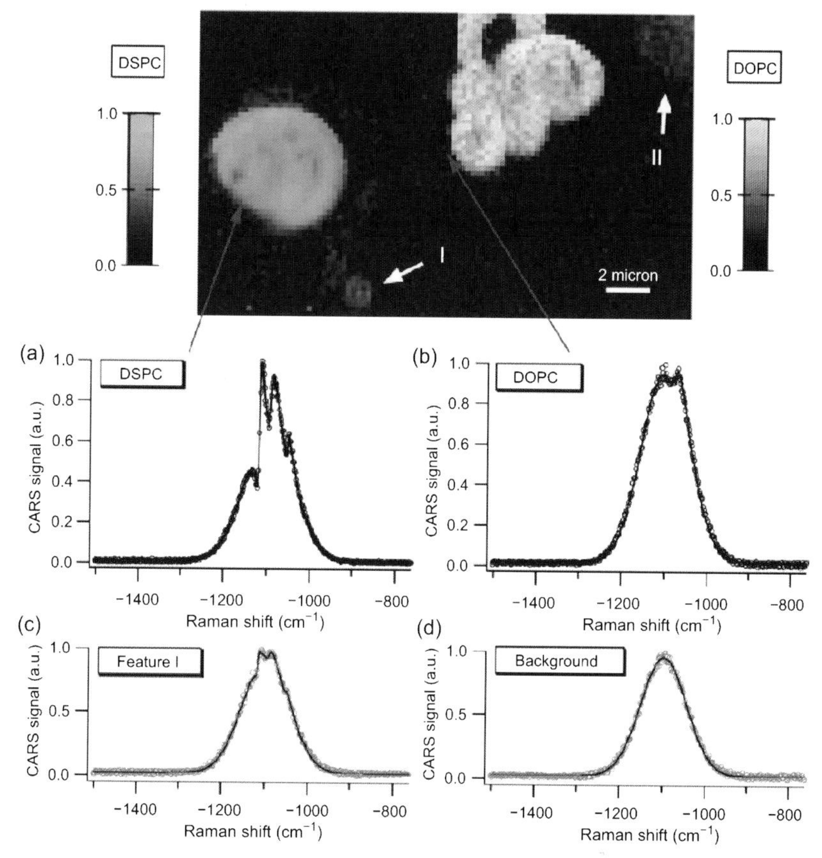

FIGURE 4.24 Multiplex CARS image of both a DSPC and a DOPC vesicle The scale bar corresponds to 2 μm. The contrast of the DSPC Raman mode at 1128 cm^{-1} is plotted in blue, whereas the contrast of the DOPC mode at 1087 cm^{-1} is plotted in yellow (color bars shown). The graphs show typical multiplex CARS spectra of DSPC (a) and DOPC (b). Graph (c) shows the CARS spectrum of the faint feature marked by I, which can be identified as DSPC and is readily discriminated from the (water) background shown in graph (d). Reprinted with permission from [152]. Copyright 2002 American Chemical Society.

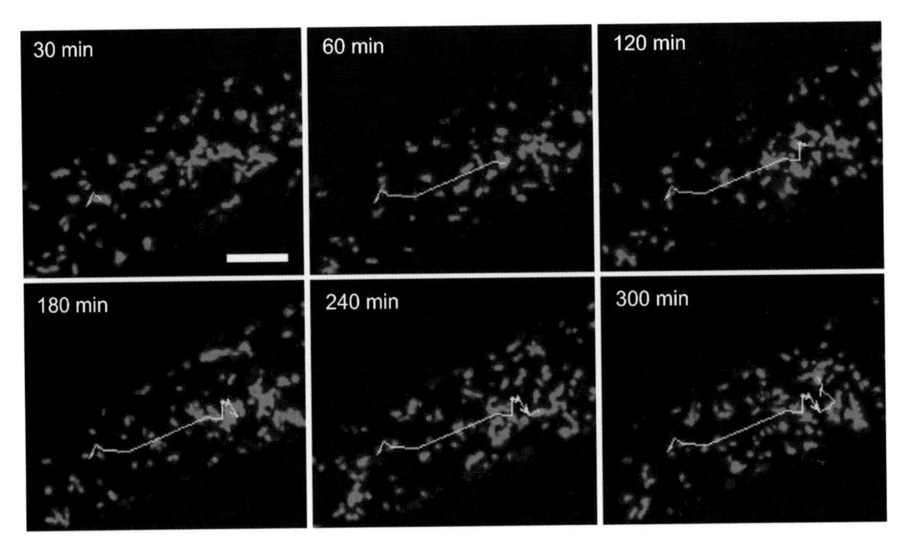

FIGURE 5.4 Tracking of a mitochondrion in a living BY-2 cell. Scale bar: 10 μm after marking a mitochondrion by two-photon excitation. The movement of the mitochondrion labeled by two-photon conversion could be tracked for 5 hours. The trajectory of the labeled mitochondrion is shown by the yellow line. Reprinted with permission from [106]. Copyright 2007 Optical Society of America.

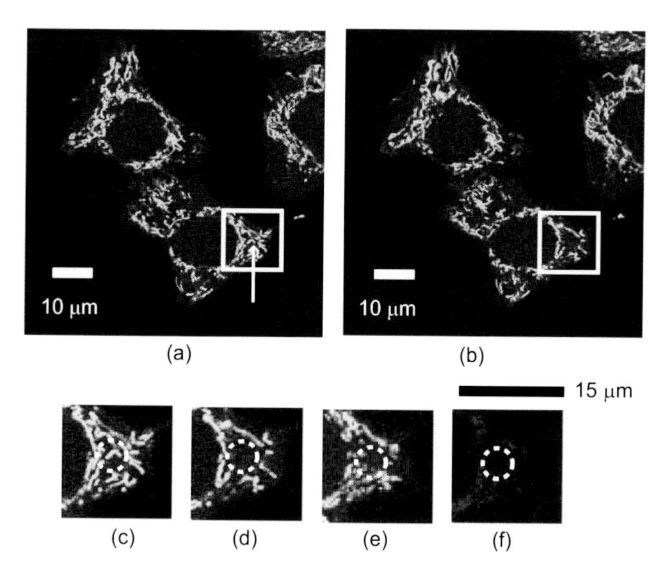

FIGURE 6.2 Nanosurgery of a single mitochondrion in a living HeLa cell with 1-kHz femtosecond laser pulses. (a) Confocal images of laser-irradiated cells before and after restaining. Yellow fluorescence shows mitochondria visualized by EYFP. Target mitochondrion is indicated by a white arrow. (c) Magnified view of square area indicated in (a) before femtosecond laser irradiation. (b) Confocal image obtained after femtosecond laser irradiation. (d) Magnified view of square area in (b). (e) Confocal image obtained by excitation with the Ar^+ laser. (f) Confocal image obtained by excitation with the He-Ne laser. Images (e) and (f) were obtained after restaining by MitoTracker Red. Dotted circles show target mitochondria. Reprinted with permission from [26]. Copyright 2004 Optical Society of America.

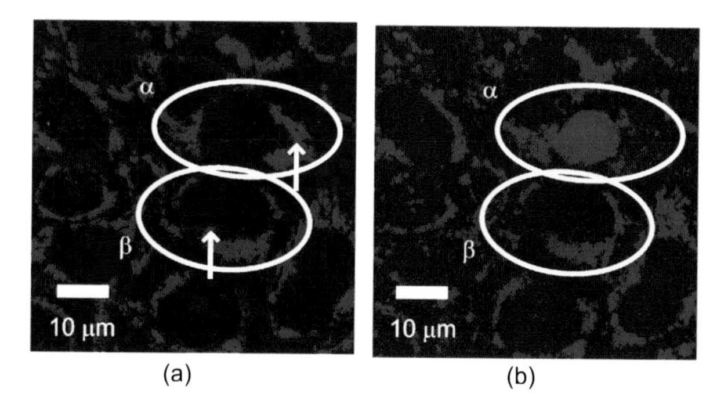

FIGURE 6.3 Confocal images of the cells (a) before and (b) after laser irradiation. Red fluorescence shows mitochondria of HeLa cells stained with MitoTracker Red. The white circles and arrows indicate individual HeLa cells and target mitochondria, respectively. Additional red fluorescence is derived from propidium iodide (PI). The laser pulses were focused inside cells α and β at energies of 7 nJ/pulse and 3 nJ /pulse, respectively. Reprinted with permission from [26]. Copyright 2004 Optical Society of America.

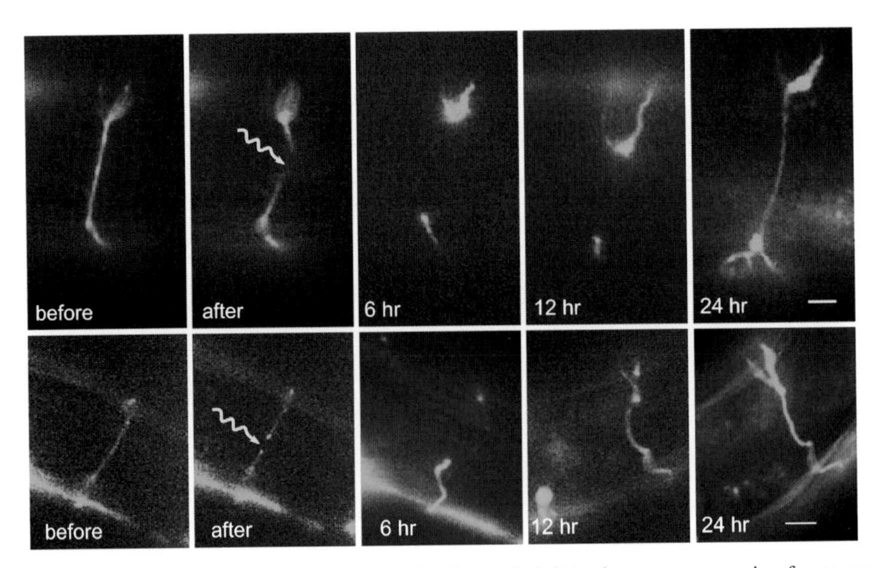

FIGURE 6.6 Femtosecond laser axotomy in *Caenorhabditis elegans* worms using femtosecond laser pulses. Fluorescence images of axons labeled with green fluorescent protein before, after. Scale bar, 5 μm. Courtesy of M. F. Yanik.

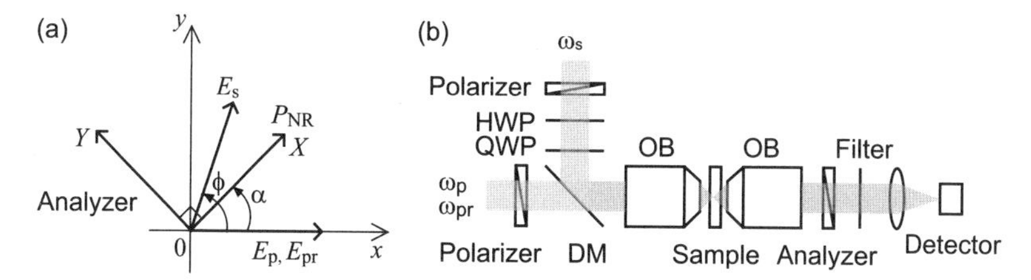

FIGURE 4.21 (a) Polarization vectors of the pump and the Stokes fields, the nonresonant CARS signal, and the analyzer polarizer. (b) Schematic of the typical P-CARS microscope: HWP, half-wave plate; QWP, quarter-wave plate; DM, dichroic mirror; OB, objective lens.

4.4.1.2 Polarization Detection The nonresonant background signal can be suppressed by taking advantage of the differences in the polarization orientation properties of the nonresonant and Raman resonant portions of the third-order polarization [134]. The pump and probe beams are linearly polarized along the x-axis, and the Stokes beam is linearly polarized along a direction at an angle of ϕ relative to the x-axis, as shown in Fig. 4.21(a). We assume that $\omega_p - \omega_s$ is resonant with a molecular vibration. The x and y components of the nonlinear polarization at frequency $\omega_{as} = \omega_p + \omega_{pr} - \omega_s$ is given by

$$P_x^{(CARS)} = \varepsilon_0 \left(\chi_{xxxx}^{(NR)} + \chi_{xxxx}^{(RR)}\right) E_p E_{pr} E_s^* \cos\phi, \tag{4.4.6}$$

$$P_y^{(CARS)} = \varepsilon_0 \left(\chi_{yxxy}^{(NR)} + \chi_{yxxy}^{(RR)}\right) E_p E_{pr} E_s^* \sin\phi, \tag{4.4.7}$$

where $\chi^{(NR)}$ and $\chi^{(RR)}$ are nonresonant and Raman resonant components of third-order susceptibility, respectively. Because the nonresonant susceptibility $\chi^{(NR)}$ is a real quantity that is independent of frequency, the direction of the nonlinear polarization arising from the nonresonant term $\chi^{(NR)}$ makes an angle

$$\alpha = \tan^{-1}\left[\frac{\chi_{yxxy}^{(NR)}}{\chi_{xxxx}^{(NR)}} \tan\phi\right], \tag{4.4.8}$$

with the x-axis. Let X and Y be the axes parallel and perpendicular to the nonresonant part of the nonlinear polarization. The nonlinear polarization components along X and Y are written as

$$P_X^{(CARS)} = \varepsilon_0 \left[\chi_{xxxx}^{(NR)} \frac{\cos\phi}{\cos\alpha} + \chi_{xxxx}^{(RR)} \cos\phi \cos\alpha \left(1 + \rho_{RR} \tan\phi \tan\alpha\right)\right] E_p E_{pr} E_s^*, \tag{4.4.9}$$

$$P_Y^{(CARS)} = -\varepsilon_0 \chi_{xxxx}^{(RR)} \cos\phi \sin\alpha \left(1 - \rho_{RR} \tan\phi / \tan\alpha\right) E_p E_{pr} E_s^*, \tag{4.4.10}$$

where $\rho_{RR} = \chi_{yxxy}^{(RR)}/\chi_{xxxx}^{(RR)}$ is the depolarization ratio of the resonant CARS field. In transparent media, the nonresonant susceptibility follows Kleinman symmetry $\chi_{xxxx}^{(NR)} = 3\chi_{yxxy}^{(NR)}$. Then, we obtain

$$3\tan\alpha = \tan\phi, \tag{4.4.11}$$

and Eqs. (4.4.9) and (4.4.10) can be rewritten as

$$P_X^{(CARS)} = \varepsilon_0 \left[\chi_{xxxx}^{(NR)} \frac{\cos\phi}{\cos\alpha} + \chi_{xxxx}^{(RR)} \cos\phi \cos\alpha \, (1 + 3\rho_{RR}) \right] E_p E_{pr} E_s^*, \tag{4.4.12}$$

$$P_Y^{(CARS)} = -\varepsilon_0 \chi_{xxxx}^{(RR)} \cos\phi \sin\alpha \, (1 - 3\rho_{RR}) \, E_p E_{pr} E_s^*. \tag{4.4.13}$$

Because the resonant susceptibility $\chi^{(RR)}$ is complex, we note immediately from Eqs. (4.4.12) and (4.4.13) that the CARS signal is elliptically polarized. Thus, by using an analyzer along Y to detect the CARS signal, we can effectively increase the resonant-to-nonresonant signal ratio. Theoretically, the signal from $P_Y^{(CARS)}$ is free of nonresonant background. In practice, there exists a residual background because of the birefringence of the optics in the beam path and the scrambling of polarization at the tight focus. For a weak resonance with $|\chi^{(RR)}| \ll \chi^{(NR)}$, the $\chi^{(RR)}$ term in Eq. (4.4.12) can be neglected. Then, we can write

$$\left(\frac{P_Y^{(CARS)}}{P_X^{(CARS)}} \right)^2 \cong \left\{ \frac{(1 - 3\rho_{RR})\, \chi_{xxxx}^{(RR)} \sin 2\alpha}{2\chi_{xxxx}^{(NR)}} \right\}^2. \tag{4.4.14}$$

According to Eq. (4.4.14), it is clear that the maximum Raman resonant-to-nonresonant signal ratio is obtained for $\alpha = 45°$. From Eq. (4.4.11), the optimal value for the angle ϕ is then 71.6°.

The schematic of typical polarization-CARS (P-CARS) microscope is shown in Fig. 4.21(b). Both the pump and the Stokes beams are linearly polarized. The pump beam is also used as the probe beam. The polarization direction of the Stokes beam is adjustable with a half-wave plate. The angle ϕ is set at 71.6°. A quarter-wave plate is used to compensate for the birefringence in the Stokes field that is induced by the dichroic mirror. The polarization direction of the detection beam is set at 135° by an analyzer. Cheng et al. successfully demonstrated background suppression by using the P-CARS technique [134]. Figure 4.22(a) shows the P-CARS spectrum of *N*-methylacetamide, a model compound containing the characteristic amide vibration of peptides and proteins [142]. The P-CARS spectrum is free of a nonresonant background and the P-CARS band positions coincide with the corresponding Raman band positions. Figure 4.22(b) and (c) shows P-CARS images of unstained epithelial cells at 1650 and 1745 cm^{-1}, respectively [142]. The main contribution to the contrast in Fig. 4.22(b) is the resonant CARS signal of the amide I band at 1650 cm^{-1}, while the vibrational contrast in Fig. 4.22(c) is largely diminished because $\omega_p - \omega_s$ is further away from the amide I band. This result confirms that P-CARS microscopy provides high vibrational contrast by suppressing the nonresonant background.

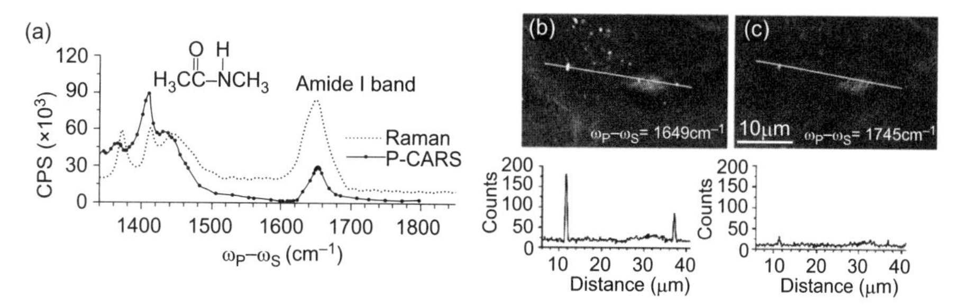

FIGURE 4.22 (a) P-CARS and spontaneous Raman spectra of pure *N*-methylacetamide liquid recorded at room temperature. (b) P-CARS image of an unstained epithelial cell with $\omega_p - \omega_s$ tuned to 1650 cm^{-1}. (c) Same as (b) but with $\omega_p - \omega_s$ tuned to 1745 cm^{-1}. Reprinted with permission from [142]. Copyright 2004 American Chemical Society.

4.4.1.3 Interferometric Detection The fundamental difference between resonant and the nonresonant nonlinear susceptibilities can be utilized to separate the resonant signal from the nonresonant signal. The nonresonant $\chi^{(NR)}$ has only a real component; $\tilde{\chi}^{(RR)}$ has both real and imaginary parts. The spectrum of the imaginary part of $\tilde{\chi}^{(RR)}$ is proportional to the spontaneous Raman spectrum. The ordinary CARS technique measures $|\tilde{\chi}^{(RR)} + \chi^{(NR)}|^2$, which combines real and imaginary components. In interferometric detection, a small CARS field from samples is mixed with a strong local oscillator (LO) field, so the imaginary part can be extracted [145,146]. Also the shot-noise limited detection can be achieved by this technique.

In the spectral interference method, a CARS field $\tilde{E}_{\text{sample}}(\omega)$ from the sample interferes with an LO field $\tilde{E}_{\text{local}}(\omega)$, which is generated by a nonlinear optical process under the nonresonant condition or is spectrally extracted from a broadband excitation light source. Here the CARS field is generated by a multiplex CARS technique where narrowband pulses are used as the pump light and the probe light and a broadband pulse is employed as the Stokes light as discussed in Section 3.2. Then, according to Eq. (3.2.14), the CARS field is described by

$$\tilde{E}_{\text{sample}}(\omega) = i\varepsilon_0 \left\{ \tilde{\chi}^{(RR)}(\omega - \omega_{\text{pr}}) + \chi^{(NR)} \right\} E_p E_{\text{pr}} \tilde{E}_s^*(\omega_p + \omega_{\text{pr}} - \omega), \quad (4.4.15)$$

where $\tilde{E}_p(\omega) = E_p\delta(\omega - \omega_p)$, $\tilde{E}_{\text{pr}}(\omega) = E_{\text{pr}}\delta(\omega - \omega_{\text{pr}})$, and $E_s(\omega)$ are the electric fields of pump, probe, and Stokes light, respectively. Then, the spectral interference signal is given by

$$\begin{aligned} I(\omega) &= \left|\tilde{E}_{\text{sample}}(\omega)\right|^2 + \left|\tilde{E}_{\text{local}}(\omega)\right|^2 \\ &\quad + i\varepsilon_0 \left\{ \tilde{\chi}^{(RR)}(\omega - \omega_{\text{pr}}) + \chi^{(NR)} \right\} E_p E_{\text{pr}} \tilde{E}_s^*(\omega_p + \omega_{\text{pr}} - \omega) \tilde{E}_{\text{local}}^*(\omega) \\ &\quad - i\varepsilon_0 \left\{ \tilde{\chi}^{(RR)*}(\omega - \omega_{\text{pr}}) + \chi^{(NR)} \right\} E_p^* E_{\text{pr}}^* \tilde{E}_s(\omega_p + \omega_{\text{pr}} - \omega) \tilde{E}_{\text{local}}(\omega) \\ &= \left|\tilde{E}_{\text{sample}}(\omega)\right|^2 + \left|\tilde{E}_{\text{local}}(\omega)\right|^2 \\ &\quad + 2\left|\tilde{E}_{\text{sample}}(\omega)\right| \left|\tilde{E}_{\text{local}}(\omega)\right| \cos\phi(\omega). \end{aligned} \quad (4.4.16)$$

Here $\phi(\omega)$ is expressed by

$$\phi(\omega) = \phi_{\text{sample}}(\omega) + \omega\tau + \phi_0(\omega) + \frac{\pi}{2}(1 - N), \tag{4.4.17}$$

where

$$\phi_{\text{sample}}(\omega) = \arctan\left(\frac{\text{Im}\left\{\tilde{\chi}^{(\text{RR})}(\omega - \omega_{\text{pr}})\right\}}{\text{Re}\left\{\tilde{\chi}^{(\text{RR})}(\omega - \omega_{\text{pr}})\right\} + \chi^{(\text{NR})}}\right). \tag{4.4.18}$$

Here $\phi_{\text{sample}}(\omega)$ indicates a spectral phase shift induced by the Raman response $\tilde{\chi}^{(\text{RR})}$, $\phi_0(\omega)$ is the relative phase delay due to the optical components in the interferometer, and τ is the temporal delay between the two pulses of the sample CARS and the LO signals. $\text{Re}\{h\}$ and $\text{Im}\{h\}$ denote the real and imaginary parts of the complex $h = \text{Re}\{h\} + i\text{Im}\{h\}$, respectively. When the LO field is generated by a nonlinear optical process under the nonresonant condition, N is 1. In contrast, when the LO field is spectrally extracted from a broadband excitation light source, N is 0. If the phase $\phi_{\text{sample}}(\omega)$ can be obtained, the real and imaginary parts of $\tilde{\chi}^{(\text{RR})}$ can be readily determined. The total phase information $\phi(\omega)$ is extracted from the spectral interferogram. To isolate $\phi_{\text{sample}}(\omega)$ from the total phase information $\phi(\omega)$, a reference measurement that provides the reference phase $\phi_{\text{ref}}(\omega) = \omega\tau + \phi_0(\omega)$ is required. In the reference measurement, a nonresonant reference sample is used because $\phi_{\text{sample}}(\omega) = 0$. The sample phase $\phi_{\text{sample}}(\omega)$ can be acquired by subtracting the reference phase $\phi_{\text{ref}}(\omega)$ from the total phase $\phi_{\text{i}}(\omega)$. Then, the real and imaginary parts can readily be calculated from the phase $\phi_{\text{sample}}(\omega)$ and amplitude $|E_{\text{sample}}(\omega)|$ [145]. The real and imaginary parts can be also obtained by the interferometric detection in the time domain [146].

In the interference method in the time domain, a CARS field $E_{\text{sample}}(t)$ is typically generated by two or three narrowband pulses, and the CARS field from the sample interferes with an LO field $E_{\text{local}}(t)$. According to Eq. (4.4.15), the CARS field is described by

$$E_{\text{sample}}(t) = i\varepsilon_0 \left\{\tilde{\chi}^{(\text{RR})}(\omega_{\text{p}} - \omega_{\text{s}}) + \chi^{(\text{NR})}\right\} E_{\text{p}}(t) E_{\text{pr}}(t) E_{\text{s}}^*(t), \tag{4.4.19}$$

where ω_{s} is the central frequency of the Stokes pulse. Then, the interference signal is written as

$$\begin{aligned} I(\tau) = {} & \left|\tilde{E}_{\text{sample}}(t)\right|^2 + \left|\tilde{E}_{\text{local}}(t)\right|^2 \\ & - 2\varepsilon_0 \left|E_{\text{p}}(t) E_{\text{pr}}(t) \tilde{E}_{\text{s}}^*(t) \tilde{E}_{\text{local}}^*(t)\right| \\ & \times \left[\text{Im}\left\{\tilde{\chi}^{(\text{RR})}(\omega_{\text{p}} - \omega_{\text{s}})\right\} \cos\varphi(\tau)\right. \\ & \left. + \left[\text{Re}\left\{\tilde{\chi}^{(\text{RR})}(\omega_{\text{p}} - \omega_{\text{s}})\right\} + \chi^{(\text{NR})}\right] \sin\varphi(\tau)\right], \end{aligned} \tag{4.4.20}$$

where

$$\varphi(\tau) = \omega_{\text{as}}\tau + \phi_0(\omega_{\text{as}}) - \frac{\pi}{2}N. \tag{4.4.21}$$

Here ω_{as} is the central frequency of the CARS field. The last two terms in Eq. (4.4.20) are the interferometric mixing terms. If $\varphi(\tau)$ is set to 0°, the resonant imaginary contributions are maximized while the nonresonant contribution vanishes. Thus, the interferometric detection in the time domain permits vibrational imaging with contrast based purely on the imaginary part of $\tilde{\chi}^{(RR)}$, which is free of the nonresonant background. In practice, the relative phase delay $\phi_0(\omega_{as})$ is modulated in order to extract the interferometric mixing terms. The modulation of the relative phase delay becomes an amplitude modulation the signal through the (heterodyne) mixing terms. By detecting the modulation signal through a lock-in amplifier, the interferometric mixing terms at an arbitrary phase can be extracted.

Evans et al. [145] and Potma et al. [146] successfully demonstrated that the interferometric detection technique can be used to extract amplitude and phase information of $\tilde{\chi}^{(RR)}$, leading to complete suppression of the nonresonant background. Figure 4.23(a) shows the amplitude and phase of the dodecane CH-stretch region. The CARS spectrum is normalized by the nonresonant CARS spectrum from the local oscillator. The real and imaginary parts of $\tilde{\chi}^{(RR)}$ calculated from the amplitude and phase are shown in Fig. 4.23(b). The CARS spectrum from the imaginary part of $\tilde{\chi}^{(RR)}$ is in good agreement with the corresponding Raman spectrum. Figure 4.23(c), (d), and (e) shows the CARS images of NIH 3T3 cells taken at 2845 cm^{-1}, the CH_2 symmetric stretching vibration of lipids, for ordinary detection, and for imaginary and real responses with the interferometric detection, respectively. A strong nonresonant background is present in Fig. 4.23(c) and (e), while the background is absent in Fig. 4.23(d). These results indicate that interferometric CARS microscopy provides a high vibrational contrast by extracting the imaginary part of $\tilde{\chi}^{(RR)}$, which is free of the nonresonant background.

4.4.1.4 Multiplex CARS Microspectroscopy In ordinary CARS microscopy, the CARS intensity including both the nonresonant contribution and the Raman resonant contribution of a single vibrational band is measured at each focal point in the samples. In multiplex CARS (M-CARS) microspectroscopy, not only the intensities but also the spectral profiles are recorded. Even though the nonresonant CARS signal is not suppressed in M-CARS microspectroscopy, the resonant and nonresonant CARS signals can be separated by analyzing the spectral profile of the CARS signal [152,154]. The spectral analysis also provides multispectral images based on various vibrational modes.

The energy level diagram of the CARS process with a multiplex excitation scheme is shown in Section 3.2. A broadband light source is used as the pump and/or Stokes pulses for simultaneously addressing a broad spectral range of various vibrational resonances. A narrowband light source is employed as the probe pulse for spectrally resolving various vibrational resonances. The spectral features of the resultant CARS signals reflect the vibrational energy levels of samples. The spectral resolution of the CARS spectrum is proportional to the spectral bandwidth of the probe pulse. In the general two-color scheme, a narrowband light source serves as the pump pulse and the probe pulse, and a broadband light source is employed as the Stokes pulse.

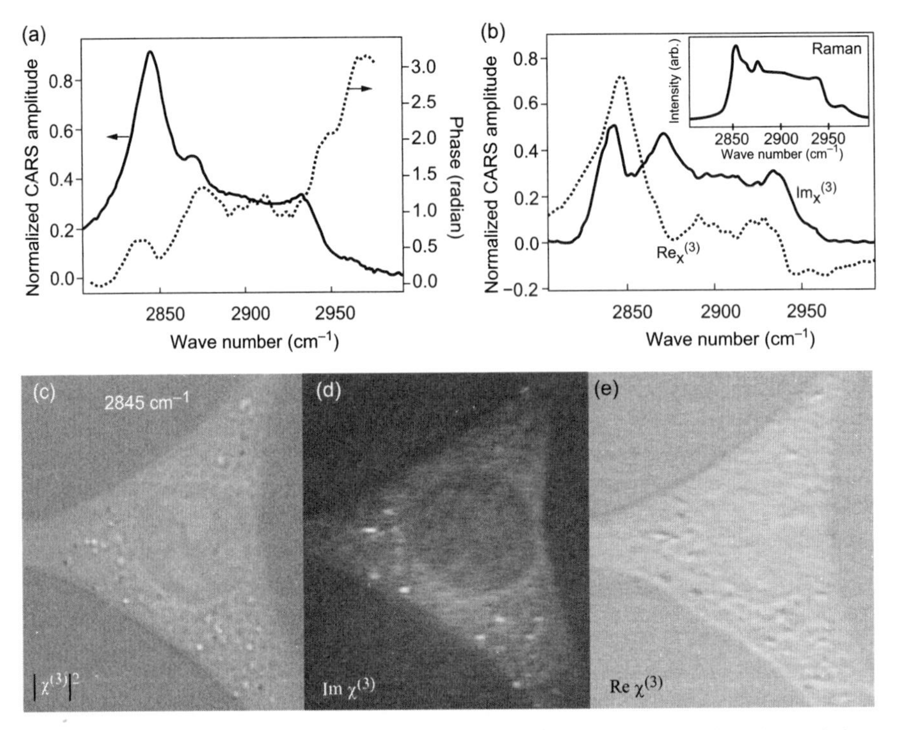

FIGURE 4.23 (a) Extracted CARS amplitude (solid curve) and phase (dotted curve) from the spectral interferograms of the CH-stretching vibrational band of dodecane. (b) The CARS spectrum reconstructed from the real (dotted curve) and imaginary (solid curve) parts of $\chi^{(RR)}$ of the CH-stretching vibrational band of dodecane. Inset, Raman spectrum of dodecane in the CH-stretch vibrational range. Reprinted with permission from [145]. Copyright 2004 Optical Society of America. (c–e) CARS imaging of live NIH 3T3 cells at 2845 cm^{-1} for the ordinary detection (c), and for the imaginary (d) and real (e) responses with the interferometric detection. Reprinted with permission from [146]. Copyright 2006 Optical Society of America.

All CARS spectra are analyzed by a curve-fitting procedure. According to Eq. (3.2.15), the CARS signal is proportional to the absolute square of the third-order nonlinear susceptibility $\chi^{(3)}$, which consists of a resonant contribution $\chi^{(RR)}$ and a nonresonant contribution $\chi^{(NR)}$,

$$I_{\mathrm{CARS}}(\omega) \propto \left|\chi^{(NR)} + \tilde{\chi}^{(RR)}(\omega - \omega_{\mathrm{pr}})\right|^2. \tag{4.4.22}$$

In the spectral range far from the one-photon resonance, $\chi^{(RR)}$ is described by

$$\tilde{\chi}^{(RR)}(\omega - \omega_{\mathrm{pr}}) \propto \sum_j \frac{R_j}{\Omega_j - (\omega - \omega_{\mathrm{pr}}) - i\Gamma_j}, \tag{4.4.23}$$

where the summation runs over all vibrational modes j. Here R_j is a real constant that contains the vibrational scattering cross section, Ω_j is the vibrational resonance frequency, and Γ_j is the HWHM of the spontaneous Raman scattering vibrational

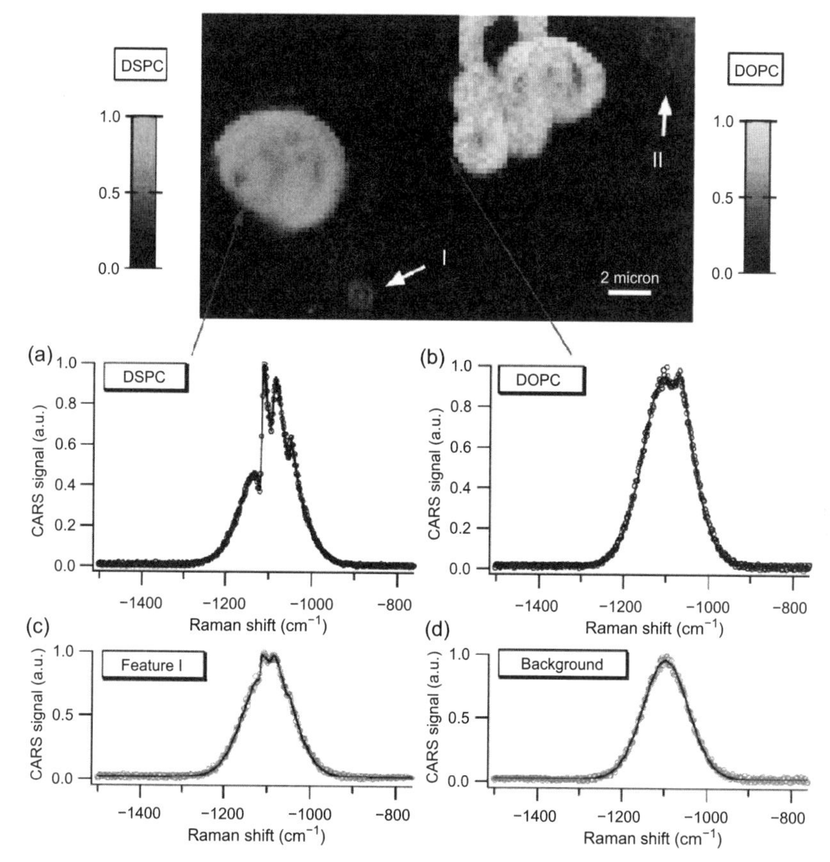

FIGURE 4.24 Multiplex CARS image of both a DSPC and a DOPC vesicle The scale bar corresponds to 2 μm. The contrast of the DSPC Raman mode at 1128 cm^{-1} is plotted in blue, whereas the contrast of the DOPC mode at 1087 cm^{-1} is plotted in yellow (color bars shown). The graphs show typical multiplex CARS spectra of DSPC (a) and DOPC (b). Graph (c) shows the CARS spectrum of the faint feature marked by I, which can be identified as DSPC and is readily discriminated from the (water) background shown in graph (d). Reprinted with permission from [152]. Copyright 2002 American Chemical Society. (*See insert for color representation of the figure.*)

transition. Since the nonresonant background is mainly generated from the solvent in a dilute sample, $\chi^{(NR)}$ is fitted by one parameter. Thus, the CARS spectra can be fitted by use of one parameter for the nonresonant background and three parameters for each vibrational transition: R_j, Ω_j, and Γ_j.

Müller and Schins have applied M-CARS microspectroscopy to extract the resonant CARS image from the CARS spectra including various Raman resonant contributions and a nonresonant contribution [152]. Figure 4.24 shows a multiplex CARS image of both a 1,2-distearoyl-sn-glycero-3-phosphocholine (DSPC; 18:0) and a 1,2-dioleoyl-sn-glycero-3-phosphocholine (DOPC; 18:1, 9-cis) vesicle, which have been discriminated on the basis of their CARS spectrum in the skeletal optical mode region [152]. The spectral data at each point were subsequently fitted to theoretical expressions for the CARS signals of the DSPC and the DOPC. It should be noted that the cross talk between the DSPC and DOPC signal is absent.

The supercontinuum (SC) generated with a narrowband laser pulse in a nonlinear fiber, such as the photonic crystal fiber (PCF) and taper fiber, and the original narrowband pulse have been well accepted to be appropriate for broadband and narrowband light sources in M-CARS microspectroscopy [155,156]. Kee et al. [155] and Kano et al. [156] generated ultrabroadband pulses with the picosecond pulses in the nonlinear fiber and applied them to the Stokes pulses in ultrabroadband ($>2500\ \mathrm{cm}^{-1}$) M-CARS microspectroscopies where the CARS spectrum can be obtained with a spectral resolution of $10\ \mathrm{cm}^{-1}$ by employing a combination of the ultrabroadband Stokes pulse and the original narrowband pulse as the pump and probe pulses. Okuno et al. developed on ultrabroadband multiplex CARS microspectroscopy using the SC generated from the PCF by seeding subnanosecond laser pulses. They further demonstrated M-CARS imaging with a high spectral resolution ($<1\ \mathrm{cm}^{-1}$) and with ultrabroadband spectral coverage ($>2000\ \mathrm{cm}^{-1}$) [157].

4.4.1.5 Time-Resolved Detection

The nonresonant signal can be suppressed by time-resolved (TR) detection based on the difference between the electronic coherence time and the vibrational coherence time. In the solvent, the electronic coherences instantaneously decay after the excitation fields are turned off. In contrast, the vibrational coherences typically persist on a picosecond time scale [158]. In TR detection, a vibrational coherence and an electronic coherence are first excited by the pump and Stokes pulse. These coherences are often excited by a single broadband pulse including pump and Stokes frequency components. Then, the vibrational coherence (nuclear wavepacket motion) must be impulsively induced by one or two pulses whose pulse durations are much shorter than the vibrational coherence time. According to Eqs. (1.1.121), (1.1.124), and (1.1.125), the sum of these coherences is expressed by

$$Q(t) = \varepsilon_0 \int_{-\infty}^{\infty} \chi^{(3)} \{(1 - f_R)\delta(t - t_1) + f_R h_{RR}(t - t_1)\} \left|E_{ps}(t_1)\right|^2 dt_1, \quad (4.4.24)$$

where $E_{ps}(t)$ is the electric field of the excitation pulse including pump and Stokes frequency components. Here the first term is the electronic coherence, and the second term is the vibrational coherence. These coherences are detected by anti-Stokes scattering $E_{as}(t)$ from a delayed probe pulse $E_{pr}(t)$ at a delay time τ. Then, the anti-Stokes field is described by

$$\begin{aligned} E_{as}^{(3)}(t, \tau) &= E_{pr}(t - \tau)Q(t) \\ &= \varepsilon_0 \chi^{(NR)} \left|E_{ps}(t)\right|^2 E_{pr}(t - \tau) \\ &\quad + \varepsilon_0 f_R \chi^{(3)} E_{pr}(t - \tau) \int_{-\infty}^{\infty} h_{RR}(t - t_1) \left|E_{ps}(t)\right|^2 dt_1, \quad (4.4.25) \end{aligned}$$

where $\chi^{(\mathrm{NR})} = (1 - f_{\mathrm{R}})\chi^{(3)}$. Because the pulse duration of the pulse $E_{\mathrm{ps}}(t)$ is much shorter than the vibrational coherence time, $|E_{\mathrm{ps}}(t)|^2$ in the second term of Eq. (4.4.25) can be approximated by $|E_{\mathrm{ps}}|^2\delta(t)e^{-i(\omega_{\mathrm{p}}-\omega_{\mathrm{s}})t}$. Then, the anti-Stokes field is written as

$$\begin{aligned} E_{\mathrm{as}}(t, \tau) \approx{}& \varepsilon_0 \chi^{(\mathrm{NR})} \left|E_{\mathrm{ps}}(t)\right|^2 E_{\mathrm{pr}}(t - \tau) \\ &+ \varepsilon_0 f_{\mathrm{R}} \chi^{(3)} \left|E_{\mathrm{ps}}\right|^2 E_{\mathrm{pr}}(t - \tau) h_{\mathrm{RR}}(t). \end{aligned} \qquad (4.4.26)$$

By illuminating the sample with the probe pulse after the electronic coherence disappears, the electronic contribution can be attenuated. Of course, to measure the vibrational contribution, the probe pulse must irradiate before the vibrational coherence vanishes. Then, the anti-Stokes field is rewritten as

$$E_{\mathrm{as}}(t, \tau) = \varepsilon_0 f_{\mathrm{R}} \chi^{(3)} \left|E_{\mathrm{ps}}\right|^2 E_{\mathrm{pr}}(t - \tau) h_{\mathrm{RR}}(t). \qquad (4.4.27)$$

Thus, the time-delayed probe pulse, whose delay time is longer than the electronic coherence time, and is shorter than the vibrational coherence time, allows efficient background suppression.

TR detection also permits multi-spectral imaging because multiple vibrational modes can be simultaneously excited by the femtosecond ultrabroadband pump and Stokes pulses. The anti-Stokes field in the frequency domain can be expressed by

$$\tilde{E}_{\mathrm{as}}(\omega, \tau) = \varepsilon_0 f_{\mathrm{R}} \chi^{(3)} \left|E_{\mathrm{ps}}\right|^2 \int_{-\infty}^{-\infty} \tilde{E}_{\mathrm{pr}}(\Omega) \tilde{h}_{\mathrm{RR}}(\omega - \Omega) e^{i\Omega\tau} d\Omega. \qquad (4.4.28)$$

If the bandwidth of the probe pulse is so narrow that multiple vibrational modes can be spectrally resolved, $\tilde{E}_{\mathrm{pr}}(\Omega)$ in the integration of Eq. (4.4.28) can be approximated by $\tilde{E}_{\mathrm{pr}}\delta(\Omega - \omega_{\mathrm{pr}})$. Then, the anti-Stokes spectrum can be written as

$$\begin{aligned} I_{\mathrm{as}}(\omega) &\propto \left|\varepsilon_0 f_{\mathrm{R}} \chi^{(3)} \left|E_{\mathrm{ps}}\right|^2 E_{\mathrm{pr}} \tilde{h}_{\mathrm{RR}}(\omega - \omega_{\mathrm{pr}}) e^{i\omega_{\mathrm{pr}}\tau}\right|^2 \\ &= \left|\varepsilon_0 \tilde{\chi}^{(\mathrm{RR})}(\omega - \omega_{\mathrm{pr}}) \left|E_{\mathrm{ps}}\right|^2 E_{\mathrm{pr}}\right|^2. \end{aligned} \qquad (4.4.29)$$

Thus, an ultrabroadband anti-Stokes spectrum without nonresonant signals, which is similar to the spontaneous Raman spectrum, can be simultaneously obtained without tuning excitation wavelengths by a spectrometer [148,149]. This technique is similar to multiplex CARS microspectroscopy, but the nonresonant contribution is not included in the anti-Stokes spectrum. In this technique, the pulse duration of the probe pulse is longer than the vibrational period and the spectral resolution of the anti-Stokes spectrum is proportional to the spectral bandwidth of the probe pulse. In contrast, if the pulse duration of the probe pulse is much shorter than the vibrational period, $E_{\mathrm{pr}}(t)$ in Eq. (4.4.27) can be approximated by $E_{\mathrm{pr}}(t) = E_{\mathrm{pr}}\delta(t)e^{-i\omega_{\mathrm{pr}}t}$. In this case, the pump and Stokes frequency components are often included in the probe pulse. Then, the intrapulse FWM effect is also induced by the probe pulse. Thus, the resultant anti-Stokes field is written by

$$\begin{aligned} E_{\text{as}}(t,\tau) &= \varepsilon_0 f_{\text{R}} \chi^{(3)} \left|E_{\text{ps}}\right|^2 E_{\text{pr}}(t-\tau) h_{\text{RR}}(t) \\ &+ \varepsilon_0 \chi^{(\text{NR})} \left|E_{\text{pr}}(t-\tau)\right|^2 E_{\text{pr}}(t-\tau) \\ &+ \varepsilon_0 f_{\text{R}} \chi^{(3)} \left|E_{\text{pr}}\right|^2 E_{\text{pr}}(t-\tau) h_{\text{RR}}(t-\tau). \end{aligned} \tag{4.4.30}$$

In a dilute sample where we can assume $\left|f_{\text{R}}\chi^{(3)}h_{\text{RR}}\right| \ll \left|\chi^{(\text{NR})}\right|$, the second term in Eq. (4.4.30) acts as an intense local oscillator. Thus, the time-resolved CARS signal can be approximated by

$$\begin{aligned} S(\tau) &\propto \int_{-\infty}^{\infty} \left|E_{\text{as}}(t,\tau)\right|^2 dt \\ &\approx \int_{-\infty}^{\infty} \varepsilon_0^2 \left|E_{\text{pr}}(t-\tau)\right|^2 \left[\left(\chi^{(\text{NR})} \left|E_{\text{pr}}(t-\tau)\right|^2\right)^2 \right. \\ &+ \chi^{(\text{NR})} f_{\text{R}} \chi^{(3)} \left|E_{\text{ps}}\right|^2 \left|E_{\text{pr}}(t-\tau)\right|^2 \left\{h_{\text{RR}}(t) + h_{\text{RR}}^*(t)\right\} \\ &+ \left. \chi^{(\text{NR})} f_{\text{R}} \chi^{(3)} \left|E_{\text{pr}}\right|^2 \left|E_{\text{pr}}(t-\tau)\right|^2 \left\{h_{\text{RR}}(t-\tau) + h_{\text{RR}}^*(t-\tau)\right\} \right] dt. \\ &\approx \int_{-\infty}^{\infty} \left(\varepsilon_0 \chi^{(\text{NR})}\right)^2 \left\{\left|E_{\text{pr}}(t-\tau)\right|^2\right\}^3 dt \\ &+ \varepsilon_0^2 \chi^{(\text{NR})} f_{\text{R}} \chi^{(3)} \left|E_{\text{ps}}\right|^2 \left(\left|E_{\text{pr}}\right|^2\right)^2 \left\{h_{\text{RR}}(\tau) + h_{\text{RR}}^*(\tau)\right\} \\ &+ \varepsilon_0^2 \chi^{(\text{NR})} f_{\text{R}} \chi^{(3)} \left|E_{\text{pr}}\right|^2 \left(\left|E_{\text{pr}}\right|^2\right)^2 \left\{h_{\text{RR}}(0) + h_{\text{RR}}^*(0)\right\}. \end{aligned} \tag{4.4.31}$$

Thus, the vibrational coherence (nuclear wavepacket motion) can be directly observed in the time domain, and appears as a sinusoidal modulation on a function of delay time of the probe pulse. Then, the Fourier transform of the measured time-resolved CARS signal results in the CARS spectrum, which is free of nonresonant background [150,151]. The spectral resolution is determined not by the spectral bandwidth of the excitation pulses, but by the inverse of maximum delay time between excitation pulses. This technique is called FT-CARS microspectroscopy. FT-CARS microspectroscopy employing femtosecond ultrabroadband pulses allows the suppression of the nonresonant signal and the measurement of an ultrabroadband CARS spectrum with high spectral resolution regardless of the pulse bandwidth [150,151].

Lee et al. [148] and Selm et al. [149] have reported ultrabroadband background-free TR-CARS microspectroscopy by employing the combination of an ultrabroadband pulse and a narrowband pulse. Figure 4.25(a) and (b) displays CARS spectra of 3-μm polystyrene (PS) beads on a glass coverslip immersed in dimethyl sulfoxide (DMSO) at positions A (PS) and B (DMSO), which were measured at delay times of 0 and 1.5 ps, respectively [148]. The ordinary multiplex detection corresponds to the zero delay scheme. At a delay time of 1.5 ps, the nonresonant background disappears, while the vibrational bands remain. Figure 4.25(c–e) shows time-resolved CARS

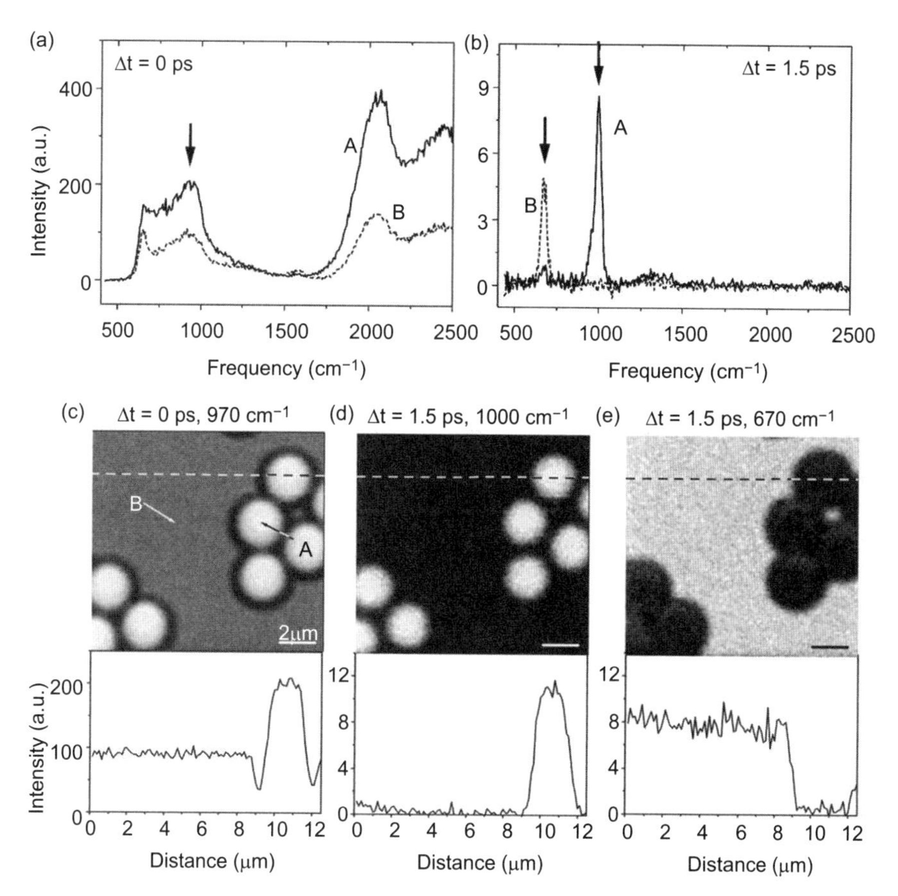

FIGURE 4.25 (a, b) CARS spectra measured at delay times of 0 ps (a) and 1.5 ps (b) at positions A and B, which correspond to PS and DMSO, respectively. (c–e) Time-resolved CARS images are constructed at the frequencies corresponding to PS and DMSO modes at different time delays. Reprinted with permission from [148]. Copyright 2008 American Institute of Physics.

images of PS beads in DMSO at delay times of 0 and 1.5 ps [148]. Even though the detection window at a delay time of 0 ps is detuned from Raman resonance of DMSO to 970 cm^{-1}, not only the resonant signal from PS beads but also the nonresonant background signal from DMSO remain as shown in Fig. 4.25(c). In contrast, in the TR detection at 1000 cm^{-1}, the nonresonant background signal from DMSO disappears and the resonant signal from PS beads becomes visible as shown in Fig. 4.25(d). On the contrary, in TR detection at 670 cm^{-1}, the resonant signal from DMSO becomes visible and the nonresonant background from the PS beads disappears as shown in Fig. 4.25(e). It should be noted that TR detection provides suppression of the nonresonant background.

Isobe et al. successfully demonstrated ultrabroadband background-free FT-CARS microspectroscopy over a bandwidth of 4000 cm^{-1} by employing ultrabroadband

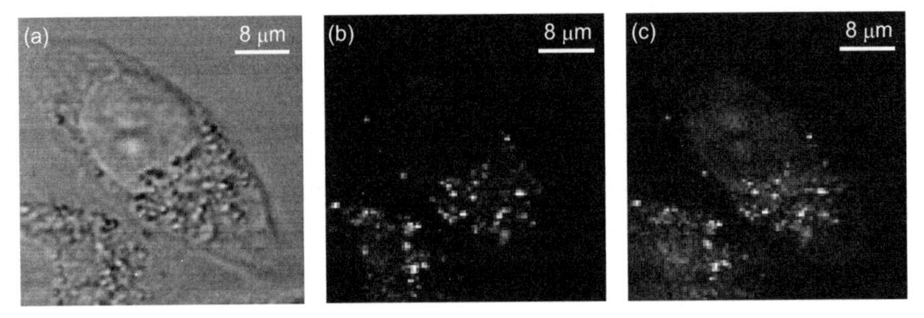

FIGURE 4.26 CARS images of an unstained HeLa cell obtained by the ordinary detection (a) and by FT-CARS microspectroscopy (b, c). (b, c) CARS images constructed from integration in the spectral bands of 2800 to 2900 cm^{-1}(b), and 2900 to 3000 cm^{-1}(c). Reprinted with permission from [151]. Copyright 2009 Optical Society of America.

pulses with a pulse duration of 5 fs [151]. Figure 4.26 shows CARS images of an unstained HeLa cell. In the ordinary detection, image contrast is determined by the refractive index based on the large nonresonant background as shown in Fig. 4.26(a). Figure 4.26(b) and (c) shows integrated spectral bands of 2800–2900 cm^{-1} and 2900–3000 cm^{-1} in FT-CARS microspectroscopy. As seen in Fig. 4.26(b) and (c), different vibrational contrasts, which is free of the nonresonant contribution, can be obtained by FT-CARS microspectroscopy.

4.4.2 Stimulated Raman Scattering Microscopy

In the stimulated Raman scattering (SRS) process, a sample is simultaneously illuminated by a pump light at ω_p and a Stokes light at ω_s whose difference frequency matches the Raman transition frequency ω_{RR}. Then, loss (ΔI_p) of the pump light at ω_p and amplification (ΔI_s) of the Stokes light at ω_s are achieved by virtue of stimulated emission of the Stokes light. The intensity loss in the pump light is called stimulated Raman loss (SRL), and the intensity gain in the Stokes light is called stimulated Raman gain (SRG). As described in Section 1.1.3, the SRL and SRG signals are proportional to the imaginary part of the Raman resonant contribution in the third-order nonlinear susceptibility, whereas the CARS signals are typically proportional to the squared modulus of the third-order nonlinear susceptibility including both the Raman resonant contribution and the nonresonant contribution. Thus, the CARS signals include a high nonresonant four-wave mixing (FWM) background, while SRS microscopy is theoretically free of the nonresonant background [159–163]. In practice, nonlinear signals that are generated by other nonlinear optical processes, including cross-phase modulation and two-photon absorption (TPA), might be mixed in SRS microscopy [161]. As shown in Section 4.4.1, the nonresonant CARS background can be suppressed by various techniques [139–157]. In particular, interferometric CARS microscopy provides vibrational imaging and contrast that is based purely on the imaginary part of the Raman resonant contribution in the third-order nonlinear susceptibility, which is free of the nonresonant background [146]. According to Eqs. (1.1.252) and (1.1.253), we see that in SRS microscopy, the induced polarizations for SRL $P_{\mathrm{SRL}} = -\mathrm{Im}\left\{\tilde{\chi}^{(\mathrm{RR})}(\omega_{\mathrm{p}} - \omega_{\mathrm{s}})\right\} I_{\mathrm{s}} E_{\mathrm{p}}$ and SRG $P_{\mathrm{SRG}} = \mathrm{Im}\left\{\tilde{\chi}^{(\mathrm{RR})}(\omega_{\mathrm{p}} - \omega_{\mathrm{s}})\right\} I_{\mathrm{p}} E_{\mathrm{s}}$ interfere constructively with the incident pump E_{p} and Stokes E_{s} fields, which act as local oscillators, respectively. Thus, we can regard SRS microscopy as a special case of interferometric CARS microscopy where the anti-Stokes field has same frequency as the pump and Stokes fields, and the pump and Stokes fields act as a local oscillator field. However, no additional external local field is required in SRS microscopy. Of course, we also need not to control the phase of the local oscillator. In addition, the phase-matching condition is automatically satisfied. Furthermore, the shot-noise-limited sensitivity of SRS microscopy is comparable to that of conventional CARS microscopy [163].

Ploetz et al. first reported SRS microscopy using multiplex detection with a photodiode array in combination with a femtosecond amplified laser system at low repetition rate [159]. Although the amplified laser pulses produce a large SRS signal, it is not suitable for bioimaging because the low repetition rate limits the image acquisition speed. This problem can be solved by applying the modulation transfer technique, which was first proposed by Tian et al. [130], to SRS microscopy. Nandakumar et al. first successfully demonstrated SRS microscopy using the modulation transfer technique [160]. Freudiger et al. subsequently reported their achieving biological imaging by SRS microscopy by employing a picosecond oscillator system at a high repetition rate [161]. Nandakumar et al. [162] and Ozeki et al. [163] also demonstrated SRS microscopy using oscillator laser systems immediately after the Freudiger et al. report. A schematic of the modulation transfer technique in SRS microscopy is shown in Fig. 4.27. As the TPA detection discussed in Section 4.3.2, the intensity of one of the two pulses is modulated at a high frequency f and the modulation transfer to the other pulse is measured with the lock-in detection at a frequency of f. By increasing the modulation frequency f, the shot-noise-limit sensitivity can be achieved. Ozeki et al. increased the modulation frequency up to 38 MHz

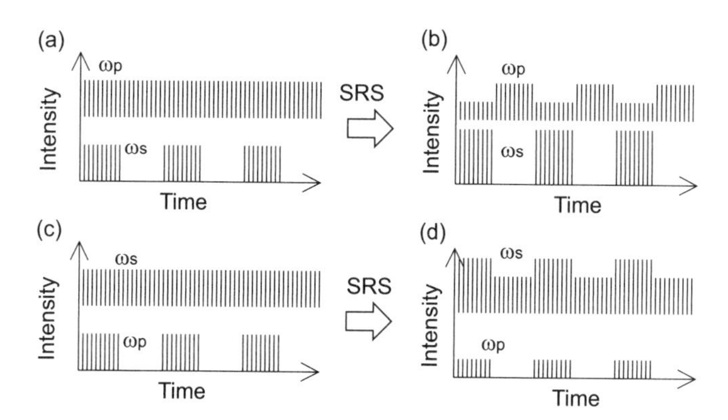

FIGURE 4.27 (a) The intensity modulation of the Stoke pulses. (b) Modulation transfer to the pump pulse intensity with loss experienced by SRS. (c) The intensity modulation of the pump pulses. (d) Modulation transfer to the Stokes pulse intensity with gain experienced by SRS.

by using subharmonically synchronized pulse lasers and achieved the shot-noise-limit sensitivity [164]. Saar et al. also demonstrated video-rate imaging by increasing the modulation frequency [165].

Now, we discuss the signal-to-noise ratio (SNR) of SRS microscopy. Assuming low incident intensities and/or a short length of the nonlinear medium, from Eqs. (1.1.255) and (1.1.256), the intensity changes of the pump and Stokes pulses by SRS are described as

$$\Delta I_{\mathrm{p}} = -\frac{\omega_{\mathrm{p}} L}{\varepsilon_0 c^2 \tilde{n}^{(1)}(\omega_{\mathrm{p}})\tilde{n}^{(1)}(\omega_{\mathrm{s}})} \mathrm{Im}\left\{\tilde{\chi}^{(\mathrm{RR})}(\omega_{\mathrm{p}} - \omega_{\mathrm{s}})\right\} I_{\mathrm{p}} I_{\mathrm{s}}, \tag{4.4.32}$$

$$\Delta I_{\mathrm{s}} = \frac{\omega_{\mathrm{s}} L}{\varepsilon_0 c^2 \tilde{n}^{(1)}(\omega_{\mathrm{p}})\tilde{n}^{(1)}(\omega_{\mathrm{s}})} \mathrm{Im}\left\{\tilde{\chi}^{(\mathrm{RR})}(\omega_{\mathrm{p}} - \omega_{\mathrm{s}})\right\} I_{\mathrm{p}} I_{\mathrm{s}}, \tag{4.4.33}$$

where

$$\chi^{(\mathrm{RR})}(\omega_{\mathrm{p}} - \omega_{\mathrm{s}}) = f_{\mathrm{R}} \chi^{(3)} \frac{\omega_{\mathrm{RR}}^2 + \Gamma_{\mathrm{RR}}^2}{\omega_{\mathrm{RR}}} \frac{1}{\omega_{\mathrm{RR}} - (\omega_{\mathrm{p}} - \omega_{\mathrm{s}}) - i\Gamma_{\mathrm{RR}}}. \tag{4.4.34}$$

Here $\mathrm{Im}\{h\}$ denotes the imaginary part of the complex $h = \mathrm{Re}\{h\} + i\mathrm{Im}\{h\}$, and I_{p} and I_{s} are the pump and Stokes intensities, respectively. The SNR of SRS microscopy is given by

$$SNR_{\mathrm{SRL}} = \frac{\Delta I_{\mathrm{p}}}{a(f) I_{\mathrm{p}} + \sqrt{I_{\mathrm{p}}}}, \tag{4.4.35}$$

Here the first term in denominator of Eq. (4.4.35) denotes the laser intensity noise of the pump light, and the second term is the shot noise of the pump light. When the Stokes intensity is modulated at a high frequency and the pump intensity is detected through the lock-in detection at f, the shot-noise-limit sensitivity can be achieved. Then, the SNR of SRS microscopy is

$$SNR_{\mathrm{SRL}} = \frac{\mathrm{Im}\left\{\tilde{\chi}^{(\mathrm{RR})}(\omega_{\mathrm{p}} - \omega_{\mathrm{s}})\right\} \omega_{\mathrm{p}} L \sqrt{I_{\mathrm{p}}} I_{\mathrm{s}}}{\varepsilon_0 c^2 \tilde{n}^{(1)}(\omega_{\mathrm{p}})\tilde{n}^{(1)}(\omega_{\mathrm{s}})}, \tag{4.4.36}$$

Next, we consider the SNR of CARS microscopy. We assume that the interaction length L is much shorter than the inverse of the wave vector mismatch Δk. Then, according to Eq. (1.1.177), the CARS intensity is expressed by

$$I_{\mathrm{CARS}} = \frac{\omega_4^2 L^2 \left|\tilde{\chi}_{\mathrm{R}}^{(\mathrm{NR})} + \tilde{\chi}^{(\mathrm{RR})}(\omega_{\mathrm{p}} - \omega_{\mathrm{s}})\right|^2 I_{\mathrm{p}} I_{\mathrm{s}} I_{\mathrm{pr}}}{16\varepsilon_0^2 c^4 \tilde{n}^{(1)}(\omega_{\mathrm{p}})\tilde{n}^{(1)}(\omega_{\mathrm{s}})\tilde{n}^{(1)}(\omega_{\mathrm{pr}})\tilde{n}^{(1)}(\omega_4)}, \tag{4.4.37}$$

where

$$\chi_{\mathrm{R}}^{(\mathrm{NR})} = (6 - 2f_{\mathrm{R}})\chi^{(3)}. \tag{4.4.38}$$

Here I_p, I_s, and I_{pr} are the pump, Stokes, and probe intensities, respectively. In a dilute sample, the CARS intensity can be approximated by

$$I_{CARS} \approx \frac{\omega_4^2 L^2 \left[\left| \tilde{\chi}_R^{(NR)} \right|^2 + 2\tilde{\chi}_R^{(NR)} \mathrm{Re}\left\{ \tilde{\chi}^{(RR)}(\omega_p - \omega_s) \right\} \right] I_p I_s I_{pr}}{16\varepsilon_0^2 c^4 \tilde{n}^{(1)}(\omega_p)\tilde{n}^{(1)}(\omega_s)\tilde{n}^{(1)}(\omega_{pr})\tilde{n}^{(1)}(\omega_4)}, \quad (4.4.39)$$

because we can assume $\left|\tilde{\chi}^{(RR)}(\omega_p - \omega_s)\right| \ll \left|\tilde{\chi}^{(NR)}\right|$. Thus, the resonant signal is expressed by

$$I_{CARS}^{(S)} = \frac{2\tilde{\chi}_R^{(NR)} \mathrm{Re}\left\{ \tilde{\chi}^{(RR)}(\omega_p - \omega_s) \right\} \omega_4^2 L^2 I_p I_s I_{pr}}{16\varepsilon_0^2 c^4 \tilde{n}^{(1)}(\omega_p)\tilde{n}^{(1)}(\omega_s)\tilde{n}^{(1)}(\omega_{pr})\tilde{n}^{(1)}(\omega_4)}, \quad (4.4.40)$$

whereas the nonresonant background is expressed by

$$I_{CARS}^{(B)} = \frac{\left| \tilde{\chi}_R^{(NR)} \right|^2 \omega_4^2 L^2 I_p I_s I_{pr}}{16\varepsilon_0^2 c^4 \tilde{n}^{(1)}(\omega_p)\tilde{n}^{(1)}(\omega_s)\tilde{n}^{(1)}(\omega_{pr})\tilde{n}^{(1)}(\omega_4)}. \quad (4.4.41)$$

Because of $I_{CARS}^{(B)} \gg I_{CARS}^{(S)}$, the shot noise and the laser $1/f$ noise derive from the nonresonant background. Therefore, the SNR of CARS microscopy becomes

$$SNR_{CARS} = \frac{I_{CARS}^{(S)}}{a(f) I_{CARS}^{(B)} + \sqrt{I_{CARS}^{(B)}}}. \quad (4.4.42)$$

Here the first term in denominator of Eq. (4.4.42) denotes the low-frequency intensity noise induced by the nonresonant background due to the $1/f$ noise of the excitation lasers, and the second term represents the shot noise of the nonresonant background. When the shot-noise-limit sensitivity where $a(f)$ is 0 is achieved, the SNR of CARS microscopy becomes

$$SNR_{CARS} = \frac{\mathrm{Re}\left\{ \tilde{\chi}^{(RR)}(\omega_p - \omega_s) \right\} \omega_4 L \sqrt{I_p I_s I_{pr}}}{2\varepsilon_0 c^2 \sqrt{\tilde{n}^{(1)}(\omega_p)\tilde{n}^{(1)}(\omega_s)\tilde{n}^{(1)}(\omega_{pr})\tilde{n}^{(1)}(\omega_4)}}, \quad (4.4.43)$$

It should be noted that the SNR of SRS microscopy is approximately equal to that of CARS microscopy under the shot-noise-limited condition, so that the laser intensity fluctuation can be eliminated completely.

Ozeki et al. have demonstrated SRS microscopy [163]. Figure 4.28(a) and (b) shows the SRS and CARS images of the polystyrene bead at a difference frequency of 2967 cm^{-1}, respectively. In the SRS image, the water does not emit any signal, so there is high contrast. In turn, the CARS image is accompanied by a large amount of nonresonant background from water. These characteristics are clearly presented in one-dimensional profiles shown in Fig. 4.28(c). Figure 4.28(d) shows the SRS

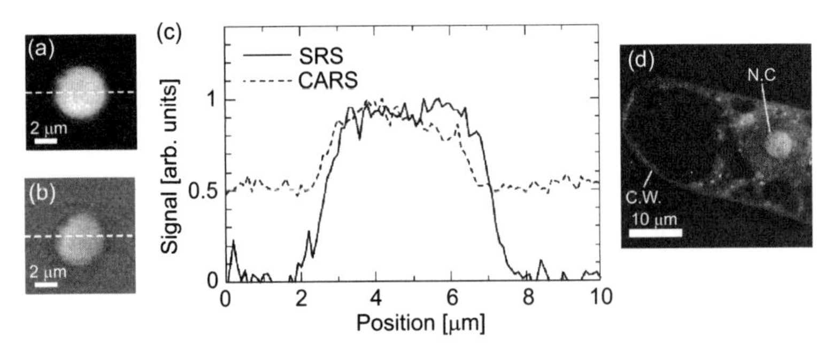

FIGURE 4.28 (a, b) SRS (a) and CARS (b) images of a polystyrene bead with a diameter of 4.5 μm. (c) Line profiles at the broken lines in (a) and (b). (d) The SRS image of an unstained tobacco BY-2 cell. Reprinted with permission from [163]. Copyright 2009 Optical Society of America.

image of a tobacco BY-2 cultured cell, which is based on vibrational contrast by CH stretching modes. The nucleus (N.C.) and cell walls (C. W.) are clearly seen.

4.5 SECOND-HARMONIC GENERATION IMAGING

Second-harmonic generation (SHG) is a second-order nonlinear optical process where two photons of frequency ω_1 are destroyed and one photon of frequency of ω_2 is generated as described in Section 1.1.2. SHG is allowed only for non-centrosymmetric media. In 1974, Hellwarth and Christensen first integrated SHG into an optical microscope to visualize the microscopic crystal structure in polycrystalline ZnSe [1]. The concept of SHG microscopy was also demonstrated in 1977 by Sheppard et al. [166]. The first biological application of SHG microscopy was imaging the endogenous collagen structure in a rat tail tendon by Freund et al. in 1986 [168]. Because structural proteins in cells and tissues, such as collagen, actomyosin complexes, and tubulin, are organized into chiral structures, and form arrays that are highly ordered and birefringent [167], significant SHG output can be obtained from the structural proteins without the additional fluorescent molecules [168–171]. Recently, SHG microscopy was applied for morphological visualizations of collagen fibrils in connective tissues [168,169], orderly arranged microtubules in mitotic spindles [170], and actomyosin lattices in muscle fiber [169,171], and polarization resolved techniques can be used to obtain orientation information about structural proteins [169], [172–175]. SHG microscopy has been also used for imaging oriented and organized structures such as amylopectin/amylase crystallized layers in starch granules [176] and cellulose fibrils in cell walls [177]. The ability to image completely off resonance has the great benefit of the virtual elimination of photonbleaching and phototoxicity, specially at longer wavelengths ($\lambda > 850$ nm) [2].

The refractive index difference between 800 nm (ω_1) and 400 nm ($2\omega_1$) in tissues is 0.03 to 0.08 [178,179]. The measured dispersion in collagen was $\tilde{n}(2\omega_1) - \tilde{n}(\omega_1) =$

0.02 [180]. Thus, it can be reasoned that perfect phase matching does not occur in biological tissues in the visible and the near infrared spectral ranges. Recently, LaComb et al. suggested that the intensity amplification of SHG in biological tissues derives from quasi-phase matching (QPM) and concluded that the SHG conversion efficiency is related not only to the fibril size but also to the packing density and order of the inter-fibril structure [180]. In general, the SHG signal is detected in the forward direction because SHG is a coherent process. However, objects with an axial size less than λ_{SHG} /10 (approx. 40 nm) exhibit not solely forward SHG signals but also backward SHG signals since destructive interference in the backward direction is incomplete [181,182]. The forward-to-backward ratio depends on the packing density and order of the inter-fibril structure as well as the fibril size [180]. In such a case, the forward-to-backward ratio can be explained by QPM theory [180].

Since SHG also arises from an asymmetric interface [183] such as cell membranes in which one leaflet has been labeled with absorbing molecules and/or fluorophores [184,185], SHG microscopy has become a powerful tool in probing membrane physiology [186]. In 1988 Huang et al. characterized SHG signals from styryl dyes, which are chiral dyes with very large non-linear optical effects, in Langmuir-Blodgett monolayers [187], and Bouevitch et al. extended this work in 1993 to determine the dependence on potential of a model membrane stained with a voltage-sensitive styryl dye on a hemispherical bilayer apparatus [188]. In 1996, Ben-Oren et al. used SHG imaging to monitor slow membrane potential responses after stimulation of photoreceptor cells by visible light [189]. Because in a membrane the presence of a chiral center relaxes the requirement of the electric dipole "selection rule" that the assembly of molecules is noncentrosymmetric, SHG from molecules with a chiral center that can sufficiently weaken the symmetry, even in cases where dye has equilibrated between the two membrane leaflets, is enhanced [184]. A strong SHG signal from achiral dyes can be obtained only if the dyes stain just one leaflet of the membrane bilayer because the dyes are applied from the external bath [185]. SHG signals can be also enhanced by use of near-resonance excitation of absorbing molecules and/or fluorophores [184,185]. Some collateral photobleaching and phototoxicity may occur in resonance-enhanced SHG. Most of the earlier work concerning SHG voltage-sensing used nonbiological systems and/or indirect methods to manipulate trans-membrane potential (TMP) [184,185,188–190]. In 2003, Millard et al. first determined the SHG voltage sensitivity of membrane-staining dyes by voltage clamping live cells in order to directly control TMP [191]. This technique has been used to image membrane potential in dendritic spines [192] and in cultured Aplysia neurons [193].

4.5.0.1 Detection of Strongly Birefringent Media According to Eq. (1.1.109), the SHG intensity by a linearly polarized pulse in the tight-focusing limit is expressed by

$$I_{\text{SHG}}(z) \propto I_1^2 \left| \int_{z_0}^{z} \frac{\chi_{\text{eff}}^{(2)}(z') e^{i \Delta k_{\text{SHG}}^{(2)} z'}}{1 + i z'/\rho_1} dz' \right|^2, \tag{4.5.1}$$

where

$$\Delta k_{\mathrm{SHG}}^{(2)} = \frac{2\omega_1 \{\tilde{n}(\omega_1) - \tilde{n}(2\omega_1)\}}{c}, \tag{4.5.2}$$

$$\rho_1 = \frac{1}{2} k_1 w_1^2. \tag{4.5.3}$$

Here $\Delta k_{\mathrm{SHG}}^{(2)}$ is the wave vector mismatch, $\chi_{\mathrm{eff}}^{(2)}$ is the second-order effective nonlinear susceptibility, I_1 is the excitation intensity, and w_1 is the focal spot radius ($1/e^2$). In the case of an infinite, uniform nonlinear medium, Eq. (4.5.1) can be rewritten as

$$I_{\mathrm{SHG}}(z) \propto I_1^2 \left| \chi_{\mathrm{eff}}^{(2)} \right|^2 \left| \int_{-\infty}^{\infty} \frac{e^{i \Delta k_{\mathrm{SHG}}^{(3)} z'}}{1 + iz'/\rho_1} dz' \right|^2, \tag{4.5.4}$$

Even if the second-order effective nonlinear susceptibility is not equal to zero, the efficiency of SHG in this limit vanishes in a normally dispersive material ($\tilde{n}(\omega_1) < \tilde{n}(2\omega_1)$) because a negative phase mismatch ($\Delta k_{\mathrm{SHG}}^{(2)} < 0$) and the Gouy phase shift of the excitation beam near focus results in complete destructive interference between the SHG signals generated before and after the focus. As shown in Section 1.1.2, strong birefringence can be used to compensate for unfavorable phase mismatches. Therefore, SHG microscopy can be used to observe strong birefringent crystals.

4.5.0.2 Detection of Interfaces Between Two Uniform Media That Differ in Nonlinear Properties When the symmetry along the optical axis is broken, destructive interference is not complete. For example, when the laser beam is focused at the position z_w, which is the distance from the interface between two media with the same linear refractive index but with different second-order nonlinear susceptibility, the SHG signal is expressed by

$$\begin{aligned}
I_{\mathrm{SHG}}(z_w) &\propto I_1^2 \left| \chi_1^{(2)} \int_{-\infty}^{z_w} \frac{e^{i \Delta k_{\mathrm{SHG}}^{(2)} z'}}{1 + iz'/\rho_1} dz' + \chi_2^{(2)} \int_{z_w}^{\infty} \frac{e^{i \Delta k_{\mathrm{SHG}}^{(2)} z'}}{1 + iz'/\rho_1} dz' \right|^2 \\
&= I_1^2 \left| \chi_1^{(2)} \int_{-\infty}^{z_w} \frac{e^{i \Delta k_{\mathrm{SHG}}^{(2)} z'}}{1 + iz'/\rho_1} dz' \right. \\
&\quad \left. + \chi_2^{(2)} \left\{ \int_{-\infty}^{\infty} \frac{e^{i \Delta k_{\mathrm{SHG}}^{(2)} z'}}{1 + iz'/\rho_1} dz' - \int_{-\infty}^{z_w} \frac{e^{i \Delta k_{\mathrm{SHG}}^{(2)} z'}}{1 + iz'/\rho_1} dz' \right\} \right|^2 \\
&= I_1^2 \delta \chi^{(2)2} \left| \int_{-\infty}^{z_w} \frac{e^{i \Delta k_{\mathrm{SHG}}^{(2)} z'}}{1 + iz'/\rho_1} dz' \right|^2 \\
&\approx I_1^2 \delta \chi^{(2)2} \rho_1^2 \left| \left[\ln \left| 1 + iz'/\rho_1 \right| \right]_{-\infty}^{z_w} \right|^2 \\
&= \frac{1}{4} I_1^2 \delta \chi^{(2)2} \rho_1^2 \left| \lim_{z \to \infty} \ln \left\{ \frac{1 + (z/\rho_1)^2}{1 + (z_w/\rho_1)^2} \right\} \right|^2,
\end{aligned} \tag{4.5.5}$$

where $\delta\chi^{(2)}$ is the difference in second-order susceptibility values. According to Eq. (4.5.5), we see signal peaks when the interface is at the beam waist position. Therefore, SHG microscopy provides the detection of interfaces between two uniform media that differ in nonlinear properties.

4.5.0.3 Detection of Objects With an Axial Size Smaller Than Wavelength/10 Let us consider SHG from objects whose size d in the axial direction is much smaller than ρ_1, which allows appreciable SHG signals. By inserting Eq. (1.1.110) into Eq. (4.5.1), we can write such SHG signals as

$$I_{\mathrm{SHG}} \propto I_1^2 \left|\chi_{\mathrm{eff}}^{(2)}\right|^2 d^2 \mathrm{sinc}^2 \left\{ \frac{\Delta k_{\mathrm{SHG}}^{(2)} d}{2} \right\}. \tag{4.5.6}$$

We see that an object with a smaller axial size generates incomplete destructive interference. Under this condition, backward SHG signals can be also obtained:

$$I_{\mathrm{BSHG}} \propto I_1^2 \left|\chi_{\mathrm{eff}}^{(2)}\right|^2 d^2 \mathrm{sinc}^2 \left\{ \frac{\Delta k_{\mathrm{BSHG}}^{(2)} d}{2} \right\}. \tag{4.5.7}$$

where $\Delta k_{\mathrm{BSHG}}^{(2)}$ is the wave vector mismatch expressed by

$$\Delta k_{\mathrm{BSHG}}^{(2)} = -\frac{2\omega_1 \{\tilde{n}(\omega_1) + \tilde{n}(2\omega_1)\}}{c}. \tag{4.5.8}$$

According to Eqs. (4.5.6) and (4.5.7), if the axial size d is much smaller than the wavelength $\lambda_1 = 2\pi c/\omega_1$, we can acquire nearly equal backward and forward SHG signals.

4.5.0.4 Quasi-Phase Matching in Biological Tissues Although SHG is a coherent process, appreciable backward SHG from collagen has been experimentally observed [182,194,195]. To give an explanation of the backward SHG signals, three hypotheses for the source of the detected backward SHG have been proposed [196]. First, the backward SHG signals arises from subsequent multiple scattering of the forward SHG in tissue. Second, the backscattering of ballistic excitation photons induces forward SHG as the backward SHG signals. Third, the backward SHG signals are directly emitted from the collagen fibrils. However, Légaré et al. demonstrated that the forward and the backward images are significantly different from each other for the imaged collagen type I tissue [195]. Their results indicate that the fraction of the forward SHG signal contributes to the overall backward SHG signal. Thus, the direct backward SHG signals are the main contributor to the overall backward SHG signals. Small fibrils with axial sizes, less than $\lambda_{\mathrm{SHG}}/10$ produce the backward signals. However, it is not sufficient to explain the forward-to-backward ratio by the small fibril size alone. Recently, direct backward SHG signals have been explained by QPM theory [180,198]. Due to the fibrillar hierarchy of collagen, which is generally described as polycrystalline in nature, QPM theory can be applied to the SHG intensity that builds up over the course of multiple fibrillar domains.

For a medium with a periodic modulation of second-order nonlinear susceptibility with a period of Λ and hence fundamental spatial frequency $K_g = 2\pi/\Lambda$, we can write the second-order nonlinear susceptibility as a Fourier series [197],

$$\chi_{\text{eff}}^{(2)}(z) = \sum_m \chi_m^{(2)} e^{iK_m z}, \tag{4.5.9}$$

where the mth spatial harmonic $K_m = mK_g$, and $\chi_m^{(2)}$ is the corresponding Fourier coefficient. For a square-wave modulation from $+\chi_{\text{eff}}^{(2)}$ to $-\chi_{\text{eff}}^{(2)}$ with duty cycle D, the Fourier coefficients are given by [197]

$$\chi_m^{(2)} = \frac{\chi_{\text{eff}}^{(2)} 2\sin(\pi D)}{m\pi}. \tag{4.5.10}$$

As we see in this equation, the largest QPM nonlinear coefficient, obtained for $m = 1$ is reduced below that of a uniform medium by a factor $2/\pi$, and for mth-order QPM a further reduction by a factor $1/m$ appears. In the case of such a periodic medium, Eq. (4.5.1) can be rewritten as

$$I_{\text{SHG}}(z) \propto I_1^2 \left| \sum_m \chi_m^{(2)} \int_{z_0}^{z} \frac{e^{i\Delta k_{\text{F}m}^{(2)} z'}}{1 + iz'/\rho_1} dz' \right|^2, \tag{4.5.11}$$

where $\Delta k_{\text{F}m}^{(2)}$ is the effective phase mismatch expressed by

$$\Delta k_{\text{F}m}^{(2)} = K_m + \Delta k_{\text{SHG}}^{(2)} = \frac{2\pi m}{\Lambda} + \frac{2\omega_1 \{\tilde{n}(\omega_1) - \tilde{n}(2\omega_1)\}}{c}. \tag{4.5.12}$$

In the case of SHG from collagen, the generation of the wave vector K_m derives from the crystalline structure of collagen fibrils [198]. Type I collagen fiber is highly organized in fibrils, exhibiting a quasi-crystalline structure as shown in Fig. 4.29 [199]. Collagen fibrils are the unit fibrils that can be observed in individual collagen fibers by electron microscopy, and are cylindrical in shape with diameters ranging from 10 to 500 nm (mean diameter about 40–80 nm) [200]. The period Λ can be related to the fibrils diameter d_1, and packing density and order of inter-fibrils structure d_2 by the following equation [198]:

$$\Lambda = d_1 + d_2. \tag{4.5.13}$$

So far we have neglected SHG in the backward direction, since in general, the wave vector mismatch for backward SHG $\Delta k_{\text{BSHG}}^{(2)} = -2\omega_1 \{\tilde{n}(\omega_1) + \tilde{n}(2\omega_1)\}/c$, is too large. Since the phase matching for backward SHG can be relaxed according to

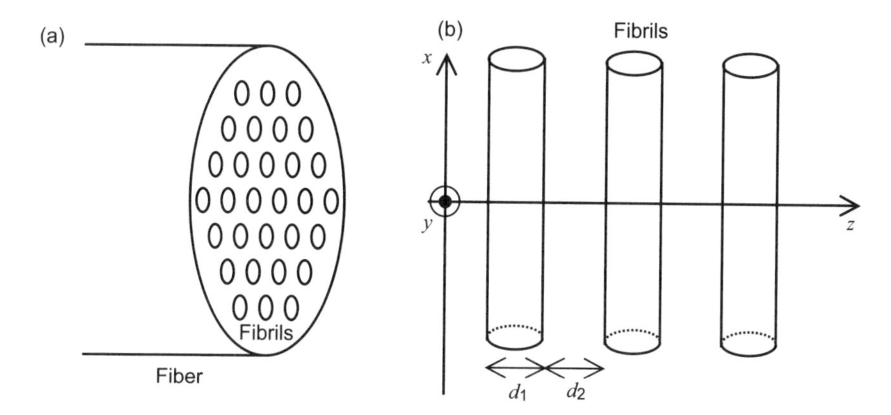

FIGURE 4.29 (a) Quasi-crystalline structural model of type I collagen fiber. (b) Simple crystalline structure model formed by fibrils at a diameter of d_1 that are separated by an interspace of d_2.

QPM, we can also treat SHG in the backward direction. The backward SHG intensity is given by

$$I_{\mathrm{BSHG}}(z) \propto I_1^2 \left| \sum_m \chi_m^{(2)} \int_{z_0}^{z} \frac{e^{i\Delta k_{Bm}^{(2)} z'}}{1 + iz'/\rho_1} dz' \right|^2, \tag{4.5.14}$$

where $\Delta k_{\mathrm{B}m}^{(2)}$ is the effective phase mismatch for backward SHG described by

$$\Delta k_{\mathrm{B}m}^{(2)} = K_m + \Delta k_{\mathrm{BSHG}}^{(2)} = \frac{2\pi m}{\Lambda} - \frac{2\omega_1 \{\tilde{n}(\omega_1) + \tilde{n}(2\omega_1)\}}{c}, \tag{4.5.15}$$

From Eqs. (4.5.12) and (4.5.15), we see that the wave vector K_m for maximizing the forward SHG signals is smaller than that for maximizing the backward SHG signals. Thus, in the case of $m = 1$, the period Λ inducing the highest forward SHG signals is larger than that inducing the highest backward SHG signals. Therefore, the forward and backward SHG efficiencies depend on not only the fibrils size but also packing density and order of inter-fibrils structure [180,198].

LaComb et al. have demonstrated the intensity amplification of SHG in biological tissues by QPM [180]. Figure 4.30(a) and (b) shows SHG images of *Valonia* cellulosein in forward and backward directions, respectively. These samples are approximately 30 μm in thickness, and the mean free path is 130 μm. In addition, the cellulose is highly regular in its structure. Thus, the contrast in the backward SHG image arises predominantly from direct quasi-coherent emission and will not contain a significant multiple-scattered contribution. The fibrils observed in the forward SHG image are long and continuous, while these fibrils frequently have a segmented appearance in the backward SHG image. If the diameter of small fibrils is less than

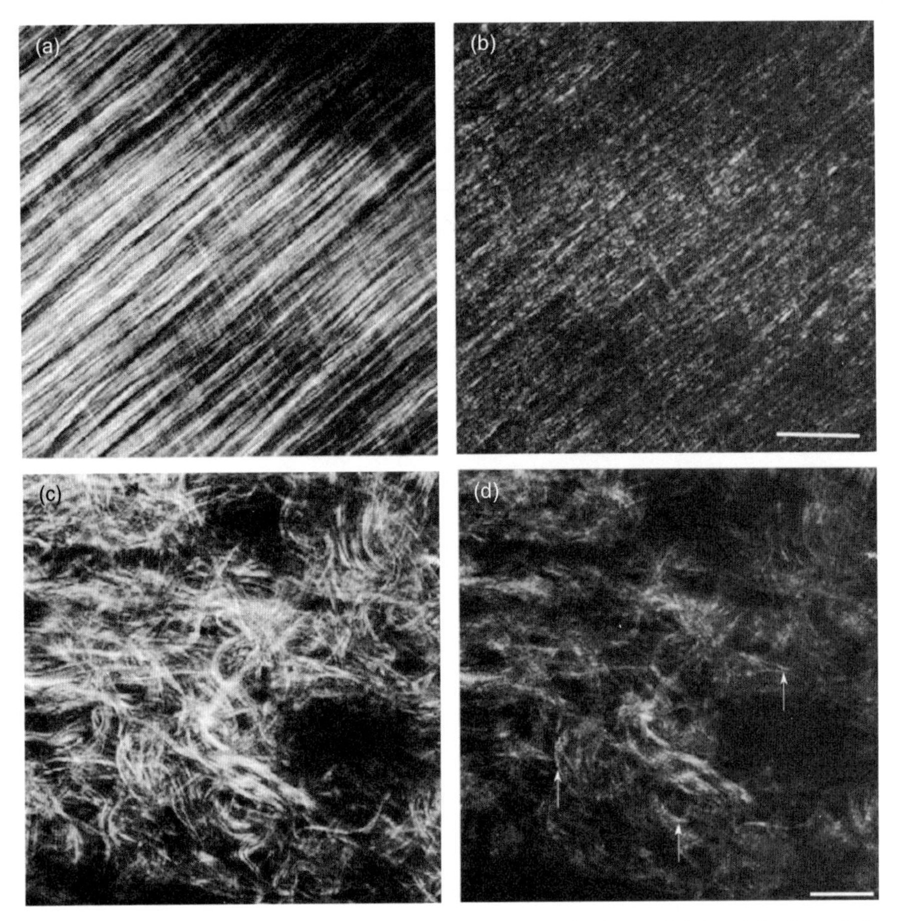

FIGURE 4.30 (a, b) SHG images of *Valonia* cellulose in forward (a) and backward (b) directions. (c, d) SHG images of oim skin in forward (c) and backward (d) directions. Scale bars is 20 μm. Reprinted with permission from [180]. Copyright 2008 Elsevier.

$\lambda_{SHG}/10$ in the absence of the QPM, the backward SHG signals are equal to the forward SHG signals and the small fibrils may appear better visualized in the backward SHG image. Thus, these results cannot be explained by treatments based solely on the fibril size. According to QPM theory, the difference in the forward and backward images can be produced by the period related to the fibrils diameter, and packing density and order of inter-fibrils structure. If the period is long enough to enhance the forward SHG signal owing to constructive interference between fibrils, destructive interference between fibrils in the backward direction diminishes the backward SHG signal and produces the segmented features. Therefore, LaComb et al. have suggested that the segmented features was produced by destructive interference due to the QPM effect. Figure 4.30(c) and (d) shows SHG images of oim skin in forward and backward directions, respectively [180]. As the arrows in Fig. 4.30(d) indicate,

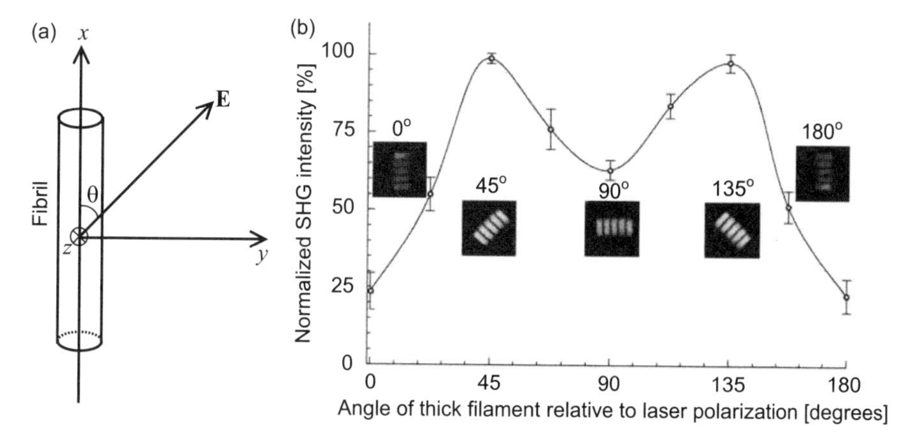

FIGURE 4.31 (a) Sketch of the fibril orientation and incident polarization angle. (b) Profile of SHG intensity versus the relative angle of scallop myofibrils to laser polarization axis. Inserts show changes of SHG intensity with rotation relative to a fixed laser polarization. Reprinted with permission from [173]. Copyright 2006 Elsevier.

the segmented morphology exists, in the backward SHG image. Because the matrix is less organized in the oim skin, the QPM is less efficient and less spatially pervasive destructive interference will occur. Therefore, based on these results, the segmented morphology is clearly due to the physical segmentation of the fibrils.

Han et al. observed corneal collagen fibrils and scleral collagen fibrils in forward and backward directions [182]. In SHG imaging of corneal collagen fibrils, the forward SHG signals arise from the fibrillar structures corresponding to the collagen bundles, which are composed of regularly packed collagen fibrils, whereas the backward SHG from the cornea is extremely weak. By contrast, the backward SHG signals from sclera are significant. According to QPM theory, these results suggest the SHG intensity to be related to the inter-fibril spacing rather than the fibril diameter [180].

4.5.0.5 Polarization-Resolved SHG We now consider the properties of the second-order nonlinear susceptibility tensor of collagen filaments. This effect is obtained as the averaged response of small collagen fibrils aligned in domains within lamellae and it reflects the direction of collagen fibrils in these domains [168,201,173]. We will assume that the fibrillar collagen exhibits a cylindrical symmetry as illustrated in Fig. 4.31(a), so that it exhibits only four independent components [173],

$$
\begin{gathered}
\chi^{(2)}_{xxx},\\
\chi^{(2)}_{xyy} = \chi^{(2)}_{xzz},\\
\chi^{(2)}_{yyx} = \chi^{(2)}_{yxy} = \chi^{(2)}_{zzx} = \chi^{(2)}_{zxz},\\
\chi^{(2)}_{yxz} = \chi^{(2)}_{yzx} = -\chi^{(2)}_{zyx} = -\chi^{(2)}_{zxy},
\end{gathered}
\tag{4.5.16}
$$

while all other components vanish. Here z is the direction of the collagen fibrils. We derive the bulk susceptibility from their molecular hyperpolarizability [173]. For molecules oriented along the molecular axes x', y', z' relative to the bulk axes x, y, z,

$$\chi_{ijk}^{(2)} = \sum_{i'j'k'} \langle \cos\phi_{ii'} \cos\phi_{jj'} \cos\phi_{kk'} \rangle \beta_{i'j'k'}, \tag{4.5.17}$$

where the angle brackets signify averaging over molecules. Here the subscripts ijk denote x, y, and z, the subscripts $i'j'k'$ denote x', y', and z', and β_{ijk} is the hyperpolarizability. For simplicity, we assume a molecule with a single preferred axis of hyperpolarizability

$$\beta_{x'x'x'} = \beta, \tag{4.5.18}$$

with all other components vanishing. Further, assuming that the molecules are distributed with a constant polar angle, $\phi_{xx'} = \phi$, and a random azimuth angle, α, we find that

$$\begin{aligned} \chi_{xxx}^{(2)} &= N\beta \cos^3\phi, \\ \chi_{xyy}^{(2)} &= N\beta \cos\phi \sin^2\phi \langle \sin^2\alpha \rangle = \tfrac{1}{2} N\beta \cos\phi \sin^2\phi, \\ \chi_{yyx}^{(2)} &= \tfrac{1}{2} N\beta \cos\phi \sin^2\phi, \\ \chi_{yxz}^{(2)} &= N\beta \cos\phi \sin^2\phi \langle \sin\alpha \cos\alpha \rangle = 0, \end{aligned} \tag{4.5.19}$$

For a cylindrically symmetric distribution of single-axis molecules, there are only two independent elements, $\chi_{xxx}^{(2)}$ and $\chi_{xyy}^{(2)} = \chi_{yyx}^{(2)}$, and their ratio, $\rho = \chi_{xxx}^{(2)}/\chi_{xyy}^{(2)}$, may be calculated from the characteristic polar angle, ϕ. Considering that light is propagating along direction z, we can write the SHG polarization as

$$\begin{aligned} P_x^{(\mathrm{SHG})} &= \varepsilon_0 \left(\chi_{xxx}^{(2)} E_x^2 + \chi_{xyy}^{(2)} E_y^2 \right) = \varepsilon_0 E_1^2 \left(\chi_{xxx}^{(2)} \cos^2\theta + \chi_{xyy}^{(2)} \sin^2\theta \right), \\ P_y^{(\mathrm{SHG})} &= \varepsilon_0 \left(\chi_{yyx}^{(2)} + \chi_{yxy}^{(2)} \right) E_x E_y = \varepsilon_0 E_1^2 \chi_{xyy}^{(2)} \sin 2\theta, \\ P_z^{(\mathrm{SHG})} &= 0, \end{aligned} \tag{4.5.20}$$

where the fundamental field is expressed by

$$\begin{aligned} E_x &= E_1 \cos\theta, \\ E_y &= E_1 \sin\theta, \\ E_z &= 0, \end{aligned} \tag{4.5.21}$$

Here θ is the angle between the direction of the collagen fibrils, x-axis, and the electric field $\mathbf{E}$ as shown in Fig. 4.31(a). According to Eq. 4.5.20, the SHG intensity can be written as

$$I_{\text{SHG}} \propto I_1^2 \left|\chi_{xyy}^{(2)}\right|^2 \left\{\left(\rho \cos^2\theta + \sin^2\theta\right)^2 + \sin^2 2\theta\right\}, \tag{4.5.22}$$

where $\rho = \chi_{xxx}^{(2)}/\chi_{xyy}^{(2)}$ reflects the anisotropy of the nonlinear response of these lamellar domains. From Eq. (4.5.22), we see that the SHG signal from collagen tissues depends on the incident polarization.

Equation (4.5.22) can be expressed as a sum of Fourier components $\cos(2m\theta)$, with $m = 0, 1, 2$ [172,174,175]. This is an efficient way to process the experimental data and determine the SHG anisotropy parameter ρ, even in the presence of optical artifacts due to diattenuation or birefringence in the propagation of the laser excitation

$$I_{\text{SHG}} = A\cos 4\theta + B\cos 2\theta + C. \tag{4.5.23}$$

The SHG anisotropy ratio ρ is then calculated as [174,175]

$$\rho = \sqrt{\frac{A+B+C}{A-B+C}}, \tag{4.5.24}$$

where we have omitted diattenuation correction. The meaningful orientation parameter O was initially introduced to describe the orientation and disorder of the molecules at the surfaces and interfaces, and it can also be defined from Eq. (4.5.19) [202,203]:

$$O = \frac{<\cos^3\phi>}{<\cos\phi>} = \frac{\rho}{2+\rho} = \cos^2\phi_e. \tag{4.5.25}$$

Regarding ϕ_e, this angle was defined in reference [202] as an effective angle corresponding to the most probable orientation of the active molecules when the distribution of molecular orientation is very narrow [202,203]. Tiaho et al. have determined the effective orientation angle ϕ_e of the harmonophores for myosin ($\phi_e \approx 62°$) and collagen ($\phi_e \approx 49°$) [169].

Figure 4.31(b) shows the relation between the SHG signals from myofibrils isolated from striated scallop adductor muscles and the relative angle of scallop myofibrils to laser polarization [173]. Inserts illustrate the effects of five different incident polarization angles θ (0°, 45°, 90°, 135°, 180°) on the emitted signal from the same field of view. An analyzer (polarizer), which is typically used to measure polarization states of the signal light, is not employed. We see that the SHG intensity from myofibrils depends on the incident polarization.

4.5.0.6 Monitoring of Membrane Potential The membrane potential in live cells can be monitored by SHG microscopy. The membranes are stained with voltage-sensitive dyes [186]. Then, the membrane-bound dye molecules are aligned in the absence of a field. Because of their placement in the membrane, a second-order nonlinear susceptibility arises from the membrane surface and induces a steady-state SHG signal. In addition to this surface effect, a second-order nonlinear susceptibility that derives from the intramembrane electric field produces a voltage sensitivity SHG signal [186]. The SHG signals can even be enhanced by exciting the dyes under the two-photon electronic resonance condition [184,185]. The polarization of the SHG with voltage sensitivity can be expressed by including a third-order term from the Taylor series as follows [204]:

$$P^{(2)}_{\mathrm{VSHG}} = \varepsilon_0 \left(\chi^{(2)}_{\mathrm{surface}} + \chi^{(3)} E_{\mathrm{DC}}\right) E_1^2 \equiv \varepsilon_0 \chi^{(2)}_{\mathrm{eff}} E_1^2, \tag{4.5.26}$$

where the effective second-order nonlinear susceptibility is written as

$$\chi^{(2)}_{\mathrm{eff}} = \chi^{(2)}_{\mathrm{surface}} + \chi^{(3)} E_{\mathrm{DC}}. \tag{4.5.27}$$

Here $\chi^{(2)}_{\mathrm{surface}}$ arises from the structural asymmetry of the interface and E_{DC} is a static electric field. The effective second-order nonlinear susceptibility has the same symmetry constraints as $\chi^{(2)}$ and is hence restricted to surfaces. Because the third-order nonlinear susceptibility is typically four to five orders of magnitude smaller than the second-order nonlinear susceptibility, the second term in Eq. (4.5.27) is usually negligible. However, since typical intramembrane electric fields are 10^5 V/m [184], this term can become significant for a cell. This mechanism is different from that of electric field induced seconds-harmonic (EFISH) generation, in which an applied field aligns a random distribution of molecules and is solely responsible for any observed second-harmonic signal.

Millard et al. have reported a SHG signal from voltage-clamped cells stained with voltage sensitive dyes, di-4-ANEPPS [191]. According to their report, the SHG signal from di-4-ANEPPS was linearly dependent on that membrane potential at voltages ranging from $+25$ mV to -100 mV. The voltage sensitivity of SHG is two to four times higher than that of fluorescence [2,191].

4.6 REFRACTIVE INDEX IMAGING

4.6.1 Third-Harmonic Generation Microscopy

Third-harmonic generation (THG) is a third-order nonlinear optical process where three photons of frequency ω_1 are destroyed and one photon of frequency of $3\omega_1$ is generated as described in Section 1.1.3. According to Eq. (1.1.159), THG intensity due to a linearly polarized pulse in the tight-focusing limit is expressed by

$$I_{\mathrm{THG}}(z) \propto I_1^3 \left| \int_{z_0}^{z} \frac{\chi^{(3)}(z') e^{i\Delta k^{(3)}_{\mathrm{THG}} z'}}{(1 + iz'/\rho_1)^2} dz' \right|^2, \tag{4.6.1}$$

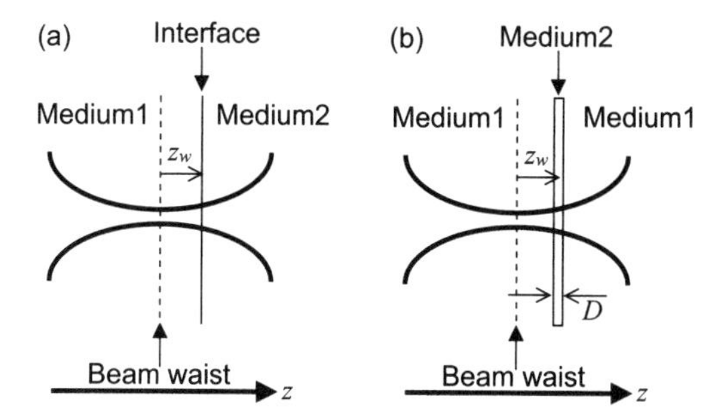

FIGURE 4.32 (a) THG near the interface between two media with the same linear refractive index but with different nonlinear susceptibility, or with the same nonlinear susceptibility but with linear refractive index. (b) THG near a thin film with a thickness $D \ll \rho_1$ embedded in a homogeneous medium.

where

$$\Delta k_{\mathrm{THG}}^{(3)} = 3\omega_1 \{\tilde{n}(\omega_1) - \tilde{n}(3\omega_1)\}/c, \tag{4.6.2}$$

$$\rho_1 = \frac{1}{2} k_1 w_1^2. \tag{4.6.3}$$

Here $\Delta k_{\mathrm{THG}}^{(3)}$ is the wave vector mismatch, I_1 is the excitation, intensity, and w_1 is the focal spot radius ($1/e^2$). In the case of an infinite, uniform nonlinear medium, this integral can be evaluated by contour integration. Unlike SHG, THG does not require asymmetry and can be obtained from any medium. However, the efficiency of THG in this limit vanishes in a normally dispersive material ($\tilde{n}(\omega_1) < \tilde{n}(3\omega_1)$) because a negative phase mismatch ($\Delta k_{\mathrm{THG}}^{(3)} < 0$) and the Gouy phase shift of the excitation beam near the focus results in complete destructive interference between the THG signals generated before and after the focus. However, when the nonlinear medium is not uniform, either in the refractive index or in the nonlinear susceptibility as shown in Fig. 4.32(a), the THG signal does not vanish [4]. Moreover, the THG signals are also observed from strongly birefringent media [206]. Thus, THG microscopy can produce images of the interfaces formed by refractive-index inhomogeneities [3,4,205], and this enables us to detect anisotropy variations [206,207] in a specimen. THG microscopy has been applied to the observation of various structures in biotissues [208–210], intracellular Ca^{2+} dynamics [211], embryos [212], hamster oral cavities [213], micrometer-sized lipid bodies [214], muscle fibers [201], and elastic fibers [215]. In THG microscopy, localized spectroscopic information can also be obtained through the technique of resonance enhancement with absorbing molecules and/or fluorophores [216].

4.6.1.1 Detection of Interfaces or Optical Inhomogeneities between Two Media That Differ in Linear or Nonlinear Properties

When an interface between two media with the same linear refractive index but with different nonlinear susceptibility is present at the beam focus of the linearly polarized pulse as illustrated in Fig. 4.32(a), the THG signal does not vanish, and significant THG output can be obtained [4]. If we assume $\Delta k^{(3)}_{\mathrm{THG}}\rho_1 \ll 2\pi$, the THG signal can be written as

$$\begin{aligned}
I_{\mathrm{THG}}(z_w) &\propto I_1^3\left|\chi_1^{(3)}\int_{-\infty}^{z_w}\frac{e^{i\Delta k^{(3)}_{\mathrm{THG}}z'}}{(1+iz'/\rho_1)^2}dz' + \chi_2^{(3)}\int_{z_w}^{\infty}\frac{e^{i\Delta k^{(3)}_{\mathrm{THG}}z'}}{(1+iz'/\rho_1)^2}dz'\right|^2\\
&= I_1^3\left|\chi_1^{(3)}\int_{-\infty}^{z_w}\frac{e^{i\Delta k^{(3)}_{\mathrm{THG}}z'}}{(1+iz'/\rho_1)^2}dz'\right.\\
&\quad\left.+\chi_2^{(3)}\left\{\int_{-\infty}^{\infty}\frac{e^{i\Delta k^{(3)}_{\mathrm{THG}}z'}}{(1+iz'/\rho_1)^2}dz' - \int_{-\infty}^{z_w}\frac{e^{i\Delta k^{(3)}_{\mathrm{THG}}z'}}{(1+iz'/\rho_1)^2}dz'\right\}\right|^2\\
&= I_1^3\delta\chi^{(3)2}\left|\int_{-\infty}^{z_w}\frac{e^{i\Delta k^{(3)}_{\mathrm{THG}}z'}}{(1+iz'/\rho_1)^2}dz'\right|^2\\
&\approx I_1^3\delta\chi^{(3)2}\frac{\rho_1^2}{1+z_w^2/\rho_1^2},
\end{aligned} \tag{4.6.4}$$

where $\delta\chi^{(3)} = \chi_1^{(3)} - \chi_2^{(3)}$ is the difference in the nonlinear susceptibility values and z_w is the distance between the interface and the beam waist. The signal peaks when the interface is at the beam waist position, and its FWHM is $2\rho_1$. The THG signal also does not vanish when the linearly polarized beam is focused near an interface between two media with the same nonlinear susceptibility but with different linear refractive index. Similarly, a thin film with a thickness $D \ll \rho_1$ embedded in a homogeneous medium will generate a THG signal with [4]

$$\begin{aligned}
I_{\mathrm{THG}}(z_w) &\propto I_1^3\left|\chi_1^{(3)}\int_{-\infty}^{z_w-D/2}\frac{e^{i\Delta k^{(3)}_{\mathrm{THG}}z'}}{(1+iz'/\rho_1)^2}dz'\right.\\
&\quad+\chi_2^{(3)}\int_{z_w-D/2}^{z_w+D/2}\frac{e^{i\Delta k^{(3)}_{\mathrm{THG}}z'}}{(1+iz'/\rho_1)^2}dz'\\
&\quad\left.+\chi_1^{(3)}\int_{z_w+D/2}^{\infty}\frac{e^{i\Delta k^{(3)}_{\mathrm{THG}}z'}}{(1+iz'/\rho_1)^2}dz'\right|^2\\
&= I_1^3\left|\chi_2^{(3)}\int_{z_w-D/2}^{z_w+D/2}\frac{e^{i\Delta k^{(3)}_{\mathrm{THG}}z'}}{(1+iz'/\rho_1)^2}dz'\right.\\
&\quad\left.+\chi_1^{(3)}\left\{\int_{-\infty}^{\infty}\frac{e^{i\Delta k^{(3)}_{\mathrm{THG}}z'}}{(1+iz'/\rho_1)^2}dz' - \int_{z_w-D/2}^{z_w+D/2}\frac{e^{i\Delta k^{(3)}_{\mathrm{THG}}z'}}{(1+iz'/\rho_1)^2}dz'\right\}\right|^2\\
&= I_1^3\delta\chi^{(3)2}\left|\int_{z_w-D/2}^{z_w+D/2}\frac{e^{i\Delta k^{(3)}_{\mathrm{THG}}z'}}{(1+iz'/\rho_1)^2}dz'\right|^2\\
&\approx I_1^3\delta\chi^{(3)2}\frac{D^2}{\left(1+z_w^2/\rho_1^2\right)^2}.
\end{aligned} \tag{4.6.5}$$

Again, we see that the signal is generated only when the film is near the focal plane as depicted in Fig. 4.32(b), now with a FWHM of $2\sqrt{\sqrt{2}-1}\rho_1 \approx 1.28\rho_1$. Note that when the symmetry along the optical axis is broken, the destructive interference is not complete and the THG signal does not vanish. Therefore, THG microscopy using a linearly polarized pulse provides the detection of interfaces or optical inhomogeneities between two media that differ in linear or nonlinear properties.

4.6.1.2 Detection of Anisotropy Variations No THG is observed from isotropic media illuminated with circularly polarized light, even in the presence of inhomogeneity [206]. The x and y components of the THG polarization induced by the circularly polarized pulse under the nonresonant condition is expressed by

$$\begin{aligned} P_x^{(\mathrm{THG})} &= \varepsilon_0 \chi_{xxxx}^{(3)} E_x^3 + \varepsilon_0 \left(\chi_{xxyy}^{(3)} + \chi_{xyyx}^{(3)} + \chi_{xyxy}^{(3)}\right) E_x E_y^2, \\ P_y^{(\mathrm{THG})} &= \varepsilon_0 \chi_{yyyy}^{(3)} E_y^3 + \varepsilon_0 \left(\chi_{yyxx}^{(3)} + \chi_{yxxy}^{(3)} + \chi_{yxyx}^{(3)}\right) E_x^2 E_y. \end{aligned} \tag{4.6.6}$$

In the case of an isotropic medium, the nonlinear susceptibility is related to

$$\chi_{xxxx}^{(3)} = \chi_{xxyy}^{(3)} + \chi_{xyyx}^{(3)} + \chi_{xyxy}^{(3)}. \tag{4.6.7}$$

By inserting Eq. (4.6.7) into Eq. (4.6.6), we obtain

$$\begin{aligned} P_x^{(\mathrm{THG})} &= \varepsilon_0 \chi_{xxxx}^{(3)} E_x \left(E_x^2 + E_y^2\right), \\ P_y^{(\mathrm{THG})} &= \varepsilon_0 \chi_{yyyy}^{(3)} E_y \left(E_x^2 + E_y^2\right). \end{aligned} \tag{4.6.8}$$

In the case of circular polarization, the x and y components of the excitation field are related to the following equation:

$$E_x = i E_y. \tag{4.6.9}$$

From Eqs. (4.6.8) and (4.6.9), we see that the THG polarization vanishes. Therefore, even if the circularly polarized pulse is focused near the interfaces or the optical inhomogeneities between two isotropic media differ in linear or nonlinear properties, no THG is observed.

However, this situation can be different for strongly birefringent media illuminated with circularly polarized light. As in standard phase matching of second-order nonlinear processes shown in Section 1.1.2, strong birefringence can be used to compensate for unfavorable phase mismatches. Because the use of the strong birefringence provides a positive phase mismatch ($\Delta k_{\mathrm{THG}}^{(3)} > 0$), the integral in Eq. (4.6.1) is not equal to zero but maximized. Let us take the example of calcite, which is a negative uniaxial crystal ($n_e < n_o$) of trigonal crystal symmetry (point group $\bar{3}m$). Here n_e is the refractive index for the extraordinary wave, which is polarized parallel to the plane containing the optic axis of the crystal and the propagation direction, and n_o is the refractive index for the ordinary wave, which is polarized perpendicular to the extraordinary wave. A positive phase mismatch can be achieved by a combination of two ordinary waves (o) and an extraordinary wave (e). Then, the third-harmonic (TH) wave is polarized parallel to the extraordinary wave and the wave vector mismatch becomes [217]

$$\Delta k_{\mathrm{THG}}^{(3)} = \frac{\omega_1 \{2\tilde{n}_o(\omega_1) + \tilde{n}_e(\omega_1) - 3\tilde{n}_e(3\omega_1)\}}{c}. \tag{4.6.10}$$

In the crystal frame (x, y, z), the circular polarized field is given by

$$\mathrm{E} = \frac{E_0}{\sqrt{2}}\mathbf{e}^{(o)} + i\frac{E_0}{\sqrt{2}}\mathbf{e}^{(e)}, \tag{4.6.11}$$

where $\mathbf{e}^{(o)}$ and $\mathbf{e}^{(e)}$ are unit vectors of the ordinary and extraordinary field and are expressed as

$$\mathbf{e}^{(o)} = (\sin\phi, -\cos\phi, 0), \tag{4.6.12}$$
$$\mathbf{e}^{(e)} = (\cos\theta\cos\phi, \cos\theta\sin\phi, -\sin\theta). \tag{4.6.13}$$

Here we neglected the walk-off angle between the wave vector $\mathbf{k}$ and the pointing vector $\mathbf{s}$. The z-axis parallel to the optical axis (c-axis), θ is the angle between the optic z-axis and the wave vector $\mathbf{k}$ and ϕ is the angle between the projection of $\mathbf{k}$ in the xy plane and the x-axis. In case of $ooe \to e$ interaction, the THG polarization in the calcite crystal excited by the circularly polarized pulse under the nonresonant condition is expressed by

$$P_x^{(\mathrm{THG})} = i\frac{\varepsilon_0 E_0^3}{2\sqrt{2}}\left\{\chi_{xxxx}^{(3)}\cos\theta\cos\phi + 3\chi_{zxxx}^{(3)}\sin\theta\cos 2\phi\right\}, \tag{4.6.14}$$

$$P_y^{(\mathrm{THG})} = i\frac{\varepsilon_0 E_0^3}{2\sqrt{2}}\left\{\chi_{xxxx}^{(3)}\cos\theta\sin\phi - 3\chi_{zxxx}^{(3)}\sin\theta\sin 2\phi\right\}, \tag{4.6.15}$$

$$P_z^{(\mathrm{THG})} = i\frac{\varepsilon_0 E_0^3}{2\sqrt{2}}\left\{-3\chi_{zyyz}^{(3)}\sin\theta - 3\chi_{zxxx}^{(3)}\left(\cos^2\phi - 3\sin^2\phi\right)\cos\phi\cos\theta\right\}, \tag{4.6.16}$$

where we used the Kleinman symmetry and the spatial symmetry for a point group of $\overline{3}m$ [218]. Because the TH wave is polarized parallel to $\mathbf{e}^{(e)}$, the THG signal is written as

$$I_{\mathrm{THG}}(z) \propto \left|\mathbf{P}^{(\mathrm{THG})}\cdot\mathbf{e}^{(e)}\right|^2 \left(2\pi\rho_1^2\Delta k_{\mathrm{THG}}^{(3)} e^{-\rho_1\Delta k_{\mathrm{THG}}^{(3)}}\right)^2, \tag{4.6.17}$$

where

$$\mathbf{P}^{(\mathrm{THG})}\cdot\mathbf{e}^{(e)} = i\frac{\varepsilon_0 E_0^3}{2\sqrt{2}}\left\{\chi_{xxxx}^{(3)}\cos^2\theta + 3\chi_{zyyz}^{(3)}\sin^2\theta + 3\chi_{zxxx}^{(3)}\sin 2\theta\cos 3\phi\right\}. \tag{4.6.18}$$

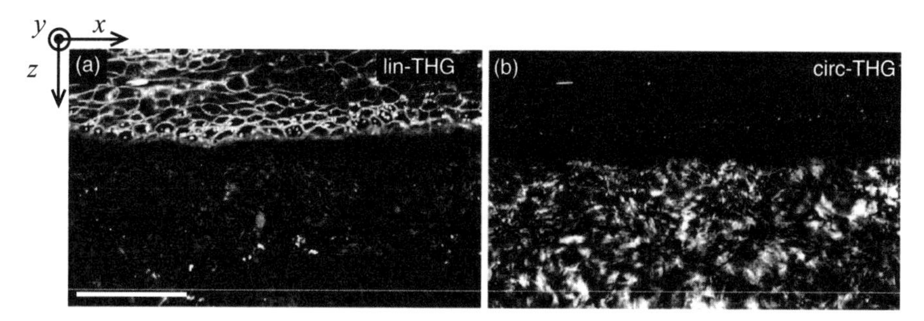

FIGURE 4.33 Polarization-sensitive THG imaging. THG imaging of a cornea with (a) linear and (b) circular incident polarization. Scale bar 100 μm. Reprinted with permission from [207]. Copyright 2010 Optical Society of America.

Moreover, when the beam is focused at an interface between a birefringent medium and an isotropic one, THG is observed even with circularly polarized illumination. This THG signal, unlike the positively phase-mismatched THG, is observed even from weakly birefringent media because a full compensation of the phase mismatch by birefringence is not required. The reason for this is that destructive interference in the weakly birefringent medium is not complete due to the symmetry breaking along the optical axis. It vanishes, however, when the focal point is located within the weakly birefringent medium. Therefore, THG microscopy with circularly polarized excitation can specifically detect anisotropy variations.

Olivier et al. demonstrated THG imaging of the anterior corneal stroma with linear incident polarization (lin-THG) and THG with circular incident polarization (circ-THG) [207]. As can be seen in Fig. 4.33(a) and (b), showing lin-THG and circ-THG images, the two images are very different. Lin-THG reveals cellular structures (epithelial cells and stromal keratocytes) and a weaker component from the stromal lamellae. Circ-THG is obtained specifically from the lamellae. This is because lamellae may be viewed as stacked slabs with alternate anisotropy directions (noted x and y). When the excitation beam (propagating along direction z) is focused near the interface between two xy lamellae, the TH wave from the first lamella emerges with a polarization state different from the one from the second lamella. Thus, destructive interference in the stromal lamellae is not complete even where birefringence is weak. To be sure, THG microscopy provides a ready means to distinguish the two contributions.

4.6.2 Nonresonant Four-Wave Mixing Microscopy

In Sections 4.3.1 and 4.4.1, it was shown that the nonresonant four-wave-mixing (FWM) signals are background signals in SPE and CARS microscopies. In fact, the nonresonant FWM (NFWM) signals can be applied to estimate three-dimensionally localized refractive index in samples because the third-order susceptibility χ^{NR} is related to the refractive indexes [218]. Third-harmonic-generation (THG) microscopy also makes it possible to three-dimensionally detect the interface between two media with different refractive indexes [4]. As shown in Section 4.6.1, the THG signal is generated only when destructive interference based on the so-called Gouy phase shift and the wave vector mismatch is incomplete. This condition cannot be satisfied in uniform nonlinear media. Therefore, it is difficult to apply THG microscopy to quantitatively evaluate refractive index. However, the relative phase between the FWM wave and its driving polarization, which originates from the Gouy phase shift, can be neglected as described in Section 1.1.3. In addition, the phase-matching condition is relaxed by using tight-focusing pulses. Thus, the FWM signal is generated in uniform nonlinear media as well as at interfaces, reflecting a local refractive index in a quantitative manner. Therefore, NFWM microscopy provides refractive index imaging [5].

Now we consider the FWM signal at a frequency of $\omega_4 = \omega_1 + \omega_2 - \omega_3$ generated by focusing into a sample ultrashort optical pulses at frequencies of ω_1, ω_2, and ω_3, which act as two pump pulses (ω_1, ω_2) and a probe pulse (ω_3), under the condition that the refractive index is independent of the frequency. The macroscopic polarization is given by

$$P = \varepsilon_0 \left\{ \chi^{(1)} E + \chi^{(3)} E^3 \right\}, \tag{4.6.19}$$

where ε_0 is the vacuum permittivity, $\chi^{(1)}$ and $\chi^{(3)}$ are the linear and the third-order nonlinear susceptibilities, respectively, and E is the electric field. Here $\chi^{(1)}$ is related to the refractive index n through the following relation:

$$\chi^{(1)} = n^2 - 1. \tag{4.6.20}$$

However, according to (1.1.186), $\chi^{(3)}$ is related to the FWM intensity by

$$I_{\mathrm{FWM}} \propto \left|\chi^{(3)}\right|^2. \tag{4.6.21}$$

The macroscopic polarization of the material P is related to the microscopic polarization of the molecule p by

$$P = Np, \tag{4.6.22}$$

where N is the number density of molecules. The microscopic polarization is expressed by

$$p = \beta E_{\mathrm{loc}} + \gamma E_{\mathrm{loc}}^3, \tag{4.6.23}$$

where β is the linear polarizability and γ is the second-order hyperpolarizability. E_{loc} is the Lorentz local field, which is given by

$$E_{\mathrm{loc}} = E + \frac{P}{3\varepsilon_0}. \tag{4.6.24}$$

These relationships lead to

$$\chi^{(1)} = \frac{L_c N \beta}{\varepsilon_0} \tag{4.6.25}$$

and

$$\chi^{(3)} = \frac{L_c^4 N \gamma}{\varepsilon_0}, \tag{4.6.26}$$

where L_c is the Lorentz's local correction factor and is given by

$$L_c = \frac{n^2 + 2}{3}. \tag{4.6.27}$$

Let us assume the following conditions: (1) The interaction of light with matter is nonresonant, namely frequencies ω_1, ω_2, $\omega_1 + \omega_2$, $\omega_1 - \omega_3$, $\omega_2 - \omega_3$, and $\omega_1 + \omega_2 - \omega_3$ are far away from the resonance frequency of the sample material. (2) The refractive index distribution is due to the variation of the number density N, while β and γ are uniform. Under the condition 1, the second-order hyperpolarizability γ relates the linear polarizability β as [218]

$$\gamma = \frac{g_f \beta^2}{\hbar \omega_r}, \tag{4.6.28}$$

where $\hbar\omega_r$ is the photon energy corresponding to the resonance frequency ω_r and g_f is a free parameter. From these equations, the relationship between the FWM intensity and the refractive index is obtained in the form

$$I_{\text{FWM}} \propto \left\{ \left(n^2 + 2\right)^3 \left(n^2 - 1\right) \right\}^2. \tag{4.6.29}$$

In practice, the refractive index distribution is determined with reference to a known refractive index because Eq. (4.6.29) is a proportional relation. Note that at least one reference point with a known refractive index has to exist at each axial position because the FWM intensity varies along the axial direction mainly due to the on-axis aberration. Then, we can obtain an 8th-order polynominal equation of n:

$$\left(n^2 + 2\right)^3 \left(n^2 - 1\right) = \sqrt{\frac{I_{\text{FWM}}}{I_{\text{FWM}}^{(\text{Ref})}}} \left(n_{\text{Ref}}^2 + 2\right)^3 \left(n_{\text{Ref}}^2 - 1\right), \tag{4.6.30}$$

where $I_{\text{FWM}}^{(\text{Ref})}$ is the FWM intensity at the reference point with a known refractive index n_{Ref}. Equation (4.6.30) can be uniquely solved numerically, for example, using the Newton's method because the left-hand side of eq. (4.6.30) is monotonically increasing for $n > 1$. Therefore, NFWM microscopy provides three-dimensional distribution of the refractive index.

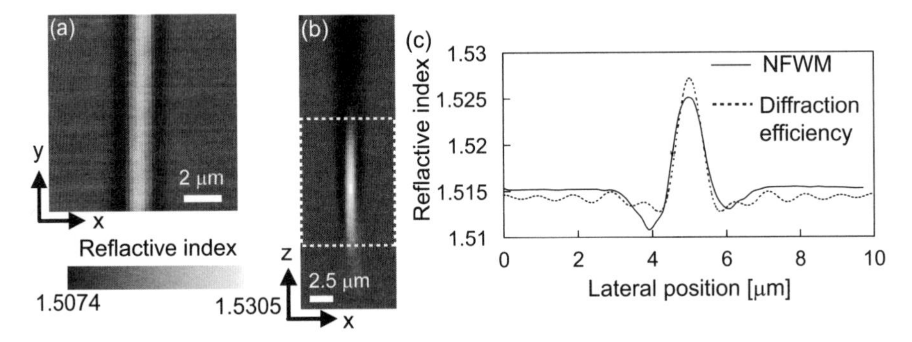

FIGURE 4.34 Refractive index distributions of an internal diffraction grating element in (a) *xy* and (b) *xz* planes. (c) Comparison of refractive index distribution (solid line) obtained by averaging over the region surrounded by the dotted line in (b) with that obtained by the diffraction efficiency (broken line). Reprinted with permission from [5]. Copyright 2008 The Japan Society of Applied Physics.

Isobe et al. demonstrated three-dimensional refractive index imaging by NFWM microscopy [5]. Figure 4.34(a) and (b) shows the measured refractive index distributions of an internal phase grating in *xy* and *xz* planes, which was fabricated in a non-alkali glass by femtosecond laser pulses. It is known that the refractive index change caused by femtosecond laser pulses is mainly due to the glass densification introduced by structural changes in the glass network and is hardly due to absorption defects [219]. This fact supports the validity of conditions 1 and 2 in the above-mentioned analysis. A high refractive index region is quite evident. Figure 4.34(c) shows the averaged refractive index profile (solid line) along the z direction over the area surrounded by the dotted line in Fig. 4.34(b) and the refractive index profile obtained by the diffraction efficiency (broken line). The two profiles are in good agreement, which confirms that NFWM microscopy allows quantitative refractive index imaging. However, the measured refractive index profiles may include artifacts that originate from (1) the refraction of the excitation beam by sample structures and (2) the power variation of NFWM signals due to the finite numerical aperture of the collector lens. The first type of artifact could be compensated for by using a heuristic method [220], and the amount of artifact could be reduced by increasing the numerical aperture of the collector lens.

REFERENCES

1. R. Hellwarth and P. Christensen, "Nonlinear optical microscopic examination of structure in polycrystalline ZnSe," *Opt. Commun.*, **12**, 318–322 (1974).
2. P. J. Campagnola and L. M. Loew, "Second-harmonic imaging microscopy for visualizing biomolecular arrays in cells, tissues and organisms," *Nat. Biotechnol.*, **21**, 1356–1360 (2003).
3. J. Squier, and M. Müller, "High resolution nonlinear microscopy: A review of sources and methods for achieving optimal imaging," *Rev. Sci. Inst.*, **72**, 2855–2867, (2001).

4. Y. Barad, H. Eisenberg, M. Horowitz, and Y. Silberberg, "Nonlinear scanning laser microscopy by third harmonic generation," *Appl. Phys. Lett.*, **70**, 922–924 (1997).
5. K. Isobe, T. Kawasumi, T. Tamaki, S. Kataoka, Y. Ozeki, and K. Itoh, "Three-dimensional profiling of refractive index distribution inside transparent materials by use of nonresonant four-wave mixing microscopy," *Appl. Phys. Express*, **1**, 022006/1–3 (2008).
6. K. Isobe, S. Kataoka, R. Murase, W. Watanabe, T. Higashi, S. Kawakami, S. Matsunaga, K. Fukui, and K. Itoh, "Stimulated parametric emission microscopy," *Opt. Express*, **14**, 786–793 (2006).
7. M. D. Duncan, J. Reintjes, and T. J. Manuccia, "Scanning coherent anti-Stokes Raman microscope," *Opt. Lett.*, **7**, 350–352 (1982).
8. A. Zumbusch, G. R. Holtom, and X. S. Xie, "Three-dimensional vibrational imaging by coherent anti-Stokes Raman scattering," *Phys. Rev. Lett.*, **82**, 4142–4145 (1999).
9. W. S. Warren, M. C. Fischer, and T. Ye, "Novel nonlinear contrast improves deep-tissue microscopy," *Laser focus world*, **43**, 99–103 (2007).
10. D. Fu, T. Ye, T. E. Matthews, B. J. Chen, G. Yurtserver, and W. S. Warren, "High-resolution in vivo imaging of blood vessels without labeling," *Opt. Lett.*, **32**, 2641–2643 (2007).
11. W. Denk, J. H. Strickle, and W. W. Webb, "Two-photon laser scanning fluorescence microscopy," *Science*, **248**, 73–76 (1990).
12. K. König, "Multiphoton microscopy in life sciences," *J. Microsc.*, **200**, 83–104 (2000).
13. W. R. Zipfel, R. M. Williams, and W. W. Webb, "Nonlinear magic: multiphoton microscopy in the biosciences," *Nat. Biotechnol.*, **21**, 1369–1377 (2003).
14. C. W. Freudiger, W. Min, B. G. Saar, S. Lu, G. R. Holtom, C. He, J. C. Tsai, J. X. Kang, and X. S. Xie, "Label free biomedical imaging with high sensitivity by stimulated Raman scattering microscopy," *Science*, **322**, 1857–1861 (2008).
15. P. Nandakumar, A. Kovalev, and A. Volkmer, "Vibrational imaging based on stimulated Raman scattering microscopy," *N. J. Phys.*, **11**, 033026/1–9 (2009).
16. Y. Ozeki, F. Dake, S. Kajiyama, K. Fukui, and K. Itoh, "Analysis and experimental assessment of the sensitivity of stimulated Raman scattering microscopy," *Opt. Express*, **17**, 3651–3658 (2009).
17. G. Y. Fan, H. Fujisaki, A. Miyawaki, R.-K. Tsay, R. Y. Tsien, and M. H. Ellisman, "Video-rate scanning two-photon excitation fluorescence microscopy and ratio imaging with cameleons," *Biophys. J.*, **76**, 2412–2420 (1999).
18. C. L. Evans, E. O. Potma, M. Puoris'haag, D. Côté, C. P. Lin, and X. S. Xie, "Chemical imaging of tissue in vivo with video-rate coherent anti-Stokes Raman scattering microscopy," *Proc. Natl. Acad. Sci. USA*, **102**, 16807–16812 (2005).
19. J. D. Lechleiter, D. T. Lin, and I. Sieneart, "Multi-photon laser scanning microscopy using an acoustic optical deflector, *Biophys. J.*, **83**, 2292–2299 (2002).
20. A. H. Buist, M. Muller, J. Squier, and G. J. Brakenhoff, "Real time two-photon absorption microscopy using multi point excitation," *J. Microsc.*, **192**, 217–226 (1998).
21. J. Bewersdorf, R. Pick, and S. W. Hell, "Multifocal multiphoton microscopy," *Opt. Lett.*, **23**, 655–657 (1998).
22. A. Egner and S. W. Hell, "Time multiplexing and parallelization in multifocal multiphoton microscopy," *J. Opt. Soc. Am. A*, **17**, 1192–1201 (2000).
23. D. N. Fittinghoff, P. W. Wiseman, and J. A. Squir, "Widefield multiphoton and temporally decorrelated multifocal multiphoton microscopy," *Opt. Express*, **7**, 273–279 (2000).

24. T. Nielsen, M. Fricke, D. Hellweg, and P. Andresen, "High efficiency beam splitter for multifocal multiphoton microscopy," *J. Microsc.*, **201**, 368–376 (2001).

25. L. Sacconi, E. Froner, R. Antolini, M. R. Taghizadeh, A. Choudhury, and F. S. Pavone, "Multiphoton multifocal microscopy exploiting a diffractive optical element," *Opt Lett.*, **28**, 1918–1920 (2003).

26. D. Oron, E. Tal, and Y. Silberberg, "Scanningless depth-resolved microscopy," *Opt. Express*, **13**, 1468–1476 (2005).

27. G. Zhu, J. van Howe, M. Durst, W. Zipfel, and C. Xu, "Simultaneous spatial and temporal focusing of femtosecond pulses," *Opt. Express*, **13**, 2153–2159 (2005).

28. K. Isobe, A. Suda, M. Tanaka, H. Hashimoto, F. Kannari, H. Kawano, H. Mizuno, A. Miyawaki, and K. Midorikawa, "Nonlinear optical microscopy and spectroscopy employing octave spanning pulses," *IEEE J. Sel. Top. Quant. Electron.*, **16**, 767–780 (2010).

29. K. Svoboda, W. Denk, D. kleinfeld, and D. W. Tank, "In vivo dendritic calcium dynamics in neocortical pyramidal neurons," *Nature*, **385**, 161–165 (1997).

30. K. Svoboda, F. Helmchen, W. Denk, and D. W. Tank, "Spread of dendritic excitation in layer 2/3 pyramidal neurons in rat barrel cortex in vivo," *Nat. Neurosci.*, **2**, 65–73 (1999).

31. F. Helmchen, K. Svoboda, W. Denk, and D. W. Tank, "In vivo dendritic calcium dynamics in deep-layer cortical pyramidal neurons," *Nat, Neurosci.*, **2**, 989–996 (1999).

32. P. Theer, M. T. Hasan, and W. Denk, "Two-photon imaging to a depth of 1000 mm in living brains by use of a Ti:Al2O3 regenerative amplifier," *Opt. Lett.*, **28**, 1022–1024 (2003).

33. D. Kobat, M. E. Durst, N. Nishimura, A. W. Wong, C. B. Schaffer, and C. Xu, "Deep tissue multiphoton microscopy using longer wavelength excitation," *Opt. Express*, **17**, 13354–13364 (2009).

34. J. M. Squirrell, D. L. Wokosin, J. G. White, and Barry D. Bavister, "Long-term two-photon fluorescence imaging of mammalian embryos without compromising viability," *Nat. Biotechnol.*, **17**, 763–767 (1999).

35. V. Kasche and L. Lindqvist, "Reactions between the Triplet State of Fluorescein and Oxygen," *J. Chem. Phys.*, **68**, 817–823 (1964).

36. L. Song, E. J. Hennink, I. T. Young, and H. J. Tanke, Photobleaching kinetics of fluorescein in quantitative fluorescence microscopy, *Biophys. J.*, **68**, 2588–2600 (1995).

37. L. Song, C. A. Varma, J. W. Verhoeven, and H. J. Tanke, "Influence of the triplet excited state on the photobleaching kinetics of fluorescein in microscopy," *Biophys. J.*, **70**, 2959–2968 (1996).

38. C. Eggeling, J. Widengren, R. Rigler, and C. A. M. Seidel, "Photobleaching of fluorescent dyes under conditions used for single-molecule detection: evidence of two-step photolysis," *Anal. Chem.*, **70**, 2651–2659 (1998).

39. C. Eggeling, A. Volkmer, and C. A. M. Seidel, "Molecular photobleaching kinetics of Rhodamine 6G by oneand two-photon induced confocal fluorescence microscopy," *ChemPhysChem*, **6**, 791–804 (2005).

40. G. H. Patterson, and D. W. Piston, "Photobleaching in two-photon excitation microscopy," *Biophys. J.*, **78**, 2159–2162 (2000).

41. J. S. Marchant, G. E. Stutzmann, M. A. Leissring, F. M. LaFerla, and I. Parker, "Multiphoton-evoked color change of DsRed as an optical highlighter for cellular and subcellular labeling," *Nat. Biotechnol.*, **19**, 645–649 (2001).

42. T.-S. Chen, S.-Q. Zeng, Q.-M. Luo, Z.-H. Zhang, and W. Zhou, High-order photobleaching of green fluorescent protein inside live cells in two-photon excitation microscopy, *Biochem. Biophys. Res. Commun.*, **291**, 1272–1275 (2002).
43. S. Kalies, K. Kuetemeyer, and A. Heisterkamp, "Mechanisms of high-order photobleaching and its relationship to intracellular ablation," *Biomed. Opt. Express*, **2**, 805–816 (2011).
44. G. Donnert, C. Eggeling, and S.W. Hell, "Major signal increase in fluorescence microscopy through dark-state relaxation," *Nat. Methods*, **4**, 81–86 (2007).
45. N. Ji, J. C. Magee, and E. Betzig, "High-speed, low-photodamage nonlinear imaging using passive pulse splitters," *Nat. Methods*, **5**, 197–202 (2008).
46. M. Born and E. Wolf, *Principles of Optics*, 5th ed., Pergamon Press, Oxford, 1975.
47. C. J. R. Sheppard and H. J. Mattews, "Imaging in high-aperture optical systems," *J. Opt. Soc. Am. A*, **4**, 1354–1360 (1987).
48. K. Itoh, W. Watanabe, H. Arimoto, and K. Isobe, "Coherence-based 3-D and spectral imaging and laser-scanning microscopy," *Proc. IEEE*, **94**, 608–628 (2006).
49. B. R. Frieden, "Optical transfer of the three dimensional object," *J. Opt. Soc. Am.*, **57**, 56–67 (1967).
50. C. J. R. Sheppard and M. Gu, "Image formation in two-photon fluorescence microscopy," *Optik*, **86**, 104–106 (1990).
51. M. Gu and C. J. R. Sheppard, "Effects on a finite-sized pinhole on 3D image formation in confocal two-photon fluorescence microscopy," *J. Mod. Opt.*, **40**, 2009–2024 (1993).
52. J. Goodman, *Introduction to Fourier Optics*, Roberts, Englewood, CO, 2005.
53. T. Higashi, S. Matsunaga, K. Isobe, T. Shimada, S. Kataoka, W. Watanabe, S. Uchiyama, K. Itoh, and K. Fukui, "Histone H2A mobility is regulated by its tails and acetylation of core histone tails," *Biochem. Biophys. Res. Commun.*, **357**, 627–632 (2007).
54. J.-P. Ritz, A. Roggan, C. Isbert, G. Müller, H. J. Buhr, and C.-T. Germer, "Optical properties of native and coagulated porcine liver tissue between 400 and 2400 nm," *Lasers Surg. Med.*, **29**, 205–212 (2001).
55. G. M. Hale and M. R. Querry, "Optical constants of water in the 200 nm to 200 μm wavelength region," *Appl. Opt.*, **12**, 555–563 (1973).
56. The Oregon Medical Laser Center, Tabulated molar extinction coefficient for hemoglobin, in water. See http://omlc.ogi.edu/spectra/hemoglobin/summary.html
57. The Oregon Medical Laser Center, Melanosome absorption coefficient. See http://omlc.ogi.edu/spectra/melanin/mua.html.
58. S. L. Jacques, D. J. McAuliffe, "The melanosome: threshold temperature for explosive vaporization and internal absorption coefficient during pulsed laser irradiation," *Photochem. Photobiol.*, **53**, 769–775 (1991).
59. P. Theer and W. Denk, "On the fundamental imaging-depth limit in two-photon microscopy," *J. Opt. Soc. Am. A*, **23**, 3139–3149 (2006).
60. C. Xu, W. Zipfel, J. B. Shear, R. M. Williams, and W. W. Webb, "Multiphoton fluorescence excitation: new spectral windows for biological nonlinear microscopy," *Proc. Natl. Acad. Sci. USA.*, **93**, 10763–10768 (1996).
61. H. Kawano, T. Kogure, Y. Abe, H. Mizuno, and A. Miyawaki, "Two-photon dual-color imaging using fluorescent proteins," *Nat. Methods*, **5**, 373–374 (2008).

62. K. Isobe, W. Watanabe, S. Matsunaga, T. Higashi, K. Fukui, and K. Itoh, "Multi-spectral two-photon excited fluorescence microscopy using supercontinuum light source," *Jpn. J. Appl. Phys.*, **44**, L167–L169 (2005).
63. K. Isobe, A. Suda, M. Tanaka, F. Kannari, H. Kawano, H. Mizuno, A. Miyawaki, and K. Midorikawa, "Multifarious control of two-photon excitation of multiple fluorophores achieved by phase modulation of ultra-broadband laser pulses," *Opt. Express*, **17**, 13737–13746 (2009).
64. S. E. Tillo, T. E. Hughes, N. S. Makarov, A. Rebane, and M. Drobizhev, "A new approach to dual-color two-photon microscopy with fluorescent proteins," *BMC Biotechnol.*, **10**, 6/1–6 (2010).
65. M. Drobizhev, N. S. Makarov, T. Hughes, and A. Rebane, "Resonance enhancement of two-photon absorption in fluorescent proteins," *J. Phys. Chem. B*, **111**, 14051–14054 (2007).
66. T. Kogure, S. Karasawa, T. Araki, K. Saito, M. Kinjo and A. Miyawaki, "A fluorescent variant of a protein from the stony coral Montipora facilitates dual-color single-laser fluorescence cross-correlation spectroscopy," *Nat. Biotechnol.*, **24**, 577–581 (2006).
67. T. Kogure, H. Kawano, Y. Abe, and A. Miyawaki, "Fluorescence imaging using a fluorescent protein with a large Stokes shift," *Methods*, **45**, 223–226 (2008).
68. K. D. Piatkevich, J. Hulit, O. M. Subach, B. Wu, A. Abdulla, J. E. Segall, and V. V. Verkhusha, "Monomeric red fluorescent proteins with a large Stokes shift," *Proc. Natl. Acad. Sci. USA*, **107**, 5369–5374 (2010).
69. D. Li, W. Zheng, and J. Y. Qu, "Two-photon autofluorescence microscopy of multicolor excitation," *Opt. Lett.*, **34**, 202–204 (2009).
70. A. Miyawaki, J. Llopis, R. Heim, J. M. McCaffery, J. A. Adams, M. Ikura, R. Y. Tsien, "Fluorescent indicators for Ca^{2+} based on green fluorescent proteins and calmodulinm," *Nature*, **388**, 882–887 (1997).
71. A. Miyawaki, O. Griesbeck, R. Heim, R. Y. Tsien, "Dynamic and quantitative Ca^{2+} measurements using improved cameleons," *Proc. Natl. Acad. Sci. USA*, **96**, 2135–2140 (1999).
72. T. Nagai, S. Yamada, T. Tominaga, M. Ichikawa, and A. Miyawaki, "Expanded dynamic range of fluorescent indicators for Ca^{2+} by circularly permuted yellow fluorescent proteins," *Proc. Natl. Acad. Sci. USA*, **101**, 10554–10559 2004.
73. M. Mank, D. F. Reiff, N. Heim, M. W. Friedrich, A. Borst, and O. Griesbeck, "A FRET-based calcium biosensor with fast signal kinetics and high fluorescence change," *Biophys. J.*, **90**, 1790–1796 (2006).
74. G. Y. Fan, H. Fujisaki, A. Miyawaki, R. K. Tsay, R. Y. Tsien, and M. H. Ellisman, "Video-rate scanning two-photon excitation fluorescence microscopy and ratio imaging with Cameleons," *Biophys. J.*, **76**, 2412–2420 (1999).
75. K.-I. Okamoto, T. Nagai, A. Miyawaki, Y. Hayashi, "Rapid and persistent modulation of actin dynamics regulates postsynaptic reorganization underlying bidirectional plasticity," *Nat. Neurosci*, **7**, 1104–1112 (2004).
76. K. Takao, K.-I. Okamoto, T. Nakagawa, R. L. Neve, T. Nagai, A. Miyawaki, T. Hashikawa, S. Kobayashi, and Y. Hayashi, "Visualization of synaptic Ca^{2+}/calmodulin-dependent protein kinase II activity in living neurons," *J. Neurosci.*, **25**, 3107–3112 (2005).

77. M. Elangovan, H. Wallrabe, Y. Chen, R. N. Day, M. Barroso, and A. Periasamy, "Characterization of one- and two-photon excitation fluorescence resonance energy transfer microscopy," *Methods*, **29**, 58–73 (2003).

78. Y. Chen and A. Periasamy, "Intensity range based quantitative FRET data analysis to localize protein molecules in live cell nuclei," *J. Fluoresc.*, **16**, 95–104 (2006).

79. R. Yasuda, C. D. Harvey H. Zhong, A. Sobczyk, L. van Aelst, and K. Svoboda, "Supersensitive Ras activation in dendrites and spines revealed by 2-photon fluorescence lifetime imaging," *Nat. Neurosci.*, **9**, 283–291 (2006).

80. S.-J. R. Lee, Y. Escobedo-Lozoya, E. M. Szatmari, and R. Yasuda, "Activation of CaMKII in single dendritic spines during long-term potentiation," *Nature*, **458**, 299–304 (2009).

81. S. Ganesan, S. M. Ameer-Beg, T. T. Ng, B. Vojnovic, and F. S. Wouters, "A dark yellow fluorescent protein (YFP)-based Resonance Energy-Accepting Chromoprotein (REACh) for Forster resonance energy transfer with GFP," *Proc. Natl. Acad. Sci. USA*, **103**, 4089–4094 (2006).

82. H. Murakoshi, S.-J. R. Lee, R. Yasuda, "Highly sensitive and quantitative FRET-FLIM imaging in single dendritic spines using improved nonradiative YFP," *Brain Cell Biol.*, *36*, 31–42 (2008).

83. C. D. Harvey, R. Yasuda, H. Zhong, and K. Svoboda, "The spread of Ras activity triggered by activation of a single dendritic spine," *Science*, **321**, 136–140 (2008).

84. C. D. Harvey, A. G. Ehrhardt, C. Cellurale, H. Zhong, R. Yasuda, R. J. Davis, and K. Svoboda, "A genetically encoded fluorescent sensor of ERK activity," *Proc. Natl. Acad. Sci. USA*, **105**, 19264–19269 (2008).

85. D. Axelrod, D. E. Koppel, J. Schlessinger, E. Elson, and W. W. Webb, "Mobility measurement by analysis of fluorescence photobleaching recovery kinetics," *Biophys. J.*, **16**, 1055–1069 (1976).

86. J. Blonk, A. Don, H. Van Aalst, and J. Birmingham, "Fluorescence photobleaching recovery in the confocal light scanning microscope," J. Microsc. **169**, 363–374 (1993).

87. E. B. Brown, E. S. Wu, W. Zipfel, and W. W. Webb, "Measurement of molecular diffusion in solution by multiphoton fluorescence photobleaching recovery," *Biophys. J.*, **77**, 2837–2849 (1999).

88. D. Mazza, F. Cella, G. Vicidomini, S. Krol, and A. Diaspro, "Role of three-dimensional bleach distribution in confocal and two-photon fluorescence recovery after photobleaching experiments," *Appl. Opt.*, **46**, 7401–7411 (2007).

89. D. W. Piston, E.-S. Wu, and W. W. Webb, "3-dimensional diffusion measurements in cells by 2-photon excitation fluorescence photobleaching recovery," *Faseb. J.*, **6**, A34-A34 (1992).

90. K. Svoboda, D. W. Tank and W. Denk, "Direct measurement of coupling between dendritic spines and shafts," *Science*, **272**, 716–719 (1996).

91. A. Majewska, E. Brown, J. Ross, and R. Yuste, "Mechanisms of calcium decay kinetics in hippocampal spines: role of spine calcium pumps and calcium diffusion through the spine neck in biochemical compartmentalization," *J. Neurosci.*, **20**, 1722–1734 (2000).

92. A. Majewska, A. Tashiro, and R. Yuste, "Regulation of spine calcium dynamics by rapid spine motility," *J. Neurosci.*, **20**, 8262–8268 (2000).

93. M. T. Hasan, R. W. Friedrich, T. Euler, M. E. Larkum, G. Giese, M. Both, J. Duebel, J. Waters, H. Bujard, O. Griesbeck, R. Y. Tsien, T. Nagai, A. Miyawaki, and W. Denk,

"Functional fluorescent Ca^{2+} indicator proteins in transgenic mice under TET control," *PLoS Biol.*, **2**, 1–13 (2004).

94. T. A. Pologruto, R. Yasuda, and K. Svoboda, "Monitoring neural activity and [Ca^{2+}] with genetically encoded Ca^{2+} indicators," *J. Neurosci.*, **24**, 9572–9579 (2004).
95. A. Sobczyk, V. Scheuss, and K. Svoboda, "NMDA receptor subunit-dependent [Ca^{2+}] signaling in individual hippocampal dendritic spines," *J. Neurosci.*, **25**, 6037–6046 (2005).
96. H. Schmidt, E. B. Brown, B. Schwaller, and J. Eilers, "Diffusional mobility of parvalbumin in spiny dendrites of cerebellar Purkinje neurons quantified by fluorescence recovery after photobleaching," *Biophys. J.*, **84**, 2599–2608 (2003).
97. H. Schmidt, B. Schwaller, and J. Eilers, "Calbindin D28k targets myo-inositol monophosphatase in spines and dendrites of cerebellar Purkinje neurons," *Proc. Natl. Acad. Sci. USA*, **102**, 5850–5855 (2005).
98. A. Grunditz, N. Holbro, L. Tian, Y. Zuo, and T. G. Oertner, "Spine neck plasticity controls postsynaptic calcium signals through electrical compartmentalization," *J. Neurosci.*, **28**, 13457–13466 (2008).
99. H. Qian, and E. L. Elson. "Analysis of confocal laser-microscope optics for 3-D fluorescence correlation spectroscopy," *Appl. Opt.*, **30**, 1185–1195 (1991).
100. R. Rigler, U. Mets, J. Widengren, and P. Kask, "Fluorescence correlation spectroscopy with high count rate and low background: analysis of translational diffusion," *Eur. Biophys. J.*, **22**, 169–175 (1993).
101. D. E. Koppel, F. Morgan, A. Cowan, and J. H. Carson, "Scanning concentration correlation spectroscopy using the confocal laser microscope," Biophys. J., **66**, 502–507 (1994).
102. K. M. Berland, P. T. So, and E. Gratton, "Two-photon fluorescence correlation spectroscopy: method and application to the intracellular environment," *Biophys J.*, **68**, 694–701 (1995).
103. P. Schwille, U. Haupts, S. Maiti, and W.W. Webb, "Molecular dynamics in living cells observed by fluorescence correlation spectroscopy with one- and two-photon excitation, " *Biophys. J.*, **77**, 2251–2265 (1999).
104. G. Alexandrakis, E. B. Brown, R. T. Tong, T. D. McKee, R. B. Campbell, Y. Boucher, and R. K. Jain, "Two-photon fluorescence correlation microscopy reveals the two-phase nature of transport in tumors," *Nat. Medicine*, **10**, 203–207 (2004).
105. D. R. Larson, Y. M. Ma, V. M. Vogt, and W.W. Webb, "Direct measurement of Gag-Gag interaction during retrovirus assembly with FRET and fluorescence correlation spectroscopy," *J. Cell. Biol.*, **162**, 1233–1244 (2003).
106. K. G. Heinze, A. Koltermann, and P. Schwille, "Simultaneous two-photon excitation of distinct labels for dual-color fluorescence crosscorrelation analysis," *Proc. Natl. Acad. Sci. USA*, **97**, 10377–10382 (2000).
107. K. G. Heinze, M. Rarbach, M. Jahnz, and P. Schwille, "Two-photon fluorescence coincidence analysis: Rapid measurements of enzyme kinetics," *Biophys. J.*, *83*, 1671–1681 (2002).
108. T. Kohl, K. G. Heinze, R. Kuhlemann, A. Koltermann, and P. Schwille, "A protease assay for two-photon crosscorrelation and FRET analysis based solely on fluorescent proteins," *Proc. Natl. Acad. Sci. USA*, **99**, 12161–12166 (2002).

109. S. A. Kim, K. G. Heinze, M. N. Waxham, and P. Schwille, "Intracellular calmodulin availability accessed with two-photon cross-correlation," *Proc. Natl. Acad. Sci. USA*, **101**, 105–110 (2004).

110. S. A. Kim, K. G. Heinze, K. Bacia, M. N. Waxham, and P. Schwille, "Two-photon crosscorrelation analysis of intracellular reactions with variable stoichiometry," *Biophys. J.*, **88**, 4319–4336 (2005).

111. K. M. Berland, "Detection of specific DNA sequences using dual-color two-photon fluorescence correlation spectroscopy," *J. Biotechnol.*, **108**, 127–136 (2004).

112. M. Collini, M. Caccia, G. Chirico, F. Barone, E. Dogliotti, and F. Mazzei, "Two-photon fluorescence cross-correlation spectroscopy as a potential tool for high-throughput screening of DNA repair activity," *Nucleic Acids Res.*, **33**, e165/1–8 (2005).

113. D. Merkle, W. D. Block, Y. Yu, S. P. Lees-Miller, and D. T. Cramb, "Analysis of DNA dependent protein kinase-mediated DNA end joining by two-photon fluorescence crosscorrelation spectroscopy," *Biochemistry*, *45*, 4164–4172 (2006).

114. M. D. Egger and M. Petráň, "New reflected-light microscope for viewing unstained brain and ganglion cells," *Science*, **157**, 305–307 (1967).

115. G. Q. Xiao, T. R. Corle, and G. S. Kino, "Real-time confocal scanning optical microscope," *Appl. Phys. Lett.*, **53**, 716–718 (1988).

116. J. B. Pawly, *Handbook of Biological Confocal Microscopy*, Springer Science and Business Media, 2006.

117. D. Shafer, "Gaussian to flat-top intensity distribution lens," *Opt. Laser Technol.*, **14**, 159–160 (1982).

118. K. H. Kim, C. Buehler, K. Bahlmann, T. Ragan, W.-C. A. Lee, E. Nedivi, E. L. Heffer, S. Fantini, and P. T. C. So, "Multifocal multiphoton microscopy based on multianode photomultiplier tubes," *Opt. Express*, **15**, 11658–11678 (2007).

119. M. E. Durst, G. H. Zhu, and C. Xu, "Simultaneous spatial and temporal focusing in nonlinear microscopy," Opt. Commun. **281**, 1796–1805 (2008).

120. E. Tal, D. Oron, and Y. Silberberg, "Improved depth resolution in video-rate line-scanning multiphoton microscopy using temporal focusing," *Opt. Lett.*, **30**, 1686–1688 (2005).

121. M. E. Durst, G. H. Zhu, and C. Xu, "Simultaneous spatial and temporal focusing for axial scanning," *Opt. Express*, **14**, 12243–12254 (2006).

122. A. Straub, M. E. Durst, and C. Xu, "High speed multiphoton axial scanning through an optical fiber in a remotely scanned temporal focusing setup," *Biomed. Opt. Express*, **2**, 80–88 (2011).

123. O. D. Therrien, B. Aubé, S. Pagés, P. De Koninck, and D. Côté, "Wide-field multiphoton imaging of cellular dynamics in thick tissue by temporal focusing and patterned illumination," *Biomed. Opt. Express*, **2**, 696–704 (2011).

124. K. Isobe, Y. Ozeki, T. Kawasumi, S. Kataoka, S. Kajiyama, K. Fukui, and K. Itoh, "Highly sensitive spectral interferometric four-wave mixing microscopy near the shot noise limit and its combination with two-photon excited fluorescence microscopy," *Opt. Express*, **14**, 11204–11214 (2006).

125. M. Yamagiwa, Y. Ozeki, T. Kawasumi, S. Kajiyama, K. Fukui, and K. Itoh, "Highly sensitive signal detection in stimulated parametric emission microscopy based on two-beam interferometry," *Jpn. J. Appl. Phys.*, **47**, 8820–8824 (2008).

126. X. Liu, W. Rudolph, and J. L. Thomas, "Photobleaching resistance of stimulated parametric emission in microscopy," *Opt. Lett.*, **34**, 304–306 (2009).

127. X. Liu, W. Rudolph, and J. L. Thomas, "Characterization and application of femtosecond infrared stimulated parametric emission microscopy," *J. Opt. Soc. Am. B*, **27**, 787–795 (2010).
128. M. Yamagiwa, G. Omura, Y. Ozeki, M. Ishii, H. M. Dang, S. Kajiyama, T. Suzuki, K. Fukui, and K. Itoh, "Dual-band stimulated parametric emission microscopy," *Jpn. J. Appl. Phys.*, **49**, 016603 (2010).
129. H. M. Dang, G. Omura, T. Umano, M Yamagiwa, S. Kajiyama,Y. Ozeki, K. Itoh, and K. Fukuia, "Label-free imaging by stimulated parametric emission microscopy reveals a difference in hemoglobin distribution between live and fixed erythrocytes," *J. Biomed. Opt.*, **14**, 040506/1–3 (2009).
130. P. Tian and W. S. Warren, "Ultrafast measurement of two-photon absorption by loss modulation," Opt. Lett., 27, 1634–1636 (2002).
131. D. Fu, T. Ye, T. E. Matthews, G. Yurtsever, and W. S. Warren, "Two-color, two-photon, and excited-state absorption microscopy," *J. Biomed. Opt.* **12**, 054004 (2007).
132. M. Hashimoto, T. Araki, and S. Kawata, "Molecular vibration imaging in the fingerprint region by use of coherent anti-Stokes Raman scattering microscopy with a collinear configuration," *Opt. Lett.*, **25**, 1768–1770 (2000).
133. J.-X. Cheng, Y. K. Jia, G. Zheng, and X. S. Xie, "Laser-scanning coherent anti-Stokes Raman scattering microscopy and applications to cell biology," *Biophys. J.*, **83**, 502–509 (2002).
134. J.-X. Cheng, L. D. Book, and X. S. Xie, "Polarization coherent anti-Stokes Raman scattering microscopy," *Opt. Lett.*, **26**, 1341–1343 (2001).
135. E. R. Dufresne, E. I. Corwin, N. A. Greenblatt, J. Ashmore, D. Y. Wang, A. D. Dinsmore, J. X. Cheng, X. S. Xie, J. W. Hutchinson, and D. A. Weitz, "Flow and fracture in drying nanoparticle suspensions," *Phys. Rev. Lett.*, **91**, 224501/1–4, (2003).
136. J.-X. Cheng, A. Volker, and X. S. Xie, "Theoretical and experimental characterization of coherent anti-Stokes Raman scattering microscopy," *J. Opt. Soc. Am. B.*, **19**, 1363–1375 (2002).
137. K. Klein, A. M. Gigler, T. Aschenbrenner, R. Monetti, W. Bunk, F. Jamitzky, G. Morfill, R. W. Stark, and J. Schlegel, "Label-free live-cell imaging with confocal Raman microscopy," *Biophys. J.*, **102**, 360–368 (2012).
138. G. J. Thomas, "Raman spectroscopy of protein and nucleic acid assemblies," *Annu. Rev. Biophys. Biomol. Struct.*, **28**, 1–27 (1999).
139. E. O. Potma, D. J. Jones, J. Cheng, X. S. Xie, and J. Ye, "High-sensitivity coherent anti-Stokes Raman scattering microscopy with two tightly synchronized picosecond lasers," *Opt. Lett.*, **27**, 1168–1170 (2002).
140. J.-X. Cheng, A. Volkmer, L. D. Book, and X. S. Xie, "An epi-detected coherent anti-Stokes Raman scattering (E-CARS) microscope with high spectral resolution and high sensitivity," *J. Phys. Chem. B*, **105**, 1277–1280 (2001).
141. A. Volkmer, J.-X. Cheng, and X. S. Xie, "Vibrational imaging with high sensitivity via epidetected coherent anti-Stokes Raman scattering microscopy," *Phys. Rev. Lett.*, **87**, 023901/1–4 (2001).
142. J.-X. Cheng, and X. S. Xie, "Coherent anti-Stokes Raman scattering microscopy: Instrumentation, theory, and applications," *J. Phys. Chem. B*, **108**, 827–840 (2004).
143. S. A. Akhmanov, A. F. Bunkin, S. G. Ivanov, and N. I. Koroteev, "Coherent ellipsometry of Raman scattering of light," *JETP Lett.*, **25**, 416–420 (1977).

144. J. L. Oudar, R. W. Smith and Y. R. Shen, "Polarization-sensitive coherent anti-Stokes Raman spectroscopy," *Appl. Phys. Lett.*, **34**, 758–760 (1979).
145. C. L. Evans, E. O. Potma, and X. S. Xie, "Coherent anti-Stokes Raman scattering spectral interferometry: determination of the real and imaginary components of nonlinear susceptibility $\chi^{(3)}$ for vibrational microscopy," *Opt. Lett.*, **29**, 2923–2925 (2004).
146. E. O. Potma, C. L. Evans, and X. S. Xie, "Heterodyne coherent anti-Stokes Raman scattering (CARS) imging," *Opt. Lett.*, **31**, 241–243 (2006).
147. A. Volkmer, L. D. Book and X. S. Xie, "Time-resolved coherent anti-Stokes Raman scattering microscopy: imaging based on Raman free induction decay," *Appl. Phys. Lett.*, **80**, 1505–1507 (2002).
148. Y. J. Lee and M. T. Cicerone, "Vibrational dephasing time imaging by time-resolved broadband coherent anti-Stokes Raman scattering microscopy," *Appl. Phys. Lett.*, **92**, 041108/1–3 (2008).
149. R. Selm, M. Winterhalder, A. Zumbusch, G. Krauss, T. Hanke, A. Sell, and A. Leitenstorfer, "Ultrabroadband background-free coherent anti-Stokes Raman scattering microscopy based on a compact Er:fiber laser system," *Opt. Lett.*, **35**, 3282–3284 (2010).
150. J. P. Ogilvie, E. Beaurepaire, A. Alexandrou, and M. Joffre, "Fourier-transform coherent anti-Stokes Raman scattering microscopy," *Opt. Lett.*, **31**, 480–482 (2006).
151. K. Isobe, A. Suda, M. Tanaka, H. Hashimoto, F. Kannari, H. Kawano, H. Mizuno, A. Miyawaki, and K. Midorikawa, "Single-pulse coherent anti-Stokes Raman scattering microscopy employing an octave spanning pulse," *Opt. Express*, **17**, 11259–11266 (2009).
152. M. Müller and J. M. Schins, "Imaging the thermodynamic state of lipid membranes with multiplex CARS microscopy," *J. Phys. Chem. B*, **106**, 3715–3723 (2002).
153. J.-X. Cheng, A. Volkmer, L. D. Book, and X. S. Xie, "Multiplex coherent anti-Stokes Raman scattering microspectroscopy and study of lipid vesicles," *J. Phys. Chem. B*, **106**, 8493–8498 (2002).
154. G. W. H. Wurpel, J. M. Schins, and M. Müller, "Chemical specificity in three-dimensional imaging with multiplex coherent anti-Stokes Raman scattering microscopy," *Opt. Lett.*, **27**, 1093–1095 (2002).
155. T. W. Kee and M. T. Cicerone, "Simple approach to one-laser, broadband coherent anti-Stokes Raman scattering microscopy," *Opt. Lett.*, **29**, 2701–2703 (2004).
156. H. Kano and H. Hamaguchi, "Ultrabroadband (>2500 cm^{-1}) multiplex coherent anti-Stokes Raman scattering microspectroscopy using a supercontinuum generated from a photonic crystal fiber," *Appl. Phys. Lett.*, **86**, 121113/1–3 (2005).
157. M. Okuno, H. Kano, P. Leproux, V. Couderc, and H. Hamaguchi, "Ultrabroadband multiplex CARS microspectroscopy and imaging using a subnanosecond supercontinuum light source in the deep near infrared," *Opt. Lett.*, **33**, 923–925 (2008).
158. A. Laubereau and W. Kaiser, "Vibrational dynamics of liquids and solids investigated by picosecond light pulses," *Rev. Mod. Phys.*, **50**, 607–665 (1978).
159. E. Ploetz, S. Laimgruber, S. Berner, W. Zinth and P. Gilch, "Femtosecond stimulated Raman microscopy," *Appl. Phys. B*, **87**, 389–393 (2007).
160. P. Nandakumar, A. Kovalev, and A. Volkmer, "Vibrational imaging and microspectroscopies based on coherent anti-Stokes Raman scattering microscopy," *8th European/French Israeli Symposium on Nonlinear and Quantum Optics*, Mo-B, 2005, http://www.weizmann.ac.il/conferences/frisno8/program.html.

161. C. W. Freudiger, W. Min, B. G. Saar, S. Lu, G. R. Holtom, C. He, J. C. Tsai, J. X. Kang, and X. S. Xie, "Label free biomedical imaging with high sensitivity by stimulated Raman scattering microscopy," *Science*, **322**, 1857–1861 (2008).

162. P. Nandakumar, A. Kovalev, and A. Volkmer, "Vibrational imaging based on stimulated Raman scattering microscopy," *N. J. Phys.*, **11**, 033026 (2009).

163. Y. Ozeki, F. Dake, S. Kajiyama, K. Fukui, and K. Itoh, "Analysis and experimental assessment of the sensitivity of stimulated Raman scattering microscopy," *Opt. Express*, **17**, 3651–3658 (2009).

164. Y. Ozeki, Y. Kitagawa, K. Sumimura, N. Nishizawa, W. Umemura, S. Kajiyama, K. Fukui and K. Itoh, "Stimulated Raman scattering microscope with shot noise limited sensitivity using subharmonically synchronized laser pulses," *Opt. Express*, **18**, 13708–13719 (2010).

165. B. G. Saar, C. W. Freudiger, J. Reichman, C. M. Stanley, G. R. Holtom, and X. S. Xie, "Video-rate molecular imaging in vivo with stimulated Raman scattering," *Science*, **330**, 1368–1370 (2010).

166. C. J. R. Sheppard, R. Kompfner, J. Gannaway, and D. Walsh, "Scanning harmonic optical microscope.," *IEEE J. Quan. Electron.*, **13E**, 100D (1977).

167. S. Inoue, "An introduction to biological polarization microscopy," *Video Microscopy*, Plenum Press, New York. 477–510 (1986).

168. I. Freund, M. Deutsch and A. Sprecher, "Connective tissue polarity: optical second-harmonic microscopy, crossed-beam summation, and small-angle scattering in rat-tail tendon," *Biophys. J.*, **50**, 693–712 (1986).

169. F. Tiaho, G. Recher, and D. Rouede, "Estimation of helical angles of myosin and collagen by second harmonic generation imaging microscopy," *Opt. Express*, **15**, 12286–12295 (2007).

170. P. J. Campagnola, A. C. Millard, M. Terasaki, P. E. Hoppe, C. J. Malone, and W. A. Mohler, "Three-dimensional high-resolution second-harmonic generation imaging of endogenous structural proteins in biological tissues," *Biophys. J.*, **81**, 493–508 (2002).

171. T. Boulesteix, E. Beaurepaire, M.-P. Sauviat, and M.-C. Schanne-Klein, "Second-harmonic microscopy of unstained living cardiac myocytes: measurements of sarcomere length with 20-nm accuracy," *Opt. Lett.*, **29**, 2031–2033 (2004).

172. P. Stoller, K. M. Reiser, P. M. Celliers and A. M. Rubenchik, "Polarization-modulated second harmonic generation in collagen," *Biophys. J.*, **82**, 3330–3342 (2002).

173. S. V. Plotnikov, A. C. Millard, P. J. Campagnola, and W. A. Mohler, "Characterization of the myosin-based source for second-harmonic generation from muscle sarcomeres," *Biophys. J.*, **90**, 693–703 (2006).

174. I. Gusachenko, G. Latour, and M.-C. Schanne-Klein, "Polarization-resolved second harmonic microscopy in anisotropic thick tissues," *Opt. Express*, **18**, 19339–19352 (2010).

175. G. Latour, I. Gusachenko, L. Kowalczuk, I. Lamarre, and M.-C. Schanne-Klein, "In vivo structural imaging of the cornea by polarization-resolved second harmonic microscopy," *Biomed. Opt. Express*, **3**, 1–15 (2012).

176. S.-W. Chu, I-H. Chen, T.-M. Liu, P. C. Chen, C.-K. Sun, and B.-L. Lin, "Multimodal nonlinear spectral microscopy based on a femtosecond Cr:forsterite laser," *Opt. Lett.*, **26**, 1909–1911 (2001).

177. R. M. Brown, A. C. Millard, and P. J. Campagnola, "Macromolecular structure of cellulose studied by second-harmonic generation imaging microscopy," *Opt. Lett.*, **28**, 2207–2209 (2003).

178. F. P. Bolin, L. E. Preuss, R. C. Taylor, and R. J. Ference, "Refractive index of some mammalian tissues using a fiber optic cladding method," *Appl. Opt.*, **28**, 2297–2303 (1989).

179. D. T. Poh, "Examination of refractive index of human epidermis in-vitro and in-vivo," *Proceedings of the International Conference on Lasers '96*, 118–125 (1996).

180. R. LaComb, O. Nadiarnykh, S. S. Townsend, and P. J. Campagnola, "Phase matching considerations in second harmonic generation from tissues: effects on emission directionality, conversion efficiency and observed morphology," *Opt. Commun.*, **281**, 1823–1832 (2008).

181. J. Mertz, and L. Moreaux, "Second-harmonic generation by focused excitation of inhomogeneously distributed scatterers," *Opt. Commun.*, **196**, 325–330 (2001).

182. M. Han, G. Giese, and J. Bille, "Second harmonic generation imaging of collagen fibrils in cornea and sclera," *Opt. Express*, **13**, 5791–5797 (2005).

183. Y. R. Shen, "Surface properties probed by second-harmonic and sum-frequency generation," *Nature*, **337**, 519–525 (1989).

184. P. J. Campagnola, M.-D. Wei, A. Lewis and L. M. Loew, "High-resolution nonlinear optical imaging of live cells by second harmonic generation," *Biophys. J.*, **77**, 3341–3349 (1999).

185. L. Moreaux, O. Sandre, M. Blanchard-Desce, and J. Mertz, "Membrane imaging by simultaneous second-harmonic generation and two-photon microscopy," *Opt. Lett.*, **25**, 320–322 (2000).

186. A. C. Millard, P. J. Campagnola, W. Mohler, A. Lewis, and L. M. Loew, "Second harmonic imaging microscopy," *Meth. Enzymol.*, **361**, 47–69 (2003).

187. J. Y. Huang, A. Lewis, and L. M. Loew, "Nonlinear optical properties of potential sensitive styryl dyes," *Biophys. J.*, **53**, 665–670 (1988).

188. O. Bouevitch, A. Lewis, I. Pinevsky, J. P. Wuskell, and L. M. Loew, "Probing membrane potential with nonlinear optics," *Biophys. J.*, **65**, 672–679 (1993).

189. I. Ben-Oren, G. Peleg, A. Lewis, B. Minke, and L. M. Loew, "Infrared nonlinear optical measurements of membrane potential in photoreceptor cells," *Biophys. J.*, **71**, 1616–1620 (1996).

190. G. Peleg, A. Lewis, M. Linial, and L. M. Loew, "Nonlinear optical measurement of membrane potential around single molecules at selected cellular sites," *Proc. Natl. Acad. Sci. USA*, **96**, 6700–6704 (1999).

191. A. C. Millard, L. Jin, A. Lewis, and L. M. Loew, "Direct measurement of the voltage sensitivity of second-harmonic generation from a membrane dye in patch-clamped cells," *Opt. Lett.*, **28**, 1221–1223 (2003).

192. M. Nuriya, J. Jiang, B. Nemet, K. B. Eisenthal, and R. Yuste, "Imaging membrane potential in dendritic spines," *Proc. Natl. Acad. Sci. USA.*, **103**, 786–790 (2006).

193. L. Sacconi, D. A. Dombeck, and W. W. Webb, "Overcoming photodamage in second-harmonic generation microscopy: Real-time optical recording of neuronal action potentials," *Proc. Natl. Acad. Sci. USA.*, **103**, 3124–3129 (2006).

194. R. M. Williams, W. R. Zipfel, and W. W. Webb, "Interpreting second-harmonic generation images of collagen I fibrils," *Biophys. J.*, **88**, 1377–1386 (2005).

195. F. Légaré, C. Pfeffer, and B. R. Olsen, "The role of backscattering in SHG tissue imaging," *Biophys. J.*, **93**, 1312–1320 (2007).
196. S. M. Zhuo, J. X. Chen, T. S. Luo, and D. S. Zou, "Multimode nonlinear optical imaging of the dermis in ex vivo human skin based on the combination of multichannel mode and Lambda mode,"*Opt. Express*, **14**, 7810–7820 (2006).
197. D. S. Hum and M. M. Fejer, "Quasi-phasematching," *C. R. Physique*, **8**, 180–198 (2007).
198. L. Tian, J. Qu, Z. Guo, Y. Jin, Y. Meng, and X. Deng,"Microscopic second-harmonic generation emission direction in fibrillous collagen type I by quasi-phase-matching theory," *J. Appl. Phys.*, **108**, 054701/1–9 (2010).
199. D. J. Prockop and A. Fertala, "The collagen fibril: the almost crystalline structure," *J. Struct. Biol.*, **122**, 111–118 (1998).
200. D. A. D. Parry and G. R. G. Barnes, "A comparison of the size distribution of collagen fibrils in connective tissues as a function of age and a possible relation between fibril size distribution and mechanical properties," *Proc. R. Soc. Lond. B*, **203**, 305–321 (1978).
201. S.-W. Chu, S.-Y. Chen, G.-W. Chern, T.-H. Tsai, Y.-C. Chen, B.-L. Lin, and C.-K. Sun, "Studies of $\chi^{(2)}/\chi^{(3)}$ tensors in submicron-scaled bio-tissues by polarization harmonics optical microscopy," *Biophys. J.*, **86**, 3914–3922 (2004).
202. A. Leray, L. Leroy, Y. Le Grand, C. Odin, A. Renault, V. Vié, D. Rouède, T. Mallegol, O. Mongin, M. H. V. Werts and M. Blanchard-Desce, "Organization and orientation of amphiphilic push-pull chromophores deposited in Langmuir–Blodgett monolayers studied by second-harmonic generation and atomic force microscopy," *Langmuir*, **20**, 8165–8171 (2004).
203. G. J. Simpson and K. L. Rowlen, "An SHG magic angle: dependence of second harmonic generation orientation measurements on the width of the orientation distribution," *J. Am. Chem. Soc.*, **121**, 2635–2636 (1999).
204. J. C. Conboy, and G. L. Richmond, "Examination of the electrochemical interface between two immiscible electrolyte solutions by second harmonic generation," *J. Phys. Chem. B.*, **101**, 983–990 (1997).
205. M. Müller, J. Squier, K. R. Wilson, and G. J. Brankenhoff, "3D microscopy of transparent objects using third-harmonic generation," *J. Microsc.*, **191**, 266–274 (1998).
206. D. Oron, E. Tal, and Y. Silberberg, "Depth-resolved multiphoton polarization microscopy by third-harmonic generation," *Opt. Lett.*, **28**, 2315–2317 (2003).
207. N. Olivier, F. Aptel, K. Plamann, M.-C. Schanne-Klein, and E. Beaurepaire, "Harmonic microscopy of isotropic and anisotropic microstructure of the human cornea," *Opt. Express*, **18**, 5028–5040 (2010).
208. D. Yelin and Y. Silberberg, "Laser scanning third-harmonic-generation microscopy in biology," *Opt. Express*, **5**, 169–175 (1999).
209. S.-W. Chu, I.-H. Chen, T.-M. Liu, C.-K. Sun, S.-P. Lee, B.-L. Lin, P.-C. Cheng, M.-X. Kuo, D.-J. Lin, and H.-L. Liu, "Nonlinear bio-photonic crystal effects revealed with multimodal nonlinear microscopy," *J. Microsc.*, **208**, 190–200 (2002).
210. V. Barzda, C. Greenhalgh, J. A. Au, S. Elmore, J. Beek, and J. Squier, "Visualization of mitochondria in cardiomyocytes by simultaneous harmonic generation and fluorescence microscopy," *Opt. Express*, **13**, 8263–8276 (2005).
211. L. Canioni, S. Rivet, L. Sarger, R. Barille, P. Vacher, and P. Voisin, "Imaging of Ca^{2+} intracellular dynamics with a third-harmonic generation microscope," *Opt. Lett.*, **26**, 515–517 (2001).

212. D. Debarre, W. Supatto, E. Farge, B. Moulia, M.-C. Schanne-Klein, and E. Beaurepaire, "Velocimetric third-harmonic generation microscopy: micrometer-scale quantification of morphogenetic movements in unstained embryos," *Opt. Lett.*, **29**, 2881–2883 (2004).
213. S.-P. Tai, W.-J. Lee, D.-B. Shieh, P.-C. Wu, H.-Y. Huang, C.-H. Yu, and C.-K. Sun, "In vivo optical biopsy of hamster oral cavity with epi-third-harmonic-generation microscopy," *Opt. Express*, **14**, 6178–6187 (2006).
214. D. Debarre, W. Supatto, A.-M. Pena, A. Fabre, T. Tordjmann, L. Combettes, M.-C. Schanne-Klein, and E. Beaurepaire, "Imaging lipid bodies in cells and tissues using third-harmonic generation microscopy," *Nat. Methods*, **3**, 47–53 (2006).
215. C.-H. Yu, S.-P. Tai, C.-T. Kung, I-. Wang, H.-C. Yu, H.-J. Huang, W.-J. Lee, Y.-F. Chan, and C.-K. Sun, "In vivo and ex vivo imaging of intra-tissue elastic fibers using third-harmonic-generation microscopy," *Opt. Express*, **15**, 11167–11177 (2007).
216. C.-H. Yu, S.-P. Tai, C.-T. Kung, W.-J. Lee, Y.-F. Chan, H.-L. Liu, J.-Y. Lyu, and C.-K. Sun, "Molecular third-harmonic-generation microscopy through resonance enhancement with absorbing dye," *Opt. Lett.*, **33**, 387–389 (2008).
217. A. Penzkofer, F. Ossig, and P. Qiu, "Picosecond third-harmonic light generation in calcite," *Appl. Phys. B*, **47**, 71–81 (1988).
218. R. Boyd, *Nonlinear Optics,* 2nd ed., Academic Press, San Diego, 2003.
219. J. W. Chan, T. R. Huser, S. H. Risbud, and D. M. Krol, "Modification of the fused silica glass network associated with waveguide fabrication using femtosecond laser pulses," *Appl. Phys. A*, **76**, 367–372 (2003).
220. T. Kawasumi, Y. Ozeki, and K. Itoh, "Analysis and compensation for artifacts in three-dimensional refractive index profiling by four-wave mixing microscopy," *Jpn. J. Appl. Phys.*, **49**, 082701/1–5 (2010).

CHAPTER 5

FUNCTIONAL IMAGING BASED ON MOLECULAR CONTROL

5.1 LOCALIZED OPTICAL MARKING AND TRACKING USING PHOTOMODULATABLE FLUORESCENT MOLECULES

Photomodulatable fluorescent molecules, which are capable of pronounced changes in their spectral properties in response to irradiation with light of a specific wavelength and intensity, provide unique possibilities for the optical marking and tracking of living cells, organelles, and intracellular molecules in a spatiotemporal manner [1–9] and for super-resolution imaging [10–15]. Photomodulatable fluorescent molecules are divided into three classes. One is irreversibly photoactivatable fluorescent molecules, which irreversibly convert from a dark (off) state to a bright fluorescent (on) state with optical stimuli (photoactivation) [16–21]. Another is irreversibly photoconvertible fluorescent molecules, whose spectral characteristics (fluorescence excitation and emission spectra) irreversibly change as a response to irradiation with light [22–33]. The last is reversibly photoswitchable fluorescent molecules, which are reversibly photoswitched between a fluorescent state and a nonfluorescent state with light irradiation [34–65]. In this section, we describe photomodulatable organic dyes, quantum dots, and fluorescent proteins, and show localized optical marking and tracking techniques using the photomodulatable fluorescent molecules.

Functional Imaging by Controlled Nonlinear Optical Phenomena, First Edition.
Keisuke Isobe, Wataru Watanabe and Kazuyoshi Itoh.

5.1.1 Photomodulatable Fluorescent Molecules

5.1.1.1 Photomodulatable Organic Dyes There are two main classes of photomodulatable organic dyes. Some examples of photomodulatable organic dyes are shown in Table 5.1 One is irreversibly photoactivatable organic dyes (also known as photocaged fluorophores) [16,17], and the other is reversibly photoswitchable organic dyes [34–49]. In caged compounds such as caged Q-rhodamine [16] and caged fluorescein [17], which are irreversibly photoactivatable organic dyes, during uncaging irradiation with UV light causes the release of a protective group with good photophysical properties but without intrinsic photoswitching ability and results in a large increase in the fluorescence intensity of the dye. Even if nonfluorescent caged fluorophores are homogeneously distributed within the system of interest, the resultant fluorescent mark is monitored with time to provide quantitative dynamic information.

On the removal of oxygen and the addition of fluorophore-specific oxidizing and reducing agents such as β-mercaptoethylamine (MEA) or glutathione (GSH), carbocyanine derivatives such as Cy5 and Alexa Fluor 647 can be reversibly photoswitched between a fluorescent *on* state and a nonfluorescent *off* state by simultaneous excitation with 514 nm (337–532 nm) and 647 nm [36,37]. The oxygen concentration can be reduced by application of an enzymatic oxygen scavenging system [66,67]. Combining Cy5 (or Alexa Fluor 647) with an activator (e.g., Alexa Fluor 405, Cy2, or Cy3) dramatically facilitates photoswitching [38,39]. Then, the fluorescent state of Cy5 is switched back by the activation with different wavelengths corresponding to the absorption wavelengths of the activators. This recovery rate depends on the close proximity of the activator [39]. Furthermore, Cy3 was found to facilitate switching of other carbocyanine derivatives, such as Cy5.5 and Cy7 [38,39].

In contrast, most rhodamine (Alexa 488, Alexa 532, Atto565, Alexa 568, Atto 590, TMR, etc.) and oxazine (Atto520, Atto655, Atto700, etc.) derivatives can be reversibly switched between a fluorescent state and a nonfluorescent state in the presence of the oxygen and thiol-containing reducing agents such as MEA or GSH [40–42], which are weaker electron donors at neutral pH [68,69]. As illustrated in Fig. 5.1(a), the photoreaction is initiated by intersystem crossing from the first-excited singlet state to the triplet state. In the presence of molecular oxygen, the triplet state is efficiently quenched via energy transfer from the dye triplet to triplet oxygen to produce singlet molecular oxygen [70]. The thiol (RSH/RS^-) competes with molecular oxygen in triplet quenching, naturally present at 200 to 250 mM concentration in aqueous solvents at room temperature, forming the thiyl and dye radical anion [71]. Efficient quenching of the triplet state under physiological conditions (pH 7–8) requires a concentration of 10 to 100 mM thiol. The radical anion is then quickly reoxidized by the oxidizing agent to repopulate the singlet ground state [71]. The produced thiyl radicals ($RS^\bullet$) scavenge molecular oxygen as well [72]. The oxygen is consumed by quenching of the triplet, dye radical, and thiyl radical with irradiation of an aqueous solution of rhodamine and oxazine derivatives in the presence of thiols. Therefore, the radical anion is formed with an efficiency limited by the amount of residual oxygen present in the sample [72]. In addition, the radical anion is immediately oxidized by re-dissolved oxygen from the headspace. The process of

TABLE 5.1 Spectroscopic Properties of Photomodulatable Organic Dyes

Fluorophore	On [nm]	Off [nm]	Pre-Color Ex [nm]	Pre-Color Em [nm]	Post-Color Ex [nm]	Post-Color Em [nm]	References
Irreversibly Photoactivatable Organic Dyes							
Caged Q rhodamine	405	NA	Dark		545	575	K. R. Gee et al., 2001 [16]
Caged fluorescein	405	NA	Dark		497	516	T. J. Mitchison et al., 1998 [17]
Reversibly Photoswitchable Organic Dyes							
Photochromic rhodamine B	375	NA	565	580	Dark		J. Fölling et al., 2007 [34]
Cy5	337–532	649	649	670	Dark		M. Heilemann et al., 2005 [36], 2008 [37]
Cy5 with Alexa Fluor 405	405	649	649	670	Dark		M. Bates et al., 2007 [38]
Cy5 with Cy2	457	649	649	670	Dark		M. Bates et al., 2007 [38]
Cy5 with Cy3	532	649	649	670	Dark		M. Bates et al., 2005 [39], 2007 [38]
Cy5.5 with activator	350–570	675	675	694	Dark		M. Bates et al., 2007 [38]
Cy7 with activator	350–570	747	747	776	Dark		M. Bates et al., 2007 [38]
Alexa Fluor 488	396	491	491	517	Dark		M. Heilemann et al., 2007 [40], S. van de Linde et al., 2011 [72]
Alexa Fluor 532	380–430	532	532	552	Dark		M S. van de Linde et al., 2009 [41], 2011 [72]
Alexa Fluor 568	380–430	572	572	600	Dark		M S. van de Linde et al., 2009 [41], 2011 [72]
Alexa Fluor 647	350–570	650	650	665	Dark		M. Bates et al., 2007 [38], M. Heilemann et al., 2008 [37]
Atto 488	400	501	501	523	Dark		S. van de Linde et al., 2011 [72]
Atto 520		516	516	538	Dark		S. van de Linde et al., 2009 [41]
Atto 532	420	532	532	553	Dark		S. van de Linde et al., 2011 [72]
Atto 565	432	564	563	592	Dark		S. van de Linde et al., 2009 [41], 2011 [72]
Atto 590		594	594	624	Dark		S. van de Linde et al., 2009 [41]
Atto 655	405	647	663	684	Dark		S. van de Linde et al., 2009 [41], 2011 [72]
Atto 700		700	700	719	Dark		S. van de Linde et al., 2009 [41]
Dy 505	386	500	500	525	Dark		S. van de Linde et al., 2011 [72]
Dy 530	380–430	535	535	556	Dark		S. van de Linde et al., 2011 [72]
Rhidamine 123	386	500	500	518	Dark		S. van de Linde et al., 2011 [72]
Rhidamine 6G	422	526	526	556	Dark		S. van de Linde et al., 2011 [72]
Tetramethyl rhodamine	405	532	555	580	Dark		T. Klein et al., 2011 [42]

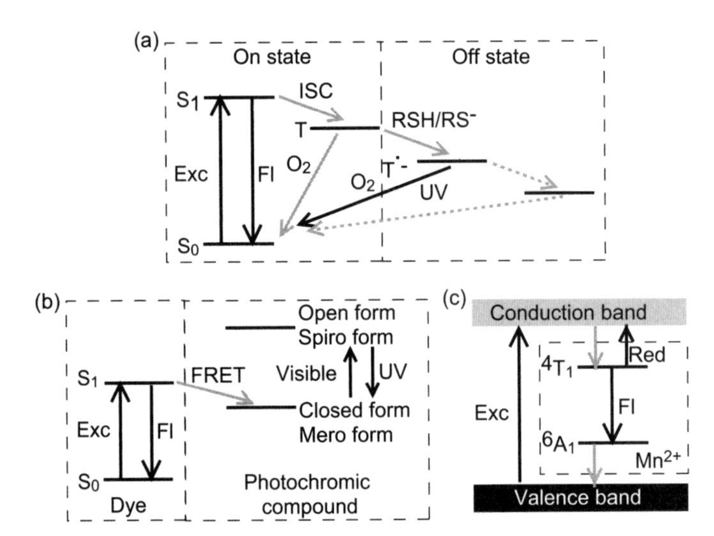

FIGURE 5.1 (a) Underlying photophysical process of reversible photoswitching of rhodamine derivatives. (b) FRET switching based on photochromic reaction of dithienylethenes derivatives and spiropirans. (c) Fluorescence photoswitching based on internal electronic transitions within Mn-doped ZnSe QD.

photo reduction by irradiation in the presence of thiols and oxidation by re-dissolving molecular oxygen is in fact reversible. Photo-excitation of the radical anion with UV irradiation also induces recovery of the fluorescent state form without any oxygen being required. Therefore, not only the photoswitching rate for the fluorescent state but also that for nonfluorescent state can be fine-tuned by the irradiation intensities at two different wavelengths [42].

Photochromic compounds such as dithienylethenes derivatives with the heterocyclic aryl group and spiropirans have been used to achieve a reversibly photoswitchable fluorescent system through addition of organic fluorescent dyes as shown in Fig. 5.1(b) [43–49]. The dithienylethenes derivatives can be reversibly converted between the open and closed ring forms, which correspond to colorless and colored modes, respectively, with different absorption properties upon exposure to UV or visible light, respectively, while the spiropirans can be reversibly switched between the spiro and mero forms with UV and visible irradiation, respectively [73,74]. The crossed form of the dithienylethene or the mero form of the spiropiran has been used as an acceptor in fluorescence resonance energy transfer (FRET) from an organic fluorescent dye serving as a donor molecule. In the open form of the dithienylethene or the spiro form of the spiropiran, however, no FRET occurs. Thus, by modulating the absorption properties of the photochromic compound with light irradiation, fluorescence from the donor fluorescent dye can be reversibly switched [43–49]. By using this photochromic FRET (pcFRET), accurate intracellular measurements of FRET efficiency can be also achieved [75,76]. The advantage of pcFRET over the other FRET assays is that it provides an accurate and repeated FRET quantification for the same FRET pair within the same live cell without the need for corrections based on reference images acquired from separate control cells.

5.1.1.2 Photomodulatable Quantum Dots Han et al. have synthesized photo-caged quantum dots (QDs) that are nonluminescent under typical microscopic illumination but can be irreversibly activated with stronger UV light [18]. Photochromic compounds such as dithienylethenes derivatives and spiropirans have been also used as an acceptor in FRET from QDs as well as from organic donors to achieve reversible photoswitching of fluorescence from the QDs donors [50–53]. Recently, Irvine et al. have developed a new reversibly photoswitchable QD without photochromic compounds, which is a manganese (Mn)-doped zinc selenide (ZnSe) QD [54]. The fluorescence photoswitching relies only on internal electronic transitions within Mn-doped ZnSe QDs as shown in Fig. 5.1(c). Initially, electrons are photoexcited with a wavelength of about 400 nm from the valence band to the conduction band of the ZnSe host, and then are subsequently transferred to the $^4T_1$ upper florescent state of the Mn^{2+} dopant within a short time (picosecond timescale). The electrons can radiatively relax from the $^4T_1$ state to the $^6A_1$ state, resulting in fluorescence emission at a wavelength of 580 nm. In the presence of activation light at a red-shifted wavelength (about 600 nm) with respect to the initial electron excitation, however, they can also be pumped to higher levels through excited-state absorption (ESA) from the $^4T_1$ state, while avoiding further excitation from the ZnSe host. Thus, fluorescence from the Mn-doped ZnSe QD can be modulated by the red-shifted activation light.

5.1.1.3 Photomodulatable Fluorescent Proteins Since 2002, various photomodulatable fluorescent proteins (FPs) have been developed by engineering existing FPs or cloning new proteins from fluorescent organisms [1–15]. Some examples of photomodulatable FPs are shown in Table 5.2. In 2002, Patterson and Lippincott-Schwartz developed a photoactivatable green fluorescent protein (PA-GFP) by improving on the wt-GFP photoconversion from a neutral to anionic species [19], and Ando et al. identified a green-to-red photoconvertible FP, Kaede from the stony coral *Trachyphyllia geoffroyi* [22]. PA-GFP exhibits up to 100-fold increase in green fluorescence when illuminated with violet or UV light [19], while Kaede displays up to nearly 2000-fold increase in red-to-green fluorescence ratio with violet or UV irradiation [22]. Violet or UV light irradiation leads to a chemical reaction within a particular FP such as cleavage of the protein backbone within the chromophore [77] or decarboxylation of Glu222, which plays a key role in the rearrangement of the hydrogen-bonding network formed around the chromophore and chromophore deprotonation [78]. This basic mechanism apparently underlies the photoactivation of PA-GFP and photoconversion of Kaede. In 2003, Chudakov et al. generated a none-to-red *kindling* fluorescent protein (KFP1) [57], which is a mutant of asFp595 [56] (also known as asulCP and asCP) discovered from the sea anemone *Anemonia sulcata*. KFP1 exhibits up to 30-fold increase in red fluorescence upon irradiation with green light, and red fluorescence can be quenched back to the initial dark state by a irradiation with blue light [57]. Both kindling and quenching of KFP1, which are due to change in spectral properties based on *cis-trans* transitions of the chromophore

TABLE 5.2 Spectroscopic Properties of Photomodulatable Fluorescent Proteins

				Pre-Color		Post-Color		
Fluorophore	Oligometric State	On [nm]	Off [nm]	Ex [nm]	Em [nm]	Ex [nm]	Em [nm]	References
Irreversibly Photoactivatable Fluorescent Proteins								
PA-GFP	Monomer	400	NA	Dark		504	517	G. H. Patterson et al., 2002 [19]
PA-mRFP1	Monomer	UV-violet	NA	Dark		578	605	V. V. Verkhusha et al., 2005 [20]
PA-mCherry1	Monomer	404	NA	Dark		564	595	F. V. Subach et al., 2009 [21]
Irreversibly Photoconvertible Fluorescent Proteins								
PS-CFP2	Monomer	400	NA	400	468	490	511	D. M. Chudakov et al., 2004 [23], Evrogen
Phamret	Tandem dimer	400	NA	458	475	458	517	T. Matsuda et al., 2008 [32]
Kaede	Tetramer	380	NA	508	518	572	582	R. Ando et al., 2002 [22]
mEos3.2	Monomer	405	NA	507	572	516	580	M. Zhang et al., 2012 [26]
KikGR	Tetramer	390	NA	507	517	583	593	H. Tsutsui et al., 2005 [27]
mKikGR	Monomer	390	NA	507	515	580	591	S. Habuchi et al., 2008 [28]
Dendra2	Monomer	UV/blue	NA	490	507	553	573	N. G. Gurskaya et al., 2006 [29], Evrogen
PSmOrange	Monomer	Blue/green	NA	548	565	636	662	O. M. Subach et al., 2011 [33]
Reversibly Photoswitchable Fluorescent Proteins								
Padron	Monomer	505	396	Dark		505	522	M. Andresen et al., 2008 [55]
KFP1	Tetramer	Green	Blue	Dark		580	600	D. M. Chudakov et al., 2003 [57]
rsCherry	Monomer	Yellow	Blue	Dark		572	610	A. C. Stiela et al., 2008 [58]
rsTagRFP	Monomer	440	567	Dark		567	585	F. V. Subach et al., 2010 [59]
mTFP0.7	Monomer	370	453	453	488	Dark		J. N. Henderson et al., 2007 [60]
Dronpa	Monomer	390	503	503	517	Dark		R. Ando et al., 2004 [61]
Dronpa-2,3	Monomer	390	487	487	514	Dark		R. Ando et al., 2007 [62]
rsFastLime	Monomer	384	496	496	518	Dark		A. C. Stiel et al., 2007 [63]
bsDronpa	Monomer	385	460	460	504	Dark		M. Andresen et al., 2008 [55]
rsEGFP	Monomer	405	491	493	510	Dark		T. Grotjohann et al., 2011 [64]
Dreiklang	Monomer	365	405	511	529	Dark		T. Brakemann et al., 2011 [65]
rsCherryRev	Monomer	Blue	Yellow	572	608	Dark		A. C. Stiela et al., 2008 [58]

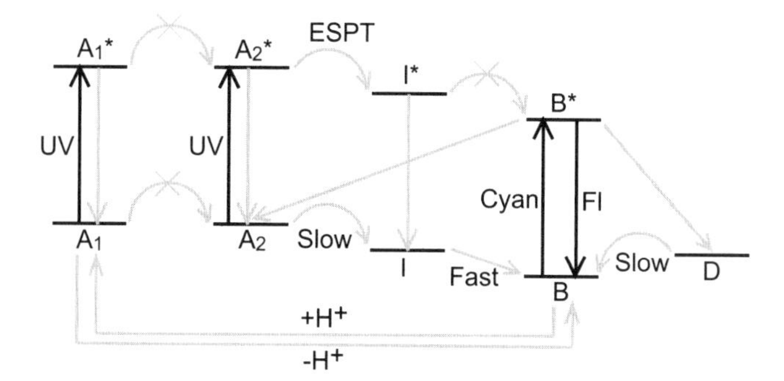

FIGURE 5.2 Reversible photoswitching scheme for Dronpa.

[79], are reversible photoswitching processes as well as asFp595. However, irradiation of KFP1 with green light of greater intensity, longer duration, or both causes irreversible kindling [57]. In 2004, Ando et al. introduced green-to-none *quenchable* FP, Dronpa, which is capable of repeated photoswitching between a fluorescent state and a nonfluorescent state using quenching and kindling light at two distinct wavelengths corresponding to deprotonated (503-nm) and protonated (390-nm) chromophore forms [61], respectively. According to the reversible photoswitching scheme for Dronpa as shown in Fig. 5.2 [80], upon irradiation of the deprotonated form (B) with cyan (503 nm) light, the excited state (B*) decays via fluorescence into the corresponding ground state (B) and into the ground states of the photoswitched protonated form (A_2) and unknown dark-state (D). Although the photoswitched form (A_2) is the protonated form of the chromophore, this form is not interconvertible with the pH-induced protonated form (A_1). On excitation of the photoswitched protonated form (A_2) with violet or UV (390-nm) light, the excited state (A_2^*) nonradiatively decays into the corresponding ground state (A_2) and can switch into the ground state of the deprotonated form (B) through excited-state proton transfer (ESPT) from the photoswitched protonated form (A_2^*) to a nonfluorescent intermediate (I*). The primary mechanism of photoswitching is thought to also arise from *cis-trans* isomerization of the hydroxybenzilidine (tyrosyl side chain) chromophore moiety that accompanies the changes in the protonation state [60,81,82].

Several none-to-red photoactivatable FPs (PA-mRFP1 [20], PA-mCherry1 [21], and PA-TagRFP (Evrogen)) based on monomeric red FPs have been developed. PA-mRFP1 is the first reported protein of this type but was characterized by rather dim fluorescence [20]. PA-mCherry1, -2, and -3 are significantly enhanced variants and was characterized by faster photoactivation, higher contrast, and better photostability [21]. PA-TagRFP (Evrogen) is a photoactivatable mutant of the bright monomeric red FP TagRFP [83]. PA-TagRFP shows better photostability than PA-mCherry.

Irreversibly photoconvertible FPs reported to date include green-to-red (Kaede [22], EosFP [24], mEosFP [24], mEos2 [25], mEos3.1 [26], mEos3.2 [26], KikGR [27], mKikGR [28], Dendra2 (Evrogen) [29], IrisFP [30] and mIrisFP [31]), cyan-to-green (PS-CFP [23], PS-CFP2 (Evrogen), and Phamret [32]), and orange-to-far-red

(PS-mOrange [33]) photoconvertible FPs. Kaede has been successfully applied as a very efficient marker for photolabeling and tracking of organelles in cells and cells in living tissues [84,85]. However, Kaede, EosFP [24], and KikGR [27] are obligate tetramers, and are thus not suitable for imaging of cellular proteins. To overcome this limitation, monomeric green-to-red photoconvertible FPs—such as mEosFP [24], mEos2 [25], mEos3.1 [26], mEos3.2 [26], mKikGR [28], Dendra2 [29], and mIrisFP [31], the best of which are currently mEos3.1, mEos3.2, and Dendra2—have been developed. These monomeric photoconvertible FPs can be also used in protein tracking [86–89]. PS-CFP2 is preferred in multicolour studies because it has the highest contrast ratio. Phamret, which is a fusion protein composed of a CFP variant (mseCFP) fused to a PA-GFP, is a FRET probe that induces a change in fluorescence emission from mseCFP to photoactivated PA-GFP [32]. Before photoactivation of PA-GFP, the excitation of the fusion protein with blue light leads to a direct emission from mseCFP. After photoactivation of PA-GFP with violet or UV light, FRET from mseCFP to PA-GFP allows a green fluorescence emission from PA-GFP.

To date, numerous reversibly photoswitchable FPs have been identified and engineered. Examples include none-to-green (Padron [55]), none-to-red (asFP595 [56], KFP1 [57], rsCherry [58] and rsTagRFP [59]), cyan-to-none (mTFP0.7 [60]), green-to-none (Dronpa [61], Dronpa-2 [62], Dronpa-3 [62], rsFastLime (also known as Dronpa-V157G) [63], Dronpa-M159T [63], bsDronpa [55], rsEGFP [64], and green form of IrisFP [30] and mIrisFP [31]), yellow-to-none (Dreiklang [65]), and red-to-none (rsCherryRev [58], and red form of IrisFP [30] and mIrisFP [31]) photoswitchable FPs. rsCherry is the first monomeric none-to-red photoswitchable FP but have limited brightness, low contrast, and complex switching behavior [58]. rsTagRFP displays higher brightness and higher contrast than rsCherry [59]. rsTagRFP has been used as an acceptor for a reversible pcFRET as well as organic photochromic compounds (dithienylethenes derivatives and spiropirans), and this technique has been applied to visualize an interaction between epidermal growth factor receptor and growth factor receptor-binding protein 2 in live mammalian cells [59]. Dronpa is currently the best-tested reversible photoswitchable FP that is monomeric, has high brightness, and can be switched on and off multiple times with high contrast. Because rsEGFP is extremely fatigue resistant and is capable of >1000 switching cycles, rsEGFP is suitable for super-resolution imaging [64]. Unlike any other photoswitchable FP, which is switched off by the same wavelength that is used for fluorescence excitation, Dreiklang can be switched on and off upon irradiation with activating light at 365 nm and quenching light at 405 nm, respectively, which are different from the wavelength for fluorescence excitation (515 nm) [65]. IrisFP [30] and mIrisFP [31] are remarkable proteins of the most complex photobehavior reported to date among FPs. These proteins are capable of irreversible photoconversion from a green- to a red-emitting state upon illumination with violet or UV light (405 nm), whereas both initial green and photoconverted red fluorescent forms are also capable of reversible photoswitching. Green IrisFP can be switched on and off upon irradiation with activating light at 405 nm and quenching light at 488 nm, respectively, while red IrisFP can be repeatedly activated at 440 nm and quenched at 532 nm [30].

5.1.2 Imaging with Localized Optical Manipulation of Photosensitive Molecules

5.1.2.1 Localized Optical Marking by Photoactivation and photoconversion with Two-Photon Excitation Irreversible photoactivation, photoconversion, and reversible photoswitching enable selective conversion or activation of fluorescence signals after optical illumination, and these are recognized as powerful tools for studying the dynamic processes of fluorescently labeled individual cells, organelles, and proteins [1–9]. A portion of photomodulatable fluorescent proteins (FPs) fused to a protein and targeted to any organelle can be selectively photoactivated within a region of interest of a live cell, and the rate of distribution of the activated fluorescent signal can be tracked. In such a way, photomodulatable FPs can help tracking protein motility [19], proteins exchange between cellular compartments [23], protein degradation [90], organelle motility [91], interaction with other organelles in fusion–fission events [92,93], and the movement of cells within cultures, living tissue, or whole organisms during development, cancerogenesis and inflammation [85,94,96], similar to the photobleaching techniques discussed in Section 2.2.4 and 4.2.2. However, the photomodulatable FP approach is more straightforward and informative, and can be applied for tracking very fast movement. This is because, instead of estimating the movement of proteins, organelles, and cells from the change in average fluorescence after photobleaching, this approach can be used to directly observe the redistribution of individual proteins, organelles, and cells labeled with the photoactivated FPs. On the one hand, reversibly photoswitchable FPs can be applied to track movement of a fused protein of interest repeatedly within the same cell, to generate multiple spatial and temporal points [97]. This allows the identification of differences in protein motility in various parts of the same cell and the monitoring of changes in motility or transport direction of a protein in response to the particular external influence or caused by internal changes of the cell state. On the other hand, irreversibly photoactivatable or photoconvertible FPs do not allow for repeated photoactivation, photoconversion, and tracking events, but instead produce a stable signal that can be tracked for a long period of time [19,22–24,29,57,91,92]. Therefore, irreversibly photoactivatable or photoconvertible FPs represent a strong alternative to traditional kinetic microscopy of protein motility in living cells, based on photobleaching methods [98].

Photoactivation, photoconversion, and photoswitching of most photomodulatable fluorescent molecules can be achieved by one-photon excitation (OPE) with violet or UV light. However, OPE leads to collateral photoconversion outside the focal volume, which limits the spatial resolution [99]. To overcome these problems, photoactivation and photoconversion by two-photon excitation (TPE) has been used [99–106].

To induce photoactivation and photoconversion by TPE, we need to know TPE spectra for photoactivation and photoconversion. TPE spectra for photoactivation in PA-GFP [100], and photoconversion in EosFP [107] and Kaede [108] have been measured by the laser-wavelength scanning method or the Fourier-transform (FT) method discussed in Chapter 3. In the case of the measurement of the TPE spectrum for photoactivation, after irradiation by the pump light for photoactivation, it is

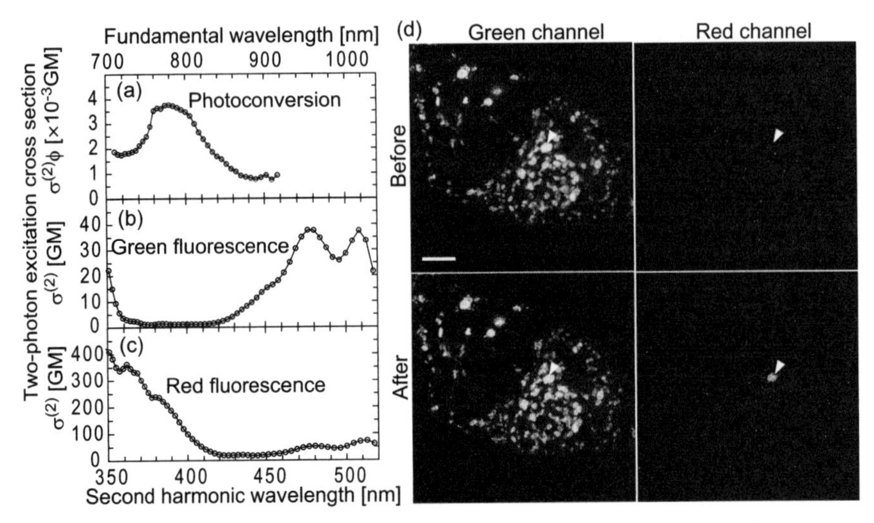

FIGURE 5.3 (a–c) TPE spectra of Kaede for (a) green-to-red photoconversion, (b) green fluorescence, and (c) red fluorescence. Reprinted with permission from [108]. Copyright 2010 Optical Society of America. (d) Selective photoconversion of mitochondria with Kaede in a fixed BY-2 cell. A single mitochondrion was photoconverted from green to red by 750-nm femtosecond laser pulses. A target mitochondrion is indicated by the red arrow. After photoconversion of the mitochondrion, time-lapse images of the green and red fluorescence using a one-photon fluorescence microscope were obtained. Scale bar: 10 μm. Reprinted with permission from [106]. Copyright 2007 Optical Society of America.

necessary to record the photoinduced spectral change with a probe light. In the laser-wavelength scanning method, because fluctuations from tuning the pump wavelength could affect the spatial overlap between the pump and probe lights, the amount of photoinduced spectral change is obtained from the fluorescence image, which is recorded by fluorescence microscopy using the probe lights [107]. In contrast, the FT method using broadband pulses allows us to measure the TPE spectrum without tuning the pump wavelength and to suppress the fluctuations in the spatial overlap between the two lights [108]. Thus, only the fluorescence signals from the focal volume, which is generated by the probe light, are used even in the measurement of TPE spectrum for photoactivation. Figure 5.3(a) to (c) shows the TPE spectra for photoconversion in Kaede, green-fluorescence in unphotoconverted Kaede, and red-fluorescence in photoconverted Kaede, which were obtained by nonlinear FT spectroscopy [108]. The TPE peak for photoconversion in Kaede was located at around a second harmonic wavelength of 390 nm. The TPE spectra for photoconversion and green fluorescence in Kaede are very similar to those for fluorescence in Sapphire and enhanced GFP (EGFP), respectively [109], which are GFP variants. The green state of Kaede shows two one-photon absorption peaks at 380 and 508 nm, corresponding to neutral and ionized forms, respectively [22]. Photoconversion is induced by excitation of the neutral form, while green-fluorescence is generated by excitation of the ionized form.

This is analogous to the photocharacterization of GFP with OPE. Sapphire molecules contain neutral chromophores with a OPE peak at 399 nm, whereas EGFP molecules have ionic chromophores with a one-photon excitation peak at 488 nm [110]. From the photocharacterization of Kaede and GFP with OPE and the similarity of the TPE spectra in Kaede and in GFP, it appears that photoconversion and green-fluorescence in Kaede on TPE are also induced by the excitation of the neutral and ionized forms, respectively. As for the photoconverted Kaede, the TPE spectrum for red-fluorescence from Kaede is similar to that from DsRed [109].

Watanabe et al. showed spatially selective labeling of a single mitochondrion by using a two-photon conversion of Kaede, and tracking the dynamics of the mitochondria [106]. In order to alter the fluorescence spectrum, femtosecond laser pulses with a wavelength of 750 nm and a repetition rate of 76 MHz were focused at the targeted mitochondrion in Fig. 5.3(d). The mitochondrion was photoconverted around the focal point. Two-photon photoconversion by the femtosecond laser irradiation enables one to perform site-specific labeling of a single organelle in three-dimensional space.

To examine the movement of a single mitochondrion, Watanabe et al. demonstrated, as an example, tracking of the movement of a single mitochondrion among several hundreds of mitochondria in a living BY-2 cell. Figure 5.4 shows a time-lapse series of confocal images of mitochondria labeled with Kaede in living BY-2 cells, obtained at 30-min intervals. This technique allows an organelle to be tracked for a long time without loss of image contrast. This indicates that labeling and tracking of an organelle using a femtosecond laser can be performed without compromising the

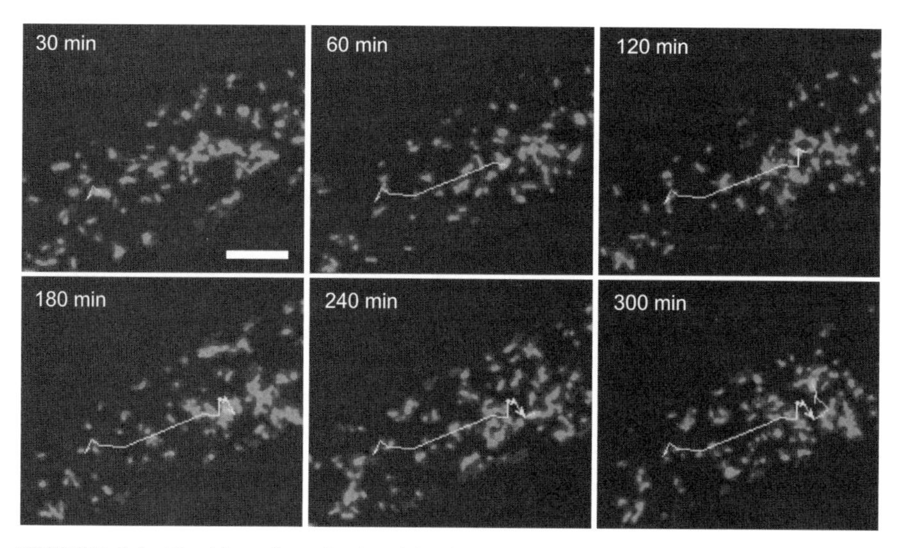

FIGURE 5.4 Tracking of a mitochondrion in a living BY-2 cell. Scale bar: 10 μm after marking a mitochondrion by two-photon excitation. The movement of the mitochondrion labeled by two-photon conversion could be tracked for 5 hours. The trajectory of the labeled mitochondrion is shown by the yellow line. Reprinted with permission from [106]. Copyright 2007 Optical Society of America. (*See insert for color representation of the figure.*)

viability. From time-lapse stacked 3D confocal images t after labeling, the dynamics of an individual mitochondrion was investgated. The velocity of the photoconverted mitochondrion was calculated to be 0.18 μm s^{-1}. The results demonstrated the ability to track the dynamics of a single mitochondrion and to reveal detailed spatial information in a living cell, such as the position and velocity. Site-specific organelle labeling enabled one to track the dynamics of a single organelle at different sites in a living cell.

5.1.2.2 Localized Photochemical Release of Caged Compounds by Two-Photon Excitation The hypothesis that changes in synaptic function are intimately linked to changes in dendritic spine structure has been tested many times. In this study, there are two important techniques. One technique is fluorescence imaging of the morphology of dendrites and spines before and after synaptic stimulation. The other is fluorescence imaging of of synaptically evoked calcium (Ca) signals by using an intracellular Ca-sensitive fluorophore, which allows identification of spines that contain the activated synapses. TPE can be used not only to achieve Ca^{2+} imaging in deep region but also to stimulate synapses via the photolysis of caged neurotransmitter such as glutamate and γ-aminobutyric acid (GABA) [111–119]. GABA is inhibitory, while glutamate is excitatory. The neurotransmitter is covalently modified with a side group that renders the neurotransmitter inert. Upon excitation of the caged neurotransmitter, the bond attaching the side group is broken, often via a hydrolysis reaction, and glutamate or GABA is effectively released [120]. With OPE, glutamate is uncaged in a large volume extending above and below the focal plane and is able to diffuse significantly before clearance. This slow and broad spatiotemporal profile of glutamate release is unable to mimic the kinetics of α-amino-3-hydroxy-5-metyl-4-isoxazole-propionic acid receptor (AMPAR)-mediated synaptic currents. AMPAR is an ionotropic transmembrane receptor for glutamate that mediates fast synaptic transmission in the central nervous system. In contrast, by uncasing glutamate with TPE, the small volume in which glutamate is released allows for rapid clearance, and the kinetics of both AMPAR-mediated synaptic currents and potentials can be matched [111,113].

Matsuzaki et al. and Ellis-Davies et al. have developed 4-methoxy-7-nitro indolinyl-L-glutamate (MNI-Glu) [111] and 4-carboxymethoxy-5,7-dinitroindolinyl-L-glutamate (CDNI-Glu) [116] for two-photon uncaging in mammalian brain slices. MNI-Glu and CDNI-Glu are photolyzed in the 710 to 730 nm range such that the response at a spine head to the uncaged glutamate appears as that evoked by single-vesicle secretion [111,116]. Because significant two-photon uncaging of MNI-Glu and CDNI-Glu falls off steeply at wavelengths roughly above 750 nm, they can be combined by using many common fluorophores with a high TPE cross section at longer wavelengths such as GFP, YFP, and DsRed. Matsuzaki et al. and Kantevari et al. have synthesized a new caged GABA derivative, which are named CDNI-GABA [118] and 7-(dicarboxymethyl)-aminocoumarin-caged GABA (N-DCAC-GABA) [119], respectively. CDNI-GABA [118] and N-DCAC-GABA [119] are effectively uncaged by TPE at 720 and 830 nm, respectively. By combining CDNI-Glu and N-DCAC-GABA, Kantevari et al. have demonstrated two-color, two-photon uncaging to fire and block action potentials from rat hippocampal CACA1 neurons in brain slices, where a two-photon-evoked AMPAR current at 720 nm and a two-photon-evoked $GABA_A$ receptor current at 830 nm have been detected [119].

5.2 MULTIFARIOUS CONTROL OF MULTIPHOTON EXCITATION BY PULSE-SHAPING TECHNIQUE

The ultrabroadband pulse technique offers a number of advantages over narrowband pulses in nonlinear optical microscopy. The ultrabroadband pulse provides simultaneous excitation of various molecules [121,122]. In addition, Silberberg's group has reported that nonlinear optical processes can be controlled by the spectral phase modulation of 10-fs ultrashort pulses whose spectra are ultrabroadband [123–125]. Two-photon absorption in atomic lines could be controlled with great precision by using sinusoidal phase modulation [124]. Several other groups have demonstrated that spectral phase modulation provides the capability to control two-photon excitation in practical two-photon excited fluorescence (TPEF) microscopy [126–131]. Dudovich et al. have also applied sinusoidal phase modulation to selective excitation in coherent anti-Stokes Raman scattering (CARS) spectroscopy, where the CARS process is induced by a single ultrabroadband pulse [125]. Spectral phase modulation techniques for single-pulse CARS microscopy have also been demonstrated by several other groups [126,132,133].

To control nonlinear optical processes, we need to know some spectroscopic information about various molecules of interest. Because nonlinear optical spectroscopies using ultrabroadband pulses, as shown in Sections 3.2 and 3.3, provide broadband excitation spectra of various molecules with a short acquisition time, nonlinear optical microscopy employing spectral phase modulation is often combined with nonlinear optical spectroscopy using ultrabroadband pulses [126]. By employing spectral phase modulation based on the excitation spectra of molecules that are obtained by nonlinear optical spectroscopy, we can control the excitation of nonlinear optical processes with different excitation spectra. Thus, various imaging modes can be switched and various image contrasts rapidly achieved without changing the experimental setup and only exchanging the phase masks. In this section, we focus on spectral phase modulation techniques for TPEF and CARS imaging.

5.2.1 Control of Two-Photon Fluorescence in Multi-Labeled Sample

Since TPEF microscopy was first reported by Denk et al. in 1990 [135], its advantages over confocal microscopy have been clearly demonstrated. TPEF microscopies combined with various fluorescence techniques have been used for investigating biological phenomena. In particular, multi-color microscopy [121,136], fluorescence resonance energy transfer (FRET) microscopy [137,138], and fluorescence cross-correlation spectroscopy (FCCS) [139] are key techniques for visualizing the movement of biomolecules and their interactions with cellar components in a living cell. By combining these techniques, the dynamic interactions of proteins and subcellular structures can be further clarified. By controlling distinctive excitation modes, these techniques can be exchanged rapidly depending on the specimen and the conditions. In FRET microscopy, only the donor fluorophore must be selectively excited without exciting any accepter fluorophores, whereas in the case of FCCS, it is necessary to excite both types of fluorophores equally. In multi-color microscopy, multi-color images are obtained by exciting all of the fluorophores simultaneously.

Ultrabroadband laser pulses provide the potential for easy and rapid switching between the excitation modes. The spectral phase modulation of ultrabroadband pulses allows control of the TPEF intensities from various fluorophores [140–143]. Provided that no intermediate resonant level is present, the TPEF intensity I_F is described by

$$I_F \propto \int \tilde{g}^{(2)}(2\omega) \left|\tilde{E}^{(2)}(2\omega)\right|^2 d\omega, \tag{5.2.1}$$

where $g^{(2)}(2\omega)$ is the two-photon excitation (TPE) spectrum of the fluorophore. Although in fact the SH fields are not generated in the TPEF process, $|E^{(2)}(2\omega)|^2$ is called the "SH power spectrum" of the excitation pulse. The SH power spectrum is expressed by [124]

$$\begin{aligned}\left|\tilde{E}^{(2)}(2\omega)\right|^2 &= \left|\int \tilde{E}(\omega+\Omega)\tilde{E}(\omega-\Omega)d\Omega\right|^2 \\ &= \left|\int |\tilde{E}(\omega+\Omega)||\tilde{E}(\omega-\Omega)| \exp[i\{\phi(\omega+\Omega)+\phi(\omega-\Omega)\}]d\Omega\right|^2,\end{aligned} \tag{5.2.2}$$

where $|E(\omega)|$ and $\phi(\omega)$ are the spectral amplitude and the spectral phase of the excitation pulse, respectively. Since $|E^{(2)}(2\omega)|^2$ can be shaped by modulating the spectral phase of the excitation pulse, the TPEF intensity can be controlled by the use of spectral phase modulation. For example, we consider the TPE process from the ground state to the excited state at an energy of $2\hbar\omega$ by the excitation pulse that is composed of the four frequency components at $\omega+\omega_1$, $\omega-\omega_1$, $\omega+\omega_2$ and $\omega-\omega_2$, all of which have equal spectral amplitude, A. We assume that in the TPE process there are two pathways connecting the initial and final states as shown in Fig. 5.5. Then, the

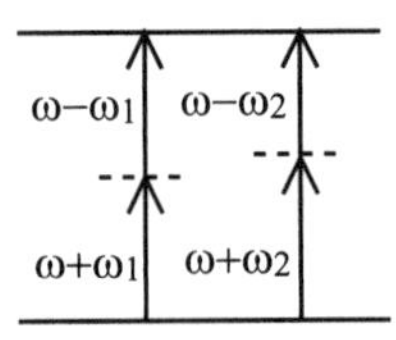

FIGURE 5.5 TPE process by the excitation pulse composed of the four frequency components at $\omega+\omega_1$, $\omega-\omega_1$, $\omega+\omega_2$, and $\omega-\omega_2$.

SH field at 2ω is given by

$$\tilde{E}^{(2)}(2\omega) = A^2 \exp\{i[\phi(\omega + \omega_1) + \phi(\omega - \omega_1)]\} + A^2 \exp\{i[\phi(\omega + \omega_2) + \phi(\omega - \omega_2)]\}. \quad (5.2.3)$$

When $\phi(\omega + \omega_1) + \phi(\omega - \omega_1) = \phi(\omega + \omega_2) + \phi(\omega - \omega_2)$, the SH field is enhanced by constructive interference, and we can obtain high TPEF intensity. In contrast, for $\phi(\omega + \omega_1) + \phi(\omega - \omega_1) = \phi(\omega + \omega_2) + \phi(\omega - \omega_2) + \pi$, the SH field is suppressed by destructive interference, and the TPEF intensity is zero. Thus, the TPE process can be controlled by modulating the spectral phase of the excitation pulse. In practice, because of the use of ultrabroadband pulses, we need to consider the TPE process by various combinations of the fundamental frequency components.

Spectral phase modulation techniques to control the TPEF intensity have been proposed by a few groups. Meshulach et al. first reported the control of two-photon absorption in atomic lines of cesium [124] by sinusoidal phase modulation. The selective excitation of organic fluorophores with different TPE spectra using sinusoidal phase modulation has been demonstrated by Walowicz et al. [140], Lozovoy et al. [141] and Dela Cruz et al. [142]. Comstock et al. proposed binary phase modulation in which the contrast ratio for selective excitation is six times greater than that achieved by sinusoidal phase modulation [143]. Kawano et al. demonstrated the reduction of the photobleaching rate by the use of a phase-optimized pulse whose SH spectrum was shaped [144]. Ogilvie et al. [127], Dela Cruza et al. [128], and Schelhas et al. [129] demonstrated that the use of sinusoidal phase modulation and binary phase modulation allows selective excitation in biological TPEF microscopy. Labroille et al. applied SH spectrum shaping based on third-order dispersion (TOD) to selective TPEF bio-imaging [131]. Isobe et al. demonstrated not only selective excitation but also simultaneous excitation, together with the control of TPEF intensities from different fluorophores with different TPE spectra independently [130]. This technique is based on the generation of a multi-wavelength pulse train by multi-level phase modulation introducing a group-delay phase. The independent control allows us to excite different fluorophores with different TPE spectra equally. Even though cells are labeled by multiple fluorophores with significantly different concentrations, the independent control provides multi-color images of FCCS signals with equal signal levels.

Inter-frame spacing is determined by summation of the imaging time and the phase modulation time. The modulation speed can be improved by selecting a modulator with a short response time. A variety of techniques for spectral phase modulation have been demonstrated to date, including the use of a liquid crystal based spatial light modulator (SLM) [145], a deformable mirror [146], acousto-optic modulators [147] and electro-optical modulators [148]. Among these experiments, electro-optical modulators have been shown to provide modulation speeds of the order of 10 ns [148]. Since this modulation rate is much higher than the frame rate in real-time imaging, the excitation mode can be switched between selective excitation and simultaneous excitation for each frame without dropping the frame rate. In addition, the TPEF intensities can be independently and arbitrarily controlled in each frame. Even if the concentrations of the various fluorophores are independently changed along with alterations in cellular morphology, we can correct the TPEF intensities by the use of intensity control for each frame. The spectral phase modulation techniques are an important tool for rapid and easy switching among multi-spectral imaging, FRET imaging, and FCCS.

In this section, we show the control of TPEF intensities from different fluorophores, all of which coexist, by using sinusoidal phase modulation, TOD phase modulation, binary phase modulation and multi-level phase modulation.

5.2.1.1 Sinusoidal Phase Modulation and Third-Order Dispersion

Sinusoidal phase modulation proposed by Meshulach et al. is expressed by [124]

$$\phi(\omega) = \delta \sin\{\gamma(\omega - \omega_{2c})\}, \tag{5.2.4}$$

where δ is the parameter to determine the spectral bandwidth of the SH spectrum, and ω_{2c} is the parameter to the central wavelength of the desired SH spectrum. γ is described by

$$\gamma = \frac{2\pi N}{\omega_{\max} - \omega_{\min}}, \tag{5.2.5}$$

where N indicates the number of peak wavelength of the SH spectrum, $\omega_{\max}$ and $\omega_{\min}$ are the maximum frequency and the minimum frequency of the excitation pulse, respectively. In sinusoidal phase modulation, $\phi(\omega_{2c} + \Omega) + \phi(\omega_{2c} - \Omega)$ is zero for all frequencies of the excitation pulse. When $\omega \neq \omega_{2c}$, $\phi(\omega + \Omega) + \phi(\omega - \Omega)$ is not zero for all frequencies of the excitation pulse. Thus, perfect constructive interference is achieved only at a frequency of $2\omega_{2c}$. In contrast, interference at the other frequencies is partially constructive or destructive. Therefore, the SH field at a frequency of $2\omega_{2c}$, $|E^{(2)}(2\omega_{2c})|^2$ is enhanced. We also consider the enhancement of $|E^{(2)}(2\omega_{2c})|^2$ from the time-spectral distribution of the modulated pulse. The time-spectral distribution of the modulated pulse is given by

$$\tau(\omega) = \frac{d\phi(\omega)}{d\omega} = \delta\gamma \cos\{\gamma(\omega - \omega_{2c})\}. \tag{5.2.6}$$

According to Eq. (5.2.6), we find that the time-spectral distribution of the modulated pulse is symmetric about the frequency ω_{2c}. Then, the frequency components around at $\omega_{2c} + \Omega$ overlap only with the frequency components around at $\omega_{2c} - \Omega$ in time. Thus, the SH field is produced only by the combination of the fundamental field at a frequency of $\omega_{2c} + \Omega$ and that at a frequency of $\omega_{2c} - \Omega$. Therefore, the SH spectrum is focused at the frequency components around at $2\omega_{2c}$.

TOD phase modulation can be written as

$$\phi(\omega) = \frac{\mathrm{TOD}}{6}(\omega - \omega_{2c})^3. \tag{5.2.7}$$

$\phi(\omega_{2c} + \Omega) + \phi(\omega_{2c} - \Omega)$ is zero for all frequencies of the excitation pulse. When $\omega \neq \omega_{2c}$, $\phi(\omega + \Omega) + \phi(\omega - \Omega)$ is not zero for all frequencies of the excitation pulse. Thus, perfect constructive interference is achieved only at a frequency of $2\omega_{2c}$. The time-spectral distribution of the pulse with the TOD is also symmetric about the frequency ω_{2c}:

$$\tau(\omega) = \frac{d\phi(\omega)}{d\omega} = \frac{\mathrm{TOD}}{2}(\omega - \omega_{2c})^2. \tag{5.2.8}$$

Thus, the application of the TOD allows us to focus the SH spectrum into the frequency components around at $2\omega_{2c}$. The spectral bandwidth of the SH spectrum is determined by the TOD. By focusing the SH spectrum into the specific spectral band where the TPE cross section of the target fluorophore is much higher than those of the other nontarget fluorophores, the selective excitation can be achieved.

Figure 5.6 shows the SH spectrum of the FTL pulse and those of the shaped pulses by sinusoidal phase modulation ($\delta = 4\pi$ and $N = 1$), and by TOD phase modulation (TOD $= 1000$ fs^3), which are calculated by using the Gaussian pulse with a spectral bandwidth of 200 nm (FWHM) at a central wavelength of 840 nm, respectively. The wavelength values in Fig. 5.6(b) and (c) correspond to ω_{2c}. The SH spectrum focusing is achieved by sinusoidal phase modulation and TOD phase modulation, and its spectral resolution is enhanced by a factor of more than 10. We see that sinusoidal phase modulation and TOD phase modulation give the modulated pulses to selectively excite to different two-photon excited states. Thus, sinusoidal

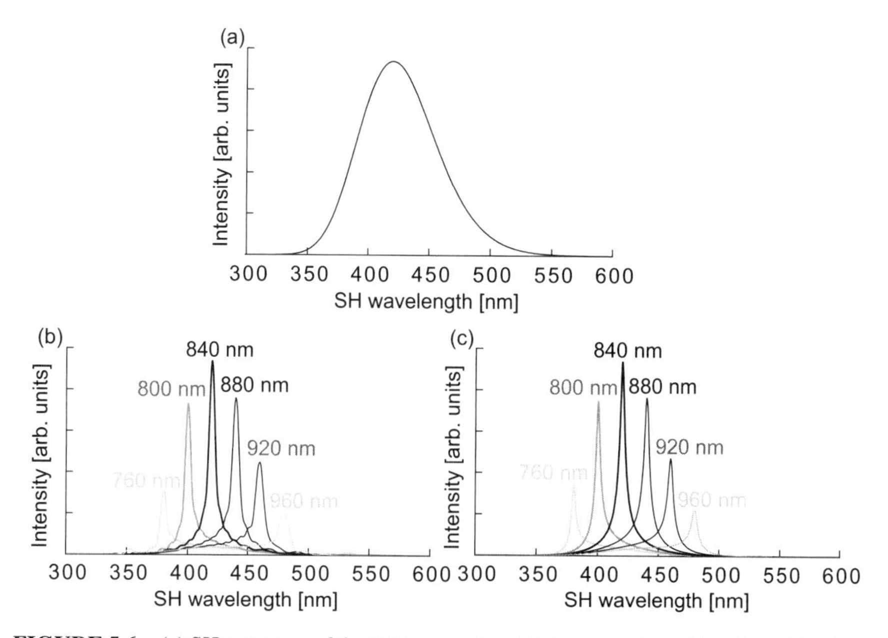

FIGURE 5.6 (a) SH spectrum of the FTL pulse. (b, c) SH spectra shaped by sinusoidal phase modulation at $\delta = 4\pi$ and $N = 1$ (b) and TOD phase modulation at 1000 fs^3(c).

phase modulation and TOD phase modulation allow the selective excitation of one fluorophore over the other fluorophores. It should be noted that the asymmetric spectral phase about the frequency ω_{2c}, $\phi(\omega_{2c} + \Omega) = -\phi(\omega_{2c} - \Omega)$ provides the SH spectral focusing at $2\omega_{2c}$.

5.2.1.2 Binary Phase Modulation Sinusoidal phase modulation and TOD phase modulation provide the pulse to achieve perfect constructive interference at the desired two-photon excited states. However, they are insufficient for use as the suppression of the TPE by destructive interference elsewhere. Thus, the background TPEF signals from the nontarget fluorophores are generated. Therefore, the contrast ratio for selective excitation by sinusoidal phase modulation or TOD phase modulation is low. To improve the contrast ratio for the selective excitation, Comstock et al. developed binary phase modulation [143]. In binary phase modulation, maximization of the SH intensity at a frequency $2\omega_{2c}$ and minimization of the background intensity at all other frequencies are simultaneously achieved by using only two phases, zero and π.

It is difficult to obtain an analytical solution of the binary phase that maximize the SH intensity at a frequency $2\omega_{2c}$ and minimize the background intensity at all other frequencies. Thus, the solution of binary phase modulation is often searched by using a simulated annealing (SA) method or a genetic algorithm (GA) method. The search space is reduced to $2^{N_{\text{SLM}}}$ by discretizing the frequency space into the number of SLM pixels, N_{SLM}. Suppose that we replace the integral in Eq. (5.2.2) with a discrete sum. The intensity of the SH signal at $2\omega_j$ is proportional to the square of the superposition of the second harmonic fields produced by the combination of the fundamental field of $\omega_j + \omega_k$ and that of $\omega_j - \omega_k$ and is expressed by

$$\tilde{S}(2\omega_j) = \left| \sum_{k=1}^{N_{\text{SLM}}} |\tilde{E}(\omega_j + \omega_k)||\tilde{E}(\omega_j - \omega_k)| \exp[i\{\phi(\omega_j + \omega_k) + \phi(\omega_j - \omega_k)\}] \right|^2, \tag{5.2.9}$$

where the fundamental field at the jth pixel of a phase modulator is $|E(\omega_j)| \exp\{i\phi(\omega_j)\}$. To further simplify the problem, the spectral amplitude is set to be 1. Then, the intensity of the SH signal at $2\omega_j$ is calculated by

$$S_{2j} = \left| \sum_{k=1}^{N_{\text{SLM}}} a_{j+k} a_{j-k} \right|^2, \tag{5.2.10}$$

where $a_j = \exp\{i\phi(\omega_j)\}$ is a binary value of 1 or -1 for $\phi(\omega_j) = 0$ or $\phi(\omega_j) = \pi$, respectively. The problem of contrast enhancement can now be formulated as finding a vector a_j such that $S_{2j} = N_{\text{SLM}}$ for $2\omega_j = 2\omega_{2c}$ and S_{2j} is minimized at all other frequencies.

5.2.1.3 Multi-Level Phase Modulation Sinusoidal phase modulation, TOD modulation, and binary phase modulation allows selective excitation of one fluorophore over the other fluorophores. However, they cannot be used for the simultaneous excitation together with the control of TPEF intensities from different fluorophores independently. To solve this problem, Isobe et al. proposed multi-level phase modulation [126,130]. The phase function of multi-level phase modulation is composed of constructive $\phi_c(\omega)$ and destructive $\phi_d(\omega)$ interference phases (which maximize and minimize the SH intensities, respectively) over the whole frequency range. $\phi_c(\omega)$ generates an FTL pulse. $\phi_d(\omega)$ is determined from the condition that results in destructive interference. The destructive interference phase depends on the fundamental spectrum. Because it is difficult to obtain an analytical solution, we determine the destructive interference phase by the use of an adaptive control technique employing a SA method or a GA method. Because the intensity of the SHG signal at $2\omega_j$ is expressed by Eq. (5.2.9), the total SHG intensity over the whole frequency range is described by

$$I_{SH} = \sum_j S(2\omega_j). \tag{5.2.11}$$

$\phi(\omega_j)$ to minimize (5.2.11) is the destructive interference phase. The broad SH spectrum obtained by the FTL pulse is divided into two spectral bands (R_1 and R_2) for two specific fluorophores. R_1 gives a high TPE cross section for one of the fluorophores, whereas R_2 gives a high TPE cross section for the other. To increase the SH intensity within R_1 while decreasing the SH intensity within R_2, $\phi_c(\omega)$ and $\phi_d(\omega)$ are applied to R_1 and the R_2, respectively. The phase function for the selective excitation is expressed by

$$\phi(\omega) = \begin{cases} \phi_c(\omega), & \omega \in R_1, \\ \phi_d(\omega), & \omega \in R_2. \end{cases} \tag{5.2.12}$$

For the simultaneous excitation of two fluorophores that exhibit different TPE spectra, we design spectral phase functions in which $\phi_d(\omega)$ is combined with an arbitrary percentage (α or β) of $\phi_c(\omega)$ within R_1 or R_2, respectively. Furthermore, we introduce a group-delay phase $\phi'(\omega - \omega_0)$ to the spectral phase function of one region (usually R_1) such that the two excitation pulses composed of fundamental spectral components in R_1 and R_2 interact independently with each of the fluorophores. The individual TPEF intensity is determined by α or β. The spectral phase is described as

$$\phi(\omega) = \begin{cases} \phi_c(\omega) + \alpha\{\phi_d(\omega) - \phi_c(\omega)\} + \phi'(\omega - \omega_0), & \omega \in R_1, \\ \phi_c(\omega) + \beta\{\phi_d(\omega) - \phi_c(\omega)\}, & \omega \in R_2. \end{cases} \tag{5.2.13}$$

5.2.1.4 Multifarious Control of TPEF by Spectral Phase Modulation A schematic of the typical experimental setup for nonlinear optical microscopy employing spectral phase modulation is shown in Fig. 5.7(a). As an ultrabroadband light source, a ultrabroadband Ti:sapphire mode-locked oscillator and a supercontinuum are typically used. Figure 5.7(b) shows the spectrum of an octave-spanning Ti:sapphire oscillator (Nanolayers, Venteon OS). In order to compensate for second-order dispersion of all the optical components, the laser pulse is passed through a

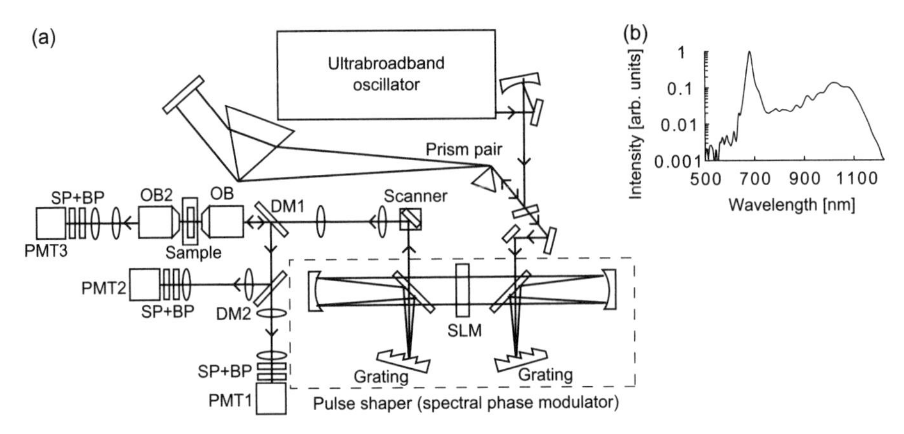

FIGURE 5.7 (a) Typical experimental setup for nonlinear optical microscopy employing spectral phase modulation. (b) Spectrum of an octave spanning Ti:sapphire oscillator.

fused silica prism pair. A grating-pair-formed pulse shaper with a spatial light modulator (SLM) is used to compensate for the higher order dispersion. The modulated pulse is launched into a laser scanning microscope. Then, the ultrabroadband pulses need to be focused into a sample by an objective lens, whose chromatic and spherical aberrations are optimized for the near-IR region. The generated signals are detected by PMTs.

To characterize the selective excitation of a single fluorophore, measurement of the SH spectrum of the excitation pulse is required. The SH spectrum of the excitation pulse $|E^{(2)}(\omega)|^2$ can be measured by obtaining the SHG spectrum using a nonlinear crystal such as a β-BBO crystal and a quartz crystal. Because the bandwidth of the phase matched SHG is narrower that that of the non–phase matched SHG, the non–phase matched SHG is typically selected for the measurement of the SH spectrum of the ultrabroadband pulse. For example, SHG from the surface of a quartz crystal cut on 90° is used as the non–phase matched SHG. Figure 5.8(a) and (c) shows the SHG spectra by the FTL pulse and the TOD shaped pulses at a TOD of 500 fs^3, both

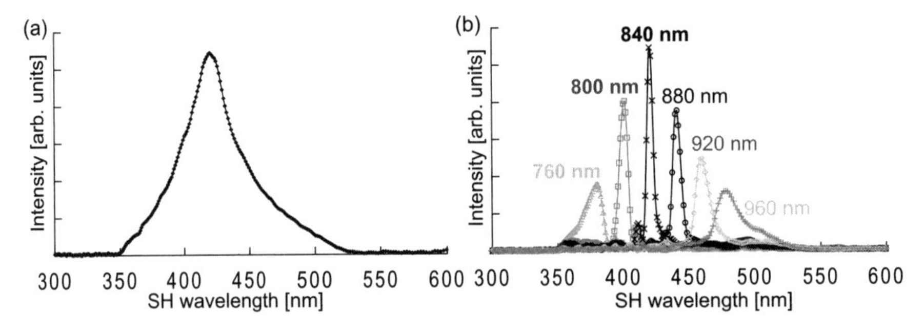

FIGURE 5.8 SHG spectra by the FTL pulse (a) and the TOD shaped pulse (b).

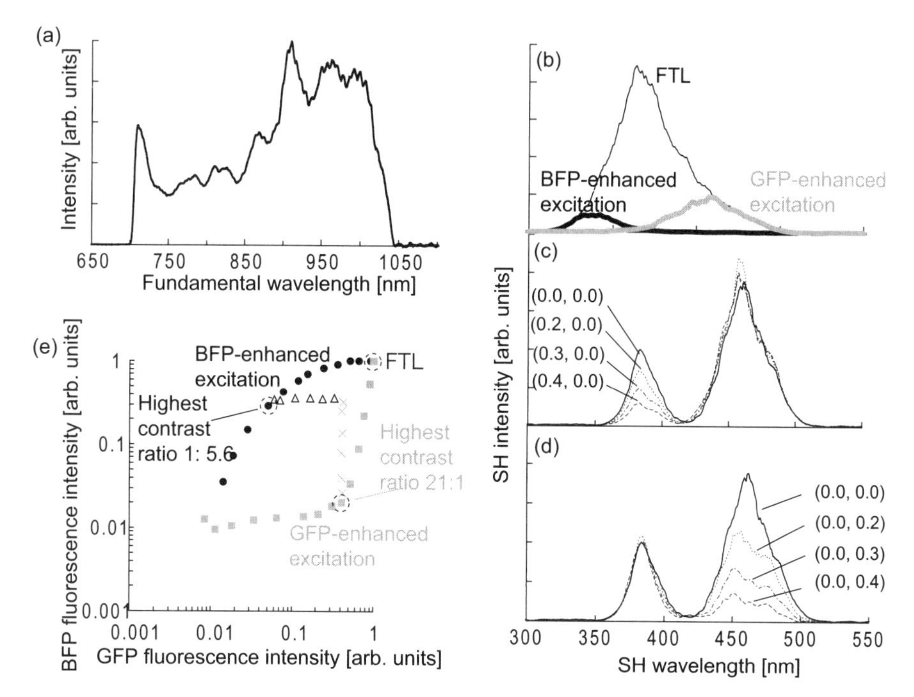

FIGURE 5.9 (a) Fundamental spectrum of the ultrabroadband excitation pulse. (b) SH spectra for selective excitations of GFP (gray) or BFP (black). (c) SH spectra for controlling TPEF intensity of BFP while keeping that of GFP constant. (d) SH spectra for controlling TPEF intensity of GFP while keeping that of BFP constant. (e) Control of TPEF intensities from GFP and BFP with selective excitations of GFP (square) or BFP (circle) and with simultaneous excitation together with control of TPEF intensity for only GFP (triangle) or BFP (cross). Reprinted with permission from [130]. Copyright 2009 Optical Society of America.

of which are obtained by using the surface of a quartz crystal cut on 90°. We can see that the bandwidth of the SHG by the TOD shaped pulse is much narrower than that by the FTL, which is broad enough to excite various fluorophores simultaneously. It should be noted that the SHG bandwidth of the TOD pulse is comparable to that of typical narrowband tuning laser pulses.

By combining multi-level phase modulation with the ultrabroadband pulse as shown in Fig. 5.9(a), Isobe et al. have demonstrated multifarious control of TPEF signals from BFP and GFP [130]. The circles and squares in Fig. 5.9(e) indicate the cases where the TPEF contrast ratios, BFP/GFP and GFP/BFP, are enhanced, respectively. The highest contrast ratios of 21:1 and 1:5.6 have achieved for GFP/BFP and BFP/GFP, respectively. The two SH spectra at the highest contrast ratios are shown in Fig. 5.9(b).We see that the product of the two highest contrast ratios is over 100. The triangles and crosses in Fig. 5.9(e) indicate the cases where the TPEF intensity of either GFP or BFP is controlled while the other is frozen. We see that multi-level phase modulation enables individual adjustment of the TPEF intensities from two

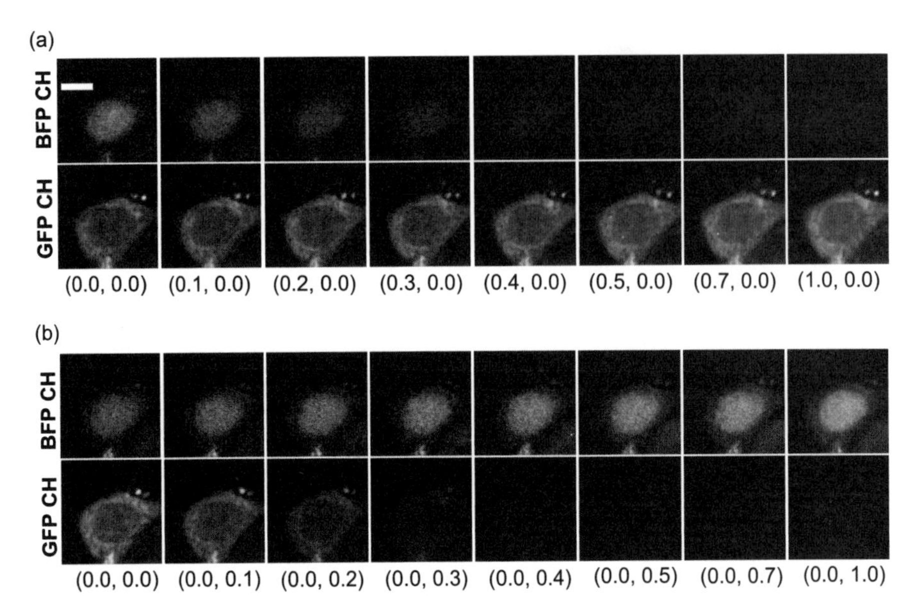

FIGURE 5.10 Dual-color images of a HeLa cell labeled with BFP (nucleus, top) and GFP (cytoplasm, bottom). (a) BFP intensity is regulated, while GFP intensity is frozen. (b) GFP intensity is regulated, while BFP intensity is frozen. Reprinted with permission from [130]. Copyright 2009 Optical Society of America.

fluorophores. Figure 5.9(c) and (d) shows the SH spectra at the spectral phase used for controlling the TPEF intensities of either one of BFP or GFP while keeping the other intensity constant. The double digits indicate the percentage parameters in the two spectral regions. The SH intensities independently decreased as the percentage parameters increased.

Figure 5.10 shows the dual-color images of a HeLa cell employing BFP (Azurite) and GFP, which are used to label the nucleus and cellular cytoplasm, respectively. We see that the contrast ratios can be freely controlled by multi-level phase modulation, and the selective excitation with a high contrast ratio and the simultaneous excitation with equal excitation rates can be easily switched only by exchanging the spectral phases.

Figure 5.11 shows FRET images of a HeLa cell using yellow-cameleon (YC3.60) to label the cellular cytoplasm [149]. Cameleon has a CFP and a YFP, and is an indicator for Ca^{2+}. CFP-enhanced excitation provides FRET imaging. Figure 5.11(a) and (b) shows TPEF images of CFP and YFP before Ca^{2+} stimulus excitation by treatment with 5 μM ionomycin, respectively. Figure 5.11(d) and (e) shows TPEF images of CFP and YFP after the Ca^{2+} stimulus excitation, respectively. FRET images before and after the Ca^{2+} stimulus excitation are presented in Fig. 5.11(c) and (f), respectively. We see that the FRET signal increases after Ca^{2+} stimulus excitation due to an increase in the Ca^{2+} concentration in the cell. Figure 5.11(g) shows the signal ratio

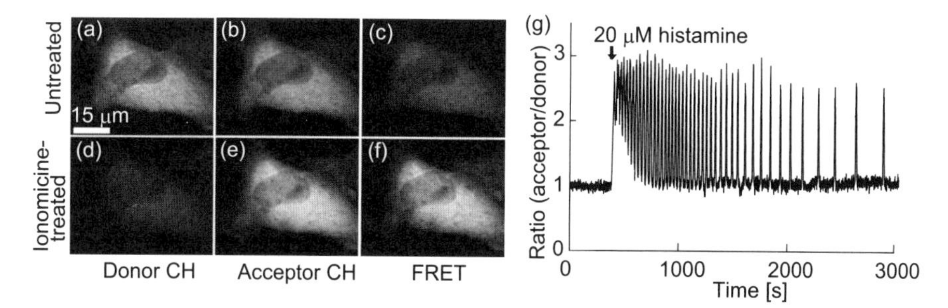

FIGURE 5.11 Ca^{2+} response in cytoplasm of a HeLa cell loaded with yellow-cameleon (YC3.60). CFP (a, d), YFP (b, e) and FRET (c, f) images with CFP-enhanced excitation before (a, b, c) and after (d, e, f) Ca^{2+} stimulus excitation by treatment with 5 μM ionomycin. (g) Signal ratio of YFP channel to CFP channel before and after Ca^{2+} stimulus excitation by treatment with 20 μM histamine. Reprinted with permission from [130]. Copyright 2009 Optical Society of America.

of the YFP channel to the CFP channel before and after Ca^{2+} stimulus excitation by treatment with 20 μM histamine. We see that Ca^{2+} oscillation is recorded with a high dynamic range. It should be noted that the selective excitation enables us to achieve FRET imaging, even if ultrabroadband pulses are used as an excitation light source.

5.2.2 Control of Vibrational Mode Excitation by a Single Broadband Pulse

CARS microscopy, which requires at least a two-color irradiation scheme, provides intrinsic vibrational contrast [150,151]. In CARS microscopy using two synchronized narrowband laser pulses, multi-spectral images based on various Raman modes have been obtained by tuning the laser wavelength. Of course, fast control techniques are required, since it takes several seconds or more to tune the wavelength of the mode-locked lasers that are currently available. In the wavelength tuning, special care must be taken to overlap between the pump and Stokes pulses spatially and temporally. The spatial overlap and/or timing jitter between the pump and Stokes pulses has been the limiting factor, severely restricting the sensitivity or the signal-to-noise ratio. These issues can be solved by three CARS microscopic techniques using only a single laser source, both of which do not require the wavelength tuning of the laser source. One is single-laser multiplex-CARS (M-CARS) microspectroscopy based on the spectral broadening in nonlinear optical fibers such as a photonic crystal fiber [152,153] and a tapered fiber [154], which have achieved by the use of one in two beams divided from a single narrowband laser. The broadband pulse from the nonlinear fiber, which is used as a Stokes pulse, is recombined with the original narrowband pump pulse in time and space. By simultaneously recording broadband CARS spectra generated by the use of the combination of the narrowband pump pulse and the broadband Stokes pulse, single-laser M-CARS microspectroscopy is realized. The spectral resolution of the M-CARS spectrum is determined by the bandwidth of the narrowband pump pulse. Kano et al. [152] and Kee et al. [154] have developed single-laser M-CARS microspectroscopy with ultrabroadband spectral coverage (>2500 cm^{-1}). Okuno et al. have developed ultrabroadband multiplex CARS microspectroscopy with a high spectral resolution (<1 cm^{-1}) and with ultrabroadband spectral coverage (>2000 cm^{-1}) [153]. Because the M-CARS spectrum includes the nonresonant CARS signals, multi-spectral images can be obtained by analyzing the profile of the M-CARS spectrum. Another technique is single-laser FT-CARS microspectroscopy obtained by using ultrabroadband pulses [155,156]. The broadband CARS spectrum, which includes no nonresonant CARS signals, can be obtained by the Fourier transform of an IAC signal. The spectral resolution is not determined by the spectral bandwidth of the excitation light source, but by the inverse of the maximum delay time between excitation pulses. From the CARS spectrum, multi-spectral images can be obtained. Ogilvie et al. have demonstrated FT-CARS microspectroscopy with a bandwidth of 1500 cm^{-1} [155]. Isobe et al. have developed FT-CARS microspectroscopy over a bandwidth of 4000 cm^{-1} [156].

The last technique is single-pulse CARS microscopy, where the CARS process is induced by a single broadband pulse with a pump and Stokes frequency components. The broadband pulse is directly delivered from an ultrabroadband oscillator [123] or is generated in the nonlinear fiber by the use of a narrowband oscillator [157]. Single-pulse CARS microscopy is more stable than the single-laser CARS microscopy because of its common-path configuration. However, in single-pulse CARS microscopy, we need to extract only the target signal of the single Raman mode at Ω_R or to suppress not only the nonresonant CARS signal but also the resonant CARS signals from nontarget Raman modes due to the simultaneous excitation of various Raman modes with energies within the pulse bandwidth. Oron et al. [158] and Dudovich et al. [125,159] have reported two techniques for single-pulse CARS microscopy using the spectral phase modulation of a broadband pulse. One is the selective excitation of a vibrational mode with wave number Ω_R by the sinusoidal phase modulation [123,125,159]. In the sinusoidal phase modulation, the nonresonant signal decreases, while the resonant signals of the Raman mode with wave number $N\Omega_R$ (where N is an integer) are maintained. Isobe et al. have demonstrated the spectral phase modulation for the selective excitation of a single Raman mode without excitation of other Raman modes and first demonstrated bio-imaging by single-pulse CARS microscopy employing spectral phase modulation [133]. Another single-pulse CARS technique is single-pulse M-CARS spectroscopy, where phase-shifting of a narrow spectral band enables effective narrow-probing of the vibrational mode [123,158,159]. The single-pulse M-CARS technique allows the same extraction of resonant CARS signals as the MCARS technique [158,159]. Von Vacano et al. [157,160] and Lim et al. [161,162] have extended the concept of single-pulse CARS spectroscopy using spectral phase modulation. They demonstrated several schemes for enhancing the sensitivity and selectivity of CARS spectroscopy. In single-pulse CARS microscopy, multi-spectral images can be acquired only by exchanging the spectral phase. In this section, we describe single-pulse CARS microscopy using spectral phase modulation.

5.2.2.1 Single-Pulse CARS Microscopy Using Sinusoidal Phase Modulation As shown in Eq. (3.2.3), the polarization of the resonant CARS process is described by

$$\tilde{P}^{(\mathrm{RR})}_{\mathrm{CARS}}(\omega) \propto \int_{-\infty}^{\infty} \tilde{E}(\omega - \Omega)\tilde{h}_{RR}(\Omega)\tilde{E}^{(2\#)}(\Omega)d\Omega, \qquad (5.2.14)$$

where $h_{\mathrm{RR}}(\omega)$ is the Raman response, $E(\omega)$ is the excitation electric field, and $E^{(2\#)}(\omega)$ is expressed by

$$\begin{aligned}\tilde{E}^{(2\#)}(\omega) &= \int_{-\infty}^{\infty} \tilde{E}(\Omega)\tilde{E}^{*}(\Omega - \omega)d\Omega \\ &= \int_{-\infty}^{\infty} |\tilde{E}(\Omega)||\tilde{E}^{*}(\Omega - \omega)| \exp[i\{\phi(\Omega) - \phi(\Omega - \omega)\}]d\Omega. \qquad (5.2.15)\end{aligned}$$

Here $|E(\omega)|$ and $\phi(\omega)$ are the spectral amplitude and the spectral phase of the excitation pulse, respectively. Although in fact the difference-frequency (DF) fields are not generated in the CARS process, $|E^{(2\#)}(\omega)|^2$ is called the "DF power spectrum," which is determined by the properties of the excitation pulses. The polarization of the nonresonant CARS process can be written as

$$\tilde{P}^{(\mathrm{NR})}_{\mathrm{CARS}}(\omega) \propto \int_{-\infty}^{\infty} \tilde{E}(\omega - \Omega)\tilde{E}^{(2\#)}(\Omega)d\Omega. \qquad (5.2.16)$$

$|E^{(2\#)}(\omega)|^2$ is determined by the interference between all frequency pairs (Ω and $\Omega - \omega$) within the excitation pulse separated by ω. By modulating the spectral phase, the DF spectrum can be focused into the Raman resonant region. Therefore, the contrast ratio between the resonant CARS and the nonresonant CARS can be enhanced by spectral phase modulation.

Dudovich et al. have applied sinusoidal phase modulation to single-pulse CARS microscopy. The sinusoidal phase modulation proposed by Dudovich et al. is expressed by

$$\phi(\omega) = \delta \sin\left(\frac{2\pi\omega}{\Omega_{\mathrm{R}}}\right), \qquad (5.2.17)$$

where δ is the parameter to determine the spectral bandwidth of the DF spectrum, and Ω_{R} is the parameter to the central wavelength of the desired DF spectrum. In sinusoidal phase modulation, if N is an integer, $\phi(\Omega) - \phi(\Omega - N\Omega_{\mathrm{R}})$ is zero for all frequencies of the excitation pulse. When $\omega \neq N\Omega_{\mathrm{R}}$, $\phi(\Omega) - \phi(\Omega - \omega)$ is not zero for all frequencies of the excitation pulse. Thus, perfect constructive interference is achieved at frequencies of $N\Omega_{\mathrm{R}}$, ($N = 1, 2, 3, \ldots$). In contrast, interference at the other frequencies is partially constructive or destructive. Therefore, the DF fields at frequencies of $N\Omega_{\mathrm{R}}$, $|E^{(2\#)}(N\Omega_{\mathrm{R}})|^2$, are enhanced. We also consider the enhancement

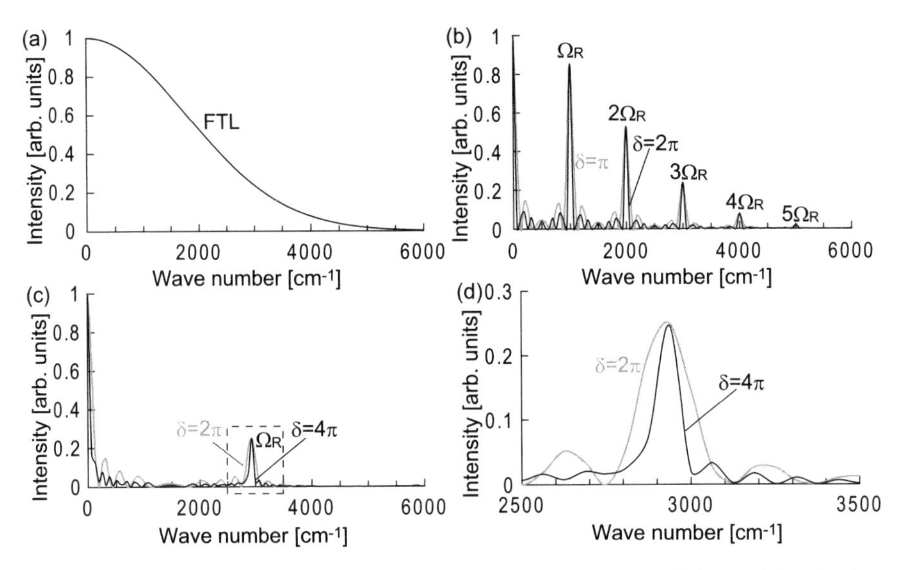

FIGURE 5.12 (a) DF spectrum of the FTL pulse. (b, c) DF spectra of the modulated pulses by sinusoidal phase modulation at $\Omega_R/2\pi c = 1000\ \text{cm}^{-1}$ (b) and $\Omega_R/2\pi c = 2930\ \text{cm}^{-1}$ (c). (d) Magnified view of (c).

of $|E^{(2\#)}(N\Omega_R)|^2$ from the time-spectral distribution of the modulated pulse. The time-spectral distribution of the modulated pulse is given by

$$\tau(\omega) = \frac{d\phi(\omega)}{d\omega} = \left(\frac{2\pi\delta}{\Omega_R}\right)\cos\left(\frac{2\pi\omega}{\Omega_R}\right). \tag{5.2.18}$$

According to Eq. (5.2.18), we find that the DF field is produced only by the combination of the fundamental field at a frequency of Ω and that at a frequency of $\Omega + N\Omega_R$ because the frequency components around at Ω overlap only with the frequency components around at $\Omega + N\Omega_R$ in time. Thus, the DF spectrum is focused at the frequency components around at $N\Omega_R$. Therefore, sinusoidal phase modulation allows the selective excitation of several vibrational modes with $N\Omega_R$.

Figure 5.12(a)–(c) shows the DF spectrum of the FTL pulse and those of the shaped pulses by sinusoidal phase modulation at $\Omega_R/2\pi c = 1000\ \text{cm}^{-1}$ and $\Omega_R/2\pi c = 2930\ \text{cm}^{-1}$, which are calculated by using the Gaussian pulse with a spectral bandwidth of 200 nm (FWHM) at a central wavelength of 840 nm, respectively. We see that the DF spectra at frequencies of $N\Omega_R$ can be enhanced by sinusoidal phase modulation. Figure 5.12(d) shows a magnified view of Fig. 5.12(c). We also see that the spectral bandwidth of the DF spectrum depends on the modulation depth δ.

5.2.2.2 Single-Pulse CARS Microscopy Using Spectral Focusing by Spectral Phase Modulation In the sinusoidal phase modulation, the nonresonant CARS signal decreases, while not only the resonant CARS signal of the target vibrational mode with $N\Omega_R$ but also that of the nontarget vibrational mode with $M\Omega_R$ (M, which is not N, is an integer) are generated. Isobe et al. proposed spectral phase modulation for the selective excitation of a single Raman mode without excitation of other Raman modes [133]. In order to excite a single vibrational mode Ω_R, the spectral phase is controlled in the following manner. The quadratic phase $\phi''(\omega-\omega_0)^2/2$ is applied over the entire spectral region and then the linear phase $\phi''\Omega_R(\omega-\omega_{max}+\Omega_R)$ is applied in the region where the frequency is smaller than $\omega_b=\omega_{max}-\Omega_R$. Here, ϕ'', ω, ω_{max} and ω_0 are the group delay dispersion (GDD), the frequency of the pulse, the maximum frequency and the central frequency of the pulse. The phase function for selective excitation of a single Raman mode without excitation of other Raman modes is described by

$$\phi(\omega)=\begin{cases}\dfrac{\phi''}{2}(\omega-\omega_0)^2+\phi''\Omega_R(\omega-\omega_b), & \omega<\omega_b=\omega_{max}-\Omega_R,\\ \dfrac{\phi''}{2}(\omega-\omega_0)^2, & \omega\geq\omega_b.\end{cases} \tag{5.2.19}$$

We consider the time-spectral distribution of the modulated pulse. The time-spectral distribution of the modulated pulse is expressed by

$$\tau(\omega)=\frac{d\phi(\omega)}{d\omega}=\begin{cases}\phi''(\omega-\omega_0)+\phi''\Omega_R, & \omega<\omega_b=\omega_{max}-\Omega_R,\\ \phi''(\omega-\omega_0), & \omega\geq\omega_b.\end{cases} \tag{5.2.20}$$

From Eq. (5.2.20), the designed time-spectral distribution of the modulated pulse is shown in Fig. 5.13. As shown in Fig. 5.13, the frequency components around at Ω overlap only with the frequency components around at $\Omega-\Omega_R$ in time. The frequency difference between the pump frequency ω_p and the Stokes frequency ω_s is constant at all times. Therefore, only a single vibrational motion is excited by focusing the DF spectrum into Ω_R. The bandwidth of the DF spectrum narrows with an increase in the GDD. This technique is similar to the spectral-focusing technique based on two chirped excitation pulses to focus their bandwidth into a narrow spectral region [163]. The difference is that the spectral focusing is not realized by using a

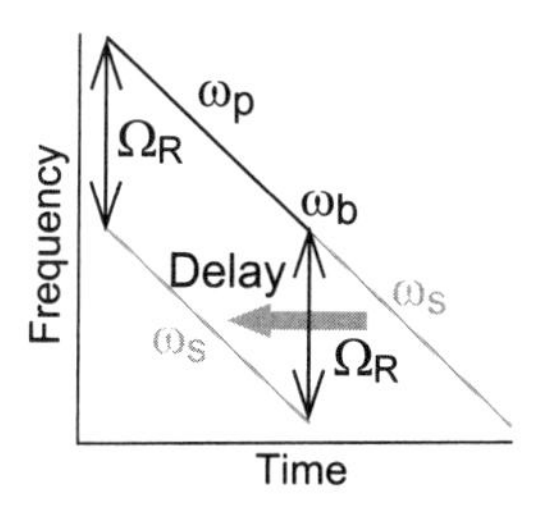

FIGURE 5.13 Time-spectral distribution of excitation pulse for selective excitation of a single Raman mode.

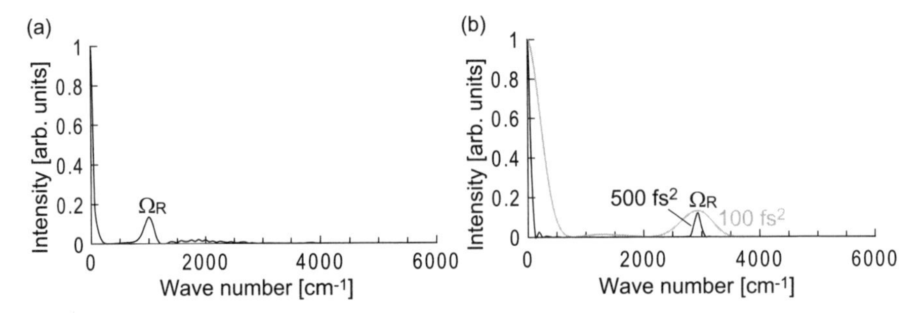

FIGURE 5.14 (a, b) DF spectra focused into at $\Omega_R/2\pi c = 1000$ cm^{-1} (a) and $\Omega_R/2\pi c =$ 2930 cm^{-1} (b) by spectral phase modulation.

pair of equally linear-chirped pulses from two synchronized lasers, but by controlling the spectral phase of a single ultrabroadband pulse.

Figure 5.14 shows the DF spectra focused into $\Omega_R/2\pi c = 1000$ cm^{-1} and $\Omega_R/2\pi c = 2930$ cm^{-1} by spectral phase modulation, which are calculated by using the FTL pulse whose DF spectrum is shown in Fig. 5.12. We see that the spectral focusing technique by spectral phase modulation allows the selective excitation of a single Raman mode at Ω_R without excitation of other Raman modes at $N\Omega_R$ ($N \geq 2$). We also see that the spectral bandwidth of the DF spectrum narrows with increasing GDD.

5.2.2.3 Single-Pulse M-CARS Spectroscopy Using Spectral Phase Modulation M-CARS spectroscopy gives ultrabroadband spectroscopic information without the wavelength tuning. However, a narrowband pulse and a broadband pulse are required. Oron et al. suggested a new method to obtain CARS spectral data, which is based on spectral phase modulation of a single broadband pulse [158]. They applied a π phase shift around the narrowband features in the shorter wavelength of the excitation pulse, which is referred to a π phase gate. The total CARS spectrum is expressed by

$$I_{\mathrm{CARS}}(\omega) \propto \left| \tilde{P}^{(\mathrm{RR})}_{\mathrm{CARS}}(\omega) + \tilde{P}^{(\mathrm{NR})}_{\mathrm{CARS}}(\omega) \right|^2, \tag{5.2.21}$$

In Fig. 5.15, we consider the CARS process generated by the π phase gated pulse whose spectral phase has shifted by π in a narrowband of about 2 nm at 720 nm as shown in Fig. 5.15(a). As shown in Fig. 5.15(b), a π phase gate hardly modifies the DF spectrum $|E^{(2\#)}(\omega)|^2$ since the energy content in a narrow spectral band part is negligible compared with the entire pulse energy. Thus, a π phase gate cannot give the selective excitation of the target Raman mode. The phase of $E^{(2\#)}(\omega)$ is also hardly changed by the π phase gate. However, the phase of the resonant CARS polarization is dramatically affected by the π phase gate. If the Raman response is expressed by

$$\tilde{h}_{\mathrm{RR}}(\Omega) \propto \frac{1}{\omega_{\mathrm{RR}} - \Omega - i\Gamma_{\mathrm{RR}}}, \tag{5.2.22}$$

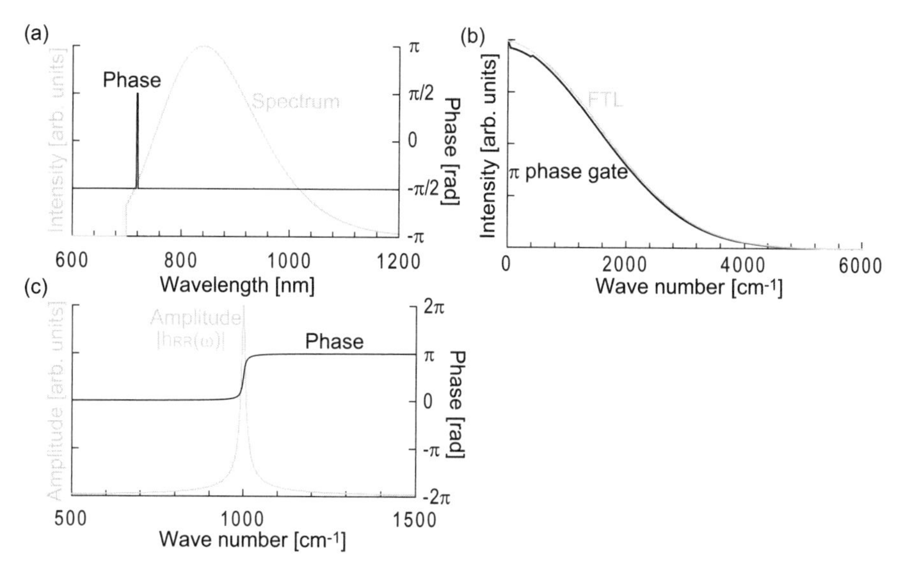

FIGURE 5.15 (a) Spectral phase and spectrum of the π phase gated pulse. (b) DF spectra for the FTL pulse (gray line) and the π phase gated pulse (black line). (c) Amplitude (gray line) and phase (black line) of the Raman response.

the Raman response induces a spectral phase shift

$$\phi_R(\Omega) = \arctan\left(\frac{\Gamma_{RR}}{\omega_{RR} - \Omega}\right). \tag{5.2.23}$$

Here ω_{RR} is the Raman resonant frequency, and Γ_{RR} is the Raman resonant bandwidth (half width at half maximum). Because the Raman response induces the sharp phase change in the narrowband part around at a resonant frequency ω_{RR} ($\omega_{RR}/2\pi c =$ 1000 cm^{-1}) as shown in Fig. 5.15(c), as is evident from Eq. (5.2.14), a π phase gate in the narrowband part around at a frequency of ω_g of the excitation pulse gives the sharp phase change in the narrowband part around at a frequency of $\omega_g + \omega_{RR}$ of the CARS polarization. In contrast, the phase of the nonresonant CARS polarization is hardly modified by the π phase gate because of a coherent sum contributed by a large portion of the pulse bandwidth. Thus, the relative phase between the resonant CARS polarization and the nonresonant CARS polarization is dramatically changed in the narrowband part around a frequency of $\omega_g + \omega_{RR}$ ($2\pi c/(\omega_g + \omega_{RR}) = 671.6$ nm) as shown in Fig. 5.16(b). Therefore, the π phase gate can be used to generate a large coherent spectral feature in the total CARS spectrum, due to constructive interference and destructive interference between two polarizations in the narrowband part around at a frequency of $\omega_g + \omega_{RR}$ as shown in Fig. 5.16(d). In addition, the small intensity variation due to the resonant signal is amplified by heterodyne detection with the strong nonresonant signal as a local oscillator.

Yet, the FTL pulse does not induce the sharp change in the relative phase between the resonant CARS polarization and the nonresonant CARS polarization as in

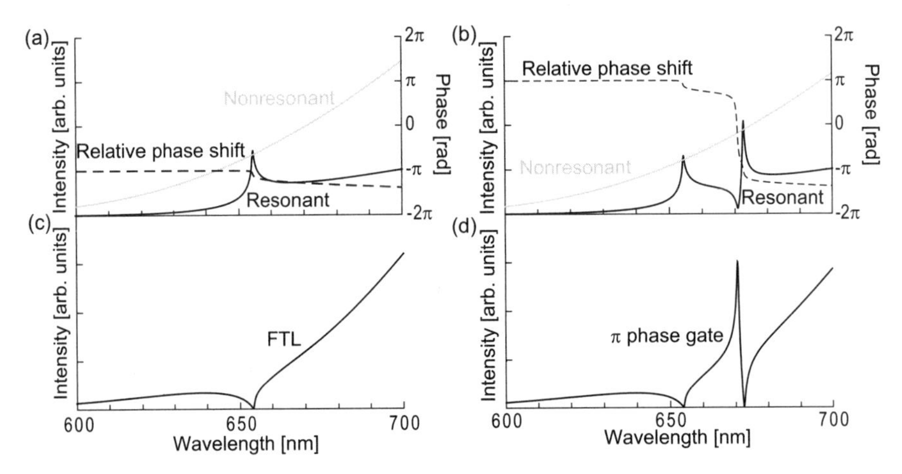

FIGURE 5.16 (a, b) Calculated spectra of the resonant CARS polarization (black line) and the nonresonant CARS polarization (gray line) for the FTL pulse (a) and π phase gated pulse (b). Also shown is the relative phase between the two (dashed line). (c, d) Calculated total CARS spectra by the FTL pulse (c) and π phase gated pulse (d).

Fig. 5.16(a). Figure 5.16(c) shows the total CARS spectrum generated by the FTL pulse. Note that a coherent spectral feature is formed in the resonant spectrum at 654 nm, which corresponds to the spectral location of the blocker on the excitation pulse, shifted by the Raman level energy, due to a transient enhancement effect. By measuring the total CARS spectrum using the π phase gate, the interference pattern between the resonant signal and the nonresonant signal can be interpreted to reveal the vibrational energy level diagram.

5.2.2.4 Multifarious Control of CARS by Spectral Phase Modulation A schematic of the typical experimental setup for nonlinear optical microscopy employing spectral phase modulation is shown in Fig. 5.15(a). To characterize the selective excitation of a single Raman mode, measurement of the DF spectrum of the excitation pulse is required. The DF spectrum of the excitation pulse $|E^{(2\#)}(\omega)|^2$ can be measured not only by obtaining the DFG spectrum using a nonlinear crystal with an IR spectrometer but also by measuring an IAC signal by nonresonant SHG as shown in Section 3.3. Isobe et al. have demonstrated the selective excitation of a single vibrational mode by the use of the spectral phase modulation of a single ultrabroadband pulse with a spectral bandwidth of 4800 cm^{-1}. Figure 5.17 shows the DF spectra for excitation pulse using spectral focusing by spectral phase modulation. Figure 5.17(a) shows the GDD dependence of the DF spectrum. As can be seen, the bandwidth of the DF spectrum narrows with increasing GDD. The achievable bandwidth of the DF spectrum is limited by the pixel number of the SLM because the maximum applied GDD is determined by the SLM pixel number. By increasing the pixel number, we can improve the bandwidth of the DF spectrum. Figure 5.17(b) shows the DF spectra obtained by using the modulated pulses to selectively excite

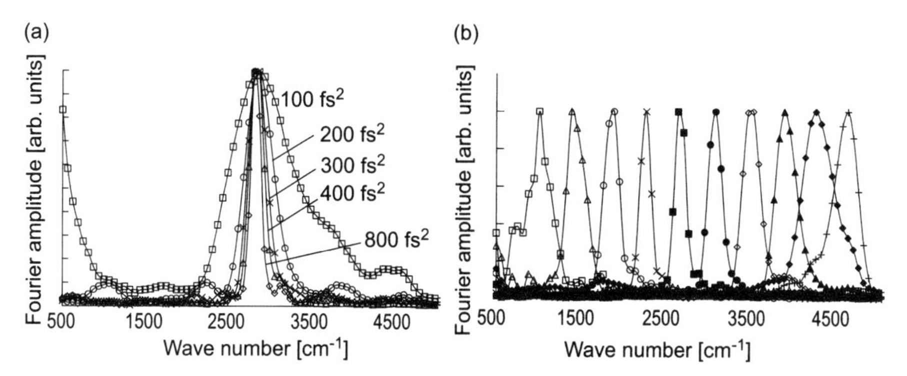

FIGURE 5.17 (a) GDD dependence of the DF spectrum. (b) Focusing the DF spectrum into various narrow spectral regions. Reprinted with permission from [133]. Copyright 2009 Optical Society of America.

different vibrational modes. From this result, we find that arbitrary vibrational modes in the laser spectrum could be selectively excited.

Figure 5.18 shows the axial distributions of the CARS signals from acetone, sandwiched between a hole-slide glass and a cover slip. The excitation wave numbers are set to 2930 and 3400 cm^{-1}, which are identified by CH stretching and nonresonant vibration, respectively. Contrast ratios of 6.5:1 and 0.6:1 are obtained for acetone/glass and glass/acetone, respectively. Note that the selective excitation of the CH stretching vibration at 2930 cm^{-1} is achieved.

Isobe et al. applied the selective excitation to vibrational bio-imaging [133]. Figure 5.19(a) and (b) shows the CARS images of an unstained HeLa cell at excitation wave numbers of 2930 and 3400 cm^{-1}, respectively. Figure 5.19(a), shows a much stronger signal for mitochondria, endoplasmic reticula, and the nucleus, which are rich in C-H, while Fig. 5.19(b) shows CARS signals not only from mitochondria,

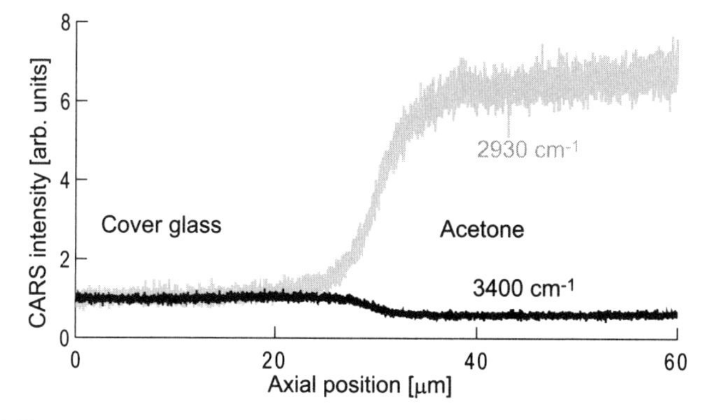

FIGURE 5.18 Intensity cross sections of acetone, sandwiched between a hole-slide glass and a cover slip, by selective excitation at 2930 cm^{-1} (black line) and 3400 cm^{-1} (gray line). Reprinted with permission from [133]. Copyright 2009 Optical Society of America.

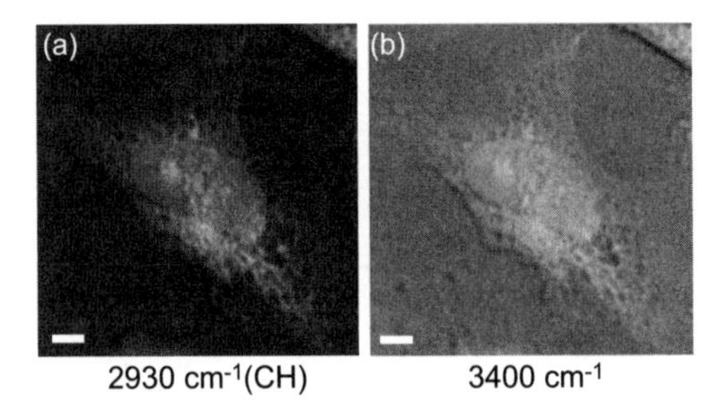

FIGURE 5.19 CARS images of an unstained HeLa cell by selective excitation at 2930 cm^{-1} (a) and 3400 cm^{-1} (b). Scale bar is 6 μm. Reprinted with permission from [133]. Copyright 2009 Optical Society of America.

endoplasmic reticulum, and the nucleus but also from water and from the cover slip around cell. As can be seen, CH vibrational imaging is successfully obtained at a wave number of 2930 cm^{-1}.

5.3 SUPER-RESOLUTION IMAGING UTILIZING NONLINEAR RESPONSE

Fluorescence microscopy has been applied to live cell imaging to investigate biological phenomena such as the expression of genes, ionic concentration changes, and membrane potential changes [165–168]. Until recently, however, spatial resolution in optical microscopy has been conceptually and practically limited by the diffraction of light. It was not until the invention of near-field scanning microscopy (NSOM) that optical imaging with sub diffraction resolution became possible [169,171]. However, NSOM measures evanescent or nonpropagating fields that exist only near the surface of samples. Thus, NSOM is limited in terms of applications. In the 1990s, the diffraction barrier as a resolution limitation for far-field fluorescence microscopy with propagating light was broken by stimulated emission depletion (STED) microscopy where the fluorophores in a bright (fluorescent) state are localized to a smaller spot than the diffraction-limited focal spot by controlling the molecular states of the fluorophores with a depletion beam [172,173]. In the control of the molecular states, saturation effects due to stimulated emission from the upper energy levels are used. The idea underlying STED microscopy has been further applied to ground-state depletion (GSD) microscopy [174–176] and has been generalized under reversible saturable optical linear fluorescence transitions (RESOLFT) microscopy [177,178]. The nonlinearity of saturation phenomena in the optical effects have been utilized in other super-resolution imaging techniques, including saturated patterned excitation microscopy (SPEM) [179], saturated structured-illumination microscopy (SSIM) [180], and saturated excitation (SAX) microscopy [181]. Some groups have developed different types of super-resolution imaging techniques based on the single-molecule localization [182–184]. Localization microscopy has managed to reconstitute full images from many thousands of localized molecules, where the localization error of single molecules can be far smaller than the optical resolution. Localization microscopy using photoswitchable fluorescent molecules is termed photoactivated localization microscopy (PALM) [182], fluorescence photoactivation localization microscopy (FPALM) [183], stochastic optical reconstruction microscopy (STORM) [184]. These techniques have already been applied to live-cell imaging [185–188].

5.3.1 Stimulated Depletion Emission Microscopy

Stimulated emission depletion (STED) microscopy was proposed in 1994 by S. W. Hell [172] and was experimentally demonstrated in 1999 [173]. The basic concept of STED microscopy is to inhibit the fluorescence process in the outer regions of the excitation beam by depleting the fluorophores in the excited state through stimulated emission (Fig. 5.20(a)) using a second beam with the intensity distribution featuring a zero at the center of the excitation beam, which is called the STED beam. The STED beam operates at wavelength that does not excite the fluorophores but efficiently induces stimulated emission (Fig. 5.20(b)). By increasing the STED beam's intensity to levels high enough to saturate the stimulated emission transition, the fluorophores remaining in the excited state are confined to around the zero

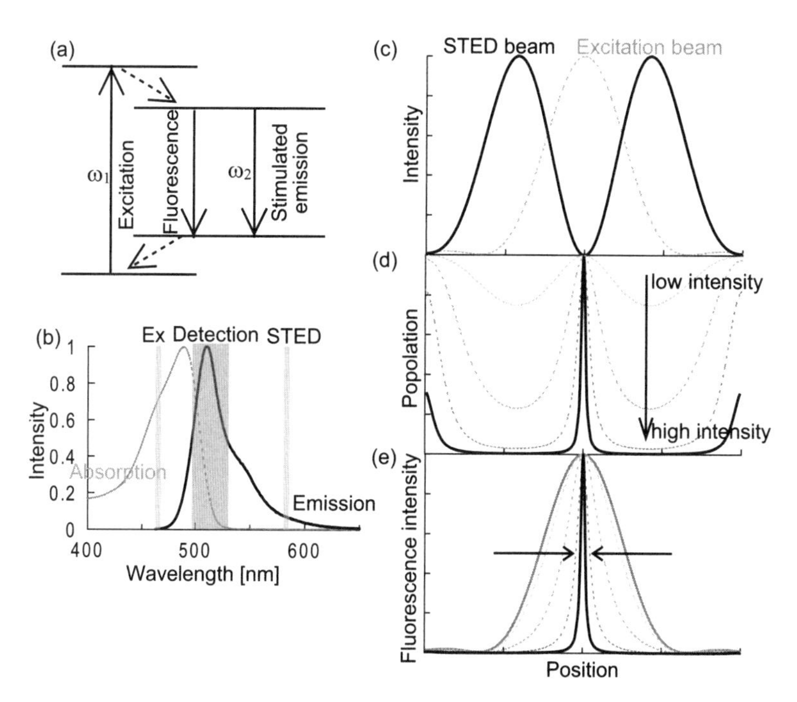

FIGURE 5.20 (a) Energy diagram for stimulated emission. (b) Spectrum including excitation, STED, and detection. (c) Intensity profiles of the excitation and STED beams. (d) Population in the excited state confined by the STED beam. (e) Effective PSF of STED microscopy.

intensity region of the STED beam as shown in Fig. 5.20(c) and (d). The point spread function (PSF) of STED microscopy is effectively narrowed by detecting fluorescence from the confined region that is squeezed as increasing the STED beam's intensity (Fig. 5.20(e)). In order to achieve a moderate resolution enhancement, STED beam intensities of >1GW/cm^2 must be routinely applied [189]. In STED microscopy, the highest resolution of 5.8 nm was achieved in 2009 by employing the color centers of nanocrystals [190], which hardly show photobleaching because the fluorescence results from point defects inside a solid matrix of the diamond [191,192]. Multi-color imaging [193] and multi-lifetime imaging [194] have been also achieved by STED microscopy. The idea underlying STED microscopy was expanded to ground-state depletion (GSD) microscopy in 1995 [174] and was generalized to a more general principle for a super-resolution imaging technique using reversible saturable optical fluorescence transitions (RESOLFT) between a bright (fluorescent) state and a dark (nonfluorescent) state in 2003 [177]. The idea underlying RESOLFT microscopy is based on the control of transitions between a bright state and a dark state using the depletion beam. For example, the control of transitions was experimentally performed by using a reversibly photoswitchable fluorescent protein, asFP595, which led to a resolution better than 100 nm at a much lower depletion intensity of 600 W/cm^2 [178].

Let us assume two fluorophore states A and B that are a bright (fluorescent) state and a dark (nonfluorescent) state, respectively. Transition A $\rightarrow$ B is induced by light, while transition B $\rightarrow$ A is caused by heat or light. For example, consider the depletion of the state A by the depletion beam via the transition A $\rightarrow$ B, whose rate is given by $k_{AB} = \sigma I/\hbar\omega$, with σ, I, and $\hbar\omega$ denoting the optical cross section of the transition, the depletion intensity, and the photon energy of the depletion light, respectively. The equilibrium population in the state A is expressed by

$$N_A(\infty) = \frac{k_{BA}}{k_{AB} + k_{BA}} = \frac{k_{BA}}{\sigma I/\hbar\omega + k_{BA}}, \tag{5.3.1}$$

where k_{BA} is the rates of transition B $\rightarrow$ A. If the depletion intensity I is much higher than the saturation intensity $I_{sat} = k_{BA}\hbar\omega/\sigma$, it follows that $N_A(\infty) \rightarrow 0$; that is, all the fluorophores end up in the state B. Provided that a spatial intensity distribution $I(r) \gg I_{sat}$ with a naught at r_0, all fluorophores end up in the state B except for those at r_0. Thus, we can generate sharp regions of the state A (Fig. 5.20(d)). With I_{max} denoting the intensity bordering the zero, the achievable resolution Δr is well approximated by [195]

$$\Delta r \approx \frac{\lambda}{2n\sin\theta\sqrt{1 + I_{max}/I_{sat}}}, \tag{5.3.2}$$

where λ, θ, and n denote the wavelength, the aperture angle of the lens, and the refractive index, respectively. The sharp regions of the state A fluorophores can be continuously squeezed down to a fluorophore size by increasing I_{max}/I_{sat}.

In order to generate the depletion beam with the intensity distribution featuring a zero at the center of the excitation beam, an additional optical element such as

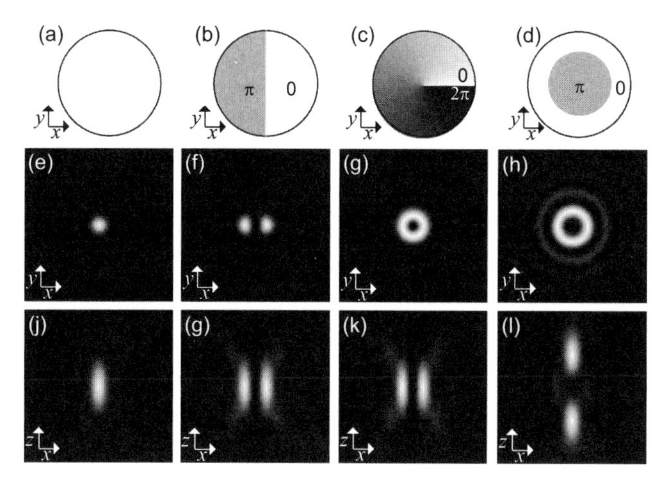

FIGURE 5.21 Phase filters (a–d) and resulting point spread functions in the xy plane (e–h) and in the xz plane (j–l).

a phase filter must be placed in the depletion beam. If a specifically spatial phase filter is placed at the entrance pupil of an objective lens, the electric fields interfere destructively in the focal center as shown in Fig. 5.21.

Figure 5.22(a) shows a schematic of the typical STED microscope. As light sources, excitation and STED pulses are required. Although picosecond pulses are typically used in STED microscopy, STED microscopy with continuous wave (CW) laser beams has been also reported [196,197]. The excitation pulse is tuned to the absorption spectrum of the fluorophore, while the STED pulse, which is synchronized with the excitation pulse, is red-shifted in frequency to the emission spectrum of the fluorophore. The STED beam passes through the spatial phase filter (PF) and is subsequently overlaid with the excitation beam using a dichroic mirror. The two pulses are focused into a sample by an objective lens (OB). A quarter-wave plate (WP) is employed to impart a circular polarization to the STED and excitation beams. The circular polarization for excitation and depletion provides a uniform effective resolution increase in the lateral directions of the focal plane of the objective lens. The fluorescence that emerges from the sample is focused thorough a confocal

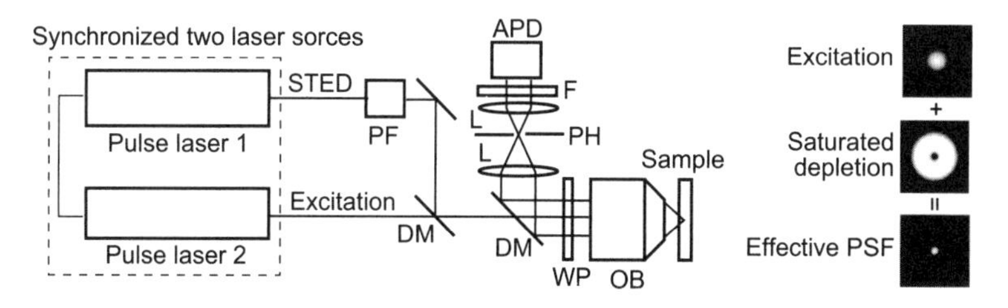

FIGURE 5.22 Typical STED microscope.

pinhole (PH) and with an avalanche photodiode (APD). The reasons for using the pinhole are the rejection of stray and ambient light and the convenient confocal axial sectioning provided. Note that confocality is not an ingredient of the concept of STED microscopy, since resolution enhancement is dominated not by the confocal pinhole but by the depletion.

Recently, Moneron and Hell [198], and Li et al. [199] have demonstrated two-photon excited fluorescence (TPEF) microscopy combined with STED microscopy. The excitation pulse is tuned to the two-photon absorption spectrum of the fluorophore, while the STED pulse is red-shifted in frequency to the emission spectrum of the fluorophore. Figure 5.23(a) and (b) shows fluorescence images of EGFP-tagged caveolin 1 (Cav1-GFP) in CHO cells obtained by ordinary TPEF microscopy and by TPEF-STED microscopy, respectively [199]. Because of the lower photon count, images obtained by STED microscopy are often smoothed or restored by various

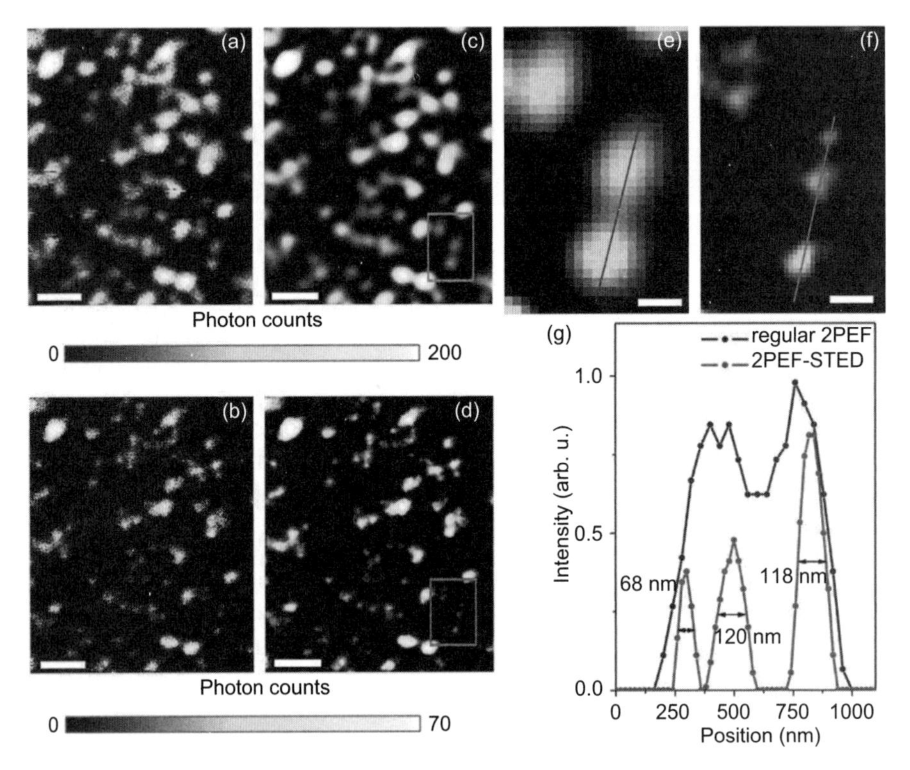

FIGURE 5.23 EGFP-tagged caveolin in a CHO cell imaged using (a) an ordinary TPEF microscope and (b) the TPEF-STED microscope. (c) Ordinary 2PEF and (d) 2PEF-STED images after image restoration by the Tikhonov–Miller filter. (e) Magnified view of the marked area for the ordinary TPEF image in (c). (f) Magnified view of the same marked area for the TPEF-STED image in (d). (g) Intensity profiles of the ordinary TPEF and TPEF-STED images, as indicated in (e) and (f). The sale bars, 1 μm (a–d) and 200 nm (e, f). Reprinted with permission from [199]. Copyright 2009 Elsevier.

filters to remove noise [200–202]. Figure 5.23(c) and (d), are the same images as in Fig. 5.23(a) and (b), respectively, after image restoration with the Tikhonov–Miller filter [203]. The small Cav1-GFP domains are presumed to be caveolar vesicles or protein-lipid chaperon complexes [199], which have diameters of 50 to 100 nm under an electron microscope [204]. Figure 5.23(e) and (f) gives magnified views of the marked areas in Fig. 5.23(c) and (d), respectively. It should be noted that TPEF-STED microscopy enables out distinguishing two caveolar vesicles located within a distance smaller than the 250-nm diffraction limit, which are observed in the ordinary 2PEF images in Fig. 5.23(e), as shown in Fig. 5.23(f) and its profile in Fig. 5.23(g). As we see Fig. 5.23(g), Li et al. has successfully obtained the fluorescence image of a single caveolar vesicle with a FWHM of 68 nm [199].

5.3.2 Saturated Excitation Microscopy

In general, one-photon excited fluorescence (OPEF) microscopy is used with relatively low excitation intensity levels in which the fluorescence intensity is linearly proportional to the excitation intensity. When the excitation intensity is sufficiently high, fluorescence emission saturates because fluorophores have a nonzero excitation lifetime, and the number of fluorophores in the focal volume is limited. Because fluorescence saturation results in a decrease in the spatial resolution, it has been avoided in OPEF microscopy [205]. However, to improve the spatial resolution, Fujita et al. employed fluorescence saturation as a nonlinear process relating emission to excitation [181]. This technique is called saturated excitation (SAX) microscopy.

The fluorescence intensity $I_f(r, t)$, emitted by an ensemble of N fluorophores, is given by [205]

$$I_f(t) = N\eta q^{(1)} \frac{\sigma^{(1)} I_{ex}(t)}{1 + I_{ex}(t)/I_{sat}}, \tag{5.3.3}$$

where η is the overall detection efficiency of the optical system, $\sigma^{(1)}$ is the one-photon absorption cross section, $q^{(1)}$ is the fluorescence quantum efficiency, $I_{ex}(t)$ is the excitation intensity, and I_{sat} is the fluorescence saturation intensity. We can now apply a Taylor expansion, so that Eq. (5.3.3) becomes

$$I_f(t) = -N\eta q^{(1)} \sigma^{(1)} I_{sat} \sum_{m=1}^{\infty} \left\{ -\frac{I_{ex}(t)}{I_{sat}} \right\}^m. \tag{5.3.4}$$

According to Eq. (5.3.4), a strong nonlinearity exists in the relation between excitation and fluorescence emission. The linear component in Eq. (5.3.4) widens the spatial profile of fluorescence emission, while the nonlinear components are confined only in the small volume near the focal spot as well as nonlinear optical microscopy. Thus, by extracting only the nonlinear components, we can improve the spatial resolution. We also find that the spatial resolution is theoretically not limited.

To achieve SAX microscopy, the nonlinear components in fluorescence saturation must be extracted. This problem has been solved by temporally modulating the excitation intensity at a frequency of f [181]. We can express the excitation intensity as $I_{ex} = I_0\{1 + \cos(2\pi ft)\}$, by which the fluorescence intensity for the mth order nonlinear component can be described as $I_f^{(m)} \propto I_0^m\{1 + \cos(2\pi ft)\}^m$. Then, the demodulated fluorescence intensity at a frequency of $mf(m = 2, 3, 4, \ldots)$ is mainly proportional to $I_0^m \cos(2\pi mft)$. Therefore, we can acquire the mth order nonlinear component by demodulating the harmonic frequency component at mf. The experimental setup for SAX microscopy can be realized by simple modification of a typical confocal fluorescence microscopy. The excitation intensity is modulated by using an acousto-optical modulator (AOM). The fluorescence signal is detected by a photomultiplier tube (PMT) with a confocal pinhole. The detected signal is demodulated by a lock-in amplifier.

In order to confirm the principle of SAX microscopy, the improvement of the spatial resolution can be calculated by using the relationship between excitation and emission intensities [206]. In this calculation, a photophysical model of fluorescence emission of rhodamine 6G molecules excited by CW laser light with a wavelength of 515 nm [181,207] was used. Figure 5.24(a) illustrates the energy diagram of a rhodamine 6G molecule with five-level electronic state model where S_0 is the ground singlet state, S_1 is the first excited singlet state, S_n is the higher excited singlet state ($n > 1$), T_1 is the lowest excited triplet state, and T_n is the higher excited triplet

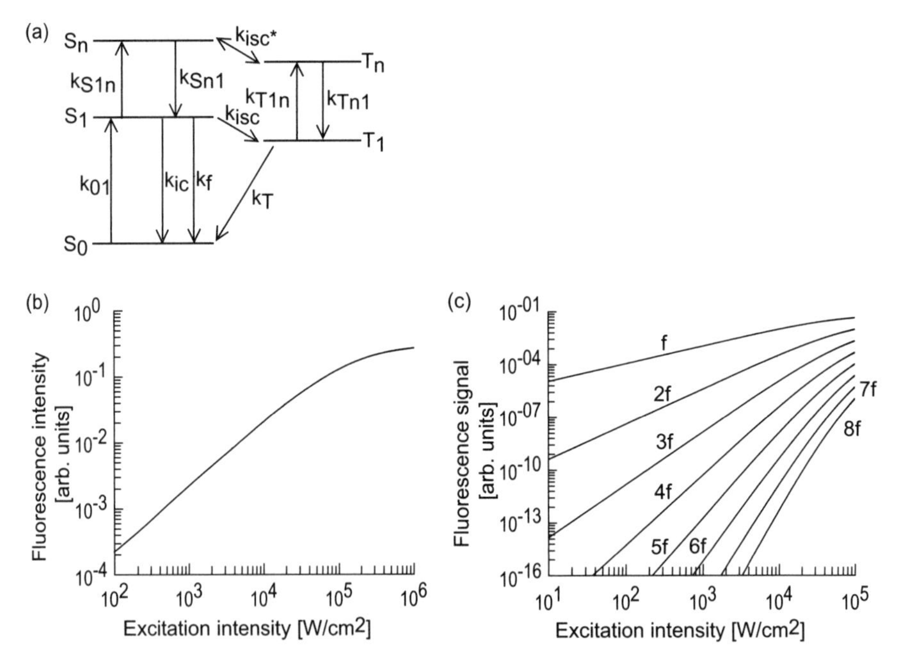

FIGURE 5.24 (a) Energy diagram of a rhodamine 6G molecule with five-level molecular electronic state model. (b) Saturation effect in fluorescence emission from rhodamine 6G molecules calculated using the five-level molecular electronic state model. (c) Relationship between the demodulated fluorescence signal and excitation intensity.

state ($n > 1$). The rate constants, k_{01} (S_0 to S_1), k_{S1n} (S_1 to S_n), and k_{T1n} (T_1 to T_n), which quantify excitation processes from the state i to the state j are quantified, are expressed by

$$k_{01} = \sigma_{01}^{(1)} I/\hbar\omega, \tag{5.3.5}$$

$$k_{\mathrm{S1n}} = \sigma_{\mathrm{S1n}}^{(1)} I/\hbar\omega, \tag{5.3.6}$$

$$k_{\mathrm{T1n}} = \sigma_{\mathrm{T1n}}^{(1)} I/\hbar\omega. \tag{5.3.7}$$

Here I is the excitation intensity, $\hbar\omega$ is the photon energy of an excited light, and $\sigma_{01}^{(1)}(= 2.22 \times 10^{-16}\mathrm{cm}^2)$, $\sigma_{\mathrm{S1n}}^{(1)}(= 0.77 \times 10^{-17}\mathrm{cm}^2)$, $\sigma_{\mathrm{T1n}}^{(1)}(= 3.85 \times 10^{-17}\mathrm{cm}^2)$ are the corresponding absorption cross section at the excitation wavelength, λ (=515 nm). Depopulation processes of each electronic excited state are characterized by the rate constants $k_{\mathrm{f}}(= 2.4 \times 10^8\ \mathrm{s}^{-1})$, k_{ic}, and $k_{\mathrm{isc}}(= 1.1 \times 10^6\ \mathrm{s}^{-1})$ for de-excitation of S_1 by fluorescence, internal conversion, and intersystem crossing to T_1, respectively, leading to an overall fluorescence lifetime of $\tau_{\mathrm{fo}} = 1/k_{\mathrm{fo}} = 1/(k_{\mathrm{f}} + k_{\mathrm{ic}} + k_{\mathrm{isc}})$ (=3.9 ns). $k_{\mathrm{T}}(= 0.49 \times 10^6\ \mathrm{s}^{-1})$, k_{Sn1} (=1/200 fs), and k_{Tn1} (=1/200 fs) denote the de-excitation rate constants of T_1, S_n, and T_n, respectively. Rate constants for the stimulated de-excitation and intersystem crossing between T_n and S_n states, k_{isc}^*, are assumed to be negligible. Furthermore, only the lowest vibronic level of each electronic state is considered; that is, fast vibrational relaxation from higher vibronic states is assumed. The time-dependent populations of each of the five levels are obtained by solving the rate equation system:

$$\frac{dS_0(t)}{dt} = -k_{01}S_0 + (k_{\mathrm{f}} + k_{\mathrm{ic}})S_1 + k_{\mathrm{T}}T_1, \tag{5.3.8}$$

$$\frac{dS_1(t)}{dt} = k_{01}S_0 - (k_0 + k_{\mathrm{S1n}})S_1 + k_{\mathrm{Sn1}}S_{\mathrm{n}}, \tag{5.3.9}$$

$$\frac{dS_{\mathrm{n}}(t)}{dt} = k_{\mathrm{S1n}}S_1 - k_{\mathrm{Sn1}}S_{\mathrm{n}}, \tag{5.3.10}$$

$$\frac{dT_1(t)}{dt} = k_{\mathrm{isc}}S_1 - (k_{\mathrm{T}} + k_{\mathrm{T1n}})T_1 + k_{\mathrm{Tn1}}T_{\mathrm{n}}, \tag{5.3.11}$$

$$\frac{dT_{\mathrm{n}}(t)}{dt} = k_{\mathrm{T1n}}T_1 - k_{\mathrm{Tn1}}T_{\mathrm{n}}. \tag{5.3.12}$$

In this calculation, we ignored the effects of photobleaching, and the fluorescence signal is proportional to the population of fluorescent molecules in the first excited singlet state. Figure 5.24(b) shows the calculated relationship between the excitation and fluorescence intensities. We can confirm from the fluorescence intensity that a linear response occurs at a low excitation intensity ($<2.0 \times 10^3$ W/cm^2) and a saturation effect at a high excitation irradiance ($>2.0 \times 10^3$ W/cm^2). Figure 5.24(c) illustrates the calculated relationships between the excitation and demodulated fluorescence intensities. The demodulated signal at a demodulation frequency of f is proportional to the excitation intensity at the low excitation intensity region. The demodulated signal at a demodulation frequency of $2f$ is proportional to the square

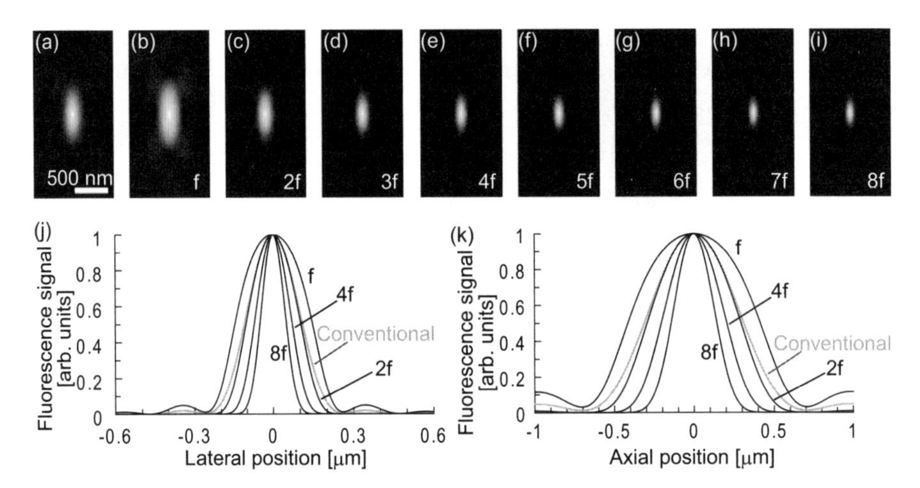

FIGURE 5.25 (a–i) The calculated excitation PSFs of conventional OPEF microscopy (a) and SAX microscopy at demodulation frequencies of f (b), $2f$ (c), $3f$ (d), $4f$ (e), $5f$ (f), $6f$ (g), $7f$ (h), and $8f$ (i). (j, k) The intensity profiles of PSFs along the lateral (j) and axial (k) directions.

of the excitation intensity. When the demodulation frequency increases further, the demodulated signal shows a higher order nonlinear response with the appearance of the saturation effect.

Figures 5.25 shows the excitation PSFs, which are calculated with an NA of 1.2 at a excitation wavelength of 515 nm, for conventional OPEF microscopy and SAX microscopy at demodulation frequencies of $f, 2f, 3f, 4f, 5f, 6f, 7f$, and $8f$ with an excitation intensity of 1.0×10^5 W/cm^2. We can confirm that the excitation PSF at a demodulation frequency of f is broadened due to the saturation effect, while those demodulated at harmonic frequencies are narrowed when the demodulation frequency increases.

The spatial resolution of SAX microscopy is theoretically not limited [206]. In practice, photobleaching effects, which are induced by high excitation intensities to induce sufficient saturated excitation [181], limit the spatial resolution in SAX microscopy. Yamanaka et al. overcame this limitation by using fluorescent nanodiamonds (FNDs) with the negatively charged nitrogen-vacancy center $(\text{N-V})^-$ as a photostable fluorescent probe for SAX microscopy [208]. The FNDs are optically transparent and capable of fluorescing from point defects [209]. The defect center absorbs strongly at 560 nm and emits fluorescence efficiently at 700 nm. The one-photon absorption cross section at the band center is in the range of 5×10^{-17} cm^2, comparable to that of a dye molecule [210]. Because the center, acting as an ion embedded in an inert solid matrix, hardly photobleaches [191], long-term observation of a single FND can be achieved [192]. Yamanaka et al. succeeded in enhancing the three-dimensional spatial resolution by demodulating the saturated fluorescence signals from the FNDs at the fourth-harmonic frequency [208]. Figure 5.26(a) and (b) shows the fluorescence images of FNDs distributed in a macrophage cell by confocal

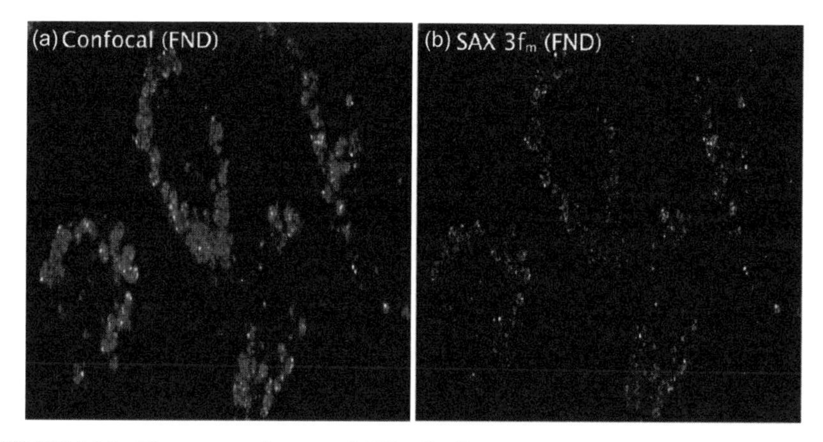

FIGURE 5.26 Fluorescence images of FNDs distributed in a macrophage cell by (a) confocal microscopy and (b) SAX microscopy with third harmonic demodulation. Reprinted with permission from [208]. Copyright 2011 Optical Society of America.

microscopy and SAX microscopy with third-harmonic demodulation [208]. Note that the spatial resolution is significantly improved by SAX microscopy.

5.3.3 Saturated Structured Illumination Microscopy

Structured illumination microscopy (SIM) was proposed in 2000 by Gustafsson. In SIM, the spatial resolution enhancement is achieved by adding the excitation and emission of Abbe's limit, and the resolution can be increased by a factor of two [211]. This can be done by wide-field OPEF microscopy by placing a fine mesh grating in the excitation light path, which introduces the highest spatial frequency during excitation, projecting a fine sinusoidal illumination pattern on the sample, which creates images in the form of moiré fringes. A series of images are used to reconstruct the image, and this composite image will include high-resolution information that is normally inaccessible. Two-dimensional SIM uses two beams of excitation light that interfere to create a sinusoidal light intensity pattern [211], while three-dimensional SIM employs three beams of excitation light to increase the axial resolution [212,213] as shown in Fig. 5.27.

To avoid degradation of the spatial resolution by fluorescence saturation, fluorescence microscopy is generally used under the condition that the emitted fluorescence

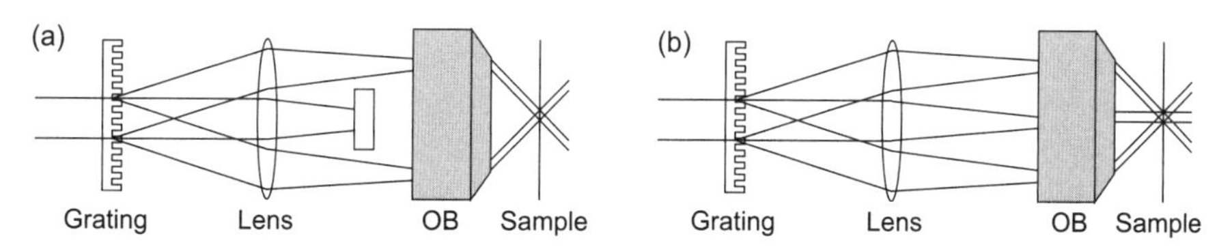

FIGURE 5.27 Experimental setup for the generation of the illumination pattern in two-dimensional SIM (s) and three-dimensional SIM (b).

is proportional to the local intensity of the illumination light. This way the image of fluorescence samples excited with a structured illumination pattern can be calculated by

$$I_{\mathrm{f}}(r,\phi) = \{C(r)I_{\mathrm{ex}}(r,\phi)\} \otimes P(r), \tag{5.3.13}$$

where $C(r)$ is the local fluorophore concentration, $P(r)$ is the PSF of the incoherent imaging system for the emitted light, and $\otimes$ denotes convolution. Here $I_{\mathrm{ex}}(r,\phi)$ is the illumination pattern of the excitation light. The spectrum in frequency space is obtained by Fourier transformation of Eq. (5.3.13):

$$\begin{aligned}\tilde{I}_{\mathrm{f}}(k,\phi) &= \int_{-\infty}^{\infty} I_{\mathrm{f}}(r,\phi)\exp(-ikr)dr \\ &= \left\{\tilde{C}(k) \otimes \tilde{I}_{\mathrm{ex}}(k,\phi)\right\}\tilde{P}(k),\end{aligned} \tag{5.3.14}$$

where $\tilde{C}(k)$, $\tilde{I}_{\mathrm{ex}}(k,\phi)$ and $\tilde{P}(k)$ are the Fourier transforms of $C(r)$, $I_{\mathrm{ex}}(r,\phi)$, and $P(r)$, respectively. The illumination pattern $I_{\mathrm{ex}}(r,\phi)$ on the sample plane is expressed as

$$\begin{aligned}I_{\mathrm{ex}}(r,\phi) &= I_{\mathrm{ex}}\{1+\alpha\cos(k_0 r+\phi)\} \\ &= I_{\mathrm{ex}}\left[1+\frac{\alpha}{2}\exp\{i(k_0 r+\phi)\}+\frac{\alpha}{2}\exp\{-i(k_0 r+\phi)\}\right],\end{aligned} \tag{5.3.15}$$

and its the Fourier transform is described by

$$\tilde{I}_{\mathrm{ex}}(k,\phi) = I_{\mathrm{ex}}\left[\delta(k)+\frac{\alpha}{2}\delta(k-k_0)\exp(i\phi)+\frac{\alpha}{2}\delta(k+k_0)\exp(-i\phi)\right]. \tag{5.3.16}$$

Substituting Eq. (5.3.16) into Eq. (5.3.14), we obtain

$$\tilde{I}_{\mathrm{f}}(k,\phi) = I_{\mathrm{ex}}\tilde{P}(k)\left\{\tilde{C}(k)+\tilde{C}(k-k_0)\exp(i\phi)+\tilde{C}(k+k_0)\exp(-i\phi)\right\}. \tag{5.3.17}$$

Here $\tilde{C}(k)\tilde{P}(k)$, $\tilde{C}(k-k_0)\tilde{P}(k)$ and $\tilde{C}(k+k_0)\tilde{P}(k)$ can be obtained from three spectra acquired at three separate phases:

$$\begin{bmatrix}\tilde{I}_{\mathrm{f}}(k,\phi_1)\\ \tilde{I}_{\mathrm{f}}(k,\phi_2)\\ \tilde{I}_{\mathrm{f}}(k,\phi_3)\end{bmatrix} = I_{\mathrm{ex}}\begin{bmatrix}1 & 0.5\alpha\exp(i\phi_1) & 0.5\alpha\exp(-i\phi_1)\\ 1 & 0.5\alpha\exp(i\phi_2) & 0.5\alpha\exp(-i\phi_2)\\ 1 & 0.5\alpha\exp(i\phi_3) & 0.5\alpha\exp(-i\phi_3)\end{bmatrix}\begin{bmatrix}\tilde{C}(k)\tilde{P}(k)\\ \tilde{C}(k-k_0)\tilde{P}(k)\\ \tilde{C}(k+k_0)\tilde{P}(k)\end{bmatrix}. \tag{5.3.18}$$

The obtained $\tilde{C}(k-k_0)\tilde{P}(k)$ and $\tilde{C}(k+k_0)\tilde{P}(k)$ are then shifted back $-k_0$ and $+k_0$ to form $\tilde{C}(k-k_0)\tilde{P}(k-k_0)$ and $\tilde{C}(k+k_0)\tilde{P}(k+k_0)$, respectively. Because these spectra enable us to extend the observable frequency region in the pattern direction, the resolution is improved accordingly in the same direction. In order to retrieve a high-resolution image, $\tilde{C}(k-k_0)\tilde{P}(k-k_0)$ and $\tilde{C}(k+k_0)\tilde{P}(k+k_0)$ are added to $\tilde{C}(k)\tilde{P}(k)$, and then their addition is inverse Fourier transformed. To retrieve an

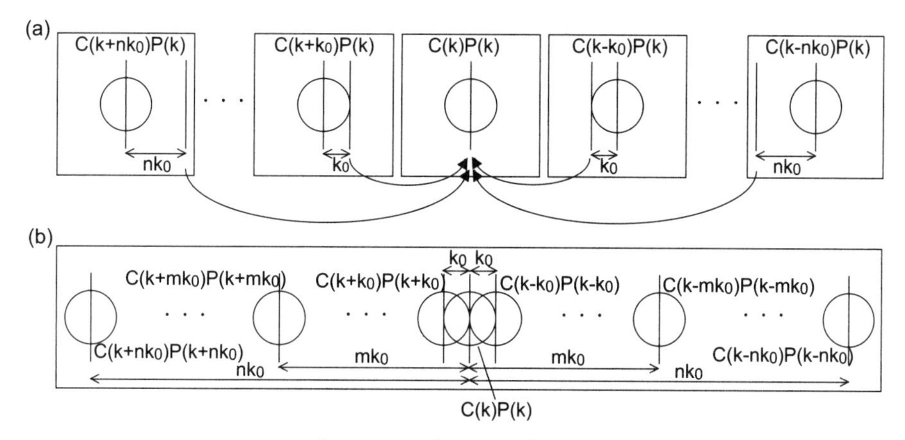

FIGURE 5.28 (a) Shifts of $\tilde{C}(k - mk_0)\tilde{P}(k)$ and $\tilde{C}(k + mk_0)\tilde{P}(k)$ to $\tilde{C}(k - mk_0)\tilde{P}(k - mk_0)$ and $C(k + mk_0)\tilde{P}(k + mk_0)$. (b) Their addition to $\tilde{C}(k)\tilde{P}(k)$ to form the final broadband spectrum.

isotropic image, further sets of $\tilde{C}(k)\tilde{P}(k)$, $\tilde{C}(k - k_0)\tilde{P}(k - k_0)$, and $\tilde{C}(k + k_0)\tilde{P}(k + k_0)$ obtained at other pattern orientations are combined before the inverse Fourier transform as shown in Fig. 5.28. Structured illumination microscopy (SIM) provides twice the spatial resolution.

The nonlinear processes can be also applied to SIM. The image of fluorescence samples with nth order nonlinearity is described by

$$I_{\mathrm{f}}^{(n)}(r, \phi) = \left\{C(r)I_{\mathrm{ex}}^{n}(r, \phi)\right\} \otimes P(r). \tag{5.3.19}$$

The spectrum in frequency space is written as

$$\tilde{I}_{\mathrm{f}}^{(n)}(k, \phi) = I_{\mathrm{ex}}\tilde{P}(k) \sum_{m=-n}^{n} a_m \tilde{C}(k + mk_0) \exp(-im\phi). \tag{5.3.20}$$

$\tilde{C}(k + mk_0)\tilde{P}(k + mk_0)$, $(m = -n, -n + 1, \ldots, n - 1, n)$ can be obtained from $2n + 1$ spectra acquired at $2n + 1$ separate phases. These spectra can enhance spatial resolution by a factor of $n + 1$. Fluorescence saturation can be used as a nonlinear process relating emission to excitation. Then, the fluorescence intensity is expressed by [205]

$$I_{\mathrm{f}}^{(\mathrm{s})}(r, \phi, t) \propto \left[C(r)\frac{\sigma^{(1)} I_{\mathrm{ex}}(r, \phi)}{1 + I_{\mathrm{ex}}(r, \phi)/I_{\mathrm{sat}}}\right] \otimes P(r), \tag{5.3.21}$$

where $\sigma^{(1)}$ is the one-photon absorption cross section and I_{sat} is the fluorescence saturation intensity. Applying a Taylor expansion, we rewrite Eq. (5.3.21) as

$$I_{\mathrm{f}}^{(\mathrm{s})}(r, \phi, t) \propto \left[C(r)\sigma^{(1)} I_{\mathrm{sat}} \sum_{m=1}^{\infty} \left\{-\frac{I_{\mathrm{ex}}(t)}{I_{\mathrm{sat}}}\right\}^{m}\right] \otimes P(r). \tag{5.3.22}$$

According to Eqs. (5.3.19), (5.3.20), and (5.3.22), it is clear that the spatial resolution is theoretically not limited in SIM when combined with fluorescence saturation. This technique is called saturated structured illumination microscopy (SSIM) or saturated patterned excitation microscopy (SPEM). In practice, the spatial resolution is limited by high irradiation intensities that are necessary to observe such nonlinear effects, which, on the one hand, require very photostable fluorophores and, on the other, careful consideration in live-cell experiments. This problem could be solved by using photoswitchable fluorophores and by extending the spatial resolution in SSIM to, where the resolution is limited only by the signal-to-noise ratio.

5.3.4 Single-Molecule Localization Microscopy

Single-molecule localization microscopy was proposed in 1995 by Betzig [214]. Although the image of a single fluorophore, which shows the PSF, is a diffraction-limited spot, the precision of determining the position of the fluorophore from its image can be much higher than the diffraction limit, as long as the image results from multiple photons emitted from the fluorophore. By recording the centers of single fluorophores from images of their PSFs with a localization precision as high as 1 nm and merging all the single-molecule positions, the spatial resolution of single-molecule localization microscopy can be enhanced. However, it is difficult to optically resolve single molecules in fluorescently labeled biological samples because a typical fluorescently labeled biological sample contains thousands or even millions of fluorophores at a high density. This barrier has been overcome by using photoactivatable fluorescent probes that can switch between a fluorescent state and a dark state. Stochastic activation to switch on individual photoactivatable molecules provides a temporal separation of single fluorophores that are not spatially resolvable from each other due to the diffraction limit. In this approach, single fluorophores within a diffraction limited region can be individually photoactivated at different time points so that they can be individually imaged, localized, and subsequently deactivated. Iterating the activation and imaging process then allows the locations of many fluorophores to be mapped and a super-resolution image to be reconstructed from these fluorophore locations as shown in Fig. 5.29. Massively parallel localization is achieved through wide-field imaging. Single-molecule localization microscopy using photoactivatable fluorescent probes was independently demonstrated by three groups and was termed photoactivated localization microscopy (PALM) [182], fluorescence photoactivation localization microscopy (FPALM) [183], and stochastic optical reconstruction microscopy (STORM) [184].

The two-dimensional localization precision of a single molecule is given by [216]

$$\Delta r_{\mathrm{loc}}^2 = \frac{s^2 + a^2/12}{F} + \frac{8\pi s^4 b^2}{a^2 F^2}, \tag{5.3.23}$$

where s is the standard deviation of the PSF (which is proportional to the conventional spatial resolution), F is the total number of photons collected (not photons per pixel), a is the size of an image pixel within the sample focal plane, and b is the background

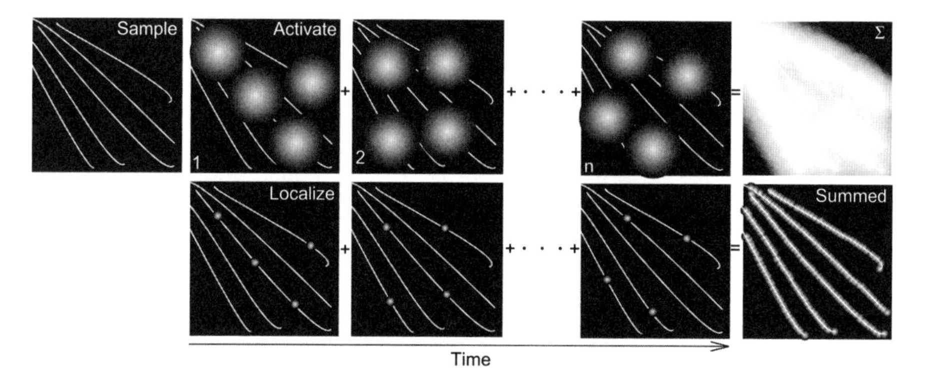

FIGURE 5.29 Principle of PALM.

noise per pixel (not background intensity). The first terms in Eq. (5.3.23) corresponds to the contribution to Δr_{loc} from shot noise in the absence of background. The second term corresponds to uncertainties introduced due to background noise, which may arise from scattered light, autofluorescence from the sample, weak fluorescence from inactive molecules, and detector electronics. For small pixel sizes and negligible background noise, Eq. (5.3.23) reduces to

$$\Delta r_{\text{loc}} = \frac{s}{\sqrt{F}}. \tag{5.3.24}$$

We then take into account the localization precision and the molecular density (or sparseness) to obtain the localization-based resolution that describes the smallest structure:

$$\Delta r^2 = \Delta r_{\text{loc}}^2 + r_{NN}^2, \tag{5.3.25}$$

where r_{NN} is the nearest neighbor distance between molecules, which depends on the molecular density [217]. If $\Delta r_{\text{loc}} \ll r_{NN}$, the resolution is limited by the molecular density. The effect of molecular density on the effective resolution can be quantified by the Nyquist criterion, which states that structural features smaller than twice that of the fluorophore-to-fluorophore distance cannot be reliably discerned. The smallest resolvable feature size is alternatively given by

$$\Delta r = \frac{2}{N_d^{1/D}}, \tag{5.3.26}$$

where N_d is the molecular density, and D is the dimension of the structure to be imaged [219].

In a typical wide-field microscope, photoactivatable molecules are activated not only in the focal plane but also in other layers because it has no optical sectioning capability. Such extraneous activation in other layers leads to higher background. Thus, it is usual in most localization microscopy to use thin biological samples in

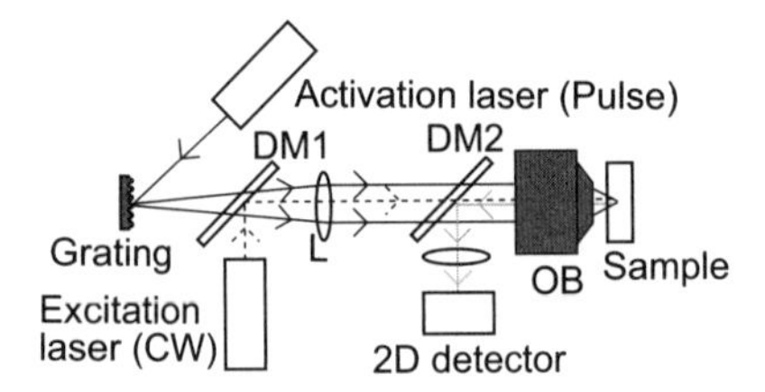

FIGURE 5.30 Experimental setup for temporal focusing localization microscope.

combination with the total internal reflection feature. However, the imaging depth is limited to a fraction of an optical wavelength. To solve this problem, localization microscopy has been combined with two-photon activation that is induced by a temporal focusing technique [220,221]. Since the temporal focusing technique enhances two-photon wide-field excitation with an optical sectioning capability, the two-photon activation is confined to a narrow region around the focal plane. Because of the axial confinement of the activation, the excitation and bleaching of unlocalized markers are substantially reduced, and this enables 3D localization imaging with high localization density in thick structures.

Figure 5.30 shows the experimental setup for the temporal focusing localization microscope. Activation laser pulses are separated into monochromatic components by a diffraction grating, and the illuminated spot on the grating is then imaged onto the sample plane using a telescope including an objective lens. This results in a pulse broadened everywhere in the sample except at the image plane, where the dispersion is compensated and the pulse duration reaches its minimum width. The temporal focusing pulses enable us to selectively activate photoactivatable fluorophores in a thin layer of the sample, thereby preventing activation of molecules in other layers. The activated molecules are imaged by using the other excitation beam, which is implemented in a one-photon absorption wide-field configuration. The temporal focusing setup is relatively simple, and can be built by using noncustomized components.

REFERENCES

1. J. Lippincott-Schwartz and G. H. Patterson, "Development and use of fluorescent protein markers in living cells," *Science*, **300**, 87–91 (2003).
2. A. Miyawaki, "Proteins on the move: insights gained from fluorescent protein technologies," *Nat. Rev. Mol. Cell Biol.*, **12**, 656–668 (2011).
3. J. C. Politz, "Use of caged fluorochromes to track macromolecular movement in living cells," *Trends Cell Biol.*, **9**, 284–287 (1999).
4. K. A. Lukyanov, D. M. Chudakov, S. Lukyanov, and V. V. Verkhusha, "Innovation: photoactivatable fluorescent proteins," *Nat. Rev. Mol. Cell Biol.*, **6**, 885–891 (2005).
5. J. N. Henderson and S. J. Remington, "The kindling fluorescent protein: a transient photoswitchable marker," *Physiology*, **21**, 162–170 (2006).
6. G. C. R. Ellis-Davies, "Caged compounds: photorelease technology for control of cellular chemistry and physiology," *Nat. Methods*, **4**, 619–628 (2007).

7. Y. Wang, J. Y.-J. Shyy, and S. Chien, "Fluorescence proteins, live-cell imaging, and mechanobiology: seeing is believing," *Annu. Rev. Biomed. Eng.*, **10**, 1–38 (2008).
8. V. Sample, R. H. Newman, and J. Zhang, "The structure and function of fluorescent proteins," *Chem. Soc. Rev.*, **38**, 2582–2864 (2009).
9. D. M. Chudakov, M. V. Matz, S. Lukyanov, and K. A. Lukyanov, "Fluorescent proteins and their applications in imaging living cells and tissues," *Physiol. Rev.*, **90**, 1103–1163 (2010).
10. S. W. Hell, "Far-field optical nanoscopy," *Science*, **316**, 1153–1158 (2007).
11. M. Fernández-Suárez and A. Y. Ting, "Fluorescent probes for superresolution imaging in living cells," *Nat. Rev. Mol. Cell Biol.*, **9**, 929–943 (2008).
12. M. Heilemann, P. Dedecker, J. Hofkens, and M. Sauer, "Photoswitches: key molecules for subdiffraction-resolution fluorescence imaging and molecular quantification," *Laser Photon. Rev.*, **3**, 180–202 (2009).
13. S. van de Linde, A. Löschberger, T. Klein, M. Heidbreder, S. Wolter, M. Heilemann, and M. Sauer, "Direct stochastic optical reconstruction microscopy with standard fluorescent probes," *Nat. Protocols*, **6**, 991–1009 (2011).
14. G. T. Dempsey, J. C. Vaughan, K. H. Chen, M. Bates, and X. Zhuang, "Evaluation of fluorophores for optimal performance in localization-based super-resolution imaging," *Nat. Methods*, **8**, 1027–1036 (2011).
15. J. Lippincott-Schwartz and G. H. Patterson, "Photoactivatible fluorescent proteins for diffraction-limited and super-resolution imaging," *Trends Cell Biol.*, **19**, 555–565 (2009).
16. K. R. Gee, E. S. Weinberg, and D. J. Kozlowski, "Caged Q-rhodamine dextran: a new photoactivated fluorescent tracer," *Bioorg. Med. Chem. Lett.*, **11**, 2181–2183 (2001).
17. T. J. Mitchison, K. E. Sawin, J. A. Theriot, K. Gee, and A. Mallavarapu, "Caged fluorescent probes," *Methods Enzymol.*, **291**, 63–78 (1998).
18. K. Y. Han, T. Mokari, C. Ajo-Franklin, and B. E. Cohen, "Caged quantum dots," *J. Am. Chem. Soc.*, **130**, 15811–15813 (2008).
19. G. H. Patterson and J. Lippincott-Schwartz, "A Photoactivatable GFP for selective photolabeling of proteins and cells," *Science*, **297**, 1873–1877 (2002).
20. V. V. Verkhusha and A. Sorkin, "Conversion of the monomeric red fluorescent protein into a photoactivatable probe," *Chem. Biol.*, **12**, 279–285 (2005).
21. F. V. Subach, G. H. Patterson, S. Manley, J. M. Gillette, J. Lippincott-Schwartz, and V. V Verkhusha, "Photoactivatable mCherry for high-resolution two-color fluorescence microscopy," *Nat. Methods*, **6**, 153–159 (2009).
22. R. Ando, H. Hama, M. Yamamoto-Hino, H. Mizuno, and A. Miyawaki, "An optical marker based on the UV-induced green-to-red photoconversion of a fluorescent protein," *Proc. Natl. Acad. Sci. USA.*, **99**, 12651–12656 (2002).
23. D. M. Chudakov, V. V. Verkhusha, D. B. Staroverov, E. A. Souslova, S. Lukyanov, and K. A. Lukyanov, "Photoswitchable cyan fluorescent protein for protein tracking," *Nat. Biotechnol.*, **22**, 1435–1439 (2004).
24. J. Wiedenmann, S. Ivanchenko, F. Oswald, F. Schmitt, C. Röcker, A. Salih, K.-D. Spindler, and G. U. Nienhaus, "EosFP, a fluorescent marker protein with UV-inducible green-to-red fluorescence conversion," *Proc. Natl. Acad. Sci. USA.*, **101**, 15905–15910 (2004).

25. S. A. McKinney, C. S. Murphy, K. L. Hazelwood, M. W. Davidson, and L. L. Looger, "A bright and photostable photoconvertible fluorescent protein," *Nat Methods*, **6**, 131–133 (2009).
26. M. Zhang, H. Chang, Y. Zhang, J. Yu, L. Wu, W. Ji, J. Chen, B. Liu, J. Lu, Y. Liu, J. Zhang, P. Xu, and T. Xu, "Rational design of true monomeric and bright photoactivatable fluorescent proteins," *Nat. Methods*, **9**, 727–729 (2012).
27. H. Tsutsui, S. Karasawa, H. Shimizu, N. Nukina, and A. Miyawaki, "Semi-rational engineering of a coral fluorescent protein into an efficient highlighter," *EMBO Rep.*, **6**, 233–238 (2005).
28. S. Habuchi, H. Tsutsui, A. B. Kochaniak, A. Miyawaki, and A. M. van Oijen, "mKikGR, a monomeric photoswitchable fluorescent protein," *PloS ONE*, **3**, e3944 (2008).
29. N. G. Gurskaya, V. V. Verkhusha, A. S. Shcheglov, D. B. Staroverov, T. V. Chepurnykh, A. F. Fradkov, S. Lukyanov, and K. A. Lukyanov, "Engineering of a monomeric green-to-red photoactivatable fluorescent protein induced by blue light," *Nat. Biotechnol.*, **24**, 461–465 (2006).
30. V. Adama, M. Lelimousinb, S. Boehmec, G. Desfondsa, K. Nienhausc, M. J. Fieldb, J. Wiedenmannd, S. McSweeneya, G. U. Nienhausc, and D. Bourgeoisa, "Structural characterization of IrisFP, an optical highlighter undergoing multiple photo-induced transformations," *Proc. Natl. Acad. Sci. USA*, **105**, 18343–18348 (2008).
31. J. Fuchs, S. Böhme, F. Oswald, P. N. Hedde, M. Krause, J. Wiedenmann, and G. U. Nienhaus, "A photoactivatable marker protein for pulse-chase imaging with superresolution," *Nat. Methods*, **7**, 627–630 (2010).
32. T. Matsuda, A. Miyawaki, and T. Nagai, "Direct measurement of protein dynamics inside cells using a rationally designed photoconvertible protein," *Nat. Methods*, **5**, 339–345 (2008).
33. O. M. Subach, G. H. Patterson, L.-M. Ting, Y. Wang, J. S. Condeelis, and V. V. Verkhusha, "A photoswitchable orange-to-far-red fluorescent protein, PSmOrange," *Nat. Methods*, **8**, 771–777 (2011).
34. J. Fölling, V. Belov, R. Kunetsky, R. Medda, A. Schönle, A. Egner, C. Eggeling, M. Bossi, and S. W. Hell, "Photochromic rhodamines provide nanoscopy with optical sectioning," *Angew. Chem. Int. Ed.*, **46**, 6266–6270 (2007).
35. J. Fölling, Vladimir B., D. Riedel, A. Schönle, A. Egner, C. Eggeling, M. Bossi, and S. W. Hell, "Fluorescence nanoscopy with optical sectioning by two-photon induced molecular switching using continuous-wave lasers," *ChemPhysChem*, **9**, 321–326 (2008).
36. M. Heilemann, E. Margeat, R. Kasper, M. Sauer, and P. Tinnefeld, "Carbocyanine dyes as efficient reversible single-molecule optical switch," *J. Am. Chem. Soc.*, **127**, 3801–3806 (2005).
37. M. Heilemann, S. van de Linde, M. Schuttpelz, R. Kasper, B. Seefeldt, A. Mukherjee, P. Tinnefeld, and M. Sauer, "Subdiffraction-resolution fluorescence imaging with conventional fluorescent probes," *Angew. Chem. Int. Ed. Engl.*, **47**, 6172–6176 (2008).
38. M. Bates, B. Huang G. T. Dempsey, and X. Zhuang, "Multicolor super-resolution imaging with photo-switchable fluorescent probes," *Science*, **317**, 1749–1753 (2007).
39. M. Bates, T. R. Blosser, and X. Zhuang, "Short-range spectroscopic ruler based on a single-molecule optical switch," *Phys. Rev. Lett.*, **94**, 108101/1–4 (2005).
40. M. Heilemann, S. van de Linde, A. Mukherjee, and M. Sauer, "Super-resolution imaging with small organic fluorophores," *Angew. Chem. Int. Ed. Engl.*, **48**, 6903–6908 (2009).

41. S. van de Linde, U. Endesfelder, A. Mukherjee, M. Schuttpelz, G. Wiebusch, S. Wolter, M. Heilemann, and M. Sauer, "Multicolor photoswitching microscopy for subdiffraction-resolution fluorescence imaging," *Photochem. Photobiol. Sci.*, **8**, 465–469 (2009).

42. T. Klein, A. Loschberger, S. Proppert, S. Wolter, S. van de Linde, and M. Sauer, "Live-cell dSTORM with SNAP-tag fusion proteins," *Nat. Methods*, **8**, 7–9 (2011).

43. M. Irie, T. Fukaminato, T. Sasaki, N. Tamai, and T. Kawai, "A digital fluorecent molecular photoswitch," *Nature*, **420**, 759–760 (2002).

44. L. Giordano, T. M. Jovin, M. Irie, and E. A. Jares-Erijman, "Diheteroarylethenes as thermally stable photoswitchable acceptors in photochromic fluorescence resonance energy transfer (pcFRET)," *J. Am. Chem. Soc.*, **124**, 7481–7489 (2002).

45. T. Fukaminato, T. Sasaki, T. Kawai, N. Tamai, and M. Irie, "Digital photoswitching of fluorescence based on the photochromism of diarylethene derivatives at a single-molecule Level," *J. Am. Chem. Soc.*, **126**, 14843–14849 (2004).

46. M. Bossi, V. Belov, S. Polyakova, and S.W. Hell, "Reversible red fluorescent molecular switches," *Angew. Chem. Int. Ed. Engl.*, **45**, 7462–7465 (2006).

47. M. Q. Zhu, L. Zhu, J. J. Han, W. Wu, J.K. Hurst, and A. D. Li, "Spiropyran-based photochromic polymer nanoparticles with optically switchable luminescence," *J. Am. Chem. Soc.*, **128**, 4303–4309 (2006).

48. T. Fukaminato, T. Umemoto, Y. Iwata, S. Yokojima, M. Yoneyama, S. Nakamura, and M. Irie, "Photochromism of diarylethene single molecules in polymer matrices," *J. Am. Chem. Soc.*, **129**, 5932–5938 (2007).

49. S. F. Yan, V. N. Belov, M. L. Bossi, and S. W. Hell, "Switchable fluorescent and solvatochromic molecular probes based on 4-amino-*N*-methylphthalimide and a photochromic diarylethene," *Eur. J. Org. Chem.*, **15**, 2503–2508 (2008).

50. L. Y. Zhu, M. Q. Zhu, J. K. Hurst, and A. D. Q. Li, "Light-controlled molecular switches modulate nanocrystal fluorescence," *J. Am. Chem. Soc.*, **127**, 8968–8970 (2005).

51. I. L. Medintz, S. A. Trammell, H. Mattoussi, and J. M. Mauro, "Reversible modulation of quantum dot photoluminescence using a protein-bound photochromic fluorescence resonance energy transfer acceptor," *J. Am. Chem. Soc.*, **126**, 30–31 (2004).

52. E. A. Jares-Erijman, L. Giordano, C. Spagnuolo, K. Lidke, and T. M. Jovin, "Imaging quantum dots switched on and off by photochromic fluorescence resonance energy transfer (pcFRET)," *Mol. Cryst. Liq. Cryst.*, *430*, 257–265 (2005).

53. L. Zhu, W. Wu, M.Q. Zhu, J. J. Han, J. K. Hurst, and A. D. Li, "Reversibly photoswitchable dual-color fluorescent nanoparticles as new tools for live-cell imaging," *J. Am. Chem. Soc.*, **129**, 3524–3526 (2007).

54. S. E. Irvine, T. Staudt, E. Rittweger, J. Engelhardt, and S. W. Hell, "Direct light-driven modulation of luminescence from Mn-doped ZnSe quantum dots," *Angew. Chem. Int. Ed. Engl.*, **47**, 2685–2688 (2008).

55. M. Andresen, A. C. Stiel, J. Fölling, D. Wenzel, A. Schönle, A. Egner, C. Eggeling, S. W. Hell, and S. Jakobs, "Photoswitchable fluorescent proteins enable monochromatic multilabel imaging and dual color fluorescence nanoscopy," *Nat. Biotechnol.*, **26**, 1035–1040 (2008).

56. K. A. Lukyanov, A. F. Fradkov, N. G. Gurskaya, M. V. Matz, Y. A. Labas, A. P. Savitsky, M. L. Markelov, A. G. Zaraisky, X. Zhao, Y. Fang, W. Tan, and S. A. Lukyanov, "Natural animal coloration can be determined by a nonfluorescent green fluorescent protein homolog," *J. Biol. Chem.*, **275**, 25879–25882 (2000).

57. D. M. Chudakov, V. V. Belousov, A. G. Zaraisky, V. V. Novoselov, D. B. Staroverov, D. B. Zorov, S. Lukyanov, and K. A. Lukyanov, "Kindling fluorescent proteins for precise in vivo photolabeling," *Nat. Biotechnol.*, **21**, 191–194 (2003).
58. A. C. Stiela, M. Andresena, H. Bocka, M. Hilberta, J. Schildea, A. Schönlea, C. Eggelinga, A. Egnera, S. W. Hella, and S. Jakobs, "Generation of monomeric reversibly switchable red fluorescent proteins for far-field fluorescence nanoscopy," *Biophys. J.*, **95**, 2989–2997 (2008).
59. F. V. Subach, L. Zhang, T. W. J. Gadella, N. G. Gurskaya, K. A. Lukyanov, and V. V. Verkhusha, "Red fluorescent protein with reversibly photoswitchable absorbance for photochromic FRET," *Chem. Biol.*, **17**, 745–755 (2010).
60. J. N. Henderson, H. Ai, R. E. Campbell, and S. J. Remington, "Structural basis for reversible photobleaching of a green fluorescent protein homologue," *Proc. Natl. Acad. Sci. USA*, **104**, 6672–6677 (2007).
61. R. Ando, H. Mizuno, and A. Miyawaki, "Regulated fast nucleocytoplasmic shuttling observed by reversible protein highlighting," *Science*, **306**, 1370–1373 (2004).
62. R. Ando, C. Flors, H. Mizuno, J. Hofkens, and A. Miyawaki, "Highlighted generation of fluorescence signals using simultaneous two-color irradiation on Dronpa mutants," *Biophys. J.*, **92**, L97–L99 (2007).
63. A. C. Stiel, S. Trowitzsch, G. Weber, M. Andresen, C. Eggeling, S. W. Hell, S. Jakobs, and M. C. Wahl, "1.8 A bright-state structure of the reversibly switchable fluorescent protein Dronpa guides the generation of fast switching variants," *Biochem. J.*, **402**, 35–42 (2007).
64. T. Grotjohann, I. Testa, M. Leutenegger, H. Bock, N. T. Urban, F. Lavoie-Cardinal, K. I. Willig, C. Eggeling, S. Jakobs, and S. W. Hell, "Diffraction-unlimited all-optical imaging and writing with a photochromic GFP," *Nature*, **478**, 204–208 (2011).
65. T. Brakemann, A. C. Stiel, G. Weber, M. Andresen, I. Testa, T. Grotjohann, M. Leutenegger, U. Plessmann, H. Urlaub, C. Eggeling, M. C. Wahl, S. W. Hell, and S. Jakobs, "A reversibly photoswitchable GFP-like protein with fluorescence excitation decoupled from switching," *Nat. Biotechnol.*, **29**, 942–947 (2011).
66. R. E. Benesch and R. Benesch, "Enzymatic removal of oxygen for polarography and related methods," *Science*, **118**, 447–448 (1953).
67. I. Rasnik, S. A. McKinney, and T. Ha, "Nonblinking and long-lasting single-molecule fluorescence imaging," *Nat. Methods*, **3**, 891–893 (2006).
68. E. Madej and P. Wardman, "The oxidizing power of the glutathione thiyl radical as measured by its electrode potential at physiological pH," *Arch. Biochem. Biophys.*, **462**, 94–102 (2007).
69. P. Wardmann, "Reduction potentials of one-electron couples involving free radicals in aqueous solution," *J. Phys. Chem.*, **18**, 1637–1755 (1989).
70. H. Görner, "Oxygen uptake induced by electron transfer from donors to the triplet state of methylene blue and xanthene dyes in air-saturated aqueous solution," *Photochem. Photochem. Photobiol. Sci.*, **7**, 371–376 (2008).
71. J. Vogelsang, R. Kasper, C. Steinhauer, B. Person, M. Heilemann, M. Sauer, and P. Tinnefeld, "A reducing and oxidizing system minimizes photobleaching and blinking of fluorescent dyes," *Angew. Chem. Int. Ed.*, **47**, 5465–5469 (2008).
72. S. van de Linde, I. Krstic, T. Prisner, S. Doose, M. Heilemann, and M. Sauer, "Photoinduced formation of reversible dye radicals and their impact on super-resolution imaging," *Photochem. Photobiol. Sci.*, **10**, 499–506 (2011).

73. M. Irie, "Diarylethenes for memories and switches," *Chem. Rev.*, **100**, 1685–1716 (2000).
74. V. I. Minkin, "Photo-, thermo-, solvato-, and electrochromic spiroheterocyclic compounds," *Chem. Rev.*, (2004) **104**, 2751–2776 (2004).
75. S. Mao, R. K. Benninger, Y. Yan, C. Petchprayoon, D. Jackson, C. J. Easley, D. W. Piston, and G. Marriott, "Optical lock-in detection of FRET using synthetic and genetically encoded optical switches," *Biophys. J.*, **94**, 4515–4524 (2008).
76. L. Giordano, T. M. Jovin, M. Irie, and E. A. Jares-Erijman, "Diheteroarylethenes as thermally stable photoswitchable acceptors in photochromic fluorescence resonance energy transfer (pcFRET)," *J. Am. Chem. Soc.*, **124**, 7481–7489 (2002).
77. H. Mizuno, T. K. Mal, K. I. Tong, R. Ando, T. Furuta, M. Ikura, and A. Miyawaki, "Photo-induced peptide cleavage in the green-to-red conversion of a fluorescent protein," *Mol. Cell*, **12**, 1051–1058 (2003).
78. J. J. Van Thor, T. Gensch, K. J. Hellingwerf, and L. N. Johnson, "Phototransformation of green fluorescent protein with UV and visible light leads to decarboxylation of glutamate 222," *Nat. Struct. Biol.*, **9**, 37–41 (2002).
79. D. M. Chudakov, A. V. Feofanov, N. N. Mudrik, S. Lukyanov, and K. A. Lukyanov, "Chromophore environment provides clues to kindling fluorescent protein riddle," *J. Biol. Chem.*, **278**, 7215–7219 (2003).
80. S. Habuchi, R. Ando, P. Dedecker, W. Verheijen, H. Mizuno, A. Miyawaki, and J. Hofkens, "Reversible single-molecule photoswitching in the GFP-like fluorescent protein Dronpa," *Proc. Natl. Acad. Sci. USA*, **102**, 9511–9516 (2005).
81. M. Andresen, M. C. Wahl, A. C. Stiel, F. Grater, L. V. Schafer, S. Trowitzsch, G. Weber, C. Eggeling, H. Grubmuller, S. W. Hell, and S. Jakobs, "Structure and mechanism of the reversible photoswitch of a fluorescent protein," *Proc. Natl. Acad. Sci. USA*, **102**, 13070–13074 (2005).
82. M. Andresen, A. C. Stiel, S. Trowitzsch, G. Weber, C. Eggeling, M. C. Wahl, S. W. Hell, and S. Jakobs, "Structural basis for reversible photoswitching in Dronpa," *Proc. Natl. Acad. Sci. USA*, **104**, 13005–13009 (2007).
83. F. V. Subach, G. H. Patterson, M. Renz, J. Lippincott-Schwartz, and V. V. Verkhusha, "Bright monomeric photoactivatable red fluorescent protein for two-color super-resolution sptPALM of live cells," *J. Am. Chem. Soc.*, **132**, 6481–6491 (2010).
84. T. Sato, M. Takahoko, and H. Okamoto, "HuC:Kaede, a useful tool to label neural morphologies in networks in vivo," *Genesis*, **44**, 136–142 (2006).
85. M. Tomura, N. Yoshida, J. Tanaka, S. Karasawa, Y. Miwa, A. Miyawaki, and O. Kanagawa, "Monitoring cellular movement in vivo with photoconvertible fluorescence protein "Kaede" transgenic mice," *Proc. Natl. Acad. Sci. USA*, **105**, 10871–10876 (2008).
86. S. Baltrusch, and S. Lenzen, "Monitoring of glucose-regulated single insulin secretory granule movement by selective photoactivation," *Diabetologia*, **51**, 989–996 (2008).
87. Z. Cvackova, M. Masata, D. Stanek, H. Fidlerova, and I. Raska, "Chromatin position in human HepG2 cells: although being non-random, significantly changed in daughter cells," *J. Struct. Biol.*, **165**, 107–117 (2009).
88. M. M. Falk, S. M. Baker, A. M. Gumpert, D. Segretain, and R. W. Buckheit III, " Gap junction turnover is achieved by the internalization of small endocytic double-membrane vesicles," *Mol. Biol. Cell*, **20**, 3342–3352 (2009).
89. K. Tachibana, M. Hara, Y. Hattori, and T. Kishimoto, "Cyclin B-cdk1 controls pronuclear union in interphase," *Curr. Biol.*, **18**, 1308–1313 (2008).

90. L. Zhang, N. G. Gurskaya, E. M. Merzlyak, D. B. Staroverov, N. N. Mudrik, O. N. Samarkina, L. M. Vinokurov, S. Lukyanov, and K. A. Lukyanov, "Method for real-time monitoring of protein degradation at the single cell level," *Biotechniques*, **42**, 446–450 (2007).

91. M. Karbowski, D. Arnoult, H. Chen, D. C. Chan, C. L. Smith, and R. J. Youle, "Quantitation of mitochondrial dynamics by photolabeling of individual organelles shows that mitochondrial fusion is blocked during the Bax activation phase of apoptosis," *J. Cell Biol.*, **164**, 493–499 (2004).

92. S. Arimura, J. Yamamoto, G. P. Aida, M. Nakazono, and N. Tsutsumi, "Frequent fusion and fission of plant mitochondria with unequal nucleoid distribution," *Proc. Natl. Acad. Sci. USA*, **101**, 7805–7808 (2004).

93. A. J. Molina and O. S. Shirihai, "Monitoring mitochondrial dynamics with photoactivatable green fluorescent protein," *Methods Enzymol*, **457**, 289–304 (2009).

94. K. A. Lukyanov, D. M. Chudakov, S. Lukyanov, and V. V. Verkhusha, "Innovation: photoactivatable fluorescent proteins," *Nat. Rev. Mol. Cell Biol.*, **6**, 885–891 (2005).

95. T. Sato, M. Takahoko, and H. Okamoto, "HuC:Kaede, a useful tool to label neural morphologies in networks in vivo," *Genesis*, **44**, 136–142 (2006).

96. D. A. Stark, and P. M. Kulesa, "An in vivo comparison of photoactivatable fluorescent proteins in an avian embryo model," *Dev. Dyn.*, 236, 1583–1594 (2007).

97. D. M. Chudakov, T. V. Chepurnykh, V. V. Belousov, S. Lukyanov, and K. A. Lukyanov, "Fast and precise protein tracking using repeated reversible photoactivation," *Traffic*, **7**, 1304–1310 (2006).

98. J. Lippincott-Schwartz, N. Altan-Bonnet, and G. H. Patterson, "Photobleaching and photoactivation: following protein dynamics in living cells," *Nat. Cell. Biol.*, **Suppl.**, S7–S14, (2003).

99. J. N. Post, K. A. Lidke, B. Rieger, and D. J. Arndt-Jovin, "One- and two-photon photoactivation of a paGFP-fusion protein in live *Drosophila* embryos," *FEBS Lett.*, **579**, 325–330 (2005).

100. M. Schneider, S. Barozzi, I. Testa, M. Faretta, and A. Diaspro, "Two-photon activation and excitation properties of PA-GFP in the 720–920-nm region," *Biophys. J.*, **89**, 1346–1352 (2005).

101. N. W. Gray, R. M. Weimer, I. Bureau, and K. Svoboda, "Rapid redistribution of synaptic PSD-95 in the neocortex in vivo", *PLoS Biol.*, **4**, e370 (2006).

102. S. R. Datta, M. L. Vasconcelos, V. Ruta, S. Luo, A. Wong, E. Demir, J. Flores, K. Balonze, B. J. Dickson, and R. Axel, "The *Drosophila* pheromone cVA activates a sexually dimorphic neural circuit," *Nature*, **452**, 473–477 (2008).

103. Y. Chen, P. J. Macdonald, J. P. Skinner, G. H. Patterson, and J. D. Müller, "Probing nucleocytoplasmic transport by two-photon activation of PA-GFP," *Microsc. Res. Tech.*, **69**, 220–226 (2006).

104. S. Ivanchenko, C. Röcker, F. Oswald, J. Wiedenmann, and U. Nienhaus, "Targeted green-red photoconversion of EosFP, a fluorescent marker protein," *J. Biol. Phys.*, **31**, 249–259 (2005).

105. H. Tsutsui, S. Karasawa, H. Shimizu, N. Nukina, and A. Miyawaki, "Semi-rational engineering of a coral fluorescent protein into an efficient highlighter," *EMBO Rep.*, **6**, 233–238 (2005).

106. W. Watanabe, T. Shimada, S. Matsunaga, D. Kurihara, K. Fukui, S. Arimura, N. Tsutsumi, K. Isobe, and K. Itoh, "Single-organelle tracking by two-photon conversion," *Opt. Express*, **15**, 2490–2498 (2007).
107. S. Ivanchenko, S. Glaschick, C. Röcker, F. Oswald, J. Wiedenmann, and G. U. Nienhaus, "Two-photon excitation and photoconversion of EosFP in dual-color 4Pi confocal microscopy," *Biophys. J.*, **92**, 4451–4457 (2007).
108. K. Isobe, H. Hashimoto, A. Suda, F. Kannari, H. Kawano, H. Mizuno, A. Miyawaki, and K. Midorikawa, "Measurement of two-photon excitation spectrum used to photoconvert a fluorescent protein (Kaede) by nonlinear Fourier-transform spectroscopy," *Biomed. Opt. Express*, **1**, 687–693 (2010).
109. H. Hashimoto, K. Isobe, A. Suda, F. Kannari, H. Kawano, H. Mizuno, A. Miyawaki, and K. Midorikawa, "Measurement of two-photon excitation spectra of fluorescent proteins with nonlinear Fourier-transform spectroscopy," *Appl. Opt.*, **49**, 3323–3329 (2010).
110. R. Y. Tsien, "The green fluorescent protein," *Annu. Rev. Biochem.*, **67**, 509–544 (1998).
111. M. Matsuzaki, G. C. R. Ellis-Davies, T. Nemoto, Y, Miyashita, M. Iino, and H. Kasai, "Dendritic spine geometry is critical for AMPA receptor expression in hippocampal CA1 pyramidal neurons," *Nat. Neurosci.*, **4**, 1086–1092 (2001).
112. M. A. Smith, G. C. R. Ellis-Davies, and J. C. Magee, "Mechanism of the distance-dependent scaling of Schaffer collateral synapses in rat CA1 pyramidal neurons," *J. Physiol.*, **548**, 245–258 (2003).
113. A. G. Carter and B. L. Sabatini, "State-dependent calcium signaling in dendritic spines of striatal medium spiny neurons," *Neuron*, **44**, 483–493 (2004).
114. A. Sobczyk, V. Scheuss, and K. Svoboda, "NMDA receptor subunit-dependent [Ca^{2+}] signaling in individual hippocampal dendritic spines," *J. Neurosci.*, **25**, 6037–6046 (2005).
115. A. Losonczy and J. C. Magee. "Integrative properties of radial oblique dendrites in hippocampal CA1 pyramidal neurons," *Neuron*, **50**, 291–307 (2006).
116. G. C. R. Ellis-Davies, M. Matsuzaki, M. Paukert, H. Kasai, and D. E. Bergles, "4-Carboxymethoxy-5,7-dinitroindolinyl-glu: an improved caged glutamate for expeditious ultraviolet and two-photon photolysis in brain slices," *J. Neurosci.*, **27**, 6601–6604 (2007).
117. M. Matsuzaki, G. C. R. Ellis-Davies, and H. Kasai, "Three-dimensional mapping of unitary synaptic connections by two-photon macro photolysis of caged glutamate," *J. Neurophysiol.*, **99**, 1535–1544 (2008).
118. M. Matsuzaki, T. Hayama, H. Kasai, and G. C. R. Ellis-Davies, "Two-photon uncaging of γ-aminobutyric acid in intact brain tissue," *Nat. Chem. Biol.*, **6**, 255–257 (2010).
119. S. Kantevari, M. Matsuzaki, Y. Kanemoto, H. Kasai, and G. C. R. Ellis-Davies, "Two-color, two-photon uncaging of glutamate and GABA," *Nat. Methods*, **7**, 123–125 (2010).
120. E. M. Callaway and R. Yuste, "Stimulating neurons with light," *Curr. Opin. Neurobiol.*, **12**, 587–592 (2002).
121. K. Isobe, W. Watanabe, S. Matsunaga, T. Higashi, K. Fukui, and K. Itoh, "Multi-spectral two-photon excited fluorescence microscopy using supercontinuum light source," *Jpn. J. Appl. Phys.*, **44**, L167–L169 (2005).

122. H. Kano and H. Hamaguchi, "Vibrationally resonant imaging of a single living cell by supercontinuum-based multiplex coherent anti-Stokes Raman scattering microspectroscopy," *Opt. Express*, **13**, 1322–1327 (2005).
123. Y. Silberberg, "Quantum coherent control for nonlinear spectroscopy and microscopy," *Annu. Rev. Phys. Chem.*, **60**, 277–292 (2009).
124. D. Meshulach and Y. Silberberg, "Coherent quantum control of two-photon transitions by a femtosecond laser pulse," *Nature*, **396**, 239–242 (1998).
125. N. Dudovich, D. Oron, and Y. Silberberg, "Single-pulse coherently controlled nonlinear Raman spectroscopy and microscopy," *Nature*, **418**, 512–514 (2002).
126. K. Isobe, A. Suda, M. Tanaka, H. Hashimoto, F. Kannari, H. Kawano, H. Mizuno, A. Miyawaki, and K. Midorikawa, "Nonlinear optical microscopy and spectroscopy employing octave spanning pulses," *IEEE J. Sel. Top. Quant. Electron.*, **16**, 767–780 (2010).
127. J. P. Ogilvie, D. Debarre, X. Solinas, J.-L. Martin, E. Beaurepaire, and M. Joffre, "Use of coherent control for selective two-photon fluorescence microscopy in live organisms," *Opt. Express*, **14**, 759–766 (2006).
128. J. M. Dela Cruza, V. V. Lozovoya, and M. Dantus, "Coherent control improves biomedical imaging with ultrashort shaped pulses," **180**, 307–313 (2006).
129. L. Schelhas, J. C. Shane, and M. Dantus, "Advantages of ultrashort phase-shaped pulses for selective two-photon activation and biomedical imaging," *Nanomedicine: Nanotechnology, Biology, and Medicine*, **2**, 177–181 (2006).
130. K. Isobe, A. Suda, M. Tanaka, F. Kannari, H. Kawano, H. Mizuno, A. Miyawaki, and K. Midorikawa, "Multifarious control of two-photon excitation of multiple fluorophores achieved by phase modulation of ultra-broadband laser pulses," *Opt. Express*, **17**, 13737–13746 (2009).
131. G. Labroille, R. S. Pillai, X. Solinas, C. Boudoux, N. Olivier, E. Beaurepaire, and M. Joffre, "Dispersion-based pulse shaping for multiplexed two-photon fluorescence microscopy," Opt. Lett. Vol. 35, pp. 3444–3446 (2010).
132. S.-H. Lim, A. G. Caster, O. Nicolet, and S. R. Leone, "Chemical imaging by single pulse interferometric coherent anti-Stokes Raman scattering microscopy," *J. Chem. Phys. B*, **110**, 5196–5204 (2006).
133. K. Isobe, A. Suda, M. Tanaka, H. Hashimoto, F. Kannari, H. Kawano, H. Mizuno, A. Miyawaki, and K. Midorikawa, "Single-pulse coherent anti-Stokes Raman scattering microscopy employing an octave spanning pulse," *Opt. Express*, **17**, 11259–11266 (2009).
134. C. Müller, T. Buckup, B. Vacano, and M. Motzkus, "Heterodyne single-beam CARS microscopy," *J. Raman Spectrosc.*, **40**, 809–816 (2009).
135. W. Denk, J. H. Strickler, and W. W. Webb, "Two-photon laser scanning fluorescence microscopy," *Science*, **248**, 73–76 (1990).
136. C. Xu, W. Zipfel, J. B. Shear, R. M. Williams, and W. W. Webb, "Multiphoton fluorescence excitation: new spectral windows for biological nonlinear microscopy," *Proc. Natl. Acad. Sci. USA.*, **93**, 10763–10768 (1996).
137. G. Y. Fan, H. Fujisaki, A. Miyawaki, R. K. Tsay, R. Y. Tsien, and M. H. Ellisman, "Video-rate scanning two-photon excitation fluorescence microscopy and ratio imaging with Cameleons," *Biophys. J.*, **76**, 2412–2420 (1999).
138. P. Allcock and D. L. Andrews, "Two-photon fluorescence: Resonance energy transfer," *J. Chem. Phys.*, **108**, 3089–3095 (1998).

139. K. G. Heinze, A. Koltermann, and P. Schwille, "Simultaneous two-photon excitation of distinct labels for dual-color fluorescence crosscorrelation analysis," *Proc. Natl. Acad. Sci. USA*, **97**, 10377–10382 (2000).
140. K. A. Walowicz, I. Pastirk, V. V. Lozovoy, and M. Dantus, "Multiphoton intrapulse interference. 1. Control of multiphoton processes in condensed phases," *J. Phys. Chem. A*, **106**, 9369–9373 (2002).
141. V. V. Lozovoy, I. Pastirk, K. A. Walowicz, and M. Dantus, "Multiphoton intrapulse interference. II. Control of two- and three-photon laser induced fluorescence with shaped pulses," *J. Chem. Phys.*, **118**, 3187–3196 (2003).
142. J. M. Dela Cruz, I. Pastirk, M. Comstock, and M. Dantus, "Coherent control through scattering tissue, Multiphoton Intrapulse Interference 8," *Opt. Express*, **12**, 4144–4149 (2004).
143. M. Comstock, V. V. Lozovoy, I. Pastirk, and M. Dantus, "Multiphoton intrapulse interference 6; binary phase shaping," *Opt. Express*, **12**, 1061–1066 (2004).
144. H. Kawano, Y. Nabekawa, A. Suda, Y. Oishi, H. Mizuno, A. Miyawaki, and K. Midorikawa, "Attenuation of photobleaching in two-photon excitation fluorescence from green fluorescent protein with shaped excitation pulses," *Biochem. Biophys. Res. Commun.*, **311**, 592–596 (2003).
145. A. M. Weiner, D. E. Leaird, J. S. Patel, and J. R. Wullert, "Programmable femtosecond pulse shaping by use of a multielement liquid-crystal phase modulator," *Opt. Lett.*, **15**, 326–328 (1990).
146. E. Zeek, K. Maginnis, S. Backus, U. Russek, M. Murnane, G. Mourou, H. Kapteyn, and G. Vdovin, "Pulse compression by use of deformable mirrors," *Opt. Lett.*, **24**, 493–495 (1999).
147. C. W. Hillegas, J. X. Tull, D. Goswami, D. Strickland, and W. S. Warren, "Femtosecond laser pulse shaping by use of microsecond radio-frequency pulses," *Opt. Lett.*, **19**, 737–739 (1994).
148. E. Frumker, E. Tal, Y. Silberberg, and D. Majer, "Femtosecond pulse-shape modulation at nanosecond rates," *Opt. Lett.*, **30**, 2796–2798 (2005).
149. T. Nagai, S. Yamada, T. Tominaga, M. Ichikawa, and A. Miyawaki, "Expanded dynamic range of fluorescent indicators for Ca^{2+} by circularly permuted yellow fluorescent proteins," *Proc. Natl. Acad. Sci. USA*, **101**, 10554–10559 (2004).
150. M. D. Duncan, J. Reintjes, and T. J. Manuccia, "Scanning coherent anti-Stokes Raman microscope," *Opt. Lett.*, **7**, 350–352 (1982).
151. A. Zumbusch, G. R. Holtom, and X. S. Xie, "Three-dimensional vibrational imaging by coherent anti-Stokes Raman scattering," *Phys. Rev. Lett.*, **82**, 4142–4145 (1999).
152. H. Kano and H. Hamaguchi, "Ultrabroadband (>2500 cm^{-1}) multiplex coherent anti-Stokes Raman scattering microspectroscopy using a supercontinuum generated from a photonic crystal fiber," *Appl. Phys. Lett.*, **86**, 121113/1–3 (2005).
153. M. Okuno, H. Kano, P. Leproux, V. Couderc, and H. Hamaguchi, "Ultrabroadband multiplex CARS microspectroscopy and imaging using a subnanosecond supercontinuum light source in the deep near infrared," *Opt. Lett.*, **33**, 923–925 (2008).
154. T. W. Kee and M. T. Cicerone, "Simple approach to one-laser, broadband coherent anti-Stokes Raman scattering microscopy," *Opt. Lett.*, **29**, 2701–2703 (2004).
155. J. P. Ogilvie, E. Beaurepaire, A. Alexandrou, and M. Joffre, "Fourier-transform coherent anti-Stokes Raman scattering microscopy," *Opt. Lett.*, **31**, 480–482 (2006).

156. K. Isobe, A. Suda, M. Tanaka, H. Hashimoto, F. Kannari, H. Kawano, H. Mizuno, A. Miyawaki, and K. Midorikawa, "Single-pulse coherent anti-Stokes Raman scattering microscopy employing an octave spanning pulse," *Opt. Express*, **17**, 11259–11266 (2009).

157. B. von Vacano and M. Motzkus, "Molecular discrimination of a mixture with single-beam Raman control," *J. Chem. Phys.*, **127**, 144514/1–4 (2007).

158. D. Oron, N. Dudovich, and Y. Silberberg, "Single-pulse phase-contrast nonlinear Raman spectroscopy," *Phys. Rev. Lett.*, **89**, 273001/1–4 (2002).

159. N. Dudovich, D. Oron, and Y. Silberberg, "Single-pulse coherent anti-Stokes Raman spectroscopy in the fingerprint spectral region," *J. Chem. Phys.*, **118**, 9208–9215 (2003).

160. B. von Vacano, T. Buckup, and M. Motzkus, "Highly sensitive single-beam heterodyne coherent anti-Stokes Raman scattering," *Opt. Lett.*, **31**, 2495–2497 (2006).

161. S.-H. Lim, A. G. Caster, and S. R. Leone, "Single-pulse phase-control interferometric coherent anti-Stokes Raman scattering spectroscopy," *Phys. Rev. A*, **72**, 041803/1–4 (2005).

162. S.-H. Lim, A. G. Caster, and S. R. Leone, "Fourier transform spectral interferometric coherent anti-Stokes Raman scattering (FTSI-CARS) spectroscopy," *Opt. Lett.*, **32**,1332–1334 (2007).

163. T. Hellerer, A. M. K. Enejder, and A. Zumbuscha, "Spectral focusing: high spectral resolution spectroscopy with broad-bandwidth laser pulse," *Appl. Phys. Lett.*, **85**, 25–27 (2004).

164. D. Oron, N. Dudovich, D. Yelin, and Y. Silberberg, "Narrow-band coherent anti-stokes raman signals from broadband pulses," *Phys. Rev. Lett.*, **88**, 063004/1–4 (2002).

165. I. Johnson, "Fluorescent probes for living cells," *Histochem. J.*, **30**, 123–140 (1998).

166. R. Y. Tsien, "The green fluorescent protein," *Annu. Rev. Biochem.*, **67**, 509–544 (1998).

167. A. Miyawaki, "Innovations in the imaging of brain functions using fluorescent proteins," *Neuron*, **48**, 189–199 (2005).

168. B. N. G. Giepmans, S. R. Adams, M. H. Ellisman, and R. Y. Tsien, "The fluorescent toolbox for assessing protein location and function," *Science*, **312**, 217–224 (2006).

169. E. H. Synge, "A suggested method for extending microscopic resolution into the ultramicroscopic region," *Phil. Mag.*, **6**, 356–362 (1928).

170. D. W. Pohl, W. Denk, and M. Lanz, "Optical stethoscopy: Image recording with resolution $\lambda/20$," *Appl. Phys. Lett.*, **44**, 651–653 (1984).

171. A. Lewis, M. Isaacson, A. Harootunian, and A. Muray, "Development of a 500 Åspatial resolution light microscope. I. Light is efficiently transmitted through $\lambda/16$ diameter apertures," *Ultramicroscopy*, **13**, 227–231 (1984).

172. S. W. Hell and J. Wichmann, "Breaking the diffraction resolution limit by stimulated emission: stimulated emission depletion microscopy," *Opt. Lett.*, **19**, 780–782 (1994).

173. T. A. Klar and S. W. Hell, "Subdiffraction resolution in far-field fluorescence microscopy," *Opt. Lett.*, **24**, 954–956 (1999).

174. S. W. Hell and M. Kroug, "Ground-state-depletion fluorescence microscopy: a concept for breaking the diffraction resolution limit," *Appl. Phys. B*, **60**, 495–497 (1995).

175. S. Bretschneider, C. Eggeling, and S. W. Hell, "Breaking the diffraction barrier in fluorescence microscopy by optical shelving," *Phys. Rev. Lett.*, **98**, 218103/1–4 (2007).

176. E. Rittweger, D. Wildanger, and S. W. Hell, "Far-field fluorescence nanoscopy of diamond color centers by ground state depletion," *Europhys. Lett.*, **86**, 14001/1–6 (2009).

177. S. W. Hell, "Toward fluorescence nanoscopy," *Nat. Biotechnol.*, **21**, 1347–1355 (2003).

178. M. Hofmann, C. Eggeling, S. Jakobs, and S. W. Hell, "Breaking the diffraction barrier in fluorescence microscopy at low light intensities by using reversibly photoswitchable proteins," *Proc. Natl. Acad. Sci. USA*, **102**, 17565–17569 (2005).

179. R. Heintzmann, T. M. Jovin, and C. Cremer, "Saturated patterned excitation microscopy—a concept for optical resolution improvement," *J. Opt. Soc. Am. A*, **19**, 1599–1609 (2002).

180. M. G. L. Gustafsson, "Nonlinear structured-illumination microscopy: wide-field fluorescence imaging with theoretically unlimited resolution," *Proc. Natl. Acad. Sci. USA*, **102**, 13081–13086 (2005).

181. K. Fujita, M. Kobayashi, S. Kawano, M. Yamanaka, and S. Kawata, "High-resolution confocal microscopy by saturated excitation of fluorescence," *Phys. Rev. Lett.*, **99**, 228105/1–5 (2007).

182. E. Betzig, G. H. Patterson, R. Sougrat, O. W. Lindwasser, S. Olenych, J. S. Bonifacino, M. W. Davidson, J. Lippincott-Schwartz, and H. F. Hess, "Imaging intracellular fluorescent proteins at nanometer resolution," *Science*, **313**, 1642–1645 (2006).

183. S. T. Hess, T. P. K. Girirajan, and M. D. Mason, "Ultra-high resolution imaging by fluorescence photoactivation localization microscopy," *Biophys. J.*, **91**, 4258–4272 (2006).

184. M. J. Rust, M. Bates, and X. Zhauang, "Sub-diffraction-limit imaging by stochastic optical reconstruction microscopy (STORM)," *Nat. Methods*, **3**, 793–796 (2006).

185. P. Kner, B. B. Chhun, E. R. Griffis, L. Winoto, and M. G. L. Gustafsson, "Super-resolution video microscopy of live cells by structured illumination," *Nat. Methods*, **6**, 339–342 (2009).

186. K. I. Willig, B. Harke, R. Medda, and S. W. Hell, "STED microscopy with continuous wave beams," *Nat. Methods*, **4**, 915–918 (2007).

187. H. Shroff1, C. G. Galbraith J. A. Galbraith, and E. Betzig, "Live-cell photoactivated localization microscopy of nanoscale adhesion dynamics," *Nat. Methods*, **5**, 417–423 (2008).

188. B. Huang, S. A. Jones, B. Brandenburg, and X. Zhuang, "Whole-cell 3D STORM reveals interactions between cellular structures with nanometer-scale resolution," *Nat. Methods*, **5**, 1047–1052 (2008).

189. B. Harke, J. Keller, C. K. Ullal, V. Westphal, A. Schönle, and S. W. Hell, "Resolution scaling in STED microscopy," *Opt. Express*, **16**, 4154–4162 (2008).

190. E. Rittweger, K. Y. Han, S. E. Irvine, C. Eggeling, and S. W. Hell, "STED microscopy reveals crystal colour centres with nanometric resolution," *Nat. Photonics*, **3**, 144–147 (2009).

191. A. Gruber, A. Dräbenstedt, C. Tietz, L. Fleury, J. Wrachtrup, and C. von Borczyskowski, "Scanning confocal optical microscopy and magnetic resonance on single defect centers," *Science*, **276**, 2012–2014 (1997).

192. S.-J. Yu, M.-W. Kang, H.-C. Chang, K.-M. Chen, and Y.-C. Yu, "Bright fluorescent nanodiamonds: no photobleaching and low cytotoxicity," *J. Am. Chem. Soc.*, **127**, 17604–17605 (2005).

193. B. R. Rankin, R. R. Kellner, and S. W. Hell, "Stimulated-emission-depletion microscopy with a multicolor stimulated-Raman-scattering light source," *Opt. Lett.*, **33**, 2491–2493 (2008).

194. J. Bückers, D. Wildanger, G. Vicidomini, L. Kastrup, and S. W. Hell, "Simultaneous multi-lifetime multi-color STED imaging for colocalization analyses," *Opt. Express*, **19**, 3130–3143 (2011).

195. V. Westphal, and S. W. Hell, "Nanoscale resolution in the focal plane of an optical microscope," *Phys. Rev. Lett.*, **94**, 143903/1–4 (2005).

196. K. I. Willig, B. Harke, R. Medda, and S. W. Hell, "STED microscopy with continuous wave beams," *Nat. Methods*, **4**, 915–918 (2007).

197. G. Moneron, R. Medda, B. Hein, A. Giske, V. Westphal, and S. W. Hell, "Fast STED microscopy with continuous wave fiber lasers," *Opt. Express*, **18**, 1302–1309 (2010).

198. G. Moneron and S. W. Hell, "Two-photon excitation STED microscopy," *Opt. Express*, **17**, 14567–14573 (2009).

199. Q. Li, S. S. H. Wu, and K. C. Chou, "Subdiffraction-limit two-photon fluorescence microscopy for GFP-tagged cell imaging," *Biophys. J.*, **97**, pp. 3224–3228 (2009).

200. K. I. Willig, R. R. Kellner, R. Medda, B. Hein, S. Jakobs, and S. W. Hell, "Nanoscale resolution in GFP-based microscopy," *Nat. Methods*, **3**, 721–723 (2006).

201. V. Westphal, S. O. Rizzoli, M. A. Lauterbach, D. Kamin, R. Jahn, and S. W. Hell, "Video-rate far-field optical nanoscopy dissects synaptic vesicle movement," *Science*, **320**, 246–249 (2008).

202. K. I. Willig, J. Keller, M. Bossi, and S. W. Hell, "STED microscopy resolves nanoparticle assemblies," *N. J. Phys.*, **8**, 1–6 (2006).

203. Wu, Q., F. Merchant, and K. Castleman, *Microscope Image Processing*, Academic Press, San Diego (2008).

204. L. Campbell, A. J. Hollins, A. Al-Eid, G. R. Newman, C. von Ruhland, and M. Gumbleton, "Caveolin-1 expression and caveolae biogenesis during cell transdifferentiation in lung alveolar epithelial primary cultures," *Biochem. Biophys. Res. Commun.*, **262**, 744–751 (1999).

205. K. Visscher, G. J. Brankenhoff, and T. D. Visser, "Fluorescence saturation in confocal microscopy," *J. Microsc.*, **175**, 162–165 (1994).

206. S. Kawano, N. I. Smith, M. Yamanaka, S. Kawata, and K. Fujita, "Determination of the expanded optical transfer function in saturated excitation imaging and high harmonic demodulation," *Appl. Phys. Express*, **4**, 042401 (2011).

207. C. Eggeling, A. Volkmer, and C. A. M. Seidel, "Molecular photobleaching kinetics of rhodamine 6G by one- and two-photon induced confocal fluorescence microscopy," *Chem. Phys. Chem.*, **6**, 791–804 (2005).

208. M. Yamanaka, Y.-K. Tzeng, S. Kawano, N. I. Smith, S. Kawata, H.-C. Chang, and K. Fujita, "SAX microscopy with fluorescent nanodiamond probes for high-resolution fluorescence imaging," *Biomed. Opt. Express*, **2**, 1946–1954 (2011).

209. G. Davies, ed. *Properties and Growth of Diamond*, Institute of Electrical Engineers: London, 1994.

210. G. Davies and M. F. Hamer, "Optical studies of the 1.945 eV vibronic band in diamond," *Proc. R. Soc. London, Ser. A*, **348**, 285–298 (1976).

211. M. G. L. Gustafsson, "Surpassing the lateral resolution limit by a factor of two using structured illumination microscopy," *J. Microsc.*, **198**, 82–87 (2000).

212. M. G. L. Gustafsson, L. Shao, P. M. Carlton, C. J. R. Wang, I. N. Golubovskaya, W. Cande, D. A. Agard, and J. W. Sedat, "Three-dimensional resolution doubling in wide-field fluorescence microscopy by structured illumination," *Biophys. J.*, **94**, 4957–4970 (2008).

213. L. Schermelleh, P. M. Carlton, S. Haase, L. Shao, L. Winoto, P. Kner, B. Burke, M. C. Cardoso, D. A. Agard, M. G. L. Gustafsson, H. Leonhardt, and J. W. Sedat, "Subdiffraction multicolor imaging of the nuclear periphery with 3D structured illumination microscopy," *Science*, **320**, 1332–1336 (2008).

214. E. Betzig, "Proposed method for molecular optical imaging," *Opt. Lett.*, **20**, 237–239 (1995).

215. A Egner, C. Geisler, C. von Middendorff, H. Bock, D. Wenzel, R. Medda, M. Andresen, A. C. Stiel, S. Jakobs, C. Eggeling, A. Schönle, and S. W. Hell, "Fluorescence nanoscopy in whole cells by asynchronous localization of photoswitching emitters," *Biophys. J.*, **93**, 3285–3290 (2007).

216. R. E. Thompson, D. R. Larson, and W. W. Webb, "Precise nanometer localization analysis for individual fluorescent probes," *Biophys. J.*, **82**, 2775–2783 (2002).

217. S. T. Hess, T. J. Gould, M. Gunewardene, J. Bewersdorf, and M. D. Mason, "Ultra-high resolution imaging of biomolecules by fluorescence photoactivation localization microscopy (FPALM)," *Methos Mol. Biol.*, **544**, 483–522 (2009).

218. H. Shroff, C. G. Galbraith, J. A. Galbraith, and E. Betzig, "Live-cell photoactivated localization microscopy of nanoscale adhesion dynamics," *Nat. Methods*, **5**, 417–423 (2008).

219. B. Huang, M. Bates, and X. Zhuang, "Super-resolution fluorescence microscopy," *Annu. Rev. Biochem.*, **78**, 993–1016 (2009).

220. A. Vaziri, J. Tang, H. Shroff, and C. V. Shank, "Multilayer three-dimensional super resolution imaging of thick biological samples," *Proc. Natl. Acad. Sci. USA*, **105**, 20221–20226 (2008).

221. A. G. York, A. Ghitani, A. Vaziri, M. W Davidson, and H. Shroff, "Confined activation and subdiffractive localization enables whole-cell PALM with genetically expressed probes," *Nat. Methods*, **8**, 327–333 (2011).

CHAPTER 6

ULTRAFAST LASER SURGERY

As we described in previous chapters of this books in multiphoton excitation microscopy ultrafast lasers can be used to image subcellular structures using without compromising viability. At higher energies, femtosecond lasers can be used to perform laser nanosurgery. The laser beam is then focused to a diffraction-limited spot to destroy or disrupt a structure within a cell and tissue [1–4]. This destructive manipulation occurs when the energy per pulse exceeds a certain threshold and induces nonlinear multiphoton absorption and avalanche ionization. In addition, the focused light is used to stimulate cellular function. In this final chapter, we descriibe such laser surgery with ultrafast lasers.

6.1 LASER CELL NANOSURGERY

6.1.1 Femtosecond Laser Surgery

An intense beam—either in the ultraviolet (UV) region or in the visible region—is tightly focused through high numerical aperture (NA) objectives until the intensity of the focal volume becomes sufficiently high to induce plasma formation. The material in the cell's focal volume could be damaged and even ablated in the submicron size regime, allowing site-specific dissection, removal, or disruption of organelles (Fig. 6.1). This technique is called laser micro/nanosurgery. Laser-mediated nanosurgery can ablate biological material inside living cells with submicrometer precision. The dissection and inactivation of subcellular organelles cells have been

Functional Imaging by Controlled Nonlinear Optical Phenomena, First Edition.
Keisuke Isobe, Wataru Watanabe and Kazuyoshi Itoh.

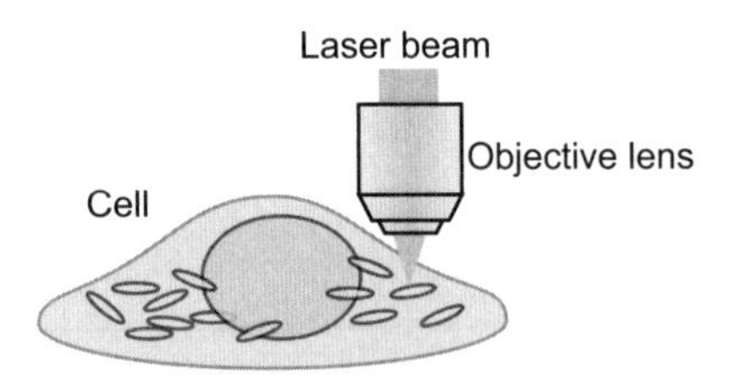

FIGURE 6.1 Schematic for laser surgery by focusing a beam.

demonstrated with submicron spatial resolution [5–12] using continuous-wave (cw) lasers and long-pulse laser pulses (nanosecond regime) in micro/nanosurgery. Lasers in the UV and visible regions present some disadvantages, namely the risk of collateral damage outside the focal volume, the risk of photodamage to living cells due to absorption, and the possible induction of oxidative stress leading to apoptosis. A focused nanosecond-pulse laser beam causes thermal damage and denaturation of the protein molecules around the laser focus. A possible hazard for the surrounding tissue is UV absorption in the DNA, which can cause mutations and genetic disease [13].

Femtosecond lasers can be used to image subcellular structures by multiphoton excitation microscopy [14] without compromising viability [15–18]. Two-photon microscopy appears to be relatively safe. It has no adverse effects on cellular metabolism, reproduction, and viability. However, above a certain energy threshold, photodamage occurs [14–16, 19]. König et al. found two types of photodamage when non-labeled cells were irradiated during scanning using femtosecond laser pulses operated at a high-repetition rate (~80 MHz) over the entire area of each cell [17]. One is a slow process probably based on two-photon excitation of endogenous absorbers, for example, by the coenzymes NAD(P)H and flavins as well as by porphyrins; excitation of these endogenous photosensitizers can result in the formation of destructive reactive oxygen species (ROS). It was reported that mitochondria and the Golgi apparatus were the major targets of NIR laser radiation [18] and the generation of ROS induced apoptosis-like death [20]. The second damage process is of immediate effect and requires high intensities; it is based on multiphoton ionization, optical breakdown phenomena, and intracellular plasma formation. Laser-exposed cells may fail to divide, become giant cells, or die through cell fragmentation. In addition, the irradiated cells may manifest membrane-barrier dysfunction, drastic alterations in the morphology of the nuclear envelope, and DNA fragmentation. It should be noted that the photodamage described above was induced when a number of laser pulses were launched into nonlabeled cells by scanning of femtosecond laser pulses over the entire area of each cell.

Focused near-infrared femtosecond lasers at higher energies can be employed as highly precise nanosurgical tools for tissues, cells, and intracellular structures [17]. Femtosecond laser surgery has been demonstrated by use of both low-repetition rate (1–250 kHz) amplified laser systems and high-repetition-rate oscillators (~ 80 MHz). König et al. proposed a novel nanosurgery tool using near-infrared femtosecond lasers with a high-repetition rate (80 MHz) to perform the dissection of chromosomes

[21,22]. The limited heat generation enables precise control of cell modification, thus avoiding peripheral thermal damage. Measurements with an atomic force microscope revealed chromosome dissection with a cut size of below 300 nm. In addition, a removal of chromosome material with a precision of 110 nm was achieved [22].

This femtosecond laser surgery was also used to produce spatially defined regions of DNA damage in live rat kangaroo cells (PtK1) and human cystic fibrosis pancreatic adenoma carcinoma cells (CFPAC-1) [23]. Spatially defined alterations in the cell nucleus with femtosecond laser are useful for studying DNA damage and repair.

6.1.1.1 Nanosurgery of Intracellular Organelles Intracellular nanosurgery is a way to alter single-cell environments and to repair/replace/modify intracellular components. The ability to affect specific organelles within living cells without directly disrupting normal cellular processes provides biologists with a large window into the functionality of various cellular constituents. Femtosecond laser pulses can be employed for nanosurgery of targeted organelles within a living cell with high spatial resolution [17]. For instance, nanosurgery could remove or replace certain sections of a damaged gene inside a chromosome, sever axons to study the growth of nerve cells, or destroy an individual cell without affecting the neighboring cells. A single organelle (cytoskeleton, mitochondrion, etc.) is completely disrupted or dissected without disturbing surface layers and affecting the adjacent organelles or the viability of both plant cells [24] and animal cells [17]. Femtosecond lasers were used for nanosurgery of organelles and structures within yeast mitotic spindles [25–33]. Figure 6.2 illustrates the nanosurgery performed on a targeted mitochondrion in a HeLa cell (a) before and (b) after irradiation with 800-nm femtosecond laser pulses at a repetition rate of 1 kHz and energy of 3 nJ/pulse with tan objective (NA, 1.4). Watanabe et al. performed an experiment on restaining of the cells, to investigate whether the absence of fluorescence intensity was due to disruption or due to bleaching of the fluorescence [26]. Restaining of the mitochondria was investigated after femtosecond laser irradiation. If the mitochondria were not disrupted, the remaining mitochondria would be detected with another indicator within the irradiation area. In contrast, after the fluorescence in mitochondria was photobleached, the mitochondria were restained. However, the YFP fluorescence in mitochondria focused by femtosecond laser pulses was not restained. The results indicated that the mitochondria were disrupted by the femtosecond laser irradiation.

The viability of the cells after femtosecond laser radiation was ascertained by using propidium iodide (PI). PI is impermeant to the membranes of living cells; it only penetrates dead or apoptotic cells with compromised membranes. The femtosecond laser pulses were focused on inside cells α and β as indicated by the arrows in Fig. 6.3(a) at different energies. After irradiation of femtosecond laser pulses, PI was added to the culture. Fig. 6.3(b) shows the confocal image of the cells after femtosecond laser irradiation and the addition of PI. At the higher energy, additional red fluorescence derived from PI was observed in the nuclei and the cytoplasm of the cell α because the plasma membrane was distorted or destroyed and then PI penetrated the cell. However, the nuclei of the cell β irradiated at lower energy were not stained with PI from the surrounding medium.

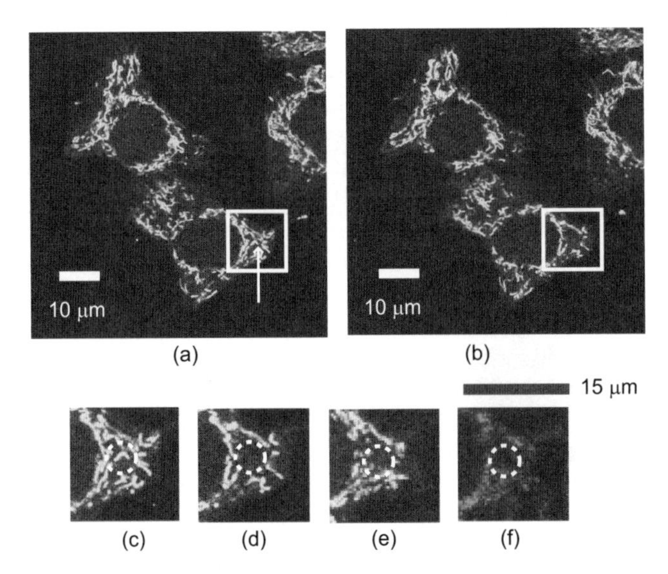

FIGURE 6.2 Nanosurgery of a single mitochondrion in a living HeLa cell with 1-kHz femtosecond laser pulses. (a) Confocal images of laser-irradiated cells before and after restaining. Yellow fluorescence shows mitochondria visualized by EYFP. Target mitochondrion is indicated by a white arrow. (c) Magnified view of square area indicated in (a) before femtosecond laser irradiation. (b) Confocal image obtained after femtosecond laser irradiation. (d) Magnified view of square area in (b). (e) Confocal image obtained by excitation with the Ar^+ laser. (f) Confocal image obtained by excitation with the He-Ne laser. Images (e) and (f) were obtained after restaining by MitoTracker Red. Dotted circles show target mitochondria. Reprinted with permission from [26]. Copyright 2004 Optical Society of America. (*See insert for color representation of the figure.*)

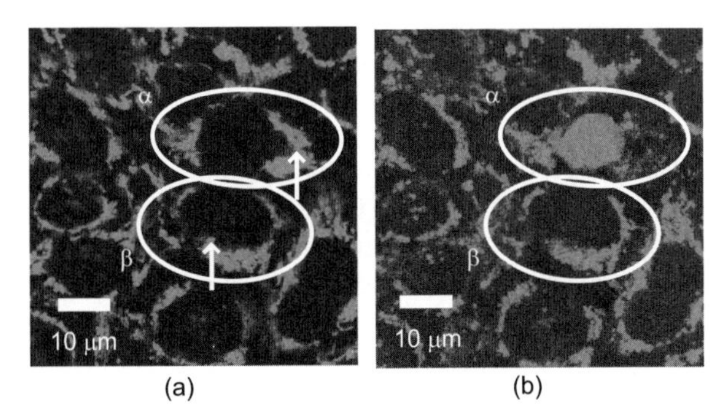

FIGURE 6.3 Confocal images of the cells (a) before and (b) after laser irradiation. Red fluorescence shows mitochondria of HeLa cells stained with MitoTracker Red. The white circles and arrows indicate individual HeLa cells and target mitochondria, respectively. Additional red fluorescence is derived from propidium iodide (PI). The laser pulses were focused inside cells α and β at energies of 7 nJ/pulse and 3 nJ /pulse, respectively. Reprinted with permission from [26]. Copyright 2004 Optical Society of America. (*See insert for color representation of the figure.*)

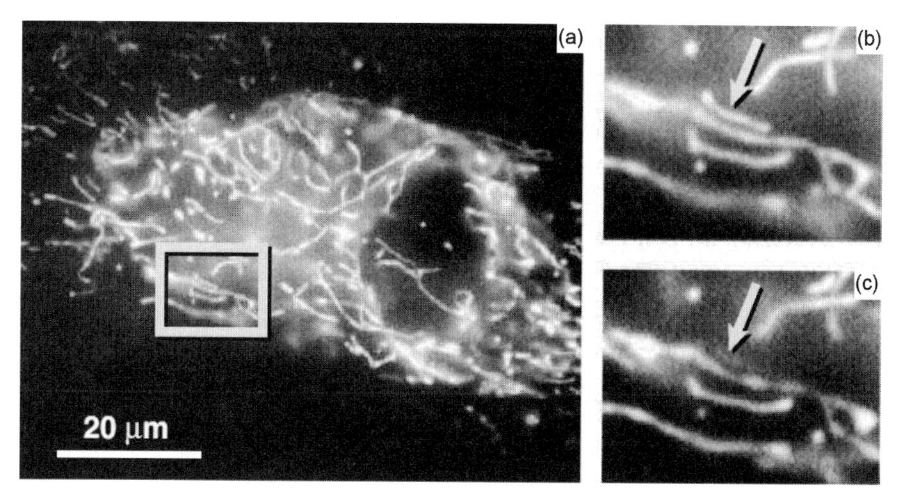

FIGURE 6.4 Nanosurgery of a single mitochondorion in a living endothelial cell. (a) Fluorescence image of mitochondria before femtosecond laser irradiation. Before (b) and after (c) laser surgery. To visualize the mitochondria, capillary endothelial cells were transfected with eYFP fused to cytochrome C oxidase. Reprinted with permission from [31]. Copyright 2005 Tech Science Press.

A single mitochondrion in living cells is disrupted through femtosecond laser irradiation without compromising the cell's viability [31]. Fig. 6.4 shows the ablation of a single mitochondrion, about 5 μm in length and separated by less than 1 μm from multiple neighboring mitochondria. After irradiating a fixed spot on the organelle with a few hundred 2-nJ laser pulses at a 1-kHz repetition rate, the entire mitochondrion disappears from the image, whereas neighboring mitochondria are not affected by the irradiation despite being only a few hundred nanometers away.

Kumar demonstrated the dissection of an individual actin filament and investigated the tension in actin stress fibers in living endothelial cells (Fig. 6.5) [33].

Femtosecond laser surgery has been demonstrated by use of high-repetition-rate laser oscillators (~80 MHz). Shimada et al. show nanosurgery of a targeted mitochondrion in a HeLa cell before and after irradiation with femtosecond laser pulses with a wavelength of 800 nm and a repetition rate of 76 MHz at an energy of 0.39 nJ /pulse [27]. Shimada et al. examined the effectiveness of the restaining method in combination with fluorescence recovery after photobleaching (FRAP) analysis in order to discern disruption or bleaching [27] (Table 6.1). An enhanced green fluorescent protein (EGFP) labeled nuclear region in a living HeLa cell was irradiated while varying the femtosecond laser energy (0.21, 0.26, and 0.39 nJ/pulse) at a wavelength of 925 nm. At the energies of 0.21 and 0.26 nJ/pulse, the fluorophore was bleached, and subsequent recovery of fluorescence in the bleached region occurred due to inward diffusion of unbleached fluorophore molecules. At the energy of 0.39 nJ/pulse, fluorescence in the focal region disappeared; however, fluorescence recovery was not observed in this case. After femtosecond laser irradiation, the same cell was restained

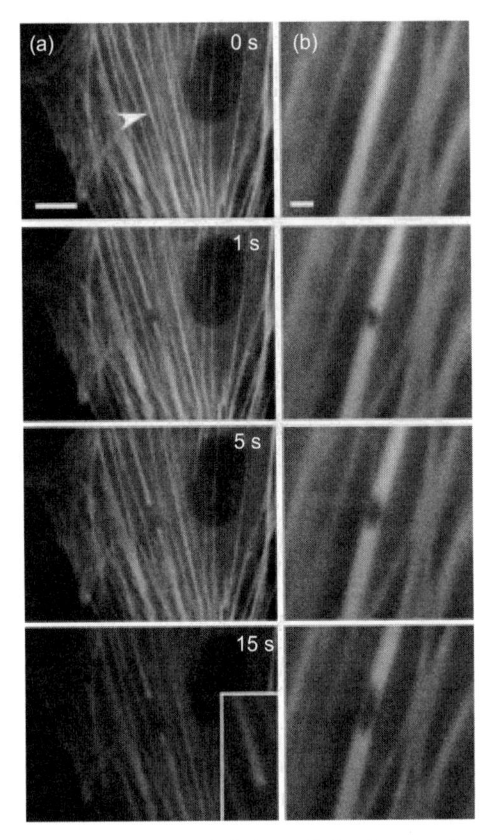

FIGURE 6.5 Dissection of stress fibers in living cells by focusing femtosecond laser pulses. (A) Severing and retraction of a single stress fiber bundle in an endothelial cell expressing EYFP-actin. Scale bar, 10 μm. (B) Strain relaxation of a single stress fiber bundle after a 300-nm hole was ablated in the fiber. Scale bar, 2 μm. Reprinted with permission from [33]. Copyright 2006 Elsevier.

with a blue fluorophore (Hoechest 33342). When the bleaching occurred, fluorescence was also observed from both EGFP and Hoechest 33342. When disruption of the nucleus occurred, no fluorescence was observed from EGFP or Hoechest 33342 in the laser-irradiated region. These results demonstrate that observation of fluorescence recovery is an indicator of the disruption of organelles in living cells. Disruption and bleaching are distinguishable using FRAP analysis and the restaining method.

TABLE 6.1 Difference Between Disruption and Bleaching

	Fluorescence after Femtosecond – Laser Irradiation	Fluorescence Recovery	Restain
Surgery	×	×	×
FRAP	×	○	○

Heisterkamp et al. confirmed the deference between bleaching of the fluorophore and disruption in fixed cells after focusing femtosecond laser pulses by transmission electron microscopy [32]. The energy threshold of ablation is at most 20% higher than the photobleaching threshold. The energies exceed 1.2 times the threshold for which fluorescence disappears to discern ablation and photobleaching.

Supatto et al. described the application of femtosecond nanosurgery in developmental biology [28]. They used femtosecond laser pulses to perform three-dimensional dissections inside live Drosophila embryos in order to locally modify their structural integrity. By tracking the outcome of the dissections by nonlinear microscopy using the same laser source, it was found that local nanosurgery can be used to modulate remote morphogenetic movements. Kohli et al. also demonstrated nanosurgery on living embryonic cells of zebrafish [34].

6.1.1.2 Mechanisms for Femtosecond Laser Surgery Femtosecond laser surgery has been demonstrated using both low-repetition-rate (1 kHz to 250 kHz) amplified laser systems and high-repetition-rate oscillators (~80 MHz). Vogel et al. proposed mechanisms for femtosecond laser nanosurgery [35, 36]. Nanosurgery at a repetition rate of 80 MHz is performed in the low-density plasma regime at pulse energies well below the optical breakdown threshold. It is mediated by free-electron-induced chemical decomposition (bond breaking) in conjunction with multiphoton-induced chemistry and is not related to heating or thermoelastic stresses. An increase in the energy gives rise to long-lasting bubbles by accumulative heating and leads to unwanted dissociation of tissue into volatile fragments. In contrast, dissection at repetition rate of 1 kHz is performed using larger (10-fold) pulse energies and relies on thermoelastically induced formation of minute tran-sient cavities with a lifespan of <100 ns.

6.1.1.3 Femtosecond Laser Nanoaxotomy Femtosecond laser surgery was applied to sever individual dendrites in sensory neurons and axons of motor neuronsin living *Caenorhabditis elegans* (*C. elegans*) roundworms to study the functional and regenerative response to nerve damage [37–39]. By only cutting a few nanoscale nerve connections (axons) inside a *C. elegans*, the backward crawl of the nematode was greatly hindered [37]. Femtosecond laser nanosurgery can control neural regrowth and allow the investigation of important biochemical and genetic pathways that are responsible for neuronal regeneration and axotomy study (Fig. 6.6).

Femtosecond laser axotomy is a versatile tool in regeneration studies when combined with microfluidic chips [40–42]. Ben-Yakar et al. demonstrated that the two-layer miocrofluidic trap allows both the immobilization of *C. elegans* and the performance of nanosurgery to sever axons and study nerve regeneration [40]. Using the nanoaxotomy chip, they discovered that axonal regeneration occurs much faster than previously described; surprisingly, the distal fragment of the severed axon regrows in the absence of anesthetics. Yanik et al. demonstrated on-chip in vivo small-animal genetic and drug screening technologies in high-throughput neural degeneration and regeneration studies [41, 42]. The high-throughput microfluidic platform allows the real-time immobilization of animals without the use of anesthesia and facilitates the

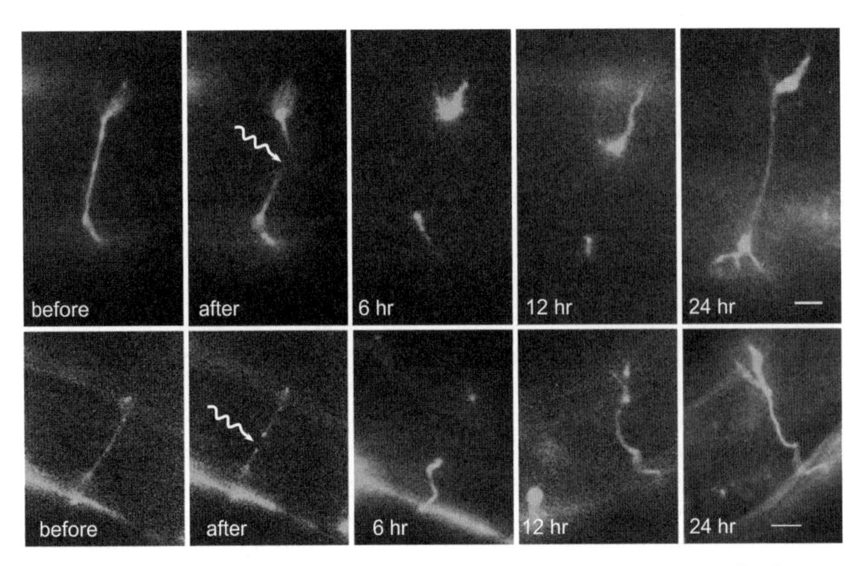

FIGURE 6.6 Femtosecond laser axotomy in *Caenorhabditis elegans* worms using femtosecond laser pulses. Fluorescence images of axons labeled with green fluorescent protein before, after. Scale bar, 5 μm. Courtesy of M. F. Yanik. (*See insert for color representation of the figure.*)

subcellular resolution multiphoton imaging on physiologically active animals. Using femtosecond laser nanosurgery and pattern recognition algorithms, subcellular precision neurosurgery can be performed in microfluidic chips on awake but immobilized animals with minimal collateral damage [43]. The ability to perform precise nanosurgery presents the potential for rapidly screening drugs and for discovering new biomolecules that affect regeneration and degeneration.

6.1.1.4 Targeted Transfection Another application of nanosurgery is cell transfection, or the creation of a small perforation in the cellular membrane and the subsequent injection of a foreign material, typically DNA. Transfection is the introduction of membrane impermeable substances such as foreign DNA into a cell and is an indispensable method for investigating and controlling the individual functions of living cells. Various gene injection techniques have been developed, such as lipofection, electroporation, sonoporation, virus vector, and particle gun injection. Laser optoperforation of individual targeted cells can be employed by directly focusing a laser beam [44, 45]. When an ultraviolet laser was used for targeted gene transfection, it was determined that laser irradiation disrupted cellular integrity. König et al. reported the targeted transportation of plasmid DNA vector pEGFP-N1 encoding-enhanced green fluorescent protein (EGFP) into Chinese hamster ovarian (CHO) cells [46] by focusing femtosecond laser pulses. Fs laser transfection was also applied to kidney epithelial (PtK2) cells of the rat kangaroo [47], canina mammary cells MTH53a [48], and stem cells [49]. Kohli and Elezzabi used femtosecond laser pulses to perform nanosurgery on living zebrafish embryos to introduce exogenous material into the embryonic cells [50].

Kohli et al. demonstrated cell isolation by nanosurgical ablation of focal adhesions adjoining epithelial cells [34]. Uchugonova et al. proposed optical cleaning of selected cells [51] by knocking out some living single stem cells within a 3D environment without causing any collateral damage. Neighbor cells can be optically destroyed while keeping the cell of interest alive. This novel method provides the possibility of controlling the development of stem cells in three dimensions, of destroying undesired cells, and of isolating stem cells of interest.

6.1.2 Plasmonic-Enhanced Nanosurgery

Local field enhancement in the near-field region of metal nanoparticles irradiated with laser pulses is a promising technique [52–54] that involves light intensity enhancement of plasmon in metal nanoparticles or fluorophores. Ben-Yakar et al. investigated plasmonic laser nanoablation of biological materials [54]. Plasmonic laser nanoablation takes advantage of the enhanced plasmonic scattering of femtosecond laser pulses in the near-field region of gold nanoparticles to vaporize various materials with a nanoscale resolution. While the use of femtosecond laser pulses ensures nonthermal tissue ablation, the use of nanoparticles improves the precision and selectivity of the ablation/surgery process. Eversole et al. showed that the technique using femtosecond laser pulses can be used to reduce the ablation threshold of cancer cell membranes labeled with anti-EGFR-coated gold nanoparticles by 7 to 8 times the original value.

6.2 PHOTODISRUPTION AND PHOTO-STIMULATION

6.2.1 Photodisruption of Tissues

6.2.1.1 Corneal Refractive Surgery The most commonly known clinical application of ultrafast laser surgery is corneal refractive surgery. Femtosecond lasers have an important application in the laser-assisted in situ leratomileusis (LASIK) technique [55–57]. Femtosecond pulses are used to make spiral cuts in the cornea with 1-μm accuracy to create lamellar flaps for laser-assisted in situ keratomileusis [56].

6.2.1.2 Optical Histology Histology is the study of the structure and function of cells and tissue by examining a thin slice of sample. Tsai et al. showed, by combining two-photon microscopy and surface laser ablation, that they are able to create a three-dimensional map of mouse cortical tissue [58]. This is a two-step process: in the first step the specimen is labeled and a thin slice is imaged through multiphoton microscopy; in the second step, the imaged section is ablated and removed, and the process is repeated. The tissue layer is imaged to a depth limited by the scattering of the excitation beam effective quantitative imaging is possible in a large number of tissue type. This iterative process of imaging and ablation is repeated as desired, resulting in a high resolution, 3D reconstruction of the desired tissue volume. A tightly focused femtosecond laser pulse is used to deposit laser energy into the endothelial cells that line a specifically targeted vessel; this causes an injury that triggers clotting or causes hemorrhage, but only in the targeted vessel. This technique allows any blood vessel, including individual arterioles, capillaries, and venules, in the top 0.5 mm of the cortex of a rodent to be selectively lesioned. Another application of femtosecond laser pulses is in imaging blood flow through cortical blood vessels and developing a model for stroke by disrupting subsurface blood vessels in living animals [59].

6.2.2 Laser-Induced Stimulation

6.2.2.1 Chromophore-Assisted Laser Inactivation Chromophore-assisted laser inactivation (CALI) is a technique that selectively inactivates proteins of interest to elucidate their in vivo functions. Specific proteins are targeted by means of antibodies attached to metallic nanoparticles or chromophores [60–62]. When the antibody-absorber conjugates are bound to the target proteins, the entire cell or group of cells is exposed to a laser beam. Protein inactivation occurs through linear absorption of the laser irradiation in the nanoparticles or chromophores, resulting in thermomechanical or photochemical destruction of the target proteins, regardless of their localization within the cell. By application of multiphoton excitation to CALI, Tanabe et al. showed that enhanced green fluorescent protein (EGFP) is an effective chromophore for inactivation of a protein's function without nonspecific photodamage in living mammalian cells [63].

6.2.2.2 Laser-Induced Stimulation Calcium is an important messenger in cells. It was found that ultrashort laser stimulation could induce Ca^{2+} wave propagation stimulation [64, 65]. Smith et al. demonstrated that the localized irradiation effects in living cells can lead to photoinduced intracellular Ca^{2+} waves [64].

6.2.2.3 Impulsive Stimulated Raman Scattering By exciting the low-frequency vibrational modes of the viral capsid through impulsive stimulated Raman scattering to a high-energy state, Tsen et al. successfully inactivated viruses such as bacteriophage M13 using femtosecond pulses centered in the visible (frequency doubled a Ti:sapphire oscillator to 425 nm) to produce destructive large amplitude vibrational modes in the M13 phage through impulsive stimulated Raman scattering [66–68].

REFERENCES

1. M. Berns and K. Greulich, *Laser Manipulation of Cells and Tissues*, Academic Press, San Diego, 2007.
2. P. S Tsai, P. Blinder, B. J. Migliori, J. Neev, Yishi Jin, J. A. Squier, and D. Kleinfeld, "Plasma-mediated ablation: an optical tool for submicrometer surgery on neuronal and vascular systems," *Curr. Opin. Biotechnol.*, **20**, 90–99 (2009).

3. K. E. Sheetza and J. Squier, "Ultrafast optics: Imaging and manipulating biological systems," *J. Appl. Phys.*, **105**, 051101 (2009).
4. C. V. Gabel, "Femtosecond lasers in biology: nanoscale surgery with ultrafast optics," *Contemporary Phys.*, **49**, 391–411 (2008).
5. R. L. Amy and R. Storb, "Selective mitochondrial damage by a ruby laser microbeam: an electron microscope study," *Science*, **150** 756–757 (1965).
6. M. W. Berns, J. Aist, J. Edwards, K. Strahs, J. Girton, P. McNeil, J. B. Rattner, M. Kitzes, M. Hammerwilson, L. H. Liaw, A. Siemens, M. Koonce, S. Peterson, S. Brenner, J. Burt, R. Walter, P. J. Bryan, D. Vandyk, J. Couclombe, T. Cahill, and G. S. Bern, "Laser microsurgery in cell and developmental biology," *Science*, **213**, 505–513 (1981).
7. M. W. Berns, W. H. Write, and R. W. Steubing, "Laser microbeam as a tool in cell biology," *Int. Rev. Cytol.*, **129**, 1–44 (1991).
8. J. Colombelli, S. W. Grill, and E. H. K. Stelzer, "Ultraviolet diffraction limited nanosurgery of live biological tissues," *Rev. Sci. Instrum.*, **75**, 472–478 (2004).
9. K. O. Greulich, *Micromanipulation by Light in Biology and Medicine*, Birkhähser, Basel, 1999.
10. V. Venygopalan, A. Guerra III, K. Hahen, and A. Vogel, "Role of laser-induced plasma formation in pulse cellular microsurgery and micromaniplation," *Phys. Rev. Lett.*, **88**, 078103 (2002).
11. A. Khodjakov, R. W. Cole, and C. L. Rieder, "A synergy of technologies: combining laser microsurgery with green fluorescent protein tagging," *Cell Motil. Cytoskeleton*, **38**, 311–317 (1997).
12. E. L. Botvinick, V. Venugopalan, J. V. Shah, L. H. Liaw, and M. W. Berns, "Controlled ablation of microtubules using a picosecond laser," *Biophys. J.*, **87**, 4203–4212 (2004).
13. M. H. Niemz, *Laser-Tissue Interactions: Fundamentals and Applications*, 3rd ed., Springer, Berlin, 1996.
14. K. König, P. T. C. So, W. W. Mantulin, B. J. Tromberg, and E. Gratton, "Cellular response to near-infrared femtosecond laser pulses in two-photon microscopes," *Opt. Lett.*, **22**, 135–136 (1997).
15. K. König, T. W. Becker, P. Fischer, I. Riemann, and K.-J. Halbhuber, "Pulse-length dependence of cellular response to intense near-infrared laser pulses in multiphoton microscopes," *Opt. Lett.*, **24**, 113–115 (1999).
16. H. J. Koester, D. Baur, R. Uhl, and S. W. Hell, "Ca^{2+} fluorescence imaging with pico- and femtosecond two-photon excitation: signal and photodamage," *Biophys. J.*, **77**, 2226–2236 (1999).
17. K. König, "Laser tweezers and multiphoton microscopes in life sciences," *Histochem. Cell Biol.*, **114**, 79–92 (2000).
18. H. Oehring, I. Riedmann, P. Fisher, K.-J. Halbhuber, and K. König, "Ultrastructure and reproduction behaviour of single CHO-K1 cells exposed to near infrared femtosecond laser pulses," *Scanning*, **22**, 263–270 (2000).
19. Y. Sako, A. Sekihata, Y. Yanagisawa, M. Yamamoto, Y. Shimada, K. Ozaki, and A. Kusumi, "Comparison of two-photon excitation laser scanning microscopy with UV-confocal laser scanning microscopy in three-dimensional calcium imaging using the fluorescence indicator Indo-1," *J. Microsc.*, **185**, 9–20 (1997).

20. U. K. Tirlapur, K. König, C. Peuckert, R. Krieg, and K. Halbhuber, "Femtosecond near-infrared laser pulse elicit generation of reactive oxygen species in mammalian cells leading to apotosis-like death," *Exp. Cell Res.*, **263**, 88–97 (2001).
21. K. König, I. Riemann, P. Fischer, and K. H. Halbhuber, "Intracellular nanosurgery with near infrared femtosecond laser pulses," *Cell. Mol. Biol.*, **45** 195–201 (1999).
22. K. König, I. Riemann, and W. Fritzsche, "Nanodissection of human chromosomes with near-infrared femtosecond laser pulses," *Opt. Lett.*, **26**, 819–821 (2001).
23. V. Gomez-Godinez, N. M. Wakida, A. S. Dvornikov, K. Yokomori, and M. W. Berns, "Recruitment of DNA damage recognition and repair pathway proteins following near-IR femtosecond laser irradiation of cells," *J. Biomed. Opt.*, **12**, 020505 (2007).
24. U. K. Tirlapur and K. König, "Femtosecond near infrared lasers as a novel tool for non-invasive real-time high resolution time-lapse imaging of chloroplast division in living bundle sheath cells of Arabidopsis," *Planta.*, **214**, 1–10 (2001).
25. U. K. Tirlapur and K. König "Femtosecond near-infrared laser pulses as a versatile non-invasive tool for intra-tissue nanoprocessing in plants without compromising viability," *Plant. J.*, **31**, 365–374 (2002).
26. W. Watanabe, N. Arakawa, S. Matsunaga, T. Higashi, K. Fukui, K. Isobe, and K. Itoh, "Femtosecond laser disruption of subcellular organelles in a living cell," *Opt. Express*, **12**, 4203–4213 (2004).
27. T. Shimada, W. Watanabe, S. Matsunaga, T. Higashi, H. Ishii, K. Fukui, K. Isobe, and K. Itoh, "Intracellular disruption of mitochondria in living HeLa cells with a 76-MHz femtosecond laser oscillator," *Opt. Express*, **13**, 9869–9880 (2005).
28. W. Supatto, D. Débarre, B. Moulia, E. Brouzés, J. L Martin, E. Farge, and E. Beaurepaire, "In vivo modulation of morphogenetic movements in Drosophila embryos with femtosecond laser pulses," *Proc. Natl. Acad. Sci. USA*, **102**, 1047–1052 (2005).
29. L. Sacconi, I. M. Tolić-NØrrelykke, R. Antolini, and F. S. Pavone, "Combined intracellular three-dimensional imaging and selective nanosurgery by a nonlinear microscope," *J. Biomed. Opt.*, **10**, 014002 (2005).
30. N. M. Wakida, C. S. Lee, E. T. Botvinick, L. Z. Shi, A. Dvornikov, and M. W. Berns, "Laser nanosurgery of single microtubules reveals location-dependent depolymerization rates," *J. Biomed. Opt.*, **12**, 024022 (2007).
31. N. Shen, D. Datta, C. B. Schaffer, P. LeDuc, D. E. Ingber, and E. Mazur, "Ablation of cytoskeletal filaments and mitochondria in live cells using a femtosecond laser nanoscissor," *Mol. Cell. Biomech.*, **2**, 17–25 (2005).
32. A. Heisterkamp, I. Z. Maxwell, E. Mazur, J. M. Underwood, J. A. Nickerson, S. Kumar, and D. E. Ingber, "Pulse energy dependence of subcellular dissection by femtosecond laser pulses," *Opt. Express*, **13**, 3690–3696 (2005).
33. S. Kumar, I. Z. Maxwell, A. Heisterkamp, T. R. Polte, T. P. Lele, M. Salanga, E. Mazur, and D. E. Ingber, "Viscoelastic retraction of single living stress fibers and its impact on cell shape, cytoskeletal organization, and extracellular matrix mechanics," *Biophys. J.*, **90**, 3762–3773 (2006).
34. V. Kohli, J. P. Acker, and A. Y. Elezzabi, "Cell nanosurgery using ultrashort (femtosecond) laser pulses: applications to membrane surgery and cell isolation," *Laser Surg. Med.*, **37**, 227–230 (2005).
35. A. Vogel, J. Noack, G. Hüttman, and G. Paltauf, "Mechanisms of femtosecond laser nanosurgery of cells and tissues," *Appl. Phys. B*, **81**, 1015–1047 (2005).

36. A. Vogel, N. Linz, S. Freidank, and G. Paltauf, "Femtosecond-laser-induced nanocavitation in water: implications for optical breakdown threshold and cell surgery," *Phys. Rev. Lett.*, **100**, 038102 (2008).
37. M. F. Yanik, H. Cinar, H. N. Cinar, A. D. Chisholm, Y. Jin, and A. Ben-Yakar, "Functional regeneration after laser axotomy," *Nature*, **432**, 822–822 (2004).
38. M. F. Yanik, H. Cinar, H. N. Cinar, A. Gibby, A. D. Chisholm, Y. S. Jin, and A. Ben-Yakar, "Nerve regeneration in *Caenorhabditis elegans* after femtosecond laser axotomy," *IEEE J. Sel. Top. Quant.*, **12**, 1283–1291 (2006).
39. F. Bourgeois, and A. Ben-Yakar, "Femtosecond laser nanoaxotomy properties and their effect on axonal recovery in *C-elegans*," *Opt. Express*, **15**, 8521–8531 (2007).
40. S. X. Guo, F. Bourgeois, T. Chokshi, N. Durr, N. J. M. Hilliard, N. Chronis, and A. Ben-Yakar, "Femtosecond laser nanoaxotomy lab-on-a-chip for in vivo nerve regeneration studies," *Nat. Methods*, **5**, 531–533 (2008).
41. C. B. Rohde, F. Zeng, R. Gonzalez-Rubio, M. Angel, and M. F. Yanik, "Microfluidic system for on-chip high-throughput whole-animal sorting and screening at sub-cellular resolution," *Proc. Natl. Acad. Sci. USA*, **104**, 13891–13895 (2007).
42. F. Zei, C. Rohde, and M. F. Yanik, "Sub-cellular precision on-chip small-animal immobilization, multi-photon imaging and femtosecond-laser manipulation," *Lab on a Chip*, **8**, 653 (2008).
43. C. B. Rohde, F. Zeng, C. Gilleland, M. Angel, and M. F. Yanik, *Proceedings of LPM2008, The 9th International Symposium on Laser Precision Microfabrication*, Tu-A-1210, 2008.
44. E. Zeira, A. Manevitch, A. Khatchatouriants, O. Pappo, E. Hyam, M. Darash-Yahana, E. Tavor, A. Honigman, A. Lewis, and E. Galun, "Femtosecond infrared laser: an efficient and safe in vivo gene delivery system for prolonged expression," *Mol. Ther.*, **8**, 342–350 (2003).
45. E. Zeira, A. Manevitch, Z. Manevitch, E. Kedar, M. Gropp, N. Daudi, R. Barsuk, M. Harati, H. Yotvat, P. J. Troilo, T. G. Griffiths, S. J. Pacchione, D. F. Roden, Z. Niu, O. Nussbaum, G. Zamir, O. Papo, I. Hemo, A. Lewis, and E. Galun, "Femtosecond laser: a new intradermal DNA delivery method for efficient, long-term gene expression and genetic immunization," *Faseb. J.*, **21**, 3522–3533 (2007).
46. U. K. Tirlapur and K. König, "Targeted transfection by femtosecond laser," *Nature*, **418**, 290–291 (2002).
47. D. Stevenson, B. Agate, X. Tsampoula, P. Fischer, C. T. A. Brown, W. Sibbett, A. Riches, F. Gunn-Moore, and K. Dholakia "Femtosecond optical transfection of cells: viability and efficiency," *Opt. Express*, **14**, 7125–7133 (2006).
48. J. Baumgart, W. Bintig, A. Ngezahayo, S. Willenbrock, H. M. Escobar, W. Ertmer, H. Lubatschovski, and A. Heisterkamp, "Quantified femtosecond laser based opto-perforation of living GFSHR-17 and MTH53 a cells," *Opt. Express*, **16**, 3021–3031 (2008).
49. A. Uchugonova, K. König, R. Bueckle, A. Isemann, and G. Tempea, "Targeted transfection of stem cells with sub-20 femtosecond laser pulses," *Opt. Express*, **16**, 9357–9364 (2008).
50. V. Kohli and A. Y. Elezzabi, "Laser surgery of zebrafish (*Danio rerio*) embryos using femtosecond laser pulses: optimal parameters for exogenous material delivery, and the laser's effect on short- and long-term development," *BMC Biotechnol.*, **8**, 7 (2008).

51. A. Uchugonova, A. Isemann, E. Gorjup, G. Tempea, R. Bückle, W. Watanabe, and K. König, "Optical knock out of stem cells with extremely ultrashort femtosecond laser pulses," *J. Biophotonics*, **1**, 463–469 (2008).
52. F. Garwe, A. Csaki, G. Maubach, A. Steinbruck, A. Weise, K. Köonig, and W. Fritzsche, "Laser pulse energy conversion on sequence-specifically bound metal nanoparticles and its application for DNA manipulation," *Med. Laser Appl.*, **20**, 201–206 (2005).
53. A. Csaki, F. Garwe, A. Steinbrueck, G. Maubach, G. Festag, A. Weise, I. Riemann,K. König, and W. Fritzsche, "A parallel approach for subwavelength molecular surgery using gene-specific positioned metal nanoparticles as laserlight antennas," *Nano Lett.***7**, 247–253 (2007).
54. A. Ben-Yakar, "Plasmonic laser nanosurgery of cells using femtosecond laser ablation in the near-field of gold nanoparticles," *LPM2008 (9th International Symposium on Laser Precision Microfabrication)*, Quebec City (2008).
55. K. König, O. Krauss, and I. Riemann, "Intratissue surgery with 80 MHz nanojoule femtosecond laser pulses in the near infrared," *Opt. Express*, **10**, 171–176 (2002).
56. T. Juhasz, H. Frieder, R. M. Kurtz, C. Horvath, J. F. Bille, and G. Mourou, "Corneal refractive surgery with femtosecond lasers," *IEEE J. Sel. Top. Quant.*, **5**, 902–910 (1999).
57. H. Lubatschowski, G. Maatz, A. Heisterkamp, U. Hetzel, W. Drommer, H. Welling, and W. Ertmer, "Application of ultrashort laser pulses for intrastromal refractive surgery," *Graf. Arch. Clin. Exp.*, **238**, 33–39 (2000).
58. P. S. Tsai, B. Friedman, A. I. Ifarraguerri, B. D. Thompson, V. Lev-Ram, C. B. Schaffer, C. Xiong, R. Y. Tsien, J. A. Squier, and D. Kleinfeld, "All-optical histology using ultrashort laser pulses," *Neuron*, **39**, 27–41 (2003).
59. N. Nishimura, C. B. Schaffer, B. Friedman, P. S. Tsai, P. D. Lyden, and D. Kleinfeld, "Targeted insult to subsurface cortical blood vessels using ultrashort laser pulses: three models of stroke," *Nat. Methods*, **3**, 99–108 (2006).
60. D. G. Jay, "Selective destruction of protein function by chromophore-assisted laser inactivation," *Proc. Natl. Acad. Sci. USA*, **85**, 5454–5458 (1988).
61. J. C. Liao, J. Roider, and D. G. Jay, "Chromophore-assisted laser inactivation of proteins is mediated by the photogeneration of free radicals," *Proc. Natl. Acad. Sci. USA*, **91**, 2659–2663 (1994).
62. Z. Rajfur, P. Roy, C. Otey, L. Romer, and K. Jacobson, "Dissecting the link between stress fibres and focal adhesions by CALI with EGFP fusion proteins," *Nat. Cell. Biol.*, **4**, 286–293 (2002).
63. T. Tanabe, M. Oyamada, K. Fujita, P. Dai, H. Tanaka, and T. Takamatsu, "Multiphoton excitation evoked chromophore-assisted laser inactivation using green fluorescent protein," *Nat. Methods*, **2**, 503–505 (2005).
64. N. I. Smith, K. Fujita, T. Kaneko1, K. Katoh, O. Nakamura, S. I. Kawata, and T. Takamatsu, "Generation of calcium waves in living cells by pulsed-laser-induced photodisruption," *Appl. Phys. Lett.*, **79**, 1208 (2001).
65. S. Iwanaga, N. Smith, K. Fujita, S. Kawata, and O. Nakamura, "Single-pulse cell stimulation with a near-infrared picosecond laser," *Appl. Phys. Lett.*, **87**, 243901 (2005).
66. K. T. Tsen, E. C. Dykeman, O. F. Sankey, N.-T. Lin, S.-W. D. Tsen, and J. G. Kiang, "Inactivation of viruses by coherent excitations with a low power visible femtosecond laser," *Virol. J.*, **3**, 79 (2006).

67. K. T. Tsen , E. C. Dykeman, O. F. Sankey, N.-T. Lin, S.-W. D. Tsen, and J. G. Kiang, "Inactivation of viruses by laser-driven coherent excitations via impulsive stimulated Raman scattering process," *J. Biomed. Opt.*, **12**, 024009 (2007).
68. K. T. Tsen, S.-W. D. Tsen, C.-L. Chang, C.-F. Hung, T.-C. Wu, and J. G. Kiang, "Inactivation of viruses with a very low power visible femtosecond laser," *J. Phys. Condens. Matter*, **19**, 322102 (2007).

INDEX

Functional Imaging by Controlled Nonlinear Optical Phenomena, First Edition.
Keisuke Isobe, Wataru Watanabe and Kazuyoshi Itoh.